Microbial Biotechnology

Jayanta Kumar Patra
Chethala N. Vishnuprasad
Gitishree Das
Editors

Microbial Biotechnology

Volume 1. Applications in Agriculture and Environment

Springer

Editors
Jayanta Kumar Patra
Dongguk University
Goyang-si, Gyeonggi-do
South Korea

Chethala N. Vishnuprasad
TransDisciplinary University
Bangalore, Karnataka, India

Gitishree Das
Dongguk University
Goyang-si, Gyeonggi-do
South Korea

ISBN 978-981-10-6846-1 ISBN 978-981-10-6847-8 (eBook)
https://doi.org/10.1007/978-981-10-6847-8

Library of Congress Control Number: 2017960428

Printed on acid-free paper

This Springer imprint is published by Springer Nature
The registered company is Springer Nature Singapore Pte Ltd.
The registered company address is: 152 Beach Road, #21-01/04 Gateway East, Singapore 189721, Singapore

Foreword

At a time when biotechnological applications are moving into innovative areas and the technology is transforming itself with quicker, cheaper, and predictable options, this publication is very timely. Many of us are aware that biotechnology has contributed immensely to our ability to deal with environmental and agricultural applications, starting from the Ananda Chakrabarty's engineered bacteria to the CRISPR-CAS9 technology.

These developments have increased our ability to access and experiment with novel combinations of genes and resulting characters. The selection of chapters in this compilation is varied providing the reader an overview of current applications and challenges in relation to biotechnology for agriculture and environment.

Particularly intriguing is the fact that the chapters, focusing on application of microbial biotechnology for environment and agriculture, deal with advanced applications such as nano-biotechnology to use of biocontrol agents with more targeted actions.

Let us congratulate the authors and editors (Dr. CN Vishnuprasad, Dr. JK Patra, and Dr. Gitishree Das) for their contribution through this volume, especially when debates continue to increase on simplified interventions for complex problems and the need to revisit some fundamentals of biotechnological processes to understand new applications.

Vice-Chancellor, TransDisciplinary University
Bengaluru, Karnataka, India
August 2017

Balakrishna Pisupati

Preface

Humans are considered as the most highly evolved species in the earth. Perhaps, microbes are placed in the lowest strata of evolution. Nevertheless, without these 'simple, tiny, invisible' creatures, earth is meaningless, nonfunctional and inhabitable. Yes, microbes are everywhere. For thousands of years, microorganisms have been part of our life in various ways. They were helping us with biotechnological processes like fermentation. Later, the advancements in microbiology and biotechnology lead to the launch of microbial biotechnology as a separate area of research and contributed dramatically to the development of the areas like agriculture, environment, biopharmaceutics, fermented foods, etc. Microbial biotechnology is one of the most influencing fields that attract researchers all over the globe. The book, *Microbial Biotechnology: Applications in Agriculture and Environment Volume 1*, is a collection of articles highlighting the recent developments in microbial biotechnology in the area of agriculture and environment.

The book is divided into two major sections: (a) 'Applications of Microbes in Agriculture' and (b) 'Applications of Microbes for the Benefit of the Environment'. These sections cover the recent developments in the applications of microorganism in various fields such as agriculture and environment. The first section of the book covers some of the emerging areas like agricultural nanotechnology emphasizing its current application and future prospects, promising applications of biofuels produced with the help of algae, advancements and application of microbial keratinase in agriculture as well as the role of *Bacillus* sp. as a biocontrol agent for rice pathogens. Besides these, this section also talks about how microbial biotechnology helps in sustainable agriculture practice. Articles discussing the role of plant growth-promoting rhizobacteria and bacterial siderophore as plant growth promoter and exploring the use of fungi and actinobacteria for sustainable agriculture are examples for this. As a component of sustainable agriculture practice, the use of microbes in detoxifying the organophosphate pesticide residues is also discussed.

The second section of the book, 'Applications of Microbes for the Benefit of the Environment', primarily focuses on the use of microbial biotechnology for the detoxification and cleaning of the environment. Aspects like bio-surfactants, biofilms, bio-detoxification of heavy metals, bioremediation of hexavalent chromium, degradation of phenol and phenolic compounds as well as microbial interaction with metals and metalloids are discussed in this section. Additionally, articles discussing the bioprospecting of endophytes for sustainable agriculture and environment as well as the concept of bioelectricity are also discussed.

This volume is the culmination of the efforts of several researchers, scientists, graduate students and postdoctoral fellows across the world who are well known in their respective areas of specialization. This book would serve as a quick reference book for the graduate and postgraduate students pursuing their study in any branch of life sciences, microbiology, health sciences and environmental biotechnology as well as researchers and scientists working in the laboratories and industries involved in research related to microbiology, environmental biotechnology and allied researches. We hope this edited volume of *Microbial Biotechnology: Application in Agriculture and Environment* would become an invaluable reference tool for researchers in their respective area of specialization.

We express our appreciation to all of the contributing authors who helped us tremendously with their contributions. We thank all of them for their time, critical thoughts and suggestions that enabled us to put together this peer-reviewed edited volume. We express our sincere gratitude to Dr. Balakrishna Pisupati, Vice Chancellor, TransDisciplinary University (TDU), Bengaluru, India, for writing the foreword for the book.

We are also thankful to Springer Nature Singapore Pte Ltd., Singapore, and their team members particularly Dr. Sue Lee, associate editor, Biomedicine, Springer Nature, South Korea, for giving us the opportunity to publish this volume. Lastly, we thank our family members for their love, support, encouragement and patience during the entire period of this work.

Dongguk University, Goyang-si, South Korea — Jayanta Kumar Patra

Gitishree Das

TDU, Bengaluru, India — Chethala N. Vishnuprasad

Contents

Contributors

Ahmed I.S. Ahmed Amity Institute of Microbial Technology, Amity University Uttar Pradesh, Noida, UP, India

Plant Pathology Unit, Plant Protection Department, Desert Research Center, Cairo, Egypt

Ovaid Akhtar Sadasivan Mycopathology Laboratory, Department of Botany, University of Allahabad, Allahabad, UP, India

Reshma R. Anilkumar Microbiology Division, Jawaharlal Nehru Tropical Botanic Garden and Research Institute, Palode, Trivandrum, Kerala, India

Nafe Aziz Amity Institute of Microbial Technology, Amity University Uttar Pradesh, Noida, UP, India

Cyanobacterial Lab, Department of Bioscience, Jamia Millia Islamia, New Delhi, India

Balasubramani S.P. School of Integrative Health Sciences, TransDisciplinary University, Bangalore, India

Gausiya Bashri Ranjan Plant Physiology and Biochemistry Laboratory, Department of Botany, University of Allahabad, Allahabad, India

Shristy Bishwakarma MITS School of Biotechnology, Bhubaneswar, Odisha, India

Mohnish Borker Department of Mechanical Engineering, National Institute of Technology Calicut, Kozhikode, Kerala, India

Gitishree Das Research Institute of Biotechnology & Medical Converged Science, Dongguk University-Seoul, Ilsandong-gu, Gyeonggi-do, South Korea

Pradeep K. Das Mohapatra Department of Microbiology, Vidyasagar University, Midnapore, West Bengal, India

Aditya Dutt Department of Biotechnology, National Institute of Technology, Raipur, Chattisgarh, India

Lekshmi K. Edison Microbiology Division, Jawaharlal Nehru Tropical Botanic Garden and Research Institute, Palode, Trivandrum, Kerala, India

E.M.H.G.S. Ekanayake National Institute of Fundamental Studies, Kandy, Sri Lanka

Sushanto Gouda Amity Institute of Wildlife Science, Amity University, Noida, Uttar Pradesh, India

Kriti Gupta Department of Biotechnology & Medical Engineering, NIT Rourkela, Rourkela, Odisha, India

Neha Gupta Department of Biotechnology & Medical Engineering, NIT Rourkela, Rourkela, Odisha, India

Amr I.M. Ibrahim Plant Pathology Unit, Plant Protection Department, Desert Research Center, Cairo, Egypt

S. Jayaraj Department of Mechanical Engineering, National Institute of Technology Calicut, Kozhikode, Kerala, India

Rout George Kerry P.G. Department of Biotechnology, Academy of Management & Information Technology, Khurda, Odisha, India

S.A. Kulasooriya National Institute of Fundamental Studies, Kandy, Sri Lanka

Sanjeet Kumar Ambika Prasad Research Foundation, Odisha, India

Nitya Kumari Department of Biotechnology & Medical Engineering, NIT Rourkela, Rourkela, Odisha, India

Anna S. Kynadi School of Biotechnology, National Institute of Technology Calicut, Kozhikode, Kerala, India

Padma Mahanti Directorate of Environment and Climate Change, Trivandrum, Kerala, India

Bibhuti Bhusan Mishra Department of Microbiology, College of Basic Science & Humanities, Orissa University of Agriculture and Technology, Bhubaneswar, Odisha, India

Bhabatosh Mitra Department of Bio-Sciences and Bio-Technology, Fakir Mohan University, Balasore, Odisha, India

Ranjan Kumar Mohapatra Environment and Sustainability Department, CSIR-Institute of Minerals and Materials Technology, Bhubaneswar, Odisha, India

Keshab C. Mondal Department of Microbiology, Vidyasagar University, Midnapore, West Bengal, India

Raghvendra Pratap Narayan Netaji Subhash Government Girls P.G. College, Lucknow, UP, India

Suman Nayak Department of Biotechnology and Biomedical Engineering, National Institute of Technology, Rourkela, Odisha, India

Suraja Kumar Nayak Department of Biotechnology, College of Engineering and Technology, Biju Patnaik University of Technology, Bhubaneswar, Odisha, India

Swapnarani Nayak Fish Genetics and Biotechnology Division, ICAR-Central Institute of Freshwater Aquaculture, Bhubaneswar, India

A. Pahari Department of Microbiology, College of Basic Science & Humanities, Orissa University of Agriculture and Technology, Bhubaneswar, Odisha, India

Chitta Ranjan Panda Environment and Sustainability Department, CSIR-Institute of Minerals and Materials Technology, Bhubaneswar, Odisha, India

Pankaj Kumar Parhi School of Applied Sciences and School of Chemical Technology, School of Biotechnology, KIIT University, Bhubaneswar, India

Parul Parihar Ranjan Plant Physiology and Biochemistry Laboratory, Department of Botany, University of Allahabad, Allahabad, India

Anuradha Patel Ranjan Plant Physiology and Biochemistry Laboratory, Department of Botany, University of Allahabad, Allahabad, India

Jayanta Kumar Patra Research Institute of Biotechnology and Medical Converged Science, Dongguk University, Goyang-si, Gyeonggi-do, South Korea

N.S. Pradeep Microbiology Division, Jawaharlal Nehru Tropical Botanic Garden and Research Institute, Palode, Trivandrum, Kerala, India

A. Pradhan Department of Microbiology, College of Basic Science & Humanities, Orissa University of Agriculture and Technology, Bhubaneswar, Odisha, India

S.K. Pradhan P.G. Department of Bioinformatics, Orissa University of Agriculture and Technology, Bhubaneswar, Odisha, India

Sheo Mohan Prasad Ranjan Plant Physiology and Biochemistry Laboratory, Department of Botany, University of Allahabad, Allahabad, India

Ram Prasad Amity Institute of Microbial Technology, Amity University Uttar Pradesh, Noida, UP, India

Department of Mechanical Engineering, Johns Hopkins University, Baltimore, MD, USA

Vanitha Ramesh Department of Genetics, Indian Academy Degree College (Autonomous), Bangalore, India

Dipak K. Sahoo Department of Microbiology, Vidyasagar University, Midnapore, West Bengal, India

Sabuj Sahoo Post Graduate Department of Biotechnology, Utkal University, Bhubaneswar, Odisha, India

Sarmistha Sarangi ICAR-Central Rice Research Institute, Cuttack, Odisha, India

Angana Sarkar Department of Biotechnology & Medical Engineering, NIT Rourkela, Rourkela, Odisha, India

Gamini Seneviratne National Institute of Fundamental Studies, Kandy, Sri Lanka

Rachana Singh Ranjan Plant Physiology and Biochemistry Laboratory, Department of Botany, University of Allahabad, Allahabad, India

M. Srinivas Department of Mechanical Engineering, National Institute of Technology Calicut, Kozhikode, Kerala, India

Pragya Srivastava Sadasivan Mycopathology Laboratory, Department of Botany, University of Allahabad, Allahabad, UP, India

T.V. Suchithra School of Biotechnology, National Institute of Technology Calicut, Kozhikode, Kerala, India

H.N. Thatoi Departments of Biotechnology, North Orissa University, Baripada, Odisha, India

Lata S.B. Upadhyay Department of Biotechnology, National Institute of Technology, Raipur, Chattisgarh, India

Chethala N. Vishnuprasad School of Integrative Health Sciences, Institute of TransDisciplinary Health Sciences and Technology, Bangalore, Karnataka, India

Mukesh Kumar Yadav Department of Otorhinolaryngology-Head and Neck Surgery & Institute for Medical Device Clinical Trials, Korea University College of Medicine, Seoul, South Korea

Korea University Guro Hospital, Seoul, South Korea

Ifra Zoomi Sadasivan Mycopathology Laboratory, Department of Botany, University of Allahabad, Allahabad, UP, India

About the Editors

Jayanta Kumar Patra, MSc, PhD is currently working as Assistant Professor at the Research Institute of Biotechnology and Medical Converged Science, Dongguk University, Ilsandong, Gyeonggido, South Korea. His current research is focused on nanoparticle synthesis by green technology methods and their potential application in biomedical and agricultural fields. He has about 11 years of research and teaching experience in the field of food, pharmacology and nano-biotechnology. Dr. Patra completed his Ph.D. (Life Sciences) from North Orissa University, India, in pharmacological application of mangrove plant bioactive compounds and PDF (Biotechnology) from Yeungnam University, South Korea. Since 2007, Dr. Patra has published around 90 papers in various national and international peer-reviewed journals of different publishing houses such as Elsevier, Springer, Taylor & Francis, Wiley, etc. and around 20 book chapters in different edited books. Dr. Patra has also authored two books, *Industrial and Environmental Biotechnology* by STUDIUM Press (India) and *Natural Products in Food: Prospects and Applications* by STUDIUM Press LLC, USA. Further, another book is under process with Apple Academic Press, Inc., Canada, CRC Press, a Taylor & Francis group. He is editorial board member of several international journals and one science magazine. Besides that, he is also a co-principal investigator of a collaborative international research project.

Chethala N. Vishnuprasad, MSc, PhD is currently working as Assistant Professor at the School of Integrative Health Sciences, TransDisciplinary University (TDU), Karnataka, India. Dr. Vishnuprasad is a postgraduate in biotechnology and Ph.D. in the area of cell biology and phytochemistry. He has more than 10 years of research and teaching experience in different universities in India and abroad. He is currently interested in pursuing transdisciplinary research in the area of Ayurveda (a traditional medical care system originated in India) and biology with a focus on understanding the concepts of Ayurveda in glucose metabolism, diabetes, obesity, etc. He is also interested in exploring the potentials of Ayurveda-specific concepts like herbal fumigation in the management of health. He has several peer-reviewed publications, conference papers and book chapters to his credit. Some of his works have been adjudged with best paper award in reputed conferences. He is also an active participant of various associations at different capacities like advisor, member, etc.

Gitishree Das, MSc, PhD is currently working as Assistant Professor at the Research Institute of Biotechnology and Medical Converged Science, Dongguk University, Ilsandong, Gyeonggido, South Korea. She has 10 years of research experience in the field of rice molecular biology, breeding and endophytic bacteria and 1 year of teaching experience. Dr. Das has completed her Ph.D. (Life Sciences) from North Orissa University, India, on rice gene pyramiding (Central Rice Research Institute, Cuttack, India) and PDF (Biotechnology) from Yeungnam University, South Korea. She has undergone many professional training in the field of biotechnology from various international and national reputed organizations. Her current research is focused on the application of endophytes and their bioactive compounds/nanoparticles in biomedical and agricultural fields. To her credit, she has published around 40 research articles in international and national reputed journals and 7 book chapters. She is editorial board member of Ambika Prasad Research Foundation. Currently, she is co-principal investigator of a collaborative international research project.

Part I
Application of Microbes in Agriculture

Chapter 1
Agricultural Nanotechnologies: Current Applications and Future Prospects

Rout George Kerry, Sushanto Gouda, Gitishree Das, Chethala N. Vishnuprasad, and Jayanta Kumar Patra

1.1 Introduction

The planet earth is inhabited by over seven billion people within the land area of 13 billion hectares, for which 99.7% food comes from the terrestrial ecosystem. As per Food and Agriculture Organization [FAO] (2017) data sheet more than 4.9 million hectares of land that constitute 37.6% of the terrestrial ecosystem is encapsulated for agricultural purposes. The fact that terrestrial land masses are the sole provider of the basic necessities of life makes every inch of landmass an important component for food source. Globally 30% of the workers are involved in farming, but in low-income countries it rises to about 60% (FAO 2017). In 2016, 54.5% of the world population were living in urban areas but by 2030, it is estimated that 60% of world's population will be shifting to urban areas (The World's Cities 2016). This shift may pose an indirect but a permanent threat as rural areas or farm lands and the farmers are the only suppliers of the food for the present generation as well as to the future generation which is expected to reach 11 billion by 2100. Even though farmers are the backbone of various countries, in India and in many other developing countries farmers face malnutrition as well as several other economic issues like

R.G. Kerry
P.G. Department of Biotechnology, Academy of Management & Information Technology, Khurda, Odisha, India

S. Gouda
Amity Institute of Wildlife Science, Amity University, Noida, Uttar Pradesh, India

G. Das • J.K. Patra (✉)
Research Institute of Biotechnology & Medical Converged Science, Dongguk University-Seoul, Ilsandong-gu, Gyeonggi-do, South Korea
e-mail: jkpatra@dongguk.edu

C.N. Vishnuprasad
School of Integrative Health Sciences, Institute of TransDisciplinary Health Sciences and Technology, Bangalore, Karnataka, India

J.K. Patra et al. (eds.), *Microbial Biotechnology*,
https://doi.org/10.1007/978-981-10-6847-8_1

doughty, high price of fertilizers, poor rainfall, less productivity etc. that even lead to suicide (NCRB ADSI annual reports 2013, 2014, 2015).

Statistics shows that currently 78% of under-privileged people worldwide are from rural areas and around 800 million people are suffering from hunger. If the current scenario is unaltered by 2030, then 653 million people will remain undernourished (FAO 2017; World Bank 2017). Although government agencies are working on improvement and development of farmers through different schemes and if the present situation still continues, then it would result in an unsustainable and irreversible imbalance in the supply of farm or agricultural products. Conventional techniques or measures taken by various government agencies across the globe are mostly restricted to providing loans, fertilizers, electricity for irrigation, improved crop seeds etc. However, all such measures are irrelevant and incapable in case of soil infertility, plant diseases and poor crop yield. Such situations continue to persist and proper scientific knowledge and technologies need to be created for agriculture activities (Srilatha 2011). Practise of monoculture or single crop plantation has also been considered as a major reason behind soil infertility and poor agricultural production. Hence, use of modern tools and techniques, science for successful plantation, good knowledge about soil, development of diseases resistant crops and production of nutrient containing crops or grains are some of the areas to be considered with immediate priority.

In the sea of the opportunities, it is impossible for the cargo to sail towards prosperity without lifting the anchor of poverty. Despite the flexibility and implementation of technology in different fields such as industrialization, biomedical research, ceramics, remote sensing, space technology, application of science and technology for sustainable agricultural development is the most important as it is the most successful and reliable source of production of any type of food, for both humans and animals. Therefore, the focus of every nation should be to find diverse and novel ways, through exploring multidisciplinary and interdisciplinary technologies such as biotechnology, nanotechnology, nano-biotechnology etc., to improve agriculture and crop production (Gul et al. 2014; Srilatha 2011). Amalgamated technology such as nanotechnology holds promising results in different aspects of agriculture through nano-formulations of agrochemicals *viz.* pesticides, insecticides, herbicide, nanobiofertilizers etc., formation of nanosensors/ nanobiosensors, crop improvement strategies, protection and identification of diseases, genetic manipulation of crop plants, improvement of health and breeding techniques of animal and poultry, post-harvest management with smarter, stronger and cost-effective techniques. (Fig. 1.1; Table 1.1) (Wang et al. 2016; Yearla and Padmasree 2016; Sekhon et al. 2014).

Nanotechnology has also been applied in identification of elite genes and their use in crop improvement to for high productivity and disease resistance (Cheng et al. 2016). Similarly, nanotechnology is also used in the health improvement and breeding of animals as well. Besides, the applications of nanotechnology also include nanosensors for accurately reporting the physiological conditions of soil and carbon based nanocarriers for targeted drug delivery, for increasing nutrient uptake, induction of better production etc. (Thornton 2010). As nanotechnology is an interdisciplinary field of science, it has vast diversity of applications in the field of sustainable development of agriculture. Henceforth, the present chapter is more

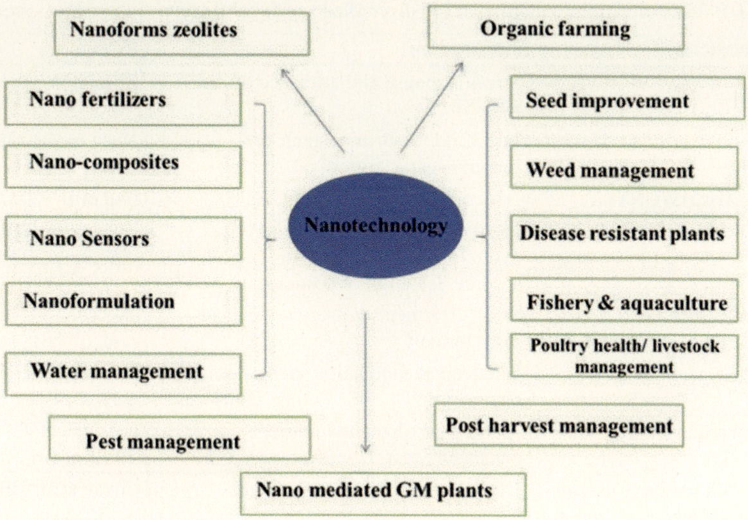

Fig. 1.1 Different fields where nanotechnology has potential application in agriculture

dedicated towards outlining the relevance of nanotechnology in the present day agricultural applications and its future prospective.

1.1.1 Nanoformulations

Nanoformulations alone cannot be considered as a single entity as they are combination of several surfactants, polymers (natural and artificial organic compounds), and metal nanoparticles in nanosize range. Pesticides made up of imidacloprid (1-(6 chloro-3-pyridinyl methyl)-N-nitro imidazolidin-2-ylideneamine), fungicides composed of hexaconazole encapsulated by chitosan nanocapsule, herbicides consisting of optimized diuron, insecticides from polycaprolactone and poly(lactic) acid nanospheres are some of the nanoformulations that are readily used in development of agriculture productivity (Adak et al. 2012; Chauhan et al. 2016; Yearla and Padmasree 2016; Boehm et al. 2003).

1.1.2 Nanofertilizers

Nanofertilizers are having growing demand in contrast to the common chemical fertilizers. Nanofertilizer includes the use of various inorganic compounds like iron oxide nanoparticles for iron deficiency, titanium dioxide nanoparticle, and silicon dioxide nanoparticles for plant growth and nourishment (Wang et al. 2016; Lu et al. 2002). Nanofertilizers have several advantages over conventional

Table 1.1 Uses of various nanoparticles in diversified agricultural sector

Nanoparticles	Applications	References
CeO_2	Stress response and tolerance of *Zea mays*	Zhao et al. (2012)
Fe_2O_3	Insecticide in Bt-transgenic and non-transgenic cotton	Nhan et al. (2015)
AgNPs/ oxMWCNTs	Insecticide	Hsu et al. (2017)
Bionanocomposite materials	Light energy conversion (e.g., photovoltaics) or biosensing (e.g., for specific detection of pesticides)	Nagy et al. (2014)
Carbon-coated Fe nanoparticles	Smart treatment-delivery systems in plant tissue	González-Melendi et al. (2008)
Cellulose nanofibers	Biomedical application such as tissue engineering	Lima et al. (2012)
Cerium oxide and titanium oxide	As nano-fertilizer and nutrient enhancement in *Hordeum vulgare* L.	Poscic et al. (2016)
Copper (II) oxide (CuO) and multiwall carbon nano-tubes (MWCNTs)	Detection of glyphosate in water	Chang et al. (2016)
Cu $(OH)_2$ nanopesticides	Enhancement of nutritional value of *Lactuca sativa*	Zhao et al. (2016)
Engineered water nanostructures	Antimicrobial platform for food safety	Pyrgiotakis et al. (2016)
Fe_2O_3	Nano-fertilizer in *Arachis hypogaea*	Rui et al. (2016)
Fullerene, C_{60} and carbon nanotubes	Increase the water retaining capacity, biomass and fruit yield in plants	Husen and Siddiqi (2014)
Gold NP-based electrochemical biosensor	Rapid and sensitive detection of plant pathogen DNA	Lau et al. (2017)
Magnesium hydroxide (Mg $(OH)_2$) nanocomposites	Efficient bioactive packaging of goods	Moreira et al. (2013)
Multi-walled carbon nanotubes (MWCNTs)	Contaminant carriers and translocate within *Brassica juncea*	Chen et al. (2015)
MWCNT and cotton CNF	Show ecotoxicological effects by inducing viability of *Chlorella vulgaris*	Pereira et al. (2014)
Nano air bubbles	Hydroponic growth of *Brassica campestris*	Ebina et al. (2013)
Nano-biosensors	Labelling products and automated storage	Ali et al. (2017)
Nanochitosan	Supports growth of *Zea mays* and also maintains soil health	Khati et al. (2017)
Nanodiagnostic kit	Plant pathogen detection	Khiyami et al. (2014)
Nanostructured lipid carriers	Antimicrobial activity	Cortesi et al. (2017)
NP-based sensors	Assessing food safety	Bulbul et al. (2015)
Optimized diuron nanoformulation	Nanoherbicide	Yearla and Padmasree (2016)

(continued)

Table 1.1 (continued)

Nanoparticles	Applications	References
Poly (γ-glutamic acid (γ-PGA) and chitosan (CS) polymers based NPs	Nanofertilizers	Pereira et al. (2016)
Protein	Nanoencapsulation of hydrophobic nutraceuticals. Improved functionalities in casein micelle (Meat processing)	Semo et al. (2007)
Quaternized chitosan-capped mesoporous silica NPs	Fungicidal activity against *Phomopsis asparagi*	Cao et al. (2016)
Silver	Control of *Colletotrichum* species	Lamsal et al. (2011)
Thermoplastic starch/ urea medleys as a matrix with hydroxyapatite as nanocomposites	Nanofertilizers	Giroto et al. (2017)
Vitamin D_3-loaded nanostructured lipid carriers (NLCs)	Fortifying food beverages	Mohammadi et al. (2017)
Zinc oxide	Transparent electronics, ultraviolet (UV) light emitters, piezoelectric devices, chemical sensors, spin electronics, enhance crop growth	Sabir et al. (2014)

chemical fertilizer due to their small size, large surface area to act and easy penetration in soil through soil porous.

1.1.3 Nanosensors

Nanosensors/ nanobiosensors are emerging as rising advantageous tools for the application in the field of agricultural research and production which have normal arrangements like ordinary sensors but only vary their size at nano scale level (Omanovic-Miklicanin and Maksimovic 2016). Nanosensors are alternative to conventional methods due to their high sensitivity, selectivity, low detection limits, fast response and small size. Nanosensors are mostly used for detecting pesticide residues and other residues of agrochemicals which is again backed up by certain morphological and economics properties such as ease of miniaturization, electrochemical, optical properties that is further simple and cost-effective (Wang et al. 2016; Cheng et al. 2016).

1.1.4 Nanomaterials

Crop improvement has also been made possible by the use of natomaterials such as carbon nanotubes and other in organic nanoparticles such as gold, SiO_2, TiO_2, ZnO which directly or indirectly helps in nutrient (element) uptake by the plant (Khot et al. 2012). Certain other metallic nanoparticles like silver and copper have shown antimicrobial properties, polymer-based copper nano-compound, silica-silver nanoparticles have been investigated with antifungal properties and for control of certain plant disease like pumpkin disease (Gul et al. 2014).

1.2 Modern Techniques Implemented for Development in Agriculture

Development and improvement of agriculture has for long been dependent on the associated biotic and abiotic components that largely influence the agricultural productivity. Some of these associated components are deteriorating due to the various forced procedure of unsustainable farming such as soil qualities (texture, water retention capacity, nutrient content etc.), crop quality (diseased crops), water availability and quality (Rajonee et al. 2017; Omanovic-Miklicanin and Maksimovic 2016; Cheng et al. 2016). Such forms of impacts are direct results of unmaintained and unsafe agricultural practices for selfish sustenance and can be overcome with the use of modern techniques. Some of these techniques involve the use of interdisciplinary science and research. Currently nanotechnology is a booming field of science like a "philosopher's stone" which is a legendary substance that have capacity to turn any inexpensive metals into gold when combined or touched. Similarly, in the field of research and development, when nanotechnology is combined with any field of science or technology, gives most spectacular and promising results, that is of gold standard. Moreover, in agricultural sector nanotechnology has revolutionized the standards of agriculture by development of soil quality, crop quality, sensing unnecessary agronomical debris, maintenance the productivity and disease progressiveness in the plants as well as or poultry animals and bards, effective and targeted gene manipulation within these farm flora and fauna of agricultural importance, and finally postharvest management with smarter, stronger, cost-effective packaging of these farm/ agricultural products (Wang et al. 2016; Khot et al. 2012; Srilatha 2011). There are many more potential applications of nanotechnology in the agriculture sector (Tables 1.1 and 1.2; Fig. 1.2), which has been subsequently discussed in this chapter.

Table 1.2 Examples of significant applications of nanotechnology in agriculture sector

	Definition	Example	Reference
Crop production			
Plant protection products	Nanocapsules, nanoparticles, nanoemulsions and viral capsids as smart delivery systems of active ingredients for disease and pest control inplants	Neem oil (*Azadirachta indica*) nanoemulsion as larvicidal agent (VIT University, IN)	Anjali et al. (2012)
Fertilizers	Nanocapsules, nanoparticles and viral capsids for the enhancement of nutrients absorption by plants and the delivery of nutrients to specific sites	Macronutrient fertilizers coated with zinc oxide nanoparticles (University of Adelaide, AU CSIRO land and water, AU Kansas State University, US)	Milani et al. (2015)
Water purification			
Water purification and pollutant remediation	Nanomaterials, e.g. nano-clays, filtering and binding to a variety of toxic substances, including pesticides, to be removed from the environment	Filters coated with TiO_2 nanoparticles for the photo catalytic degradation of agrochemicals in contaminated waters (University of Ulster, UK)	McMurray et al. (2006)
Diagnostic			
Nanosensors and diagnostic devices	Nanomaterials and nanostructures (e.g. electrochemical lyactive carbon nanotubes, nanofibers and fullerenes) that are highly sensitive bio-chemical sensors to closely monitor environmental conditions, plant health and growth	Pesticide detection with a liposome-based nano-biosensor (University of Crete, GR)	Vamvakaki and Chaniotakis (2007)
Plant breeding			
Plant genetic modification	Nanoparticles carrying DNA or RNA to be delivered to plant cells for their genetic transformation or to trigger defense responses, activated by pathogens.	Mesoporus silica nanoparticles Transporting DNA to transform plant cells (Iowa State university, US)	Torney et al. (2007)
Nanomaterials from plant			
Nanoparticles from plants	Production of nanomaterials through the use of engineered plants or Microbes and through the processing of waste agricultural products	Nanofibres from wheat straw and soy hulls for bio-nanocomposite production (Canadian Universities and Ontario Ministry of Agriculture, Food and Rural Affairs, CA)	Alemdar and Sain (2008)

Source: Parisi et al. (2015)

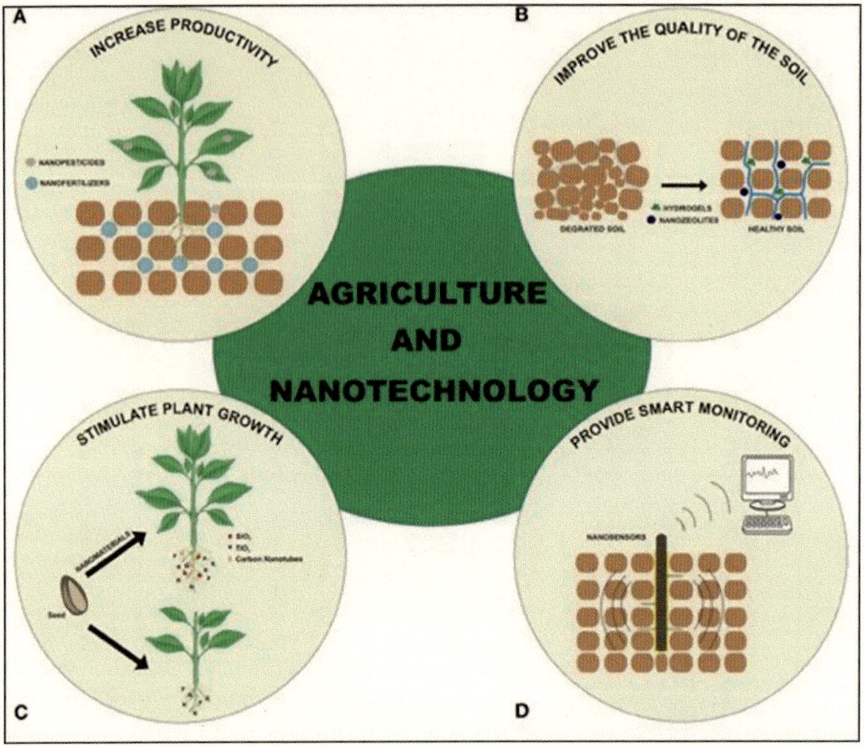

Fig. 1.2 Application of nanotechnology in agriculture. (**a**) for increasing productivity, (**b**) for improvement of the quality of soil, (**c**) for stimulating plant growth and (**d**) for providing smart monitoring (Source: Fraceto et al. 2016)

1.2.1 Nano-Formulations of Agrochemicals

Nanoformulation is a mixture of different particles which gives rise to a single particle of nano range. The mixture in general is composed of several surfactants which stabilizes the formulation either by ionic or non-ionic interaction, polymers (natural and artificial organic compounds), and metal nanoparticles that serves as a carrier or container of the active component of the nanoformulation (Sekhon et al. 2014). Together with the surfactant, the carrier and the active is referred to as a nanoformulation. But when the active component in the nanoformulation is agronomically important compound or substance then the whole formulation can be referred to as nanoagrochemical (Adak et al. 2012; Boehm et al. 2003). The nature of the nanoagrochemical depends on the function or the purpose of the type of agronomically active compound or substance they withhold or composed, for instance the formulation of nanopesticides, nanoherbicides, nanoinceticides and nanofertilizers.

1.2.2 Nanotechnology in Pesticides

Currently nanopesticides that are copper based are widely being used, in comparison to conventional pesticides, in agriculture particularly in organic farming. Zhao et al. (2016) studied metabolomics application of Cu $(OH)_2$ nanopesticides on the nutritional value of *Lactuca sativa* and along with the help of certain sophisticated high-throughput instrumentation and technology such as Gas chromatography-Time-of-Flight-Mass Spectrometry (GC–TOF-MS) based metabolomics approach and Partial Least Squares-Discriminant Analysis (PLS-DA). They also emphasized that the deposition of copper in the vascular bundle of leaves for foliar application. These levels of copper deposition were recommended guidelines including increased concentration of certain minerals specifically potassium as well as upregulation of physiologically important proteins, vitamins, and phytohormones. Optimized Diuron Nano Formulation (ODNF) is a stable nanoherbicide made from diuron (1,1-dimethyl, 3-(3′,4′-dichlorophenyl) urea) and stem lignin as a matrix of *Leucaena leucocephala* for control release of this nanoformulation. Diuron (1,1-dimethyl, 3-(3′,4′-dichlorophenyl) urea) is herbicide, inhibits plant photosynthesis and growth. Diuron activity can be further increased by the help of nanotechnology. Due to its small size, inhibiting the metabolic functions of the seedlings until 16 weeks without losing its activity successfully leads to chlorosis and mortality. ODNF not only function as herbicide but also protects the active ingredient from microbial degradation, UV damage (Yearla and Padmasree 2016). Hexaconazole based *Controlled-released Nitrogen Fertilizers* (CRNF) are the nanoformulation used for the controlled release of hexaconazole. It consists of a fungicide called hexaconazole encapsulated by chitosan nanocapsule that is a combination of chitosan and is synthesized from naturally occurring chitin by partial N-deacetylation with alginate and tripolyphosphate. It is a fungicide that is extensively used for controlling fungal pathogens on various crops mainly against *Rhizoctonia solani*. Besides its fungicidal activity hexaconazole show herbicidal activity and also has major effects including reduction in shoot length, leaf area, and whole dry weight in the case of some *Plectranthus* spp. (Chauhan et al. 2016). Nanostructured Lipid Carriers (NLC) is a nano formulation known for encapsulating natural molecules having antimicrobial activity such as plumbagin, hydroquinon, eugenol, alpha-asarone, and alpha-tocopherol. The NLC were prepared by melting and ultrasonication method, characterized by Cryo-TEM for morphology and SdFFF for dimensional distribution and active encapsulation yields. The efficiency of the system could be mainly described by its efficiency in controlling the phytopathogen. The NLC produced by using blends of solid (e.g. triglycerides) and liquid (e.g. tricaprylin) lipids at ambient temperatures increase solubility and enhances the efficiency of natural antimicrobial molecules. It only gives protective encapsulation for delivery and increases solubility leads to rapid penetration (Cortesi et al. 2017).

It has been previously acknowledged these nano-agroformulations showing more promising results compared to traditional or conventional formulations. Bifenthrin, a pyrethroid insecticide has also been developed, which is neurotoxic to both insects and mammals but its activity dynamically modulated in its nanoformulated form. In a case study conducted by Kah et al. (2016) showed that nanoformulations had remediating effect on its soil absorption and degradation. But the ability of controlled release is not clear. Therefore, further investigation is necessary to evaluate the efficiency of these nano-formulations both on biotic and abiotic components of the parameters influencing agricultural practices.

1.2.3 Nanotechnology in Fertilizers

Presently the synthesis and application of nanoparticles have taken a discreet turn in various field of agricultural science, but when particularly considered to the dynamic nature of soil and its fertility, nanoparticleshave certainly proved their potential. Nanoparticles based fertilizers or in a single term nano-fertilizers such as phosphate, nitrogen, iron, zinc, titanium, aluminium, copper, and silver based nano-fertilizers have tremendously shifted the goal of sustainable agriculture to the next higher level (Malik and Kumar 2014). Some of these nano-fertilizers have been effectively used to improve agricultural products along with sustaining the biotic and abiotic factors associated with farming. Recently a greenhouse experiment was conducted to evaluate the impact of these nano-fertilizers on the production of total phenolic content, and antioxidant activity of rice. In a particular study the efficacy of nano-fertilizers (FRR-CF+FRR-NF) on plant at different stages starting right form seedling, tillering and till panicle initiation stage was evaluated. It was found that FRR-CF+FRR-NF significantly enhanced the plant height, chlorophyll content and the number of reproductive tillers, panicles, and spikelets (Benzon et al. 2015). Nano iron is another example of nanofertilizer, important nanoformulation against Fe deficiency and show Fe fertilization effect in agricultural applications.

Treatment of iron oxide nanoparticles (Fe_2 O_3 NPs) to soil with different concentration showed significant physiological changes in content of soluble sugar and protein, content of chlorophyll and malondialdehyde (MDA), and activity of antioxidant enzymes of watermelon leaves. Proper concentration of Fe_2O_3 NPs has the ability to improve iron deficiency chlorosis and enhance growth of watermelon plant. Studies show that chlorophyll content increases upon exposure to iron-based nanoparticles and thus the growth of certain plant such as *Lactuca sativa* could be significantly increased (Wang et al. 2016). Poly (γ-glutamic acid (γ-PGA) and chitosan (CS) polymers based nanoparticles are currently exploited for their advancement in drug delivery system as a carrier of the phytohormone (gibberellic acid) that regulates plant growth. In general gibberellic acid has several uses in the field such as improvement in the germination and development, plants along with enhancement in their productivity and quality. But upon encapsulating the same phytohormone has higher efficiency than normal phytohormones resulting in increase of leaf area and induction of root development. This phenomenon has been studied in *Phaseolus vulgaris* where γPGA/CS-nano encapsulation of the hormone gibberellic acid as thenano particles favoured the germination of the plant (Pereira et al. 2016).

In another study phosphorus (P) and potassium (K) incorporated nano-fertilizer was prepared, characterized by X-ray diffraction (XRD) and its efficiency was evaluated in a pot culture experiment with *Ipomoea aquatic*.

Zeolite a microporous, aluminosilicate minerals was used for nano-fertilizer synthesis by modification with certain surfactants. The study further concluded that nano-fertilizers efficiently contributed to the growth of the plant as well as to the accumulation P and K. Post-impact of the application of nano-fertilizers on soil showed improved pH, moisture, cations-exchange capacity and micro-nutrients retention capacity (Rajonee et al. 2017). It has been assumed that nano-fertilizers could play an indispensable effective role in increasing agronomic productivity without hampering the natural longitivity agronomic associated factors.

1.2.4 Application of Nanosensors/Nanobiosensors in Sustainable Agriculture

Sensors are the interface between the real and the virtual/ digital world. They are slowly and steadily amalgamating with versatile emerging technologies in inter or multidisciplinary fields of science and research to dynamically transcend the current evaluation of the magnificent nature along with the miraculous life within it. Commonly a sensor can be defined as a physical instrument used to measure physical properties, record, track, indicate or otherwise respond to it (Schneider et al. 2015). The three basic components of a sensor are detector, amplifier and transducer (Arlett et al. 2011; Rai et al. 2012). Based on the origin, type and the mode of detection these sensors can be segregated into different groups such as photo sensors, electrical sensors, chemical sensors or biological sensors (Rai et al. 2012; Turner 2013). Moreover, all these sensors can be further called as nano-sensors if their size falls within the nano range (Rai et al. 2012). Due to the small size, nano-sensors could be more efficiently exploited in heterogeneous filed for the benefit of mankind. Currently the use of these nano-sensors in the field of agriculture has made influential embellishments that forced the present thinkers to rethink the whole traditional processes of framing and adapt this spectacular piece of technology and integrate it into the farming (Omanovic-Miklicanin and Maksimovic 2016). Henceforth the nanotechnology based nano-sensors are employed in detection of residues of agrochemicals, crop improvement, protection and identification of diseases (Khot et al. 2012; Wang et al. 2016; Cheng et al. 2016).

1.2.4.1 Application in Detection of Residues of Agrochemicals

Agrochemical residues are those chemical components of the pesticides or the pesticide itself which gets accumulated in the field or the crops harvested from the field (Bhandari 2014). Food and Drug Administration (FDA) have detected 1045–1603 chemicals designated as pesticide residues (FDA 2005). In another FDA report of 2014 fiscal year a total sample of 6638 were analysed of both human and animal foods, where 705 pesticides and industrial chemicals were detected by the FDA (FDA 2014).Despite the significance of agrochemicals in efficient farming, they leave a trail of their despicable virtue that devours the fruitfulness sustainable agriculture. The over use of these chemicals have serious impact on the surrounding or the whole ecological habitat which may lead to immediate but chronic, catastrophic and irreversible modulations on the sustenance of agricultural productivity. A simple but yet effective remedy for stabilizing this patronizing deteriorative phenomenon is possible by the means of certain miniature devices based on nanotechnology (Bhandari 2014). These devices could efficiently detect the residues and simultaneously neutralize the threat and report.

These devises are generally called as sensors or more specifically nano-sensors because of the miniature size. They offer profound serviceability by their high sensitivity, super selectivity, fast responses and low detection limits. Nano-sensors have already been used for detection of some of the pesticide residues such as methyl parathion, parathion, fenitrothion, pirimicarb, dichlorvos and paraoxon (Khot et al. 2012). In the present scenario, an efficient mode of sensing such as luminescence had significantly increased the serviceability of these nano-sensors. Vasimalai and John (2013) had developed and studied a luminescent sensor using chitosan capped silver nanoparticles as a fluorophore for detection of malathion. The yellow color of the nanoparticle formulation changed to brown and a sharp decrease in the absorbance along the redshift. The nano-sensor was again applied for the determination of malathion in fruits and water samples the result of which further supported and validated by HPLC.

Recently a turn-off sensor was developed for measuring glyphosate using amalgamated form of copper (II) oxide (CuO) and multiwall carbon nano-tubes (MWCNTs), the efficiency of which was indicated by the decreased catalytic activity of the CuO/MWCNTs. The sensor showed promising efficiency in detection of glyphosate in water (Chang et al. 2016). Another marvelous detection method of dimethoate by the use of newly developed oxidized multi-walled carbon nanotubes (oxMWCNTs) modified with silver nanoparticle (AgNPs) having peroxidase-like activity further proved the extent to which nanotechnology could be exploited. Excellency of these synthesized nano-sensor's(AgNPs/ oxMWCNTs) peroxidase-like activity was verified by hydrogen peroxide-Amplex red system and it was found that, the catalytic activity of AgNPs/ oxMWCNTs decreased in the presence of dimethoate, because the insecticide effectively interacted with AgNPs of the nano-formulation (Hsu et al. 2017). Similarly nano-biosensors are an advanced eco-friendly wing of the nano-sensors, as they include a biological organic moiety as a variant component in their formulation (Rai et al. 2012). In a more simple way it can be said that these nano-biosensors are equipped with utilitarian bio-components such as enzymes, proteins and nucleic acids, which could be perceptible via diversified means within plants (Misra et al. 2013). Da Silva et al. (2013) demonstrated for the first time the implementation of atomic force spectroscopy in the detection of enzyme-inhibiting herbicides. They developed and characterized a nano-biosensor based on atomic force microscopy tip functionalized with acetolactate synthase enzyme which was used for detection of an herbicide metsulfuron-methyl. Nano-biosensors are a growing technology of for simple, sensitive, selective, and rapid detection methods and with the present circumstances it could be speculated that one day this multidisciplinary piece of technology will certainly change the course of history.

1.2.4.2 Application in Crop Improvement

Crop improvement is the primary task of any agricultural inputs which also shows significant and promising results. This has also been achieved using nanotechnology, where enhanced sensing technologies through nano-sensors/ nano-biosensors, have up regulated the management and conservation of input in crops in the current agricultural practices. In present situation, the foremost limiting factor for crop improvement includes biotic and abiotic stress or the combined stress which adversely affects the productivity of the plant (Singh et al. 2016; Pandey et al. 2017). The biotic stress encompasses different plant disease due to viruses, bacteria and funguses and secondly the abiotic stress, which is mainly due to up or down regulation of certain protein or growth factors, again the min influencer for which are temperature, pressure, pH, moisture content, imbalance in ion and nutrient concentration (Pandey et al. 2017). In the present century, the population explosion has laid to uncontrolled pollution of air, water and land has enhanced the destabilization of the abiotic factors resulting in increased stress (Crippa et al. 2016). Moreover, the stingy nature of mankind has laid the founding for unethical experimentation which resulted in origin of new genetically modified resistance organisms causing unprecedented diseases in both plants and animals (Maghari and Ardekani 2011).

Precocious detection of plant disease ahead of time has impelled present researchers to look for nanotechnology based remedies for crop plants against biotic stress through the utilization of autarchic nano-sensors which are linked to GPS system for real time monitoring the status or the condition of the soil and the crops (Misra et al. 2013). DNA detection is one of the basic research methods that have been widely exploited science the beginning. Despite that, it is still a sophistication to detect unambiguous sequence of DNA or low abundance genes in biological sample with specificity and sensitivity. Recently surface enhanced Raman scattering (SERS) technology based bio-sensing platform has been studied. Here a target DNA (tDNA) accelerates self-orientation of gold nanoparticles (AuNPs) probes on DNA which is in the form of nanowires for signal amplification in DNA detection based on hybridization chain reaction. This technique can be used for any biological sample (Chen et al. 2014). Optical nano-biosensors based on fluorescence resonance energy transfer (FRET) have been studied by Bagheri et al. (2017) where they detected tropane alkaloids as anti-cholinergic agents in both natural and transgenic hairy root extract of *Atropa belladonna*. They formulated a sensor of cadmium telluride quantum dots with M2 muscarinic receptor and tioglycoyic acid as capping agent along with scopolamine-rhodamine 123 conjugate.

Abiotic stress on the other hand is another speed breaker that hinders the progressivity of sustainable crop improvement. Nutritional scarcity and heavy metal accumulation leading to generation of ROS which further results in senescence, must be detected at an early stage (Meena et al. 2017). The task which has been achieved to a greater extent by nano-sensors is the primary consideration of present research (Zaytseva and Neumann 2016). Taher et al. had developed a new solid-phase extraction method using oxidized multiwalled carbon nanotubes in concentrated HNO_3 which was again modified with 2-(5-bromo-2-pyridylazo)-5-diethylaminophenol for extraction, pre-concentration, and electrothermal atomic absorption spectrometric determination in real sample which were at ng/L level (Taher et al. 2014). Choi and Gilroy (2014) developed Fluorescence Resonance Energy Transfer (FRET) based biosensor to report the level of stress hormone specifically abscisic acid within the plant cell of *Arabidopsis thaliana* in real-time. Sooner or later the nanoform of this biosensor will be developed and during that time no stones will be left unturned.

1.2.5 Genetic Manipulation of Crop Plants through Nano-Devices for Disease Diagnosis and Improvement

Genetic manipulation of plants means modification of the chromosomal or extra chromosomal DNA directly or indirectly, which leads to the formation of genetically modified organisms (GMOs) or genetically modified plants or crops (Zhang et al. 2016). Since 1859, when Charles Darwin published the first edition of "On the Origin of Species" till now when Nanocarriers for carrying desired gene to the preferred site with efficiency and controlled release, the research is constantly booming (Zhang et al. 2016; Wu et al. 2017). But the beginning was always humble in case of its application for the razing poverty. One of the finest pioneered works in this sector is "Golden rice" that had been designed to express high level of beta-carotene or a cost-effective and efficient way to deliverer dietary source of Vitamin A (Oliver 2014). Plants resistance against various parasites was developed and adapted that had previously been modified through genetic engineering conjugated with nanotechnology. In some studies biological pesticides like abamectin (Abm) which has been modified through nanotechnology and genetic manipulation to improve its poor mobility and nematicidic activity that was previously due to restricted area of protection around the developing root system. Technically Abm's physical chemistry was manipulated by encapsulating it with *Red clover necrotic mosaic virus* (RCNMV) that resulted in the formation of an efficient plant virus nanoparticles (PNV) delivery system (Cao et al. 2015).

Nutrient deficiency is another important problem that has been easily mitigated by use of nanocarriers that directly fulfills the requirement of plant nutrient requirement. One such study was conducted on cereals in zinc deficient soil where the deficiency of the micronutrient was ameliorated by the application of nanotechnology. Here zinc deficiency in the cereal grains were enhanced by zinc complexes chitosan nanoparticles which was synthesized by using tri-polyphosphate as cross-linker (Deshpande et al. 2017). Crop quality can be further improved by local translocation of small RNAs between cells. This phenomenon has been experimented in tomato which was grafted into goji (*Lycium chinense*) that reveals the hidden activity of miRNAs in regulation and expression arrangements within a distance grafting system (Khaldun et al. 2016). The point where nanotechnology meets plant biotechnology is like science-fiction meeting reality, where nanocarriers are used in gene delivery to plant cells. Nanocarriers such as carbon nanotubes immobilized with cellulase had been studied for its ability to deliver DNA effectively (Fouad et al. 2008). One of the important biomedical systems, named calcium phosphate (CaP) having diversified application including delivering plasmid DNA, is a widely used non-viral gene delivery method. The efficiency of CaP nanoparticles in delivering pBI121 harboring GFP by 35S promoter-encoding plasmid DNA into tobacco cells have been evaluated (Ardekani et al. 2014). Currently the application only involves the delivery of bioactive compounds such as proteins and genetic materials or certain drugs to the animal cells. Again there is a long way to go in the path where nanodevices will be efficiently used for the delivery of genetic material to plant cells (Menaa 2015).

1.2.6 Nanocomposites/Nano-Biocomposites in Agricultural Development

A nanocomposite is a multiphase solid material with a dimension of less than 100 nm composed of nano-sized ceramics or other natomaterials either organic or inorganic with a capacity of substantially enhancing the composite property of the matter containing these nanocomposites (Bogue 2011; Barahuie et al. 2013). Unlike nanocomposites, polymeric nanocomposites are intensely cross-linked, that gives rise to the exquisite properties characteristically high stiffness, strength, creep, chemical corrosion resistance and elevated temperature tolerability. Further the polymer nanocomposite property depends upon diversified parameters such as shape, adhesive natures and dispersive phase (Duggal et al. 2016). Polymers shows better biocompatibility properties when they are of biological origin such as proteins/ enzymes that have excellent electro catalytic activity. The activity which is a result of charge distribution within the inter or intra molecular residues of the protein which together with the matrix gives rise to a versatile form of biocompatible nanocomposites with indefinite possibilities (Jamir and Mahato 2016).

These nanocomposites/nanobiocomposites are currently being exploited for the development in agricultural activity in a diversified way directly or indirectly. Though there has not yet been any direct study on how nanocomposites effect or interact with crops, but it can be assumed that the side effect is negligible because every risk in science is stepping forward in the path of enlightenment in knowledge of understanding the scientific phenomenon little bit closer. Presently the application of nanocomposites is focused or in other words it can also be said that it is limited only to certain application like food packaging, preservation, restoration of dynamic property of soil and/ or nano-biofertilizers, biosensors/ nano-biosensors (Sekhon 2014). A report proves the development of efficient bioactive packaging by nutraceuticals inspired pectin- magnesium hydroxide (Mg $(OH)_2$) nanocomposites. Here a bioactive edible film based on pectin was developed as a dietary scaffold and nanoplates of Mg $(OH)_2$ as the reinforcing filler. The nanocomposite morphological characterization was carried out with the help of atomic force microscopy and Fourier transform infrared spectroscopy (Moreira et al. 2013). Food preservation could be effectively enhanced by nanocomposite synthesized from chitosan, nano-cellulose fiber and thyme oils.

The nanocomposite affected the moisture content of the preserved fruits which was further followed by decrement in weight and total sugar content, while acidity was unaltered. Overall it can be said that the coating with nanocomposite leads to increased shelf life and reduction in fungal growth (Nabifarkhani et al. 2015). Soil is the most vital component of the sustainable agriculture and therefore conserving its dynamic nature in itself is a great task. Though presently fertilizers are been used for maintaining its stability, its adverse effect is unavoidable. Therefore, nanofertilizers are the current applied techniques to reduce the adverse effect of the fertilizers. One of such study is being done by using nanocomposite for improving phosphate and urea intake of the soil. The nanocomposite was prepared from urea or extruded thermoplastic starch/ urea medleys as a matrix with hydroxyapatite (Giroto et al. 2017). There are many other application such as nano-clay, biochar-nanoparticle as nanocomposite and many other polymer based nanocomposite that have already being used for the development of agriculture but only at experimental level. The present effort should be to bring these versatile magnificent technologies out in to the field to each and every farmer so that the productivity would never cease.

1.2.7 Application of Nanotechnology in Hydroponic

Hydroponics is a precocious technology where plants were artificially grown in a liquid solution containing all of the required nutrients. This marvelous technology named hydroponic system is in extensive use in the present decade to study and understand the biotic and abiotic stress response to plant (Nguyen et al. 2016). Now a days in supermarkets the displayed fruits and vegetables are grown hydroponically of which the most common crops are tomatoes, cucumbers, sweet peppers, melons, lettuce, strawberries, herbs, eggplant, and chilies (Sekhon 2014). This technique was actually able to accelerate the kinetics of growth in the plant with respect to time that the plant normally required for its growth. The phenomenon was studied successfully in the Electro-Hydroponic culture system where the application of an electric field with varying intensity of direct current at galvanostatic regime (50–12.5 mA) along with required nutrient solution for the effective, alternative and interesting harvest technique for the growth of *Lactuca sativa* (Fuentes-Castaneda et al. 2016). The technique can further be advanced with the help of nanotechnology by the implementation of nano-pesticides which is sometimes a necessary, nanofilters, nano-preservatives for maintenance of the adequate moisture content of seeds. Currently nano bubbles play a major role in the process of seed germination which is again a vital step for plant development. These nano bubbles ultimately help to hydrate the seedling and improve its metabolism and ultimately their growth enhancement.

This prepares a wider sphere for growth of hydroponic plants and hence, improves agro ecosystem. This whole process takes place by increasing the OH^- concentration in the water which forms shells and hence, no space for further gas dissolution (Ushikubo et al. 2010). Again these nano-bubbles generally increase the oxygen concentration in the air nano bubble reactor. In a study conducted for time duration of 4 weeks, on hydroponic growth of *Brassica campestris* shows that nanobubble has a great impact on the co-cultured organisms like fishes. An air nanobubble reactor which continuously releases air nano bubbles improves the growth of sweet fishes and rainbow trout. Nano air bubbles for 3 weeks in case of sweet fishes and 4 weeks in case of rainbow trout is shows better growth (Ebina et al. 2013).The impact of titanium dioxide nanoparticles (TiO_2 NPs) on growth and survival of the nitrogen fixing bacterium *Rhizobium trifolii* in the symbiotic root nodule of the plant *Trifolium pretense* along with the growth of the plant in the hydroponic system was investigated through eco-toxicological tests. The results indicated that about 21% of the TiO_2 NPs treated plants were devoid of the nodulation and at elevated concentration of NPs resulted in the impaired *R. trifolii* as well as the growth of *T. pretense* plant (Moll et al. 2016). Nanotechnology in these ways could reduce the energy requirements, time and overall cost and soil exploitation and pave a path for sustainable agriculture.

1.2.8 Application of Nanotechnology in Organic Agriculture

The biological or ecological agriculture in conjugation with classical feasible farming methods with advanced sustainable farming technology which is commonly known as organic agriculture is one of the preferred nutrient source. It focuses mainly on rotating crops, natural management of pests, variation in crops, diversity in livestock, and conservation and maintenance of soil quality along the addition of compost and green manures (Reganold and Wachter 2016). Organic agriculture enhances the health of agro-ecosystem in diversified ways and means. Now, incorporating nanotechnology to it would not only improve the quality of crops but also give a boost to the livestock. Several companies are predicting that the application of nanotechnology would uplift the economy of food industry and hence, termed as 'agrifood nanotechnology'. Firstly, nanotechnology is applied to the food and animal feed in form of colouring, vitamins, or flavours within nanocapsules. Secondly, nanotechnology has applications in fertilizers which lessen its utility in crops (Jahanban and Davari 2014). At present nanoparticles are used in animal feeds as colouring, flavours and nutrition/ vitamins in the form of nanocapsules, which because of nano size could able to dissolve readily and efficiently in beverages (Huang et al. 2015). Moreover nano-formulations of fertilizers of biotic origin shows more efficiency in comparison to conventional fertilizers and lastly biopolymers with integrated nanoparticles proves to be compostable and more kinetically stable than other biopolymers (Jahanban and Davari 2014). As International Federation on Organic Agriculture Movements (IFOAM) had rejected the use of nanotechnology and nanomaterials in organic agriculture, therefore research regarding the use of nanoparticles in organic agriculture is somewhat limited (Sekhon 2014).

1.3 Application of Nanotechnology in Poultry and Animal Health

It was previously known that nanotechnology is venturing in every field of science and research and leaving behind spectacular trails of new scientific possibilities. These new panoramas of nanotechnology in combination with other derived fields of science such as molecular biology, animal biotechnology and to some extent clinical biotechnology have invigorated nearly every sector of veterinary and animal sciences by stunningly modulating the synergetic applications in relation to sustainable poultry production (Muktar et al. 2015). There are discrete forms of nanomaterials that could be effectively used for disease diagnosis, treatment, supplementation in animal nutrition, efficient animal breeding, safe reproduction, and value-added poultry products (Thornton 2010). Some of the commonly used nanomaterials includes metallic nanoparticles, quantum dots, single walled and multi-walled carbon nanotubes, fullerenes, liposomes and dendrimers (Sekhon 2014). Despite the serviceability of nanotechnology and its exploitations in major innovations, it is still developing and assumed to hinder environment and its components directly or indirectly. Hence, the research of the present decade is more concerned to develop nanomaterial with ameliorated toxicity and spread awareness of the benefits as well as potential risk (Thornton 2010).

Nutrition delivery is the primary concern of any livestock industry and nanotechnology is slowly advancing in the concern sector. A study was conducted to evaluate the impact of different level of nano chromium picolinate (nanoCrPic) on egg quality, mineral retention and mineral accumulation within the tissue of chickens. It was found that the supplemental nano CrPic could effectively enhance egg quality, withholding of chromium and zinc ions, and increased concentration of mineral accumulation in the liver (Cr, Ca, P) and yolk (Ca), and in eggshell (Ca) (Sirirat et al. 2013). Nano-polymers in conjugation to nutrients could be highly efficient for delivery to the gastrointestinal compartment. Because it is evident from previous studies that the nanoparticle with their versatile morphological, physical and chemical properties could easily overcome the extreme alkaline and acidic condition of the gut (Ban et al. 2015). Biocidic applicability could be easily attained by fabricating magnetic nanoparticles around gold nanoparticles by means of photothermal lysis of photogenic bacterium *Salmonella typhi* is one of the growing technique is investigated by the researchers all over the world. It is a simple technique where after the absorption of light by the nanohybrid, it from its electromagnetic energy conversion generates heat in the surrounding medium that quickly shifts the temperature to an intolerable extent for the pathogen and results in their lysis (Ramasamy et al. 2016).

Detection and diagnosis through luminescence is an expanding field of medical science. Presently real-time imaging by a probe, named ratiometric mitochondrial cysteine-selective two-photon fluorescence was developed by using the biothiol reaction site of acrylate moiety and a merocyanine as a fluorophore is widely studied for their applicability in diagnostics (Niu et al. 2016). Nanoparticle-based antibodies or lectins can be used as surface markers for the removal of aberrant spermatozoa. A study conducted on the ability to remove defective spermatozoa by nanoparticle-based magnetic purification method from bull semen was proven to be efficient in improving sperm sample viability, fertilizing ability both *in vitro* and *in vivo*. The two types of nanoparticles used in the present research include an antibody against ubiquitin and another nanoparticle coated with lectin PNA. No side effect was observed in all of the 466 healthy inseminated animals or their offspring (Odhiambo et al. 2014). Nanocarrier mediated delivery, on the other hand, is another fast growing and promising technique in reproductive biology because of the potentiality in the improvement of the safety and efficiency of existing methodologies that includes *in vitro* and *in vivo* experimental gene therapy and sperm-mediated gene transfer. A pioneering study on the use of mesoporous silica nanoparticles with the symmetry of the pores that were hexagonal in structure and surface functionalized with polyethileneimine and aminopropyltriethoxysilane and facultatively packed with two common types of bioactive compounds such as nucleic acid or proteins for intact association with sperm without showing any negative effect upon the primary parameters of sperm functions (Barkalina et al. 2014). Nanoparticles are commercialized and their development is in continues progression, their spectacular properties are being manipulated and optimized on the basis of particular function in which they could be used, but still then in the case of animal production nanotechnology is still in its infancy in certain applications (Hill and Li 2017).

1.4 Application of Nanotechnology in Post-Harvest Management of Agricultural Goods

Fresh Fruits and Vegetables (FFV) are the imperative origins of vital vitamins such as vitamin A and C along with minerals like potassium and other ions for eudaimonia. The problem is that, these are perishable living products that required an extra coordinated attention by the producers after harvest, but due to unawareness of proper consideration of the necessity condition results in the loss of harvested FFVs unexpectedly (Mahajan et al. 2014).Post-harvest management refers to mitigate the unexpected measurable loss of both quality and quantity of harvested food crops in storage, packaging, transportation, processing and appropriate preparation before consumption. Paucity in appropriate skill and technology post-harvest management such as maintenance of the optimum temperature for longer time period without the loss of important nutritional values and proper packaging techniques to avoid several unexpected short comings that might hamper the food security that has already caused alarming threat of high level of poverty among the developing countries (James and Zikankuba 2017). Classical management system is inefficient in maintaining all the adequate condition simultaneously because of high energy cost of both labor and equipment and scaling up of the products, deficiency of edible materials with required properties, intensive investment, autonomous harmonized regulatory jurisdiction or laws, sluggish consumer acceptance owing to noticed alliance with radioactivity, difficulty to reach unreachable sites for treatment of fresh products within calyx and wax areas and elevated concentration of the chemical agents used may induce health hazards (Mahajan et al. 2014).

In the present scenario, some of these drawbacks could be alleviated by the application of nanotechnology. Currently modulating the growth and development of microorganisms by the use nanomaterials like graphene oxide that is in the form of nanosheet, generation of efficient packaging covers or films made up of nano-objects like nanorods, nanotubes, and nanowires that can also be called as nanofiber are slowly commercializing because of their inhibiting the entry of gases and the harmful rays (Palmieri et al. 2017; Wyser et al. 2016). Strength, quality and the morphological beauty of the packaging material is also further advanced by the application of nanotechnology (Pradhan et al. 2015). Further nano-biosensors are used for labeling products and are considered as the primary step in an automatically controlled storage (Ali et al. 2017). In this way the application of nanotechnology in the processes of altering the sophistications that previously were limiting developments in the field of effectively of the post-harvest management.

1.5 Conclusion and Future Prospects

Future could be alarming, but it is this fear that propels the mankind to constantly focus on eradicating or mitigating the possible expected cause that might ignite such alarming effect. Researchers are increasingly, trying to unfold the vast mysteries which are seemingly hidden in the earth as well as in the distant stars within visible universe as well as beyond it in the near future. The more mankind enlightens themselves with the understanding of the nature, the greater will be the understanding of our insignificant character. Humans have much to learn and much to understand from this unexplored and unique universe as well as from ourselves and our needs. As mankind is the present known smartest creature of the earth, the responsibilities in conserving the nature should be also done in a smarter way. But instead exploitation of the dispensable values of the Mother Nature is currently on the verge of extinction. Still there is hope. An antidote which we call as science is referred to as good servant but bad master. Present unsustainable harvest technique demands for more developed ways and means to combat or tackle the upcoming wave of poverty that might one day take over the world. Some of the future technique that might be a silver lining for agricultural development includes the following;

The indispensable abiotic yet dynamic component of agriculture is soil, which is slowly losing its dynamic properties, but could be restored efficiently by the application of nanotechnology (Malik and Kumar 2014). Nano-biosensors are currently a simple but effective way of detecting the deficiency in the soil by colour, light and heat (Wang et al. 2016). After detection the next task is the neutralization of the problem that too is possible with the help of nanotechnology. Basing the specific type of deficiency of the soil, nanocarriers could be developed which could effectively deliver the required supplement of the soil (Malik and Kumar 2014; Deshpande et al. 2017). Once the soil is enriched, plants are the vital component to which the focus shifts. The linked problems include insects and parasites that could also be more efficiently dealt with the application of nano-pesticides, nano-insecticides, nano-fungicides etc. (Yearla and Padmasree 2016). Further, genetic engineered crop plants could be developed by the application of molecular biology in combination with nanocarriers could result in the development of disease resistance plants, plants with essential vitamins, proteins, hormones etc. (Ardekani et al. 2014; Oliver 2014; Zhang et al. 2016; Wu et al. 2017). Lastly effectual mode of management of the agricultural products till it reach the end user is a very important task. Therefore promising measures should be taken to manage these products for which competent mode of preservation, packaging, transport and delivery is required. Nanotechnology could further used for the improvement of the post-harvest management of the agricultural goods (Palmieri et al. 2017; Wyser et al. 2016). Apart from these applications nanotechnology have touched every aspect of sustainable agricultural development in the present as well as the future. It is not late when nanotechnology will be regarded as the central axis in overthrowing the poverty through its application in sustainable agricultural development in the current century.

Acknowledgements Authors are grateful to Dr. Leonardo F. Fraceto of Department of Environmental Engineering, São Paulo State University, Sorocaba, Brazil for giving permission to include their Fig. 1.1 (as Fig. 1.2) and to Dr. Claudia Parisi of European Commission, Joint Research Centre (JRC), Institute for Prospective Technological Studies (IPTS), Edificio Expo, C/ Inca Garcilaso3, 41,092 Seville, Spain for giving permission to include their table (as Table 1.2) in the current chapter. We are also grateful to Frontiers in Environmental Science and Nano Today journal for permission to include the figure and table under creative commons open access licence.

References

Adak T, Kumar J, Shakil NA, Walia S (2012) Development of controlled release formulations of imidacloprid employing novel nano-ranged amphiphilic polymers. J Environ Sci Health B 47(3):217–225

Alemdar A, Sain M (2008) Isolation and characterization of nanofibers from agricultural residues: wheat straw and soy hulls. Bioresour Technol 99(6):1664–1671

Ali J, Najeeb J, Ali MA, Aslam MF, Raza A (2017) Biosensors: their fundamentals, designs, types and most recent impactful applications: a review. J Biosens Bioelectron 8:1

Anjali CH, Sharma Y, Mukherjee A, Chandrasekaran N (2012) Neem oil (*Azadirachta indica*) nanoemulsion--a potent larvicidal agent against *Culex quinquefasciatus*. Pest Manag Sci 68(2):158–163

Ardekani MRS, Abdin MZ, Nasrullah N, Samim M (2014) Calcium phosphate nanoparticles a novel non-viral gene delivery system for genetic transformation of tobacco. Int J Pharm Pharm Sci 6(6):605–609

Arlett JL, Myers EB, Roukes ML (2011) Comparative advantages of mechanical biosensors. Nat Nanotechnol 6:203–215

Bagheri F, Piri K, Mohsenifar A, Ghaderi S (2017) FRET-based nanobiosensor for detection of scopolamine in hairy root extraction of *Atropa belladonna*. Talanta 164:593–600

Ban C, Park SJ, Lim S et al (2015) Improving flavonoid bioaccessibility using an edible oil-based lipid nanoparticle for oral delivery. J Agric Food Chem 63:5266–5272

Barahuie F, Hussein MZ, Hussein-Al-Ali SH et al (2013) Preparation and controlled-release studies of a protocatechuic acid-magnesium/aluminum-layered double hydroxide nanocomposite. Int J Nanomedicine 8:1975–1987

Barkalina N, Jones C, Kashir J et al (2014) Effects of mesoporous silica nanoparticles upon the function of mammalian sperm *in vitro*. Nanomedicine 10:859–870

Benzon HRL, Rubenecia MRU, Ultra VU Jr, Lee SC (2015) Nano-fertilizer affects the growth, development, and chemical properties of rice. Int J Agric Res 7(1):105–117

Bhandari G (2014) An overview of agrochemicals and their effects on environment in Nepal. Appl Ecol Environ Sci 2(2):66–73

Boehm AL, Martinon I, Zerrouk R et al (2003) Nanoprecipitation technique for the encapsulation of agrochemical active ingredients. J Microencapsul 20(4):433–441

Bogue R (2011) Nanocomposites: a review of technology and applications. Assem Autom 31:106–112

Bulbul G, Hayat A, Andreescu S (2015) Portable nanoparticle-based sensors for food safety assessment. Sensors (Basel) 15(12):30736–30758

Cao J, Guenther RH, Sit TL et al (2015) Development of abamectin loaded plant virus nanoparticles for efficacious plant parasitic nematode control. ACS Appl Mater Interfaces 7(18):9546–9553

Cao L, Zhang H, Cao C et al (2016) Quaternized chitosan-capped mesoporous silica nanoparticles as nanocarriers for controlled pesticide release. Nanomaterials (Basel) 6(7):126

Chang YC, Lin YS, Xiao GT et al (2016) A highly selective and sensitive nanosensor for the detection of glyphosate. Talanta 161:94–98

Chauhan N, Gopal DM, Kumar R, Kim K-H, Kumar S (2016) Development of chitosan nanocapsules for the controlled release of hexaconazole. Int J Biol Macromol 97:616–624

Chen K, Wu L, Jiang X et al (2014) Target triggered self-assembly of Au nanoparticles for amplified detection of *Bacillus thuringiensis* transgenic sequence using SERS. Biosens Bioelect 62:196–200

Chen G, Qiu J, Liu Y et al (2015) Carbon nanotubes act as contaminant carriers and translocate within plants. Sci Rep 2015(5):15682

Cheng HN, Klasson KT, Asakura T, Wu Q (2016) Nanotechnology in agriculture. In: Cheng HN, Doemeny L, Geraci CL, Schmidt DG (eds) Nanotechnology: delivering on the promise, vol 2, ACS symposium series, vol 1224. American Chemical Society, Washington, DC, pp 233–242

Choi WG, Gilroy S (2014) Plant biologists FRET over stress. eLife 3:e02763

Cortesi R, Valacchi G, Muresan X-M et al (2017) Nanostructured lipid carriers (NLC) for the delivery of natural molecules with antimicrobial activity: production, characterization and in vitro studies. J Microencapsul 34:63–72

Crippa P, Castruccio S, Archer-Nicholls S et al (2016) Population exposure to hazardous air quality due to the 2015 fires in Equatorial Asia. Sci Rep 6(37074):1–9

Da Silva ACN, Deda DK, Da Roz AL et al (2013) Nanobiosensors based on chemically modified *afm* probes: a useful tool for metsulfuron-methyl detection. Sensors 13:1477–1489

Deshpande P, Dapkekar A, Oak MD et al (2017) Zinc complexed chitosan/TPP nanoparticles: a promising micronutrient nanocarrier suited for foliar application. Carbohydr Polym 165:394–401

Ebina K, Shi K, Hirao M et al (2013) Oxygen and air Nanobubble water solution promote the growth of plants, fishes, and mice. PLoS One 8(6):e65339

FAO (2017) The future of food and agriculture – trends and challenges, Rome

FDA (2005) Glossary of pesticide chemicals. http://www.fda.gov/

FDA (2014) Pesticide residue monitoring program fiscal year report. https://www.fda.gov/downloads/Food/.../Pesticides/UCM546325

Fouad M, Kaji N, Jabasini M et al (2008) Nanotechnology meets plant biotechnology: carbon nanotubes deliver DNA and incorporate into the plant cell structure, 12th international conference on miniaturized systems for chemistry and life sciences, 2008, San Diego, CA, USA

Fraceto LF, Grillo R, de Medeiros GA et al (2016) Nanotechnology in agriculture: which innovation potential does it have? Front Environ Sci 4(20). https://doi.org/10.3389/fenvs.2016.00020.

Fuentes-Castaneda O, Dominguez-Patiño ML, Dominguez-Patino J et al (2016) Effect of electric field on the kinetics of growth of lettuce (*Lactuca sativa*) in a hydroponic system. J Agric Chem Environ 5:113–120

Giroto AS, Guimaraes GGF, Foschini M, Ribeiro C (2017) Role of slow-release nanocomposite fertilizers on nitrogen and phosphate availability in soil. Sci Rep 7(46032):1–11

González-Melendi P, Fernandez-Pacheco R, Coronado MJ et al (2008) Nanoparticles as smart treatment-delivery systems in plants: assessment of different techniques of microscopy for their visualization in plant tissues. Ann Bot 101(1):187–195

Gul HT, Saeed S, Khan FZA, Manzoor SA (2014) Potential of nanotechnology in agriculture and crop protection: a review. Appl Sci Bus Econ 1(2):23–28

Hill EK, Li J (2017) Current and future prospects for nanotechnology in animal production. J Anim Sci Biotechnol 8:26

Hsu CW, Lin ZY, Chan TY et al (2017) Oxidized multiwalled carbon nanotubes decorated with silver nanoparticles for fluorometric detection of dimethoate. Food Chem 224:353–358

Huang S, Wang L, Liu L et al (2015) Nanotechnology in agriculture, livestock, and aquaculture in China: a review. Agron Sustain Dev 35:369–400

Husen A, Siddiqi KS (2014) Carbon and fullerene nanomaterials in plant system. J Nanobiotechnol 12:16

Jahanban L, Davari M (2014) Organic agriculture and nanotechnology. In: Rahmann G, Aksoy U (eds.) Proceedings of the 4th ISOFAR scientific conference. 'Building Organic Bridges', at the Organic World Congress, Istanbul, Turkey

James A, Zikankuba V (2017) Post harvest management of fruits and vegetable: a potential for reducing poverty, hidden hunger and malnutrition in sub-Sahara Africa. Cogent Food Agric 3:1312052

Jamir A, Mahato M (2016) A review on protein based nanobiocomposite for biosensor application. Rev Adv Sci Eng 5:109–122

Kah M, Weniger AK, Hofmann T (2016) Impacts of (nano) formulations on the fate of an insecticide in soil and consequences for environmental exposure assessment. Environ Sci Technol 50:10960–10967

Khaldun BM, Huang W, Haiyan L et al (2016) Comparative profiling of miRNAs and target gene identification in distant-grafting between tomato and *Lycium* (Goji Berry). Front Plant Sci 7(1475):1–18

Khati P, Chaudhary P, Gangola S, Bhatt P, Sharma A (2017) Nanochitosan supports growth of *Zea mays* and also maintains soil health following growth. 3 Biotech 7(1):81

Khiyami MA, Almoammar H, Awad YM et al (2014) Plant pathogen nanodiagnostic techniques: forthcoming changes? Biotechnol Biotechnol Equip 28(5):775–785

Khot LR, Sankaran S, Maja JM et al (2012) Applications of nanomaterials in agricultural production and crop protection: a review. Crop Prot 35:64–70

Lamsal K, Kim SW, Jung JH et al (2011) Application of silver nanoparticles for the control of Colletotrichum species *in vitro* and pepper anthracnose disease in field. Mycobiology 39(3):194–199

Lau HY, Wu H, Wee EJH et al (2017) Specific and sensitive isothermal electrochemical biosensor for plant pathogen DNA detection with colloidal gold nanoparticles as probes. Sci Rep 7:38896

Lima R, Feitosa LO, Maruyama CR et al (2012) Evaluation of the genotoxicity of cellulose nanofibers. Int J Nanomedicine 7:3555–3565

Lu CM, Zhang CY, Wen JQ et al (2002) Research of the effect of nanometer materials on germination and growth enhancement of *Glycine max* and its mechanism. Soybean Sci 21(3):168–171

Maghari BM, Ardekani AM (2011) Genetically modified foods and social concerns. Avicenna J Med Biotech 3(3):109–117

Mahajan PV, Caleb OJ, Singh Z, Watkins CB, Geyer M (2014) Postharvest treatments of fresh produce. Phil Trans R Soc A 372:20130309

Malik S, Kumar A (2014) Approach for nano-particle synthesis: using as nano-fertilizer. Int J Pharm Res Bio-Sci 3(3):519–527

McMurray TA, Dunlop PSM, Byrne JA (2006) The photocatalytic degradation of atrazine on nanoparticulate TiO2 films. J Photochem Photobiol A Chem 182:43–51

Meena KK, Sorty AM, Bitla UM et al (2017) Abiotic stress responses and microbe-mediated mitigation in plants: the omics strategies. Front Plant Sci 8(172):1–25

Menaa F (2015) Genetic engineering and nanotechnology: when science-fiction meets reality! Adv Genet Eng 4(2):1000128

Milani N, Hettiarachchi GM, Kirby JK et al (2015) Fate of zinc oxide nanoparticles coated onto macronutrient fertilizers in an alkaline calcareous soil. Plos One 10:e0126275. https://doi.org/10.1371/journal.pone.0126275

Misra AN, Misra M, Singh R (2013) Nanotechnology in agriculture and food industry. Int J Pure Appl Sci Technol 16(2):1–9

Mohammadi M, Pezeshki A, Abbasi MM et al (2017) Vitamin D3-loaded nanostructured lipid carriers as a potential approach for fortifying food beverages; in vitro and in vivo evaluation. Adv Pharm Bull 7(1):61–71

Moll J, Okupnik A, Gogos A et al (2016) Effects of titanium dioxide nanoparticles on red clover and its rhizobial symbiont. PLoS One 11(5):e0155111

Moreira FKV, Camargo LAD, Marconcini JM, Mattoso LHC (2013) Nutraceutically inspired pectin–$Mg(OH)_2$ nanocomposites for bioactive packaging applications. J Agric Food Chem 61:7110–7119

Muktar Y, Bikila T, Keffale M (2015) Application of nanotechnology for animal health and production improvement: a review. World Appl Sci J 33(10):1588–1596

Nabifarkhani N, Sharifani M, Garmakhany AD et al (2015) Effect of nano-composite and Thyme oil (*Tymus Vulgaris*) coating on fruit quality of sweet cherry (*Takdaneh* Cv) during storage period. Food Sci Nutr 3(4):349–354

Nagy L, Magyar M, Szabo T et al (2014) Photosynthetic machineries in nano-systems. Curr Protein Pept Sci 15(4):363–373

National Crime Records Bureau (NCRB) (2013) Accidental Deaths & Suicides in India (ADSI) annual reports. http://ncrb.nic.in/StatPublications/ADSI/PrevPublications.htm

National Crime Records Bureau (NCRB) (2014) Accidental Deaths & Suicides in India (ADSI) annual reports. http://ncrb.nic.in/StatPublications/ADSI/PrevPublications.htm

National Crime Records Bureau (NCRB) (2015) Accidental Deaths & Suicides in India (ADSI) annual reports. http://ncrb.nic.in/StatPublications/ADSI/ADSI2015/ADSI2015.asp

Nguyen NT, McInturf SA, Mendoza-Cózat DG (2016) Hydroponics: a versatile system to study nutrient allocation and plant responses to nutrient availability and exposure to toxic elements. J Vis Exp 113:e54317

Nhan LV, Ma C, Rui Y (2015) The effects of Fe_2O_3 nanoparticles on physiology and insecticide activity in non-transgenic and Bt-Transgenic cotton. Front Plant Sci 6:1263

Niu W, Guo L, Li Y et al (2016) Highly selective two-photon fluorescent probe for ratiometric sensing and imaging cysteine in mitochondria. Anal Chem 88(3):1908–1914

Odhiambo JF, DeJarnette JM, Geary TW et al (2014) Increased conception rates in beef cattle inseminated with nanopurified bull semen. Biol Reprod 91(4):97
Oliver MJ (2014) Why we need GMO crops in agriculture. Mo Med 111(6):492–507
Omanovic-Miklicanin E, Maksimovic M (2016) Nanosensors applications in agriculture and food industry. Bull Chem Technol Bosnia Herzegovina 47:59–70
Palmieri V, Bugli F, Lauriola MC et al (2017) Bacteria meet graphene: modulation of graphene oxide nanosheet interaction with human pathogens for effective antimicrobial therapy. ACS Biomater Sci Eng 3(4):619–627
Pandey P, Irulappan V, Bagavathiannan MV, Senthil-Kumar M (2017) Impact of combined abiotic and biotic stresses on plant growth and avenues for crop improvement by exploiting physio-morphological traits. Front Plant Sci 8(537):1–15
Parisi C, Vigani M, Rodriguez-Cerezo E (2015) Agricultural nanotechnologies: what are the current possibilities? Nano Today 10:124–127
Pereira MM, Mouton L, Yepremian C et al (2014) Ecotoxicological effects of carbon nanotubes and cellulose nanofibers in Chlorella vulgaris. J Nanobiotechnol 12:15
Pereira AES, Sandoval-Herrera IE, Zavala-Betancourt SA, Oliveira HC, Ledezma-Perez AS, Romero J, Fraceto LF (2016) Polyglutamic acid/chitosan nanoparticles for the plant growth regulator gibberellic acid: characterization and evaluation of biological activity. Carbohydr Polym 157:1862–1873
Poscic F, Mattiello A, Fellet G et al (2016) Effects of cerium and titanium oxide nanoparticles in soil on the nutrient composition of barley (*Hordeum vulgare* L.) Kernels. Int J Environ Res Public Health 13(6):577
Pradhan N, Singh S, Ojha N et al (2015) Facets of nanotechnology as seen in food processing, packaging, and preservation industry. Biomed Res Int 2015:1–17
Puggal S, Dhall N, Singh N, Litt MS (2016) A review on polymer nanocomposites: synthesis, characterization and mechanical properties. Indian J Sci Technol 9(4):1–6
Pyrgiotakis G, Vedantam P, Cirenza C et al (2016) Optimization of a nanotechnology based antimicrobial platform for food safety applications using engineered water nanostructures (EWNS). Sci Rep 6:21073
Rai V, Acharya S, Dey N (2012) Implications of nanobiosensors in agriculture. J Biomater Nanobiotechnol 3:315–324
Rajonee AA, Zaman S, Huq SMI (2017) Preparation, characterization and evaluation of efficacy of phosphorus and potassium incorporated nano fertilizer. Adv Nanopart 6:62–74
Ramasamy M, Kim S, Lee SS, Yi DK (2016) Recyclable photo-thermal nano-aggregates of magnetic nanoparticle conjugated gold nanorods for effective pathogenic bacteria lysis. J Nanosci Nanotechnol 16:555–561
Reganold JP, Wachter JM (2016) Organic agriculture in the twenty-first century. Nat Plants 2:1–8
Rui M, Ma C, Hao Y et al (2016) Iron oxide nanoparticles as a potential iron fertilizer for peanut (*Arachis hypogaea*). Front Plant Sci 7:815
Sabir S, Arshad A, Chaudhari SK (2014) Zinc oxide nanoparticles for revolutionizing agriculture: synthesis and applications. Sci World J 2014:925494
Schneider J, Börner D, Rosmalen PV, Specht M (2015) Augmenting the senses: a review on sensor-based learning support. Sensors 15:4097–4133
Sekhon BS (2014) Nanotechnology in agri-food production: an overview. Nanotechnol Sci Appl 7:31–53
Sekhon H, Ennew C, Kharouf H, Devlin J (2014) Trustworthiness and trust: influences and implications. J Mark Manage 30(3-4):409–430
Semo E, Kesselman ED, Livney DYD (2007) Casein micelle as a natural nano-capsular vehicle for nutraceuticals. Food Hydrocoll 21:936–942
Singh A, Singh S, Prasad SM (2016) Scope of nanotechnology in crop science: profit or loss. Res Rev J Bot Sci 5(1):1–4
Sirirat N, Lu JJ, Hung AT, Lien TF (2013) Effect of different levels of nanoparticles chromium picolinate supplementation on performance, egg quality, mineral retention, and tissues minerals accumulation in layer chickens. J Agric Sci 5:150–159

Srilatha B (2011) Nanotechnology in agriculture. J Nanomed Nanotechnol 2(123):7. https://doi.org/10.4172/2157-7439.1000123

Taher MA, Mazaheri L, Ashkenani H et al (2014) Determination of nickel in water, food, and biological samples by electrothermal atomic absorption spectrometry after preconcentration on modified carbon nanotubes. J AOAC Int 97:225–231

The World's Cities (2016) United Nations, Department of Economic and Social Affairs, Population Division. Data booklet (ST/ESA/ SER.A/392), pp 1–26

Thornton PK (2010) Livestock production: recent trends, future prospects. Philos Trans R Soc B 365:2853–2867

Torney F, Trewyn BG, Lin VS, Wang K (2007) Mesoporous silica nanoparticles deliver DNA and chemicals into plants. Nat Nanotechnol 2(5):295–300

Turner APF (2013) Biosensors: sense and sensibility. Chem Soc Rev 42:3184–3196

Ushikubo FY, Furukawa T, Nakagawa R et al (2010) Evidence of the existence and the stability of nano-bubbles in water. Colloids Surf A Physicochem Eng Asp 361:31–37

Vamvakaki V, Chaniotakis NA (2007) Pesticide detection with a liposome-based nano-biosensor. Biosens Bioelectron 22(12):2848–2853

Vasimalai N, John SA (2013) Biopolymer capped silver nanoparticles as fluorophore for ultrasensitive and selective determination of malathion. Talanta 115:24–31

Wang Y, Sun C, Zhao X et al (2016) The application of nano-TiO_2 photo semiconductors in agriculture. Nanoscale Res Lett 11:529

World Bank (2017) Enabling the business of agriculture. World Bank, Washington, DC. https://doi.org/10.1596/978-1-4648-1021-3

Wu Z, Xu XL, Zhang JZ et al (2017) Magnetic cationic amylose nanoparticles used to deliver survivin-small interfering RNA for gene therapy of hepatocellular carcinoma *in vitro*. Nanomat 7(110):1–13

Wyser Y, Adams M, Avella M et al (2016) Outlook and challenges of nanotechnologies for food packaging. Packag Technol Sci 29:615–648

Yearla SR, Padmasree K (2016) Exploitation of subabul stem lignin as a matrix in controlled release agrochemical nanoformulations: a case study with herbicide diuron. Environ Sci Pollut Res Int 23(18):18085–18098

Zaytseva O, Neumann G (2016) Carbon nanomaterials: production, impact on plant development, agricultural and environmental applications. Chem Biol Tech Agri 3(17):1–26

Zhang C, Wohlhueter R, Zhang H (2016) Genetically modified foods: a critical review of their promise and problems. Food Sci Human Wellness 5(3):116–123

Zhao L, Peng B, Hernandez-Viezcas JA et al (2012) Stress response and tolerance of *Zea mays* to CeO_2 nanoparticles: cross talk among H_2O_2, heat shock protein and lipid peroxidation. ACS Nano 6(11):9615–9622

Zhao L, Huang Y, Hannah-Bick C et al (2016) Application of metabolomics to assess the impact of $Cu(OH)_2$ nanopesticide on the nutritional value of lettuce (*Lactuca sativa*): enhanced Cu intake and reduced antioxidants. NanoImpact 3(4):58–66

Chapter 2
Antimycotic Role of Soil *Bacillus* sp. Against Rice Pathogens: A Biocontrol Prospective

Suraja Kumar Nayak, Swapnarani Nayak, and Bibhuti Bhusan Mishra

2.1 Introduction

Application of agrochemicals for protection of crops against various fungal phytopathogens is a current trend around the globe, although their toxicity on the environment is highly studied. Many countries have restricted the use of fungicides & pesticides because of higher price and resistant pathogenic fungal strains (Rahman 2013). Technological advances in agriculture leading to effective utilisation of natural resources. Microbial antagonists are one of them. Biocontrol through using antagonistic microorganisms has emerged as an alternative to reduce the use of agrochemicals. An overwhelming suitable alternative for the control of bedevil is the use of bacteria with antagonistic effects against phytopathogens (Thomashow 1996). Bacteria might interact close proximate with the host plant thus could be effective biocontrol agents in sustainable agriculture production. Soil is an important treasure to isolate and select bacteria with biocontrol activities (Bloemberg and Lugtenberg 2001). Soil is the most essential resource and only source of minerals and water to the living organisms and harbours innumerable microbes which are essential for maintenance of soil fertility vis-à-vis crop productivity. Macromolecular soil substrates gone through degradation, predation, feeding and nutrients absorption by microbes have drawn attention on in soil biogenic chemical processes

S.K. Nayak (✉)
Department of Biotechnology, College of Engineering and Technology, Biju Patnaik University of Technology, Bhubaneswar, Odisha, India
e-mail: surajnayak3@gmail.com

S. Nayak
Fish Genetics and Biotechnology Division, ICAR-Central Institute of Freshwater Aquaculture, Bhubaneswar, Odisha, India

B.B. Mishra
Department of Microbiology, College of Basic Science & Humanities, Orissa University of Agriculture and Technology, Bhubaneswar, India

J.K. Patra et al. (eds.), *Microbial Biotechnology*,
https://doi.org/10.1007/978-981-10-6847-8_2

and thus releases a small number and highly diversified secondary metabolites. Biotic interactions dominating soil biology differ from those in other systems because of the dominating role of sessile organisms and the lack of autotrophy in soil (chemolithoautotrophs being an interesting but not significant exception). Soil microbial diversity has been exploited for many years. Most natural products of economic value such as antibiotics or other organic chemicals, pharmaceuticals are derived from soil microbial strains (Daniel 2004). Independent of soil composition microorganisms distributed inside and outside of microaggregates in soil (Hattori 1988). This heterogeneity leads to a high diversity of soil microorganisms and a wide variety of microbial niches.

Soil bacteria exhibit antifungal properties due to various enzymes production as may be a part of their lytic system, enables to intake hyphae as substrate for sustenance (De Boer et al. 1998). Because of antibiotics production most of them have been used as biocontrol agents against phytopathogenic fungi (Yilmaz et al. 2005). Reports on antifungals opined that half of antifungals endorsed after 1994 are of biological origin.

Rice (*Oryza sativa* L.) fungal disease are not only serious concern in Asian continent including countries of Bangladesh, China, Indonesia, India, Japan, Korea, Malaysia, Philippines, Sri Lanka, Thailand and Vietnam but also in other continents (Rabindran and Vidhyasekaran 1996). Rice serves as the staple food for half of population worldwide. In humid tropical and subtropical conditions, mycelia spread in the soil externally and on the diseased plant parts sclerotia (resting fungus structures with a thickwalled, melanized rind enclosing tissue rich in stored nutrients) are produced. The efficacy of antifungals and host pathogenic susceptibility is evaluated by the interaction of cultural and environmental conditions in case of root rot & spot and blast disease (Dean et al. 2005). Also fungi have developed resistance against both chemical treatments and genetic "disease resistance" reported by agronomists in some cultivars. A number of isolates give some control of the fungal diseases (*Rhizoctonia* sp., *Pythium* sp.) on crops in the field. Antagonists are organisms with traits enabling themselves to interfere with pathogen growth, division, survival and infection by various direct or indirect mechanisms.

Some beneficial soil bacteria (especially belonging to the genera *Bacillus* and nearly related *Paenibacillus*) living in and around plant roots are of particular interest in that context. *Bacillus* sp. are widespread in nature, nonpathogenic and innoxious to humans and other animals but harmless to plants (Acea et al. 1988). These bacteria excretes antimicrobial compounds *in vitro*, including the (lipo) peptides antibiotics (Stelle and VlamiM de Souza 2002) and antifungal protein (Liu et al. 2007). Moreover *Bacillus* sp. offers advantages over others against fungal phytopathogens due to endospores and secretion of broad-spectrum antibiotics. The endospores forming ability facilitates resistant to the elevated temperature & high concentrations of chemicals, long-term storage and commercialization. Few of the well documented *Bacillus* sp. are engaged to soil fertility and optimization of plant nutrition apart from that also involved in the production of bacterial phytohormones and the solubilization of phosphate minerals (de Freitas et al. 1997).

Successful and effective biological control antagonists requires a clear understanding of the complex regulation of disease suppression in answer to biotic and abiotic factors and knowledge of the dynamics and composition of plant and surrounding soil associated bacterial communities. An attempt has been made in the present chapter to use soil *Bacillus* sp. as biocontrol agent combating rice fungal pathogens.

2.2 Soil Microbial Diversity Emphasizing on *Bacillus* sp.

Soil environment is considered as one of the most appropriate environment for microbial growth and development (Cavalcanti et al. 2006). The occurrence and importance of antimicrobial microorganisms in soil were demonstrated in late 70's. The biological inhibition of microorganisms occurs in all types of soil (both treated and non treated) for any kind of phytopathogens. Moreover, for any particular case in conducive soil there is a continuous synthesis of structural analog help in suppression of specific group of pathogens. The suppressiveness of biological origin, are may not be transferrable in high dilution to a favourable condition to produce antibiotics in soil (Baker 1987). The advantageous effect of soil bacteria enlisted as nitrogen fixation, phytohormones production, synthesis of various antistress enzymes, siderophore, solubilization potassium and zinc and disease control by suppression or killing of phytopathogens (Kumar et al. 2012). Moreover, the antagonistic bacteria produces wide range of secondary metabolites due to various secretion mechanisms and are effectuate various biochemical reactions with atypical lytic enzymes (Das et al. 2006).

Generally 2–5% of rhizosphere bacteria are plant growth promoting rhizobacteria (PGPR) involved in synthesis of various kinds of bacteriocins and protein exotoxins, which are biologically active peptide moieties with fungicidal mode of action (Riley and Wertz 2002). While many antibiotics are existed, efforts to discover new are still continues. Thus, numerous species of *Streptomyces*, *Penicillium* and *Bacillus* have been studied due to their inherent properties of antibiotics production (Brock and Madigan 1991) with relevant to agriculture realm. Amongst all, spore-forming bacilli have an edge over the non-spore formers such as *Pseudomonas* sp., and others. *Bacillus* sp. are ubiquitously found in soil and most of from them have proteolytic activity and the ability to disintegrate proteins (Aslim et al. 2002), excellent colonization proficiency with a superlative sporulation and diseases suppression for which they proved as candidates for potential fungal biocontrol agents. In addition, they are easy to cultivate and produce metabolites of non pathogenic origin, found to be cheaper and more effective. Non pathogenic *Bacillus* sp. were found to colonize in the soil as well as root surface, increase the plant growth and destroyes fungal mycelia, sclerotia (Basha and Ulaganathan 2002). Members of *Bacillus* sp. are able to produce more than two dozens of antibiotic compounds with varied chemical properties (Stein 2005), amongst them large studied are of peptide properties (Mannanov and Sattarova 2001; Stein 2005). Additionally, phospholipid derivatives (i.e., Bacilysocin) and other groups of antibiotics were also produced by this soilborn species (Tamehiro et al. 2002).

Plenty of investigators have been engaged to isolate various terrestrial *Bacillus* strains and identify their bioactive inhibitory compounds (Lisboa et al. 2006) and its potential to produce multistructure inhibitory compounds (Stein 2005). *Bacillus* sp. can produce agents with different chemical properties from peptide group. Several members from the genus *Bacillus*, including *Paenibacillus* and *Brevibacillus*, produce various antimicrobial substances, e.g. antibiotics. *Brevibacillus* (erstwhile *Bacillus brevis*) is characterized by its capability to produce antimicrobial substances. Furthermore, a wide range of antibiotic compounds are produced by *B. licheniformis*, *B. pumilus*, *B. circulans*, *B. cereus*, *Brevibacillus aterosporus*, *Paenibacillus* (erstwhile *Bacillus polymyxa*) and others (Abdalla et al. 2014).

In the mid nineteenth century research was more focused on finding and isolating soil microorganisms antagonistic to plant pathogens that caused specific crops diseases (Santoyo et al. 2012). The potential of *B. subtilis* to produce antibiotics has been recognized for 50 years. Peptide antibiotics represent the predominant class. However, the complete spectrums of antibiotic activities by different *B. subtilis* strains by systematic studies are rare (Pinchuk et al. 2002). Being capable of producing more than 70 different antibiotics, *B. subtilis* is one of the major producer amongst all and inhibiting *Cercospora beticola* (Lindow and Brandl 2003), *Colletotrichum gloeosporioides* (Collins and Jacobsen 2003), *Pseudocercospora purpurea* (Eeden and Korsten 2006) and *Rhizoctonia solani* (Kai et al. 2007).

2.3 Emerging Rice Fungal Pathogens

Soil fertility parameters can be trapped through the disease proliferation (Texas Agric. Exp. Stn 1996). Pathogenic fungi are primary causes of grain loss, and some of them produce compounds toxic to humans. Amongst these fungi, a number of soil borne phytopathogens seriously affects the rice, causing devastating effects (Chaiharn et al. 2009) (Fig. 2.1). These are also considered as economically important (Aye and Matsumoto 2010), responsible for severe economic losses in harvest yields. However, upto 80% of crops plants are infected by anyone of these pathogens infection (Leslie and Summerell 2006) and reduce the quality of staple foods (Jens Laurids et al. 2013). Mycelium presence inside the plant tissue or debris are also exhibiting pathogens survival and serves as mean of an initial inoculum for establishment of a new infection (Ali et al. 2016). Rice diseases due to various phytopathogens appear to be proliferating at exponential increasing rates (Table 2.1). In addition, the diseases profile on rice has changed with time due to evident changes in global climatic conditions, diversified variety and cultivation practices as region specific. Many diseases earlier considered as minor have become threatening (Laha et al. 2009). More than 70 diseases have been reported to occur on rice out of which 36 fungal, 21 viral, 06 bacterial and 06 nematode diseases have been recorded (Ou

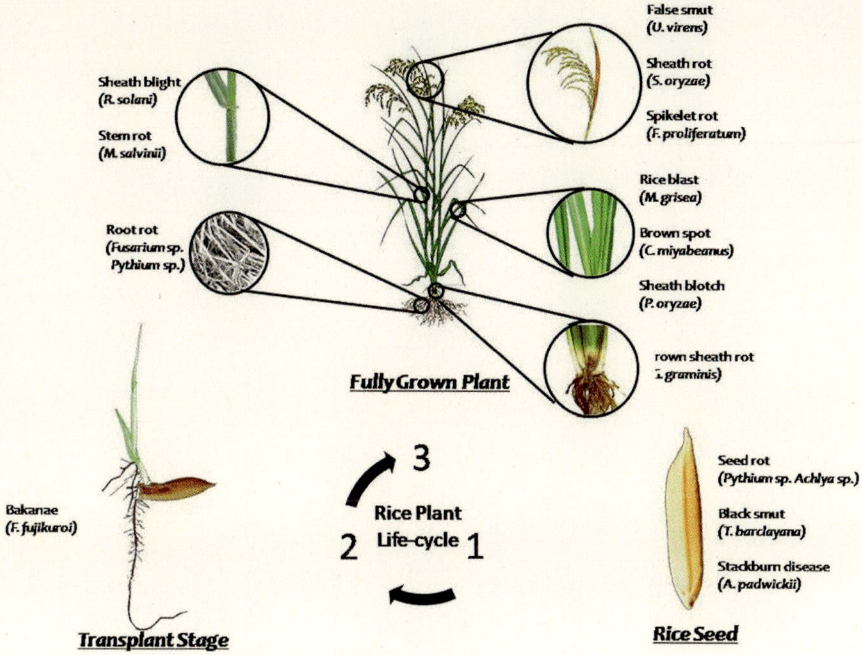

Fig. 2.1 Pathogens infecting different parts of rice

1985). Of these, blast (*Magnaporthe grisea*; anamorph: *Pyricularia grisea*), sheath blight (*Rhizoctonia solani* Kuhn; teleomorph: *Thanetophorus cucumeris* (Frank) Donk) and false smut (*Ustilaginoidea virens*; teleomorph: *Villosiclava virens*) are the major fungal diseases and causes remarkable quantitative and qualitative losses. In addition, bakanae or foot rot (*Fusarium moniliforme*; synonym: *Gibberella fujikuroi*), sheath rot **(*Nigrospora oryzae*; synonym: *Sarocladium oryzae*; teleomorph: *Khuskia oryzae*),** brown spot (*Cochliobolus miyabeanus*; synonym: *Drechslera oryzae*), stem rot (*Magnaporthe salvinii*; synonym: *Leptoshaeria salvinii*), crown sheath rot (*Gaeumannomyces graminis*; erstwhile *Ophiobolus oryzinus*, *O. graminis*) r**oot rot or feeder root necrosis** (*Fusarium* sp., *Pythium* sp., *P. dissotocum*, *P. spinosum*) have worsen the scenario worldwide (Sharma and Bambawale 2008; Singh 2012). Unexpected diseases such as glume discoloration (*D. oryzae*, *S. oryzae*, *F. moniliforme*, *Alternaria padwickii*, *Epicoccum* sp., *Curvularia* sp.) and kernel smut (*Tilletia barclayana*), **spikelet rot disease or pecky rice** (*F. proliferatum*, *Bipolaris australiensis*, *C. lunata*, *A. alternate*; erstwhile: *A. tenuis*), seedling blight or damping off (*Cochliobolus lunatus*; anamorf: *C. lunata*) are projected to become more important as location-specific diseases and are becoming significant problems where they were erstwhile regarded as of minor importance (Krishnaveni et al. 2015).

Table 2.1 Fungicidal activity of soil *Bacillus* sp. against rice phytopathogens

Disease	Causative agent	Inhibiting/Affecting Part	Symptoms	Antagonistic organism	References
Sheath blight	*Rhizoctonia solani* *Thanatephorus cucumeris* (teleomorph)	Stem (sclerotial infection during tillering stage)	1. Irregular lesions with dark brown margin and grayish inner colors on sheath and leaf blades. 2. Poor filling of the grains and emergence of panicles.	*B. amyloliquefaciens*	Shrestha et al. (2016)
Rice blast (leaf and collar) disease	*Magnaporthe grisea* (anamorph), *Pyricularia grisea* synonym *P. oryzae*)	Above ground parts as leaf, collar, node, neck, small portion of panicle and sometimes leaf sheath	White to gray-green elliptical or spindle-shaped lesions or spots, with red to brownish borders on leaves, neck and panicles.	*B. firmus* *B. subtilis*	Suryadi et al. (2011) Taguchi et al. (2003)
False smut	*Ustilaginoidea virens* (teleomorph *Villosiclava virens*)	Grains in panicle	1. Greenish, smooth velvety spore balls (chlamydospores) around the grains in the panicle. 2. Chlamydospores contain ustiloxins (mycotoxins), toxic to animals and human.	*B. subtilis*	Yan et al. (2014)
Brown spot	*Cochliobolus miyabeanus* (erstwhile *Helminthosporium oryzae*) *Bipolaris oryzae*(Breda de Haan)	Coleoptiles, leaves, leaf sheath, panicle branches, glumesand spikelets.	Lesions on the leaves, initially small dots or oval spots or circular eye shaped appear light brown or grey centre surrounded by reddish brown in the periphery.	*B. vallismortis*	Krishnaveni et al. (2015) Chalkley (2010) Mwalyego et al. (2011)
Bakanae	*Fusarium fujikuroi* (syn. *Gibberella fujikuroi*)	Seedlings	Seedling elongation, slight pale yellowing at the transplanting stage	*B. cereus*	Kazempour and Elahinia (2007)

Root rot, Or, Feeder root necrosis	*Fusarium* sp., *Pythium* sp., *P. dissotocum*, *P. spinosum*	Roots	Roots exhibit dark brown to black discoloration and necrosis. Further roots decayed leads to absorption of nutrient disrupted, leaves turn yellow and the plants lack sturdiness.	*B. subtilis* *B. pumilus*	Cawoy et al. (2011)
Sheath rot	***Sarocladium oryzae, Nigrospora oryzae,* (teleomorph) *Khuskia oryzae***	Young panicles covered by leaf sheaths in the primary part	Irregular spots with brown margins and grey centres or greyish brown throughout.	*B. subtilis*	Shamsi et al. (2010) Bigirimana et al. (2015)
Stem rot	*Magnaporthe salvinii* (syn. *Leptoshaeria salvinii*)	Culm	1. A small black lesion (sclerotia) appears at the outer leaf covers after mild tiller stage. On progression the tiller killed. 2. Dark lesions develop on the stems at the water line due to lodging, unfilled panicles and chalky grains.	*B. pumilus* *B. subtilis*	Webster et al. (1981) Venkateswarlu et al. (2015) Jayaprakashvel and Mathivanan (2009)
Crown (black) sheath rot	*Gaeumannomyces graminis* (erstwhile *Ophiobolus oryzinus*, *O. graminis*; syn. *Rhaphidophora graminis*)	Roots, crown and lower parts of culm	1. Black to dark brown diffuse lesions on the sheath close to the water line. 2. Perithecial necks protruding from the upper surface with a thick fungal mat between the leaf sheath and the culm.	*B. cereus*	Renwick et al. (1991) Peixoto et al. (2013)

(continued)

Table 2.1 (continued)

Disease	Causative agent	Inhibiting/Affecting Part	Symptoms	Antagonistic organism	References
Seed rot and Seedling diseases/ Damping off	*Pythium* sp., *Achlya* sp.,	Seeds and seedlings	1. Fungal mycelium emerges from cracks in the seed glumes or from infected seedling's plumule. 2. Infected seed encircled in a dark circular spot on the soil surface. 3. Primary leaves are stunted. 4. Leaves and sheaths become yellow or chlorotic, leads to delayed in development	*Bacillus* sp.	Chun and Schneider (1998) Espino et al. (2015) Whipps and Lumsden (1991)
Bunt of rice, or Black smut, or Kernel smut	***Tilletia barclayana***	Seed grain	1. Diseased grains are filled with black powder. 2. Endosperm replaced partially or completely by a black mass of smut spore consists of black mass of chlamydospores. 3. All or part of individual kernels near or at maturity replaced	*B. pumilus*	El-kazzaz et al. (2015) Espino et al. (2015)

Sheath blotch	***Pyrenochaeta oryzae***	Leaf sheath, sometimes the leaf blade and glumes	1. The blotches are ovaloid, long (1inch) and brownish at first and the centre turns grey orgreyish brown gradually and dotted with the black fruiting bodies of the fungus. 2. The margin remains brown.	*B. subtilis*	Patel et al. (2006) Shamsi et al. (2010) Cawoy et al. (2011)
Stackburn disease or Seed discoloration, seed rot, sheath-rotting pathogen	*Alternaria padwickii* *Trichoconis padwickii* (erstwhile *Trichoconiella padwickii*)	Seeds	1. Oval to circular lesions on leaves (3-10 mm dia), grey to white having small, black dots in center as sclerotia in older infected parts. 2. Discoloured, shrivelled and brittle grains in severe infection.	*B. cereus*	Chalkley (2010) Shubha et al. (1992) Krishnaveni et al. (2015)
Spikelet rot disease or Pecky rice or Kernel spotting (In USA)	*Fusarium proliferatum*, *Bipolaris australiensis*, *C. lunata*, *Alternaria alternate* (erstwhile *A. tenuis*)	Panicle and grain	Reddish discoloration of grain at flowering and milk stage, and which changed to brown and black at the ripening and fully mature stage.	*Bacillus* sp. *B. subtilis* *B. licheniformis*	Huang et al. (2011) Abdalla et al. (2014) Soleimani et al. (2005)

2.3.1 Sheath Blight

Rhizoctonia solani Kuhn (teleomorph: *Thanetophorus cucumeris* (Frank) Donk), is responsible for Sheath blight disease which is one of the most destructive and economically important rice diseases round the globe mostly in tropical and temperate regions (Ou 1985). *R. solani* is a soil borne fungal plantpathogen having a vast host range. Primary inoculum of the pathogen, viz. sclerotia, builds up between growing seasons and is the source of contamination in the subsequent rice leads to substantial remarkable loss in yields (Kanjanamaneesathian et al. 1998). High nitrogen rates, plant density and favorable abiotic conditions including high relative humidity, low light and cloudy days favour the disease during crop growth as early heading and grain-filling stages (Dath 1990).

Crop alteration will be ineffective for disease management with low host resistance. Fungal sclerotia resist in soil and plant debris for prolonged period and its management requires judicious fungicidal treatment (Dath 1990). However, indiscriminate fungicide application increases the risk of the emergence of fungicide-resistant phytopathogens (Shrestha et al. 2016).

2.3.2 Rice Blast Disease

Blast disease of rice is caused by *Magnaporthe grisea* (anamorph: *Pyricularia oryzae* Sacc. synonym *P. oryzae* Cav.), which is distributed worldwide and occurs nearly in 100 countries including yield losses between 1–100% in Japan (Kato, 2001), 70% in China, 21–37% in Bali, Indonesia (Suprapta and Khalimi 2012) and 30–50% in South America and South-East Asian countries as Sri Lanka, Bangladesh, Vietnam, Philippines, India and Thailand (Baker et al. 1997). The disease occurrence increased due to the use of intensive agronomic practices as excess N_2 fertilization in the field and humid climate. It appears at three parts of crop plant (on the leaf, node and neck) causing heavy to total loss in yield. The yield reduction due to this disease may be as high as 75% or more (Ou 1985; Srinivas Prasad et al. 2011).

Under moist conditions, mycelia develop and spread in soil and sclerotia are produced on contaminated parts. The fungus spreads to the upper foliage by lesions on entire tillers starting from the water line up to the flag leaf. The disease occurrence is initially soil borne, subsequent spread is foliar. The approximate yield losses range from 2.5% to 50% (Rabindran and Vidhyasekaran 1996). *P. oryzae* develops new races quickly resulting in invading the rice resistance.

2.3.3 False Smut

Rice false smut disease, caused by the pathogenic ascomycete *Ustilaginoidea virens* (Cooke) Takahashi (teleomorph: *Villosiclava virens*; synonym: *Claviceps oryzae-sativae* Hashioka) is one of the most acute and devastating diseases (Talbot and Foster 2001). This fungus is considered to be a biotrophic parasite as the hyphae extend into the central vascular tissues (Tang et al. 2013) and causes no harm during growth. The hyphae penetrates at the upper parts of the floral organs to colonize internally the spikelet and roots at the seedling stage and at the earlier booting stage the filaments of stamen. At a later stage of *U. virens* colonization, the basal area of the floral region marked as the presence of a dense mass of mycelia due to the promotive microenvironment in terms of nutrients and moisture content. *U. virens* do not infects the reproductive organs (Mebeaselassie et al. 2015). The fungal hyphae seldom penetrate through the plant cell wall. The disease causes huge economical downfall. Normally, this disease appears more in upland than in low-land rice areas. It is well understood that whenever a spell of wet weather coincides with the heading time, false smut disease appears.

2.3.4 Brown Spot

Brown spot of rice caused by *Cochliobolus miyabeanus* erstwhile *Helminthosporium oryzae* (anamorph: *Bipolarisoryzae* Breda de Hann; Syn: *Drechslera oryzae*), is documented as a significant biotic limiting factor in rice production of the world, particularly under semi-dry situations (Ou 1985) and is a seed borne disease. The disease was more significant and made history during the 'Great Bengal Famine' in 1942–43 in India. The approx. yield loss varies from 26% to 50% as reported (Chakrabarti 2001).

The disease occurs more or less every year in mild to severe form in many upland and rain-fed lowland rice-growing areas. Besides causing brown spot on the leaves, the fungus is also responsible for grain discoloration, another important setback in paddy-growing areas. It involves in production of a host-specific toxin. In spite of everything we have known about the pathogen, brown spot remains as important but not a major rice production constraint (Krishnaveni et al. 2015).

2.3.5 Bakane

Bakanae caused by *Fusarium fujikuroi* Nirenberg is a seed borne disease and extensively distributed in Asian continent. Application of nitrogen fertilisers spreads the disease. The spores are carried out by physical factors like water and wind. Temperature within 30–35 °C favors disease development (Nyvall 1999). It is transmitted primarily by seed.

The pathogen has a wide host range and is widespread throughout the world. *F. fujikuroi* induces seedling elongation, foot rot, seedling rot, grain sterility and grain discoloration on rice (Ou 1985). Often there are symptoms of infected seedling except minor yellowing at the transplant stage (Kim 1981). In India, bakanae or foot rot disease is caused by the fungus *F. moniliforme*(teleomorph: *Gibberella fujikuroi*), which is known to cause severe damage in the states of Tamil Nadu, Andhra Pradesh, eastern districts of Uttar Pradesh and in Haryana.

2.3.6 Root Rot or Feeder Root Necrosis

Root rots are probably one of the most common but illtreated diseases of rice. Due to multiple factors like extreme environmental conditions, insect feeding (by rice water weevil larvae) and infections of other pathogens causes this kind of disorders.

Fertilizer usually reduces the aboveground symptoms – although actual nutrient use is poor. Rice water weevil control greatly reduces root rots. Though it is a minor disease of rice but very less work has been done so far on this.

2.3.7 Sheath Rot

Sheath rot is one of the principal diseases of rice but is considered as a minor and geographically limited disease in yesteryears. Major pathogens associated with rice sheath rot are *Sarocladium oryzae* (Sawada) W. Gams and D. Hawksw (erstwhile *Acrocylindrium oryzae*) and *Fusarium* sp. belonging to the *F. fujikuroi* complex which recognized three other species (including *F. verticillioides*, *F. moniliforme* and *F. proliferatum*). The symptoms are found in all rice-growing lands especially in lowland areas (Pearce et al. 2001) all over the world (Sakthivel 2001). Apart from that due to crop intensification practices like use of semi-dwarf and photoperiod-insensitive cultivars, applications of maximum N_2 fertilizers and increase in plant density have worsen the scenario. In addition to this intercountry transport of planting material also increase the disease spreading (Bigirimana et al. 2015).

S. oryzae is being very frequent in rainy seasons (Mew and Gonzales 2002). It has till now been reported in the countries (CABI 2007) as Argentina, Australia, Bangladesh, Brazil, Brunei Darussalam, Burundi, Cameroon, China, Côte d'Ivoire, Gambia, India, Indonesia, Japan, Kenya, Madagascar, Malaysia, Mexico, Nepal, Niger, Nigeria, Pakistan, Philippines, UAE, Senegal, Sri Lanka, Tajikistan, Tanzania, Thailand, Uzbekistan, USA, Vietnam and Venezuela. Sheath rot development intensified at 20–30 °C temperature and 65–85% relative humidity. Depending on the pathogenic conditions the losses varies in between 20–85% (Sakthivel 2001).

S. oryzae produces secondary metabolites helvolic acid and cerulenin play major role in pathogenesis (Ghosh et al. 2002; Ayyadurai et al. 2005). The mechanism involved infiltration of plant tissues with the two metabolites causes disruption in ion exchange and increase the electrolyte leakage susceptibility (Sakthivel et al. 2002). Chlorophyll biosynthesis was retarded by Helvolic acid, a tetracyclic triterpenoid (Ayyadurai et al. 2005). Cerulenin is a hexaketide amide, inhibiting the malonyl-ACP: acyl-ACP condensation step along with fatty acid synthesis for inhibition of polyketide synthesis (Omura 1976). The fungus also produces hydrolytic and oxidative enzymes that play major role in disease outbreak (Joe and Manibhushanrao 1995; Pearce et al. 2001).

2.3.8 Stem Rot

Rice plants are infected by many phytopathogens causing various diseases, resulting in reduce quality of the crop and yield. Stem rot disease in rice is provoked by *Sclerotium oryzae* Cattaneo (teleomorph: *Magnaporthe salvinni* Cattaneo Krause & Webster; syn. *Leptosphaeria salvinii* Cattaneo) (Tsuda et al. 1982). It is one of the notable rice diseases in India and is prevalent in the states like Andhra Pradesh, Haryana, Punjab and Uttarakhand. It affects to the extent of up to 75% yield loss annually (Kumar et al. 2003). The disease is induced by *S. oryzae*, is a facultative polyphagous parasite found in soil. Though the fungus is found in plant debris, soil borne inoculum is more important for infection and disease. The first symptoms can be seen after the mid-tillering stage in the crop field. When the fungus penetrates into the inner leaf sheaths it causes rotting and infection of the culm, resulting in lodging, unfilled panicles, chalky grains and tillers die under severe conditions.

2.3.9 Crown (Black) Sheath Rot

The fungus *Gaeumannomyces graminis* (Sacc.) Arx & D. Olivier var. *graminis* (anamorph: *Gaeumannomyces hyphopodioides* sp. with lobed hyphopodia), causing crown (black) sheath rot of rice was reported first in upland rice and subsequently spreaded to irrigated rice of North Eastern part of South America (Prabhu and Filippi 2002; Nunes 2008). The morphological features that discriminate the varieties are the lobed hyphopodia and the ascospore sizes (Mathre 1992). The hyphopodia are produced from mycelium at the infected culms base or in culture medium (Walker 1981).

This has worldwide distribution in temperate and dry regions. Its development is favoured by moist, cool soils ranging from 12–20 °C. Soils deficient in nitrogen, phosphorous and copper can increase the severity.

2.3.10 *Seed Rot and Seedling/Damping off Diseases*

Seed rot or damping off of rice is most serious disease and prevalence in seasons of low temperature and the development of seedlings is uneven or delayed. In this physiological disorder seedlings undergrown by low temperature or poor aeration effects are more prone to attacked by water molds and other fungi. Sophisticate transplantations (by machine) avail extremely favourable conditions with high seed density, high temperature and high humidity for disease occurrence (Ohata 1981). Ungerminated seeds and diseased seedlings are usually infected with various fungi. The members are either from *Saprolegniaceae* or *Pythiaceae* Family. The causative organisms are basically *Pythium* sp. and *Achlya* sp. *Pythium* sp. grows well at temp. 25–30 °C. So their damage is very little while *A. klebsiana* isolates were pathogenic at low as well as high temperature. Thus the plants under warm condition produce half infected and favourable for vigorous growth.

Pythium aphanidermatum and *P. helicum* were strongly pathogenic to rice *in vitro* but other *Pythium* species were moderate. *Pythium* sp. grow well in relative moderate temperature and infect through the wounded part of seeds, roots and coleptiles. Thus the rot disease is principally due to *Pythium* sp.

Water sown rice conditions are ideal for the growth of the fungi. *Achlya* sp. was slightly pathogenic but more destructive than other, if seedlings have grown at lower temperature. *A. klebsiana* appears to infect the seed coat and endosperm. The mechanism involved in failure of seed to germinate and failure of seedlings to emerge through the water may be due to a weakening of the seedling through utilisation of energy source (endosperm) by fungus, instead of direct damage to the germinating embryo.

2.3.11 *Bunt of Rice or Black Smut or Kernel Smut*

Kernel smut is usually considered a inconsequential disease of rice. Caused by *Tilletia barclayana* (Brefeld) Sacc. & Syd. *Neovossia barclayana* Bref.; *N. horrida* (Tak.) Padw. & Kahn. It is prevalence in most rice producing countries including Burma, China, India, Indonesia, Japan, Korea, Mexico, Philippines, Senegal and the USA. In Australia the disease is absent. In southern U.S. rice growing areas, it is more persistence during rainy seasons and in areas of fields receiving more than adequate amount of N_2 fertilizer (Slaton et al. 2004).

Wind borne spread of kernel smut of rice spores infects and reinfects rice panicles both within a crop and onto neighbouring crops. The fungus causing kernel smut is not known to be transmitted by systemic infection, but to develop from a local inoculation at anthesis by airborne secondary sporidia (Whitney and Frederiksen 1975). Kernel smut of rice is not initiated from a systemic infection but from a local inoculation at anthesis.

In *Neovossia*, chlamydospores are simple and are produced singly at the ends of special, fertile hyphae. Spores of both genera germinate by promycelia. The rice kernel smut fungus has fragmentary appendages that are occasionally evident on younger spores, indicating that the spores were borne singly and terminally. Short and medium rice varieties have lower incidence of the disease due to their small floret opening at flowering. In rice growing areas of California the spores are widespread and no effort has been made to restrict overspread seed lots. Rice-growers can chose to grow tolerant cultivars; however, these cultivars generally yield less than the highly susceptible ones (Slaton et al. 2004).

2.3.12 Sheath Blotch

Sheath blotch caused by *Pyrenochaeta oryzae* Shirai ex Miyake, was first observed in Japan, 1910 and considered as sheath brown spot. It has since been reported from all over the commonwealth rice growing continents. In Asia Myanmar, China, India, Philippines, Thailand, Malaysia, Japan, Bangladesh in Africa, Sierra Leone, Swaziland in Australia, Fiji and Solomon island, in America it's found in some parts of Arkansas and Florida. Punjab, Gujarat (Patel et al. 2006) and Chhattisgarh states of India are showing susceptibility to the disease pathogens.

Another species *P. nipponica* Hara had also been found to infect the rice leaves; it has smaller pycinidia, setae and conidia than the *P. oryzae* (Datnoff and Jones 1992). The fungus did not sporulate on PDA medium. All rice cultivars were susceptible to sheath blotch. The disease are infecting the sheaths and leaves, develop reddish brown to brown, circular to irregular lesions. Pycnidia were maximally observed on infected tissue, it also found in the healthy tissue in minimum amount. Sheath blotch is least concern to rice and scanty information is available.

2.3.13 Stackburn Disease or Seed Discoloration

Rice grains may be infected by various organisms before or after harvest, causing seed discoloration. The discoloration may appear externally on the glumes and internally on the kernals, or on both. The causative organisms involved with grain discoloration are principally *Alternaria padwickii* (Ganguly) M.B. Ellis, ***C. miyabeanus* and *D. oryzae*. In addition, *G. fujikuroi*, *Nigrospora* sp., *Epicoccum* sp., *Curvularia* sp. and *Phoma* sp. are also plays role in the incidence**. It occurs primarily in tropical **hot and humid climatic** regions (CMI 1984) of Africa, Asia (southern), Oceania, South America (Rodriguez and Nass 1990). **These microorganisms have various effects on the grains. Deterioration of stored seeds and grains caused by a variety of fungi is a persistent problem in the Indian storage system due to climatic situations**. Seed borne incidence was favoured by heavy rainfall, relative humidity and minute variations in temperature (Abul Khair et al. 1988) in premature stage. The pathogen enters to seeds via the glumes and infect the kernel. In the "stackburn", spots on glumes are light brown to white or light pink or reddish-brown, with a dark dense border (Groth 1992). The fungus survives as sclerotia in soil or crop debris and mycelium in host tissues. Due to the presence of pathogen, seed treatment applications reduce disease incidence.

2.3.14 Rice Spikelet Rot Disease

Spikelet rot disease (SRD), for instance, has become an emerging rice disease in South-East Asian subcontinent, especially in China, and the yield reduction increasing regularly over few years. It appears to be influenced by a number of factors, including the familiarization of large-panicle, non-glutinous plant type found in both modern inbred varieties and hybrids. *F. proliferatum*, *B. australiensis*, *C. lunata* and *A. tenuis* are the causative agents. These can grow in wide range of mild environmental (temperature, pH) conditions, range of carbon as well as nitrogen sources proves their existence with various trophisms. SRD causes grain discoloration as well as grain malformation. Windy and rainy conditions promote the disease severity. The pathogens utilizes host nutrients for growth and development with symptoms of dark reddish discoloration at flowering and milk stage.

Different rice varieties and hybrids seem to show strong difference in susceptibility against the disease (Huang et al. 2011). Usually, closed-panicle plant types, as round-grained non-glutinous rice inbreeds and hybrids, seems to be quickly and severely infected. Monovarieties, growing a single susceptible variety over large areas are exhibiting less resistance against the disease. Moreover, the pathogens that cause SRD also produce fungal toxins. *F. proliferatum* produces fumonisin (Stankovic et al. 2007), while *A. tenuis* produces tenuazonic acid, alternariol, altertoxin-I and alternariolmonomethyl ether (Panigrahi and Dallin 1994). These mycotoxins are harmful to humans and animals.

2.4 Molecular Approaches for Pathogenesis

Vast array microorganisms are causative agents of various diseases in crops, which results in serious production loss worldwide each year. In evolution plants have undergone disease resistance capability. Genetically, plant disease resistance is divided into two kinds, qualitative or complete resistance and quantitative or partial resistance, based on the speed and strength of plant response to pathogens invasion and proliferation (Kou and Wang 2010). Qualitative resistance can be strengthened through presence of a single major disease resistance (MR) gene and is mostly pathogen community-specific. While multiple gene complexes are involved in quantitative resistances as well as nonspecific to pathogens. Mostly the qualitative resistance has been widely targeted for field application due to its maximal resistance and easy manipulation.

Pathogen invade plant tissue can be resisted by the host plant through a double line defence innate immune system. One is denoted as pathogen-associated molecular pattern (PAMP)-triggered immunity (PTI) while the other is effector-triggered immunity (ETI) (Jones and Dangl 2006). In plant–pathogen interactions pathogen-produced PAMPs &/or damage in plants created by host peptides or cell wall fragments released during pathogen invasion recognized by cell membrane-localized plant pattern recognition receptors (PRRs) to initiate PTI. This array of self defense response is frequently weak and acts as pathogen species-nonspecific or community-nonspecific, and thus it is also recognized as basal resistance (Thomma et al. 2011). Resistance pathogens can surpass PTI using effectors that they secrete into plant cells. Plants carrying resistance (R) proteins can initiate ETI by direct or indirect perception of specific effectors. ETI is pathogen community-specific and generally exerts a maximum resistance, and thus it is also called community-specific or gene-for-gene resistance. Qualitative resistance can be regulated by either of R genes functioning in ETI and PRR genes functioning in PTI. Genes initiating qualitative resistance are denoted by MR.

2.4.1 Common Structural Features of MR Genes

Approximately 20 types of MR genes have been identified, and which shares common structural features (Sacco and Moffett 2009). However, the cloned MR gene (dominant) of angiosperms encodes cytoplasmic nucleotide-binding (NB)-leucine-rich repeat (LRR) proteins (Moffett 2009). These genes mediate resistance to various types of pathogens including fungi. NB-LRR proteins directly or indirectly interact with specific pathogen effectors secreted into host cells to initiate ETI (Eitas and Dangl 2010). A large number of MR genes only encode a few types of proteins in most of the well-studied plant-pathogen pathosystems, suggesting that the initiation of host qualitative resistance to different pathogens in these pathosystems appears to share similar mechanisms (Zhang and Wang 2013).

2.5 Active Soil Bacteria Against Rice Phytopathogens

Control of plant diseases by other living organisms is explained as biological control (Trigiano et al. 2004). These usually involve interactions between the pathogen and host. Because of its inherent potential to inhibit various phytopathogens causing different crop plant diseases involved in a blend of diverse modes of action and the possibilities to be combined enhanced effects with other control methods, the use soil bacterial strains as biological control agents have received a great attention (Kondoh et al. 2001; Shoda 2000). Several *Bacillus* sp. with excellent colonization capacity and versatility towards phytopathogens proves them as promising bacterial biocontrol agent and sporulation capability assures their relative abundance in the environment. In addition, the bountiful soil bacteria play an important role in fungal biocontrol through destroying chitin, a major fungal cell wall constituent. Altogether, *Bacillus* species also offer several advantages by facilitates increased shelf life and utilisation. *Bacillus* sp. belongs to class *Bacilli*, order *Bacillales*, family *Bacillaceae*.

Rhizospheric *Bacillus* sp. worthy to be used as biological control agents against most immerging phytopathogens. One of the most convincing properties contributing to that suggestion is the ability to modify attachment to different surfaces for survival of the *Bacillus* cells in their habitat (Stein 2005). In green house and field setups treatment of plant diseases with these bacteria have been successfully carried out. Subsequently they can antagonize fungal pathogens for nutrients and niche, secreting fungitoxic metabolites and by stimulating plant immune system (Lecle're et al. 2005). *Bacillus* is examined for potential to increase the yield, growth and nutrition of other crop plants under organic growing conditions (Orhan et al. 2006).

Various *Bacillus* sp. have been reported to be reservoir of bioactive compounds of phospholipids, polyketides and (lipo)peptides in nature (Tamehiro et al. 2002). Surfactin, fengycin and iturins are some proteins expresses their significance involvement in phytopathogenic disease control and maximum agricultural output (Emmert and Handelsman 1999; Romero et al. 2004). However, these are amphiphilic, membrane active multi-peptides with a β-amino fatty acid and strong antimycotic property (Thimon et al. 1995) and also composed of C_{13} to C_{18} lengths of the fatty acids (Toure et al. 2004). Specifically Iturin A exhibits a strong antibiotic activity by broad antifungal spectrum, proving it an ideal strong biocontrol agent with the aim to reduce agrochemical uses (Hsieh et al. 2008). As targeting the fungal membrane Iturin A disrupts it and creating transmembrane ion channels, which permit the release of vital cellular electrolytes (Hsieh et al. 2008). As strong biosurfactants, surfactins also have synergistic effects with iturin A. Reductions in the severity of fungal diseases by *Bacillus* representatives, are *B. amyloliquefaciens*, *B. subtilis*, *B. mycoides*, *B. cereus*, *B. pumilus*, *B. pasteurii* and *B. sphaericus*. While 8% of genome of *B. subtilis* and *B. amyloliquefaciens* involved in wide array of antimicrobial compounds synthesis. The organisms have employed vast array of multienzyme complex system for antifungal and plant growth promotion (Probanza et al. 2002; Gutiérrez Mañero et al. 2003). Such antibiotic weaponry and excellent soil colonisation explains the strong biocontrol ability of *Bacilli* both controlled and natural field conditions and its success as marketed product. Furthermore, the focus is on mechanisms of disease suppression and the degree of reduction in soil has not been studied systematically.

2.5.1 Mechanism

The fungitoxicity of *Bacillus* sp. against some devastating and economically important rice pathogens permits for the development of biocontrol agents for strategistic use (Harman et al. 2004). In order to understand its control mechanism, the host-pathogenic interaction must be understood. The multiple mechanisms involved are as follows;

2.5.1.1 Antibiosis

It is carried out in between microbial interactions involving low molecular or is mediated by the particular or non-particular metabolite of microbial origin, by enzyme, volatile organic compounds and other toxic substance (Limon et al. 2004; El-Ghaouth et al. 2004). It is achieved through

(i) Antibiotic

These are the microbial secondary metabolites lethal at very low concentration and kill others. For example, *B. amyloliquefaciens* and *B. methylotrophicus* are synthesizing antimicrobial secondary metabolites of types lantibiotics, phospholipids and polyketides responsible for the suppression of *R. solani*. Meanwhile, several strains of *Lysinibacillus* sp. (erstwhile *Bacillus*) closest to *L. sphaericus* and *L. xylanilyticus* were shown to have antagonism on food borne and fungal pathogens possibly through bacteriocins production (Ahmad et al. 2014).

(ii) Volatile organic compounds (VOCs) production

Microorganisms produces functional VOCs as extracellular metabolite, interferes with the growth and proliferation of fungal pathogens. Chemically these are lipophilic with low mol mass (<300 Da). Bacterial VOCs retards pathogens growth by reduction in both mycelial germination and spore formation. Few bacterial volatiles are CO_2, NH_3 or HCN. NH_3 has been known to toxic for many photosynthetic organisms, as it destroys the photosystem II O_2 evolving complex (Drath et al. 2008). Bacterial CO_2 could be a carbon source for chemolithoautotrophic microorganisms and green plants and promotes growth. *Bacillus* sp. having potent pathogen suppressing attributes by production of HCN (Kumar et al. 2012).

2.5.1.2 Hyperparasitisms/Mycoparasitism

Direct parasitism or destruction of pathogen or its propagules by other microorganisms is termed as a hyperparasitism. It is one of the most direct type growth retardation with no supportive effort from other organism or environment. Bacteria that are parasitic on other pathogenic fungi are mycoparasites. Hyperparasites are facultative parasites, obligate pathogens and predators. For example, *B. cereus* and *B. subtilis* along with some fungi hyperparasitized the mycelium of *Pythium* sp. Hyperparasitisms increases with synergistic culture rather monoculture (Bankole and Adebanjo 1998). Sclerotia are metabolically active and very resistant to invasion by microorganisms. While the microsclerotia may readily infect without extensive mycelial growth, on contact with developing roots as of *S. sclerotiorum*. The interaction in between antagonist *Bacillus*, phytopathogens, plant and environment determines the success or failure of annoyance. In addition bacterial predations are moistly communal and non-particular and exhibited in auxotrophic conditions.

2.5.1.3 Stress Effects

When the abiotic conditions for the growth and development of rice exceeds indiscriminately leading to stress, and more susceptible to pathogenic attack. Pathogens may be directly affected by such stress, which weakens them and makes more vulnerable to antagonists, thus decreasing disease severity. When sclerotia germinate to budding mycelium or to sporangium, nutrient leakage is increased and become susceptible for antagonist attack. Sclerotia may be stressed by the soil environment and made more vulnerable to antagonistic microflora thus favours the decay by antagonists. In some instances by decreasing the host plant stress the disease prevalence may be diminished.

2.5.1.4 Competition

A general believe is that the competition between pathogens and nonpathogens for nutrients, space and other resources is important advantage in biocontrol. However, it is more critical for soil borne pathogens. In view of this, nutrient chellation and rapid colonisation used as protection for plants through non pathogenic endo and exophytes. Prevention in infection sites (excess nutrients) accumulation for competitors; or modifying physiological conditions as rate of oxygen diffusion, water potential (Ψ), pH and other of the host are surplus role for growth limitation. Side by side they produce the metabolites that are effective in suppression of pathogens. These microbes colonize the sites where water and carbon containing nutrients are available, such as exit points of secondary roots, damaged epidermal cells, nectaries or secreting glands and utilize the root mucilages.

Competition for Fe, an essential micronutrient has also been shown to important in disease biocontrol. Soil *Bacillus* sp. is generally considered to be aggressive competitors, fast growers and rapidly colonize substrates to exclude pathogens. As Fe is required for spore germination and hyphal growth, Fe competition in alkaline soils may be a limiting factor (Costa and Loper 1994). Reports have demonstrated that siderophore, iron chelating molecules synthesised in *B. firmus* plays a role in pathogen suppression (Chaiharn et al. 2009). Siderophore are the secondary metabolites produced in late exponential stage. It also have reported *B. firmus* and other *Bacillus* strains produce hydroxamate and catachol type siderophores at low iron conditions and infected by *Alternaria* sp., *F. oxysporum*, *P. oryzae* and *Sclerotium* sp. in rice. The ability for siderophore production and utilisation confers an ecological advantage for soil borne bacteria to colonizing in the soil.

2.5.1.5 Induced Systemic Resistance (ISR)

Induced Systemic Resistance (ISR) is a mechanism in which bacteria enhances plant defence systems to overcome biotic challenges (Kloepper et al. 2004). From genera *Bacillus*, it is interesting to note the diversity of species can induce systemic resistance in plants. *B. amyloliquefaciens*, *B. subtilis*, *B. pasteurii*, *B. cereus*, *B. pumilus*, *B. mycoides* and *B. sphaericus*, to name a few (Ryu et al. 2004; Kloepper et al. 2004; Ongena et al. 2007; Rudrappa et al. 2010; Niu et al. 2011). The (lipo)peptides, being produced and actively involved in ISR for plants (Ongena et al. 2007). ISR network regulated basically on the signalling molecules of jasmonic acid (JA), ethylene (ET) and a often nitric oxide(NO). It does not involve the pathogen related proteins or salicylic acid (SA). In case of *B. cereus* and *B. amyloliquefaciens*, VOCs are responsible for the ISR. The signal transduction leading to ISR requires receptivity to both JA and ET. Methyl jasmonate (MJA) and the ET precursor i.e., 1-amino-cyclopropane-1-carboxylate (ACC) are effective in inducing resistance against pathogenic microflora (Thomma et al. 2001). JA induces defence-related proteins e.g., proteinase and thionins while several pathogenesis-related (PR) gene were activated due to ET. They act synergistically in stimulating elicitor-induced PR gene expression and systematically induce defence responses. ISR was independent of SA but induced by ET and JA dependent pathways in *B. subtilis* GB03 (Ryu et al. 2004).

2.5.1.6 Mycolytic Enzymes (Fungal Cell Wall Lytic Enzymes) Production

Bacteria able to synthesize enzymes like chitinases, lipases, proteases and β-1,3-glucanases are detrimental for phytopathogens and adds on improvement of the biocontrol efficacy (Whipps 2001). Plant resistance to pathogenic fungi also enhanced by various reaction pathways including accumulation of β-1,3-glucanase enzymes and hydrolytic chitinases (Boller 1985). Chitinases degrade chitin, major component of fungal cell wall and act synergistically with the β- 1,3-glucanases to seize growth. The enzymes are less expressed in leaves and highly in roots and seeds. *F. oxysporum* antagonised by *B. subtilis in vitro* (Chen et al. 2009). The strain was evaluated for production of chitinase, β-1,3-glucanase, indole-3-acetic acid (IAA) and other metabolites with the selected medium. The enzymes causes vacuolation, swelling and lysis of fungal hyphae. *B. cereus* generate and discharge a chitinase of 36 kDa which significantly protect against *R. solani* (Pleban et al. 1997).

β-1,3-glucanases targets β-1,3 glucan and cleaves of β-1,3 glucosidic bonds in fungal cell wall. β-1,3- glucanase, have 25–36 KDa of molecular weight, resistant to proteases and have extracellular secretion (Bol et al. 1990). *In vitro* glucanase activity of *B. subtilis* against rice blast was demonstrated (Leelasuphakul et al. 2005). Both the mycolytic enzymes act synergistically for maximum effort.

2.5.2 Physiology

In regular interaction in between the plants and soil microbes some microbial biomolecules are denoted as Microbe Associated Molecular Patterns (MAMPs) are detected. These includes β-glucan, (lipo)polysaccharides, chitin, elongation factor (HrpZ), ergosterol, flagellin (bacterial), NEP1-like proteins (NLPs) and oomycete-derived Pep13. Helps in triggering of primary plant innate immunity, contribute to cease infection before the pathogen gains a hold on host. Alterations in physiochemical biomolecules as changes in fluxes of H^+, K^+, Cl^- and Ca^{2+} ions across the plasma membrane, activation of cyclic nucleotide gated channels, signalling systemic resistance molecules production and stomatal closure results in plants after elicitation by MAMPs (Muthamilarasan and Prasad 2013) and helps in changes the defense mechanisms to suppress propagation of pathogens.

Members of the *Bacillaceae* produce a wide variety compounds differs in length and fatty acid chains, like subtilin and sublancin of ribosomal origin, others such as bacilysin, rhizocticins, difficidin and (lipo)peptides are formed by nonribosomal peptide synthetases and/or polyketide synthases (Leclère et al. 2005). The biocompounds are amphiphilic cyclic peptides comprises of α or β-amino acids branched to unique β-amino/hydroxy fatty acid. Fungitoxicity increases with increase in carbon atom as C_{17} homologues are 20-fold more active in comparison to C_{14}. The *srf* operon encoding subunits of surfactin synthetase catalyzes the thiotemplate mechanism for incorporation of the seven amino acids into the surfactin (lipo)peptide. The mycosubtilin gene cluster consists of four open reading frames, designated fenF, mycA, mycB, and mycC, regulated by the promoter, P_{myc} (Duitman et al. 1999). The mycosubtilin peptide moiety synthesis was carried out by three *myc* genes encoding seven modules. The fatty acid and polyketide synthases are structurally homologous to N-terminus of *mycA*.

The mycocidal effect of lipopeptides not only rely on their chemical structure but also on the sterol content of the plasma membrane in the target phytopathogens, due to a buffering effect of sterols on the increase in membrane fluidity caused by the long chain fatty acids (Latoud et al. 1990; Avis and Bélanger 2002).

2.6 Mode of Action

The wider use of biological control agents signifies its ability to affect plant health by notable suppressing different plant infections involving various modes of action (Sharma et al. 2009) and the probable synergism with other control methods and processes (Correa et al. 2009). As a result from specific and non-specific biointeractions, biological control can be considered as a positive one. *Bacillus* sp. is the best contender for bacterial biocontrol agents (Schallmey et al. 2004). Mutualism, protocooperation, commensalism, neutralism, competition, amensalism, parasitism, and predation are various interactions occur in nature for biocontrol. In general the modes of action have been counted important in the biological control of fungi with *Bacillus* sp. are (A) preventative colonization; (B) antagonism through toxic metabolites (Hallmann 2001). Each of these mechanisms affects one or more of the weak interactions links (Fig. 2.2) determining which stage in the life-cycle of a pathogen is the ideal objective for biological control.

Competition between the phytopathogenic fungi and the soil *Bacillus* sp. for space and nutrients is ever present, for the simple fact that they occupy the same soil

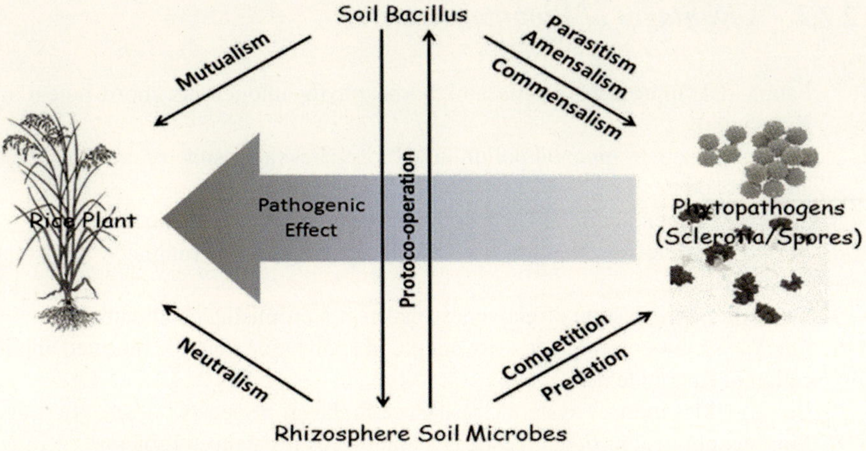

Fig. 2.2 Different interactions of soil bacteria with phytopathogens

ecological niche concomitantly. However, this is the basic mode of action, the levels of control would always be closely related to bacterial population density and degree of prevalence in soil, or to the location of bacterial colonisation in relation to the phytopathogen feeding site.

The activation of host defences also have a role against antagonistic microorganisms (Droby and Chalutz 1994). Direct interaction of the *Bacillus* sp. with the pathogen hyphae does not tacit higher antagonistic activity, whereas *in vitro* extracellular enzyme (β-1,3-glucanase, chitinases) activity might account for higher levels of armamentship. In few instances the pathogens are inhibited by metabolites through obstructing protein synthesis in lieu of nucleic acid synthesis. A better understanding of the antagonists' mode(s) of action is an important prerequisite both for improving their performance and to establish screening criteria in the search for emerging isolates.

2.7 Advantages and Limitations

The advantage of *Bacillus* sp. based products is their long shelf-life at room temperature, due to endospore production. It is relatively easy for microorganism to ferment with economical production compared to fungal chemical control agents. Since no major direct effect exerted on pathogen and, efficacy lowered in high pathogens innocula in soil or against very competitive soil borne pathogens. In several *Bacillus* sp. the highest efficacy is observed on growing seeds or budding plantlets, because of the combination of direct antibiosis, induced resistance and growth promotion.

2.7.1 *Advantages of Bioantagonists*

1. Fungicidal approach towards soil borne phytopathogens is more potent in *Bacillus* sp.
2. Circumvent environmental pollution in biospheres (atmospheres, hydrospheres and lithospheres).
3. Avoids detrimental effects on beneficial organisms and maintains balance.
4. Very high biocontrol capacity by integrating with fungicide resistant antagonist.
5. Durability, long lasting effectiveness makes it a promising candidate.
6. *Bacillus* sp. biocontrol agents are simple in application than the repeated application of fungicide needs.
7. Biopesticides induces systemic resistance in the crop species.
8. Most economical *in lieu* to pesticides and avoids resistance problems.
9. Substantial effective against mycelia and sclerotia as of *Bacillus* sp. on *R. solani.*
10. In case of less severe incidence biopesticides are degraded gradually causing no harm to yield.

2.7.2 *Limitations*

1. Host specificity in biocontrol agents exhibits non target deleterious effects through non-direct interactions (Dean and Ragan 2003).
2. In host shifting (expansion of host range) attack native organisms.
3. Living organisms are difficult to store (few require cold storage), handle, apply, take longer time to act and loss of viability.
4. Continuous availability of free moisture helps to prevalence the disease.
5. Environmental parameters (sub)optimal for growth and development of an antagonist can reduce the effectiveness of biocontrol.
6. Often the degree of control varies from place to place and time to time, destination of pathogen in the host and aetiology of the diseases involved.
7. Concerning the environmental safety a few legal and ethical issues are with the release of genetically modified (engineered) organisms into the environment.

2.8 Effects on Soil Health

Agrochemicals primarily used for influencing beneficial soil microbial community with management of soil environment. All over the world, most of the crop fields are applied with the N and P fertilizers. With this level of inputs going onto the soil it is important to understand and manage the effects these chemicals on the soil environment.

There are many different chemical fertilizers, fungicides, pesticides which are directly affecting and inhibiting native soil microbes and their community activity associated with the ecological processes (Grant and Wu 2008). Soil microbiota mediates 80–90% of the biotic processes in soil, its role in terms of nutrient recycling, disease suppressiveness (control) etc. Additionally, the substantial application of fungicides against foliar disease influences production of VOCs in the aboveground parts, as well as the production of some antimicrobial photochemical in rhizosphere (Cruz et al. 2012). The relationship between soil quality and microbial diversity is not well defined. It is widely accepted that the total soil microbial community including the slumbering microbes with functional redundancy that maintains soil health and homeostasis. A commonly used indicator of soil status is the determination of diversity and structure of microbial communities.

Soil characteristics and plant diversity influences the function and structure of the microbial community (Butenschoen et al. 2011), but also respond to minute alterations in environmental conditions, including different aspects driven by anthropogenic interventions such as biocontrol of soil borne pathogens, biofertilization, rotation, intercropping, tillage and chemical or organic inputs and others). Commonly soil *Bacillus* sp. as biocontrol agents have minor to transient effect on the soil inhabiting microflora and microfauna due to pathogen specific interaction. These are not bioaccumulate, existed for limited time span and are biodegradable. Thus good strategic implementation is required for the reduction of soil inoculum potential and improvement in plant and soil health. Moreover, after inoculation it changes the balance between the resident soil microbial communities by altering the microecosystem. Biocontrol agents inhibit the growth of plant pathogens, increase soil disease suppressiveness, have the ability to ameliorate some abiotic and physiological stresses and enhance nutrient uptake in plants, improved plant growth. Biocontrols application on soil attenuates the microbial community because of a buffer effect of vegetation, which aids recovery of the originally existed microbial population. Increased soil microbial biomass, active other biocontrol bacterial diversity & enzyme activities and also promotes higher in C-biomass compared to chemical fertilizer applied conditions (Bajsa et al. 2013). The impact of commercial formulations of biocontrols on soil microbes should be trailed in the field, since excipients used might have effects which are unobserved during inoculation *in vivo*. Fortunately, the use of *Bacillus* as biocontrol of rice phytopathogens on a large scale as is happening nowadays is a clear indication of the safety and usefulness.

2.9 Conclusions

The native as well as exotic soil microbial communities influence plant health, nutrition and ecosystem functioning and nutrient recycling. Therefore, it is imperative to understand the relationship between soil microbial communities and agricultural practices. Mostly plants are infected by soil fungi residing in the rhizosphere. Rice, as the essential food for half of the global population and to protect it from devastating different phytofungal pathogens possessing different level of pathogenicity is a matter of utmost concern. The scene is worsening by addition of chemical fertilizers that pollute and weakens the environment, make the pathogen resistant. Soil *Bacillus* sp. provides an alternate and ecofriendly roadway to mitigate the complication in a smoother path. By various mechanisms, mode of action and different physiology it is pertinent to mention here that genus *Bacillus* always combat against different rice phytopathogens coherently as described in this chapter.

References

Abdalla SA, Algam SAA, Ibrahim EA, El Naim AM (2014) In vitro screening of bacillus isolates for biological control of early blight disease of tomato in Shambat soil. World J Agric Res 2(2):47–50

Acea MJ, Moore CR, Alexander M (1988) Survival and growth of bacteria introduced into soil. Soil Biol Biochem 20:509–515

Ahmad V, Iqbal AN, Haseeb M, Khan MS (2014) Antimicrobial potential of bacteriocin producing Lysini bacillus jx416856 against foodborne bacterial and fungal pathogens isolated from fruits and vegetable waste. Anaerobe 27:87–95. https://doi.org/10.1016/janaerobe201404001

Ali HZ, AbdulRahman AQA, Abdullah AA, Saood HM (2016) Prescreening of pathogenicity of rice pathogens prior to biological control assay under greenhouse conditions. Asian J Sci Tech 07(02):2416–2422

Aslim B, Saúlam N, Beyatli Y (2002) Determination of some properties of Bacillus isolated from soil. Turk J Biol 26:41–48

Avis TJ, Bélanger RR (2002) Mechanisms and means of detection of biocontrol activity of Pseudozima yeast against plant-pathogenic fungi. Fed Eur Microbiol Soc Yeast Res 2:5–8

Aye SS, Matsumoto M (2010) Genetic characterization by Rep-PCR of Myanmar isolates of *Rhizoctonia* spp causal agents of rice sheath diseases. J Plant Pathol 92:255–260

Ayyadurai N, Kirubakaran SI, Srisha S, Sakthivel N (2005) Biological and molecular variability of *Sarocladium oryzae* the sheath rot pathogen of rice (*Oryza sativa* L.) Curr Microbiol 50:319–323. https://doi.org/10.1007/s00284-005-4509-6

Bajsa N, Morel MA, Brana V, Castro-Sowinski S (2013) The effect of agricultural practices on resident soil microbial communities: focus on biocontrol and biofertilization. In: de Bruij FJ (ed) Molecular microbial ecology of the rhizosphere, vol 2. Wiley, Hoboken

Baker B, Zambryski P, Staskawicz B, Dinesh-Kumar SP (1997) Signaling in plant- microbe interactions. Science 276:726–733

Baker KF (1987) Evolving concepts of biological control of plant pathogens. Annu Rev Phytopathol 25:67–85

Bankole SA, Adebanjo A (1998) Efficacyof some fungal and bacterial isolatesin controlling wet rot disease of cowpea caused by *Pythium aphanidermatum*. J Plant Protect Tropics 11:37–43

Basha S, Ulaganathan K (2002) Antagonism of Bacillus species (strain BC121) towards *Curvularia lunata*. Curr Sci 82(12):1457

Bigirimana VP, Hua GKH, Nyamangyoku OI, Höfte M (2015) Rice sheath rot: an emerging ubiquitous destructive disease complex. Front Plant Sci 6:1066. https://doi.org/10.3389/fpls201501066

Bloemberg GV, Lugtenberg BJJ (2001) Molecular basis of plant growth promotion and biocontrol by rhizobacteria. Curr Opin Plant Biol 4:343–350

Bol JF, Linthrost HJM, Cornelissen BJC (1990) Plant pathogenesis related proteins induced by virus infection. Annu Rev Phytopathol 28:113–138

Boller T (1985) Induction of hydrolases as defense reaction against pathogens. In: Key JL, Kosuge TA, Liss R (eds) Cellular and molecular biology of plant stress. Alan R Liss, New York, pp 247–262

Brock TD, Madigan MT (1991) Biology of microorganisms, 6th edn. Prentice-Hall International, Englewood Cliffs

Butenschoen O, Scheu S, Eisenhauer N (2011) Interactive effects of warming soil humidity and plant diversity on litter decomposition and microbial activity. Soil Biol Biochem 43:1902–1907

Cavalcanti MA, Oliveira LG, Fernandes MJ, Lima DM (2006) Filamentous fungi isolated from soil in districts of the Xingo'region. Braz. Acta Bot Bras 20(4):831–837

Cawoy H, Bettiol W, Fickers P, Ongena M (2011) Bacillus-based biological control of plant diseases pesticides in the modern world. In: Stoytcheva M (ed) Pesticides use and management. InTech open science, Rijeka, pp 274–302

Chaiharn M, Chunhaleuchanon S, Lumyong S (2009) Screening siderophore producing bacteria as potential biological control agent for fungal rice pathogens in Thailand. World J Microbiol Biotechnol 25:1919–1928

Chakrabarti NK (2001) Epidemiology and disease management of brown spot of rice in India. In: Sreenivasaprasad S, Johnson R (eds) Major fungal diseases of rice: recent advances. Springer, Dordrecht, pp 293–306

Chalkley D (2010) Invasive Fungi Stackburn seedling blight leaf spot of rice -Alternaria padwickii Systematic Mycology and Microbiology Laboratory ARS USDA

Chen F, Wang M, Zheng Y, Luo J, Yang X, Wang X (2009) Quantitative changes of plant defense enzymes and phytohormones in biocontrol of cucumber Fusarium wilt by *Bacillus subtilis* B579. World J Microbiol Biotechnol 26:675–684

Chun SC, Schneider RW (1998) Sites of infection by pythium species in rice seedlings and effects of plant age and water depth on disease development. Phytopathology 88(12):1255–1261

CMI (1984) Distribution maps of plant diseases. Map No 314, 4th edition. CAB International, Wallingford

Collins DP, Jacobsen BJ (2003) Optimizing a Bacillus subtilis isolate for biocontrol of sugar beet Cercospora leaf spot. Biol Conserv 26(2):153–161. https://doi.org/10.1016/S1049-9644(02)00132-9

Correa OS, Montecchina MS, Berti MF, Fernandez Ferrari MC, Pucheu NL, Kerber NL, Garcia AF (2009) *Bacillus* amyloliquefaciens BNM122 a potential microbial biocontrol agent applied on soybeen seeds cause a minor impact on rhizosphere and soil microbial communities. Appl Soil Ecol 41:185–194

Costa JM, Loper JE (1994) Characterization of siderophore production by the biological control agent *Enterobacter cloacae*. MPMI 7:440–448

Crop Protection Compendium (2007) CAB International, Wallingford

Cruz AF, Hamel C, Yang C et al (2012) Phytochemicals to suppress fusarium head blight in wheat-chickpea rotation. Phytochemistry 78:72–80

Daniel R (2004) The soil metagenome – a rich resource for the discovery of novel natural products. Curr Opin Biotech 15:199–204

Das S, Lyla PS, Khan SA (2006) Marine microbial diversity and ecology: importance and future perspectives. Current Sci 90(10):1325–1335

Dath AP (1990) Sheath blight disease of rice and its management. Associate Publishing Company, New Delhi, p 152

Datnoff LE, Jones DB (1992) Sheath blotch of rice: a disease new to Americas. Plant Dis 76:1182–1184

De Boer W, Paulien JA, Gunnewiek K, Lafeber P, Janse JD, Spit BE, Woldendorp JW (1998) Antifungal properties of chitinolytic dune soil bacteria. Soil Biol Biochem 30(2):193–203

de Freitas JR, Banerjee MR, Germida JJ (1997) Phosphate-solubilizing rhizobacteria enhance the growth and yield but not phosphorus uptake of canola (*Brassica napus* L.) J Biol Fertil Soils 24(4):358–364. https://doi.org/10.1007/s003740050258

Dean EP, Ragan MC (2003) Indirect effects of host-specific biological control agents. Trends Ecol Evol 18:456–461

Dean RA, Talbot NJ, Ebbole DJ, Farman ML, Mitchell TK, Orbach MJ, Thon M, Birren BW (2005) The genome sequence of the rice blast fungus *Magnaporthe grisea*. Nature 434:980–986. https://doi.org/10.1038/nature03449

Drath M, Kloft N, Batschaer A, Marin K, Novak J, Forchhammer K (2008) Ammonia triggers photodamage of photosystem II in the cyanobacterium Synechocystissp Strain PCC 6803. Plant Physiol 147:206–215

Droby S, Chalutz E (1994) Mode of action of biocontrol agents of postharvest diseases. In: Wilson CL, Wisniewski ME (eds) Biological control of postharvest diseases – theory and practice. CRC Press, Boca Raton, pp 63–75

Duitman EH, Hamoen LW, Rembold M, Venema G, Seitz H, Saenger W, Bernhardt F, Schmidt M, Ulrich C, Stein T, Leenders F, Vater J (1999) The mycosubtilin synthetase of *Bacillus subtilis* ATCC 6633: a multifunctional hybrid between a peptide synthetase an amino transferase and a fatty acid synthase. Proc Natl Acad Sci USA 96:13294–13299

Eeden MV, Korsten L (2006) A predictive model for biological control of cercospora spot: effect of nutrient availability. South African Avocado Growers' Association Yearbook. pp 48–52

Eitas TK, Dangl JL (2010) NB-LRR proteins: pairs pieces perception partners and pathways. Curr Opin Plant Biol 13:472–477

El-Ghaouth A, Droby S, Wilson CL, Wisniewski M, Smilanick J, Korsten L (2004) Biological control of postharvest diseases of fruits and vegetables. In: Khachatourians GG, Arora DK (eds) Applied mycology and biotechnology: Agriculture and food production. Elsevier Science, Amsterdam, pp 11–27

El-kazzaz MK, Salem EA, Ghoneim KE, Elsharkawy MM, El-Kot GAN, Kalboush ZA (2015) Biocontrol of Tilletia barclayana the causal agent of Kernel Smut disease in rice. Egypti J Biol Pest Control 25(3):535–544. Proceeding of 4th International Conference, ESPCP2015, Cairo, Egypt, 19–22 October 2015

Emmert EAB, Handelsman J (1999) Biocontrol of plant disease: a (gram-) positive perspective. FEMS Microbiol Lett 171:1–9

Espino LA, Barlow VM, Greer CA, Godfrey LD, Lawler SP (2015) Revised continuously UC IPM Pest Management Guidelines Rice UC ANR Publication, 3465, Oakland, California, pp 27–28. (http://www.ipmucanreduanredu/PMG/selectnewpestrice.html)

Ghosh MK, Amudha R, Jayachandran S, Sakthivel N (2002) Detection and quantification of phytotoxic metabolites of Sarocladium oryzae in sheath rot-infected grains of rice. Lett Appl Microbiol 34:398–401

Grant C, Wu R (2008) Enhanced-efficiency fertilizers for use on the Canadian Prairies. Crop Manag 7(1)

Groth D (1992) Stackburn. In: Webster RK, Gunnell PS (eds) Compendium of rice diseases. American Phytopathological Society, St Paul, pp 18–19

Gutiérrez Mañero FJ, Probanza A, Ramos B, Colon Flores JJ, Lucas Garcia JA (2003) Effects of culture filtrates of rhizobacteria isolated from wild lupine on germination growth and biological nitrogen fixation of lupine seedlings. J Plant Nutr 26:1101–1115

Hallmann J (2001) Plant interactions with endophytic bacteria. In: Jeger MJ, Spence NJ (eds) Biotic interactions in plant-pathogen associations. CAB International, London, pp 87–119

Harman GE, Howell CR, Vitarbo A, Chet I, Lorito M (2004) Trichoderma species eopportunistic avirulent plant symbionts. Nat Rev Microbiol 2:43–56

Hattori T (1988) Soil aggregates as microhabitats of microorganisms. Biol Fertil Soils 6:189–2037

Hsieh FC, Lin TC, Meng M, Kao SS (2008) Comparing methods for identifying Bacillus strains capable of producing the antifungal lipopeptide iturin A. Curr Microbiol 56:1–5

Huang SW, Wang L, Liu LM, Tang SQ, Zhu DF, Savary S (2011) Rice spikelet rot disease in China-1 characterization of fungi associated with the disease. Crop Prot 30:1–9

Jayaprakashvel M, Mathivanan N (2009) Biological control and its implications on rice diseases management. In: Role of biocontrol agents for disease management in sustainable agriculture, pp 440–455

Jens Laurids S, Elina A, Ulf T, Henriette G, Teis Esben S (2013) Production of fusarielins by Fusarium. Int J Food Microbiol 160:206–211

Joe Y, Manibhushanrao K (1995) Biochemical constituents of differentially virulent isolates of *Sarocladium* spp causing sheath rot disease of rice. Z Pflanzenkr Pflanzenschutz 102:291–298

Jones JD, Dangl JL (2006) The plant immune system. Nature 444:323–329

Kai M, Effmert U, Berg G, Piechulla B (2007) Volatiles of bacterial antagonists inhibit mycelial growth of the plant pathogen *Rhizoctonia solani*. Arch Microbiol 187(5):351–360. https://doi.org/10.1007/s00203-006-0199-0

Kanjanamaneesathian M, Kusenwiriyawong C, Pengnoo A, Nilratana L (1998) Screening of potential bacterial antagonists for control of sheath blight in rice and development of suitable bacterial formulations for effective application. Australas Plant Pathol 27:198–206

Kato H (2001) Rice blast disease. Pesticide Outlook 12:23–25

Kazempour MN, Elahinia SA (2007) Biological control of Fusarium fujikuroi the causal agent of Bakanae disease by rice associated antagonistic bacteria. Bulgarian J Agr Sci 13:393–408

Khair A, Shetty SA, Shetty HS, Safeeulla KM (1988) The incidence and distribution of Trichoconiella padwickii in Karnataka in relation to environmental factors. Int J Trop Plant Dis 6(1):107–113

Kim CK (1981) Ecological studies of bakanae disease of the rice cause by Gibberella fujikoroi Korean. J Plant Prot 20:146–151

Kloepper JW, Ryu CM, Zhang S (2004) Induced systemic resistance and promotion of plant growth by Bacillus spp. Phytopathology 94:1259–1266

Kondoh M, Hirai M, Shoda M (2001) Integrated biological and chemical control of damping-off caused by Rhizoctonia solani using Bacillus subtilis lXB14-C and flutolanil. J Biosci Bioeng 91:173–177

Kou Y, Wang S (2010) Broad-spectrum and durability: understanding of quantitative disease resistance. Curr Opin Plant Biol 13:181–185

Krishnaveni D, Ladhalakshmi D, Laha GS, Prakasam V, Asma Jabeen Mangrauthia SK, Srinivas Prasad M (2015) Role of natural products in disease management of rice. In: Ganesan S (ed) Sustainable crop disease management using natural products. CABI Press, London, pp 144–159

Kumar A, Singh R, Jalali BL (2003) Management of stem rot of rice with resistance inducing chemicals and fungicides. Indian. Phytopathology 56:266–269

Kumar P, Dubey RC, Maheshwari DK (2012) Bacillus strains isolated from rhizosphere showed plant growth promoting and antagonistic activity against Phytopathogens. Microbiol Res 167(8):493–499

Laha GS, Reddy CS, Krishnaveni D, Sundaram RM, Srinivas Prasad M, Ram T, Muralidharan K, Viraktamath BC (2009) Bacterial blight of rice and its management. Technical Bulletin No 41 DRR (ICAR) Rajendranagar, Hyderabad, India, p 37

Latoud C, Peypoux F, Michel G (1990) Interaction of iturin A a lipopeptide antibiotic with Saccharomyces cerevisiaecells: influence of the sterol membrane composition. Can J Microbiol 36:384–389

Lecle're V, Be'chet M, Adam A, bastien Guez J-S, Wathelet B, Ongena M, Thonart P, Gancel F, Marle'ne C, Philippe J (2005) Mycosubtilin overproduction by Bacillus subtilis BBG100 enhances the Organism's antagonistic and biocontrol activities. Appl Environ Microbiol 71:4577–4584

Leelasuphakul W, Sivanunsakul P, Phongpaichit S (2005) Purification characterization and synergistic activity of ß −1 3-glucanase and antibiotic extract from an antagonistic *Bacillus subtilis* NSRS 89-24 against rice blast and sheath blight. Enzym Microb Technol 38:990–997

Leslie JF, Summerell BA (2006) The fusarium laboratory manual. Blackwell, London

Limon MC, Chacon MR, Mejías R, Delgado-Jarana J, Rincón AM, Codón AC, Benítez T (2004) Increased antifungal and chitinase specific activities of Trichoderma harzianum CECT 2413 by addition of a cellulose binding domain. Appl Microbiol Biotechnol 64(5):675–685

Lindow SE, Brandl MT (2003) Microbiology of the Phyllosphere. Appl Environ Microbiol 69(4):1875–1883. https://doi.org/10.1128/AEM.69.4.1875-1883.2003

Lisboa MP, Bonatto D, Bizani D, Henriques JAP, Brandelli A (2006) Characterization of a bacteriocin-like substance produced by *Bacillus amyloliquefaciens* isolated from the Brazilian Atlantic forest. Int Microbiol 9:111–118

Liu YF, Chen ZY, Ng TB, Zhang J, Zhou MG, Song F, Lu F, Liu Y (2007) Bacisubin an antifungal protein with ribonuclease and hemagglutinating activities from *Bacillus subtilis* strain B-916. Peptides 28:553–559. https://doi.org/10.1016/j.peptides.2006.10.009

Mannanov RN, Sattarova RK (2001) Antibiotics produced by Bacillus bacteria. Chem Nat Compd 37:117–123

Mathre DE (1992) Gaeumannomyces. In: Singleton LL, Mihail JD, Rush CM (eds) Methods for research on soil borne Phytopathogenic Fungi. APS Press, St Paul, pp 60–63

Mebeaselassie A, Luoye L, Aiqing F, Xiaoyuan Z, Jianxiong L (2015) Development of GFP-expressing Ustilaginoidea virens strain to study fungal invasion and colonization in rice spikelets. S Afr J Bot 97:16–24. https://doi.org/10.1016/jsajb201411013

Mew TW, Gonzales P (2002) A handbook of rice seedborne Fungi. Science Publishers, Enfield

Moffett P (2009) Mechanisms of recognition in dominant R gene mediated resistance. Adv Virus Res 75:1–33

Muthamilarasan M, Prasad M (2013) Plant innate immunity: an updated insight into defense mechanism. J Biosci 38:1–17

Mwalyego FS, Kayeke JM, Mghogho RM (2011) Important diseases in rice production: symtoms damage and management ASARECA Rice project Tanzania, pp 1–11

Niu DD, Liu HX, Jiang CH, Wang YP, Wang QY, Jin HL, Guo GH (2011) The plant growth-promoting rhizobacterium B. cereus AR156 induces systemic resistance in Arabidopsis thaliana by simultaneously activating salicylate-and jasmonate/ethylene-dependent signalling pathways. Mol Plant Microbe Interact 24:533–542

Nunes CDM (2008) Ocorrência das doenças mal-do-pé (Gaeumannomyces graminis) e mancha parda (Dreschlerasp) na cultura de arroz Pelotas RS Embrapa Clima Temperado (Comunicado Técnico 205)

Nyvall RF (1999) Field crop diseases. Iowa State University Press, Ames, p 1021

Ohata K (1981) Change in the outbreak of rice diseases in mechanized transplant cultures. Jap Pest Inform 38:9–12

Omura S (1976) The antibiotic cerulenin a novel tool for biochemistry as an inhibitor of fatty-acid synthesis. Bacteriol Rev 40:681–697

Ongena M, Adam A, Jourdan E, Paquot M, Brans A, Joris B, Arpigny JL, Thonart P (2007) Surfactin and fengycin lipopeptides of Bacillus subtilis as elicitors of induced systemic resistance in plants. Environ Microbiol 9:1084–1090. https://doi.org/10.1111/j.1462-2920.2006.01202.x

Orhan E, Esitken A, Ercisli S, Turan M, Sahin F (2006) Effects of plant growth promoting rhizobacteria (PGPR) on yield growth and nutrient contents in organically growing raspberry. Sci Hortic 111:38–43

Ou SH (1985) Rice diseases, 2nd edn. Commonwealth Mycological Institute, Kew

Panigrahi S, Dallin S (1994) Toxicity of the Alternaria spp metabolites tenuazonic acid alternariol altertoxin-I and alternariol monomethyl ether to brine shrimp -(Artemia salina L) larvae. J Sci Food Agric 66:493–496

Patel RV, Chauhan HL, Sabalpara AN, Patel SJ (2006) Occurrence of sheath blotch of rice in South Gujarat. Indian Phytopath 59(1):123

Pearce DA, Bridge PD, Hawksworth DL (2001) Species concept in *Sarocladium* the causal agent in sheath rot in rice and bamboo blight. In: Sreenivasaprasad S, Johnson R (eds) Major fungal diseases of rice: recent advances. Springer, Dordrecht, pp 285–292

Peixoto CN, Ottoni G, Filippi Marta CC, Silva-Lobo VL, Prabhu AS (2013) Biology of *Gaeumannomyces graminis* var. *graminis* isolates from rice and grasses and epidemiological aspects of crown sheath rot of rice. Trop Plant Pathol 38(6):495–504

Pinchuk IV, Bressollier P, Sorokulova IB, Verneuil B, Urdaci MC (2002) Amicoumacin antibiotic production and genetic diversity of *Bacillus subtilis* strains isolated from different habitats. Res Microbiol 153:269–276

Pleban S, Chernin L, Chet I (1997) Chitinolytic activity of an endophytic strain of *Bacillus cereus*. Appl Environ Microbiol 25:284–288

Prabhu AS, Filippi MC (2002) Ocorrência do mal-do-pé causado por Gaeumannomyces graminis var graminis uma nova enfermidade em arroz no Brasil. Fitopatol Bras 27:417–419

Probanza A, Lucas Garcia JA, Ruiz Palomino M, Ramos B, Gutierrez Manero FJ (2002) *Pinus pinea* L. seedling growth and bacterial rhizosphere structure after inoculation with PGPR Bacillus(*B licheniformis* CECT5106 and *B pumilus* CECT5105). Appl Soil Ecol 20:75–84

Rabindran R, Vidhyasekaran P (1996) Development of a formulation of Pseudomonas iluorescens PfALR2 for management of rice sheath blight. Crop Prot 15:715–721

Rahman MM (2013) Insecticide substitutes for DDT to control mosquitoes may be causes of several diseases. Environ Sci Pollut Res 20:2064–2069

Renwick A, Campbell R, Coe S (1991) Assessment of in vivo screening systems for potential biocontrol agents of *Gaeumannomyces graminis*. Plant Pathol 40:524–532

Riley MA, Wertz JE (2002) Bacteriocins: evolution ecology and application. Annu Rev Microbiol 56:117–137

Rodríguez HA, Nass H (1990) Diseases of rice in Venezuela Revista de la Facultad de Agronomía Universidad Central de Venezuela No 39:130–134

Romero D, Pérez-García A, Rivera ME, Cazorla FM, de Vicente A (2004) Isolation and evaluation of antagonistic bacteria to-wards the cucurbit powdery mildew fungus *Podosphaera fusca*. Appl Microbiol Biotechnol 64:263–269

Rudrappa T, Biedrzycki ML, Kunjeti SG, Donofri NM, Czymmek KJ, Pare PW, Bais HP (2010) The rhizobacterial elicitor action induces system resistance in *Arabidopsis thaliana*. Integr Biol 3:130–138

Ryu CM, Murphy JF, Mysore KS, Kloepper JW (2004) Plant growth-promoting rhizobacteria systemically protect *Arabidopsis thaliana* against cucumber mosaic virus by a salicylic acid and NPR1-independent and jasmonic acid-dependent signaling pathway. Plant J 39:381–392

Sacco MA, Moffett P (2009) Disease resistance genes: form and function. In: Bouarab K, Brisson N, Daayf F (eds) Molecular plant-microbe interactions. CABI Press, London, pp 94–141

Sakthivel N (2001) Sheath rot disease of rice: current status and control strategies. In: Sreenivasaprasad S, Johnson R (eds) Major fungal diseases of rice: recent advances. Springer, Dordrecht, pp 271–283

Sakthivel N, Amudha R, Muthukrishnan S (2002) Production of phytotoxic metabolites by *Sarocladium oryzae*. Mycol Res 106:609–614

Santoyo G, Mosqueda M, Orozco C, Govindappa M (2012) Mechanisms of biocontrol and plant growth-promoting activity in soil bacterial species of Bacillus and Pseudomonas: a review. Biocontrol Sci Tech 22(8):855–872

Schallmey M, Singh A, Ward OP (2004) Developments in the use of Bacillus sp. for industrial production. Can J Microbiol 50:1–17

Shamsi S, Naher N, Chowdhury P, Momtaz MSTS (2010) Fungal diseases of three aromatic rice (*Oryza Sativa* L.) J Bangladesh Acad Sci 34(2):163–170

Sharma OP, Bambawale OM (2008) Integrated management of key diseases of cotton and rice integrated Management of Plant. Pest Dis 4:271–302

Sharma RR, Singh D, Singh R (2009) Biological control of postharvest diseases on fruits and vegetables by microbial antagonists: a review. Biol Control 50:205–221

Shoda M (2000) Bacterial control of plant diseases. J Biosci Bioeng 89:515–521

Shrestha BK, Karki HS, Groth DE, Jungkhun N, Ham JH (2016) Biological control activities of Rice-associated Bacillus sp strains against sheath blight and bacterial panicle blight of rice. PLoS One 11(1):e0146764. https://doi.org/10.1371/journalpone0146764

Shubha K, Shetty SA, Shetty HS, Karanth NGK (1992) Interactions between seed-borne bacteria and fungi of paddy. Int J Trop Plant Dis 10(1):125–130

Singh R (2012) Integrated disease management in rice. In: Padmavathi C, Sreedevi B, Ladhalakshmi D, Arun Kumar, Sampath Kumar M (eds) ICAR Sponsored Winter School on new frontiers in integrated pest management in rice and rice based cropping systems. DRR, Hyderabad, pp 256–262

Slaton NA, Gbur EE, Cartwright RD, DeLong RE, Norman RJ, Brye KR (2004) Grain yield and kernel smut of rice as affected by preflood and midseason nitrogen fertilization in Arkansas. Agron J 96:91–99

Soleimani MJ, Shamsbakhsh M, Taghavi M, Kazemi SH (2005) Biological control of stem and root rot of wheat caused by Bipolaris spp by using antagonistic bacteria fluorescent Pseudomonas and Bacillus sp. J Biol Sci 5(3):347–353

Srinivas Prasad M, Madhav MS, Laha GS, Ladhalakshmi D, Krishnaveni D, Satendra Kumar M, Balachandran SM, Sundaram RM, Aruna Kanthi B, Madhan Mohan K, Ratna Madhavi K, Kumar V, Viraktamath BC (2011) Rice blast disease and its management DRR. Technical Bulletin No 57/2011, p 56

Stankovic S, Levic J, Petrovic T, Logrieco A, Moretti A (2007) Pathogenicity and mycotoxin production by *Fusarium proliferatum* isolated from onion and garlic in Serbia. Eur J Plant Pathol 118:165–172

Stein T (2005) Bacillus subtilis antibiotics: structures syntheses and specific functions. Mol Microbiol 56(4):845–857

Stelle JM, VlamiM de Souza JT (2002) Antibiotic production by bacterial biocontrol agents. Antonie Van Leeuwenhoek 81:379–407. https://doi.org/10.1016/j.biocontrol.2009.05.021

Suprapta DN, Khalimi K (2012) Pengembangan agen hayati untuk mengendalikan penyakit blas memacu pertumbuhan dan meningkatkan hasil tanaman padi Laporan Penelitian Riset Invensi Udayana Universitas Udayana Denpasar, p 42

Suryadi Y, Susilowati DN, Putri KE, Mubarik NR (2011) Antagonistic activity of indigenous Indonesian bacteria as the suppressing agent of rice fungal pathogen. J Internat'l Environ Appl Sci 6(4):558–568

Taguchi Y, Hyakumachi M, Horinouchi H, Kawane F (2003) Biological control of rice blast disease by Bacillus subtilis IK-1080. Jpn J Phytopathol 69:85–93

Talbot N, Foster A (2001) Genetics and genomics of the rice blast fungus Magnaporthe grisea: developing an experimental model for understanding fungal diseases of cereals. Adv Bot Res 34:263–287. https://doi.org/10.1016/S0065-2296(01)34011-9

Tamehiro N, Okamoto-Hosoya Y, Okamoto S, Ubukata M, Masa Hamada M, Naganawa H, Ochi K (2002) Bacilysocin a novel phospholipid antibiotic produced by *Bacillus subtilis* 168. Antimicrob Agents Chemother 46:315–320

Tang YX, Jin J, Hu DW, Yong ML, Xu Y, He LP (2013) Elucidation of the infection process of Ustilaginoidea virens (teleomorph: Villosiclava virens) in rice spikelets. Plant Pathol 62:1–8. https://doi.org/10.1111/j1365-305920120 2629x

Texas Agricultural Experiment Station (1996) Symptoms of rice diseases. Texas Agricultural Experiment Station, Beaumont

Thimon L, Peypoux F, Wallach J, Michel G (1995) Effect of the lipopeptide antibiotic iturin A on morphology and membrane ultra struture of yeast cell. FEMS Microbiol Lett 128:101–106

Thomashow LS (1996) Biological control of plant root pathogens. Curr Opin Biotechnol 7:343–347

Thomma BP, Nurnberger T, Joosten MH (2011) Of PAMPs and effectors: the blurred PTI-ETI dichotomy. Plant Cell 23:4–15

Thomma BP, Penninckx IA, Broekaert WF, Cammue BP (2001) The complexity of disease signaling in Arabidopsis. Curr Opin Immunol 13:63–68

Toure Y, Ongena M, Jacques P, Guiro A, Thonart P (2004) Role of lipopeptides produced by *Bacillus subtilis* GA1 in the reduction of grey mould disease caused by *Botrytis cinerea* on apple. J Appl Microbiol 96:1151–1160

Trigiano RL, Windham MT, Windham AS (2004) Plant pathology: concepts and laboratory exercises, vol 413. CRC Press, Boca Raton

Tsuda M, TWakiTaga M, Ueyama A (1982) Ascocarp production of Magnaporthe salvinii in culture. Trans Br Mycol Soc 78(3):515–519

Venkateswarlu N, Sireesha O, Aishwayra S, Vijaya T, Sriramulu A (2015) Isolation screening of rhizosphere fungi antagonistic to rice stem rot disease pathogen. Catt Asian J Pharm Clin Res 8(5):54–57

Walker J (1981) Taxonomy of take-all fungi and related genera and species. In: Asher MJC, Shipton PJ (eds) Biology and control of take-all. Academic Press, London, pp 15–74

Webster RK, Wick CM, Brandon DM, Hall DH, Bolstad J (1981) Epidemiology of stem rot disease of rice: effects of burning vs soil incorporation of rice residue. Hilgardia 49(3):1–12

Whipps JM (2001) Microbial interactions and biocontrol in the rhizosphere. J Exp Bot 52:487–511

Whipps JM, Lumsden RD (1991) Biological control of Pythium species. Biocontrol Sci Tech 1:75–90

Whitney NG, Frederiksen RA (1975) Kernel smut of rice. Texas Agric Exp Sta MP-123

Yan L, Xue-mei Z, De-qiang L, Fu H, Pei-song H, Yun-liang P (2014) Integrated approach to control false smut in hybrid rice in Sichuan Province China Rice. Science 21(6):354–360

Yilmaz M, Soran H, Beyatli Y (2005) Antimicrobial activities of some Bacillus spp strain isolated from the soil. Microbiology 161:127–131

Zhang H, Wang S (2013) Rice versus Xanthomonas oryzae pv oryzae: a unique pathosystem. Current Opin Plant Biol 16:188–195

Chapter 3
Role of Plant Growth Promoting Rhizobacteria in Reclamation of Wasteland

Ifra Zoomi, Raghvendra Pratap Narayan, Ovaid Akhtar, and Pragya Srivastava

3.1 Introduction

Since the arrival of civilization, the productive top layer of soil has been degrading due to innumerable environmental and ecological causes but the last few decades have witnessed a tremendous increase in this process resulting from an accelerated growth of human activities, leading to wasteland formation. Wasteland development is a problematic issue all over the globe and initiated by number of anthropogenic activities that resulted in loss of soil fertility and yield. The major cause of the wasteland development is deforestation, inappropriate agronomic practices (unbalanced use of fertilizers, treated or untreated wastewater irrigation), nutrient deficiency or excess and industrialization (Dregne 2002). Wasteland developers can directly lead to loss of soil organic carbon, nutrient contents, water holding capacity of soil and below-ground biodiversity and indirectly leads to habitat loss (Gisladottir and Stocking 2005). Thus, the reclamation of degraded land is regarded as an effective measure to accommodate the increasing demand of space for living and development (Wang et al. 2014).

Reclamation of wasteland by physical and chemical method often resulted in many serious environmental problems, including loss of soil biodiversity, deterioration of soil structure, nutrient imbalance and costly. Therefore, an effort are required to figure out substitute, groundbreaking, and ecofriendly selections to overcome from costly and non-ecofriendly methods. In this context, microbial community plays an important role in ecosystem functioning such as decomposition and nutrient cycling (Grayston and Prescott 2005; Belimov and Wenzel 2009) and is therefore

I. Zoomi • O. Akhtar • P. Srivastava
Sadasivan Mycopathology Laboratory, Department of Botany, University of Allahabad, Allahabad, UP, India

R.P. Narayan (✉)
Netaji Subhash Government Girls P.G. College, Lucknow, UP, India
e-mail: narayan.raghvendra@gmail.com

J.K. Patra et al. (eds.), *Microbial Biotechnology*,
https://doi.org/10.1007/978-981-10-6847-8_3

an integral component of determining reclamation success (Macdonald et al. 2012). Among microbial community plant growth-promoting rhizobacteria (PGPR) naturally occurring in soil may represent a viable alternative to reduce costs and environmental impacts in reclamation programs (Dary et al. 2010). PGPR stimulate the plant development by an assortment of mechanisms that includes, synthesis of phytohormones (Patten and Glick 2002), synthesis of ACC deaminase enzyme (Glick 2014), besides this increasing the accessibility of macro-nutrients like nitrogen (Compant et al. 2005), phosphorus, iron and other micro-elements (Costa and Loper 1994; Rodriguez and Fraga 1999; Trends in biology). Therefore, the current book chapter manifests the PGPR-induced marked acceleration, particularly in the success of the restoration of disturbed and degraded lands and their presence has been found critical for the regeneration of natural ecosystems.

3.2 Types of Wastelands

According to National Wasteland Development Board (NWDB) wastelands can be categorized into (1) Culturable and (2) Unculturable.

3.2.1 Culturable Wastelands

Culturable wastelands are those lands which have the potential to restore or reclaims in later stage. These lands remain unproductive for various reasons such as water logging, salinity, drought, heavy metal contamination and water stress etc.

3.2.2 Unculturable Wastelands

Unculturable wastelands are unfertile land and cannot be put to important uses. Nevertheless, certain areas of unculturable wasteland can be rehabilitated into grassland which can improve the ecological balance by protecting soil erosion and maintain the moisture of the soil. These lands are deserted areas near snowline with high degree to slope, etc.

3.3 Major Factors of Wasteland Development

3.3.1 *Heavy Metals*

Metals having densities more than 5 g/cm^3 are generally called as heavy metals and metalloids (Oves et al. 2012). Heavy metal contamination is a major problem all over the world and activities such as, industrialization, urbanization and intensive agricultural practices plays a major role in wasteland development.

3.3.2 *Drought*

Drought is a constant period of dry meteorological conditions and distresses almost all climatic regions of the biosphere (Wilhite 2000). The main cause of drought is worldwide climate alteration, i.e. Escalating temperature that changed the soil moisture.

3.3.3 *Overgrazing of Grasslands*

Overgrazing decreases heather cover (vegetation) of top soil. According to a report by the Forestry Commission, approximately 40% of threatened grassland areas suffer from livestock grazing and fodder removal.

3.3.4 *Salinization*

The process of increasing the salt content in the soil is called salinization. It is a serious environmental problem and reported in many parts of the world, mostly in arid and semi-arid regions (Giri et al. 2003). The major cause of salinization are high temperature, low rainfall and improper irrigation practices.

3.3.5 *Flooding*

Flooding is a common abiotic factor that negatively distress the development of plants. Lack of oxygen in roots is the key outcome of flooding.

3.3.6 Deforestation

Deforestation is the removal of a tree or stands that results in permanent conversion of forests in wasteland. The major causes of deforestation are shifting cultivation, converting forest into agricultural land and urban use.

3.4 Approaches for Reclamation

Reclamation of wasteland is considered as a thought-provoking job with respect to technical complexity and cost (Mahar et al. 2016). A number of approaches to reclaim the degraded soil are being used over the years. These approaches are mainly categorized as biological, chemical and physical (Chen et al. 2000; Lim et al. 2014). Physical approaches cause loss in microbial diversity, negatively affect the soil physical properties and costly process. Similarly, chemical approach also requires intensive labor and generate secondary pollution and costly (Jegatheesan et al. 2016; Mahar et al. 2016). Thus, physical and chemical approaches for reclamation of wasteland is costly for both trade and industry and an ecological point of view. On the other hand, biological approach is useful as it is ecofriendly, natural process, inexpensive and publically accepted (Mani and Kumar 2014; Ullah et al. 2015a, b; Kang et al. 2016). Biological approaches are also preferred over physical and chemical approach because it utilizes solar energy and conserve natural soil properties (Kang et al. 2016).

3.5 Plant Growth Promoting Rhizobacteria (PGPR)

Microorganisms are cosmopolitan in their distribution. Most of the soil encompasses a massive assortment of microorganisms, including actinomycetes, bacteria, fungi, protozoa and algae. According to, Alexander (1991) a typical gram of soil contains ~9×10^7 bacteria, 4×10^6 actinomycetes, 2×10^5 fungi, 3×10^4 algae, 5×10^3 protozoa and 3×10^1 nematodes. However, numbers of these microorganisms may greatly vary from one soil to another (Glick 2014). Kaymak (2010) reported that among rhizospheric microorganisms, bacteria are the most abundant. Numerous bacterial genera such as *Pseudomonas*, *Bacillus*, *Enterobacter*, *Azospirillum*, *Klebsiella*, *Burkholderia*, *Variovorax*, *Serratia* and *Azotobacter* cause a noticeable influence on plant development (Nadeem et al. 2014). These beneficial bacteria are termed as plant growth-promoting rhizobacteria (PGPR) by Kloepper and Schroth (1981). PGPR classify into two categories (1) Bacterial strains that can accomplish at least two of the three conditions, for instance, hostile growth rate, stimulating plant growth and biological control (Vessey 2003) and (2) on the basis of the degree of association of PGPR with the plant root cells. On the basis of the degree of association, PGPR further classified into extracellular plant growth promoting rhizobacteria (ePGPR) and intracellular plant growth promoting rhizobacteria (iPGPR) (Martinez-Viveros et al. 2010). The ePGPRs possibly occurs in the rhizospheric region of the plant, including the external surface of a plant root (rhizoplane) and root cortex. In contrast to this, iPGPRs generally found inside the root nodules. The iPGPR includes ePGPRs, *Micrococcous*, *Agrobacterium*, *Arthrobacter*, *Azotobacter*, *Azospirillum*, *Caulobacter*, *Burkholderia*, *Chromobacterium*, *Flavobacterium*, *Erwinia*, *Pseudomonas*, *Bacillus*, and *Serratia* (Gray and Smith 2005; Bhattacharya and Jha 2012). Whereas, iPGPR comprises of endophytes and *Frankia* species, both of which can forms symbiotic association with the higher plants and contain nitrogenase enzyme to fix atmospheric Nitrogen (Verma et al. 2010). Some endophytes belongs to family rhizobiaceae generally gram positive and pleuromorphic forms, for instance, *Allorhizobium*, *Azorhizobium*, *Bradyrhizobium*, *Mesorhizobium* and *Rhizobium* (Bhattacharya and Jha 2012).

3.5.1 PGPR in Reclamation of Wasteland

Soil degradation as a result of salinity, heavy metals, drought and flooding are major disablement to optimal utilization of land resources. Numbers of plant species have been recommended for reclamation or rehabilitation of different degraded soils (Padilla and Pugnaire 2006). However, there are various limiting factors like excess of salts, deficiency of nutritional elements, low moisture, initiation and elongation of roots is inhibited that affect the plant growth and establishment from seedling to maturity. For reclaiming wastelands into usable lands, there is a need to put efforts to restrict the mortality rate and improves the plant growth. Thus, the interaction among soil microorganisms and rhizosphere is considered as an important for the establishment and survival of plants under harsh conditions. The role of soil microbes in accelerating the reclamation process in the wastelands has been recognized and emphasis has been given on the use of microbial inoculants in revegetation programs (Dash and Gupta 2011; Ashraf et al. 2017). PGPR is gaining importance in the reclamation of degraded soils. They have markedly increased the success of restoration of disturbed and degraded lands and their presence has been found critical for the regeneration of natural ecosystems (Fig. 3.1). PGPR has been reported from degraded soils of India, such as *Rhizobium, Bradyrhizobium, Azotobacter, Azospirillum, Pseudomonas* and *Bacillus* has been (Tilak et al. 2005; Upadhyay et al. 2009). The adaptability of these PGPR is due to aggressive growth rate, synthesis of osmoregulator, siderophore production (Nadeem et al. 2014). For instance, Belimov et al. (2001) isolated PGPR strain from the rhizoplane of pea (*Pisum sativum* L.) and Indian mustard (*Brassica juncea* L.) grown in different soils and heavy metals contaminated sewage sludge. The isolated genera and species were *Pseudomonas brassicacearum*, *Pseudomonas putida*, *Pseudomonas oryzihabitans*, *Pseudomonas marginalis*, *Pseudomonas* sp., *Alcaligenes xylosoxidans*, *Alcaligenes* sp., *Variovorax paradoxus*, *Bacillus pumilus*, and *Rhodococcus* sp.

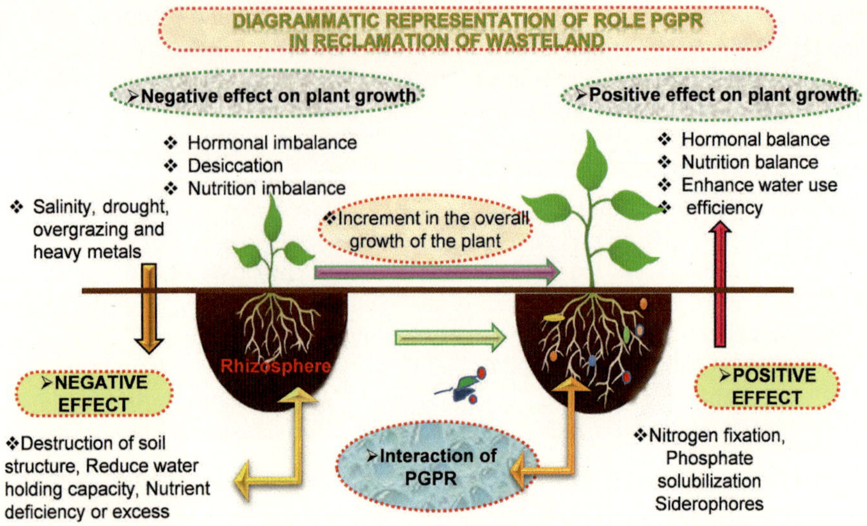

Fig. 3.1 Diagrammatic representation of role of PGPR in reclamation of wasteland

Egamberdiyeva, (2007) reported that *Pseudomonas alcaligenes* PsA15, *Bacillus polymyxa* BcP26 and *Mycobacterium phlei* MbP18 were capable to withstand in high temperatures and salinity. However, plant growth promoting rhizobacterial strains can also improve the growth and development of the plant against salinity, drought, flooding, and heavy metal toxicity and, consequently, assist the plants to endure under stressful environments (Belimov et al. 2001; Mayak et al. 2004a, b; Nadeem et al. 2007; Zahir et al. 2008; Sandhya et al. 2009; Glick 2010; Ma et al. 2011; Nadeem et al. 2014). For instance, twenty PGPR strains were isolated from the rhizospheric region of maize (*Zea mays* L.) plant of salt-affected fields. Among them S5 (*Pseudomonas syringae*), S14 (*Enterobacter aerogenes*), and S20 (*Pseudomonas fluorescens*) were the most operative strains for stimulating the development and productivity of maize plants (Nadeem et al. 2007).

PGPR possibly stimulate the plant growth and establishment by two (directly and indirectly) major ways (Glick et al. 1995). According to Kloepper et al. (1987), direct way includes synthesis of beneficial compound and its uptake by host plant or assisting in resource acquisition (N, P and other macronutrients) from the soil environment. PGPR known to assist the plant growth by nitrogen fixation and synthesizing siderophores and phytohormones in order to escalate its bioavailability (Glick 1995; Kloepper et al. 1989; Patten and Glick 2002; Nadeem et al. 2014). In addition to this, PGPR also benefit the host plants by synthesizing ACC-deaminase enzymes, and chitin degrading enzyme (chitinase) and also by the excreting exopolysaccharides, rhizobitoxine, etc. that helps in soil aggregation (Glick et al. 2007; Sandhya et al. 2009). Indirect way of stimulation of plant growth occurs when PGPR inhibits or reduce the damaging effect of plant pathogens (Glick and Bashan 1997; Bhattacharyya and Jha 2012).

3.5.2 Mechanisms of PGPR for Establishment of Plants under Stress Conditions

PGPR are naturally occurring soil microbes that inhabit roots and facilitate plant development either indirect or direct mechanism (Glick 1995) under degraded soil conditions. Some of these mechanisms include enhanced nutrient solubility and bioavailability, bioremediation of metal contaminated soils and synthesis of phytohormones and ACC deaminase enzyme has been discussed below.

3.5.2.1 Synthesis of Phytohormones and ACC Deaminase Enzyme

Phytohormones play a key role in stimulating plant growth and establishment under different environmental conditions. Phytohormones are small organic molecules such as ethylene, auxins, cytokinins and gibberlic acid. Among phytohormones ethylene plays a beneficial role in plant growth, such as root initiation and elongation, nodule formation, ageing, abscission and ripening, beside this, ethylene also involves in stress signaling (Wang et al. 2002; Ludwig et al. 2005). In response to stress environmental condition, plants produce high level of ethylene which causes root growth inhibition, inhibits nodule formation, indole acetic acid (IAA) transport, promotes senescence and abscission (Jackson 1985; El-Iklil and Benichou 2000). In response to ethylene stress, PGPR produce 1 aminocyclopropane-1-carboxylate (ACC) deaminase enzyme (EC: 4.1.9.9.4) that catalyze the cleavage of 1 aminocyclopropane-1-carboxylate (ACC) (precursor of ethylene) into ammonia (NH_3) and alpha- ketoglutaric acid (Arshad et al. 2007). Number of workers reported that PGPR containing the ACC deaminase enzyme activity (Belimov et al. 2001; Ma et al. 2003; Mayak et al. 2004a; Madhaiyan et al. 2006; Shaharoona et al. 2006; Saravanakumar and Samiyappan, 2007; Nadeem et al. 2007; Rodriguez-Diaz et al. 2008) and reduces several types of stress posed by degraded land (Arshad et al. 2007, 2008; El-Howeity and Asfour 2012; Arora et al. 2012). PGPR containing ACC deaminase enzyme stimulates the plant growth against flooding, the presence of organic toxicants, the presence of a heavy metals, salinity; phytopathogens and drought (Grichko and Glick 2001; Mayak et al. 2004a, b; Reed and Glick 2005; Saravanakumar and Samiyappan 2007; Farwell et al. 2007; Zhuang et al. 2007; Gurska et al. 2009).

In addition to this, PGPR are also synthesize Indole acetic acid (IAA) and protect the plants in adverse environmental conditions (Frankenberger and Arshad 1995; Remans et al. 2008; Ahmed and Hasnain 2014). Egamberdieva (2009) reported that IAA increased root/shoot biomass of wheat seedling exposed to elevated levels of salt. Similarly, *Sinorhizobium meliloti* DR-64 increased the tolerance by producing IAA in nodulated *Medicago truncatula* plant against salt tress (Bianco and Defez 2009). In addition to this, PGPR also synthesize cytokinin which can promote the growth and development of plants under both stressed and unstressed environmental conditions (Garcia et al. 2001; Ortiz Castro et al. 2009). PGPR improves the plant growth under different environmental conditions by stimulating or synthesis of phytohormones and ACC deaminase enzyme depicted in Table 3.1.

Table 3.1 Some examples of PGPR in synthesis of ACC deaminase and phytohormones

Stress	PGPRS	Mechanisms	Host plant	References
Salinity	*Achromobacter piechaudii* ARV8	Increased ACC deaminase activity stimulated the plant growth	Tomato (*Lycopersicon esculentum* L.)	Mayak et al. (2004a)
Salinity	*Pseudomonas fluorescens* *P. aeruginosa* *P. stutzeri*	Increased ACC deaminase activity stimulated the plant growth	Tomato (*Lycopersicon esculentum* L.)	Tank and Saraf (2010)
Salinity	*Pseudomonas putida* N21, *Pseudomonas aeruginosa* N39, *Serratia proteamaculans* M35	Increased ACC deaminase activity stimulated the plant growth	Wheat (*Triticum aestivum* L.)	Zahir et al. (2009)
Salinity	*Pseudomonas putida*	Increased ACC deaminase activity stimulated the plant growth	Canola (*Brassica napus* L.)	Cheng et al. (2007)
Salinity	*Pseudomonas fluorescens* TDK1	Increased ACC deaminase activity	Groundnut (*Arachis hypogaea* L.)	Saravanakumar and Samiyappan (2007)
Salinity	*Pseudomonas syringae* S5, *Pseudomonas fluorescens* S20 *Enterobacter aerogenes* S14	Increased ACC deaminase activity	Maize (*Zea mays* L.)	Nadeem et al. (2007)
Salinity	*Achromobacter piechaudii* ARV8	Increased ACC deaminase activity	Tomato (*Lycopersicon esculentum* L.)	Mayak et al. (2004b)
Salinity	PGPR (Mk1, *Pseudomonas syringae*; Mk20, *Pseudomonas fluorescens*; and Mk25, *Pseudomonas fluorescens* biotype G) and *Rhizobium phaseoli* strains M1, M6, and M9	Increased ACC deaminase activity	Mung bean (*Phaseolus radiate* L.)	Ahmad et al. (2011)

(continued)

Table 3.1 (continued)

Stress	PGPRS	Mechanisms	Host plant	References
Salinity	*Raoultella planticola* Rs-2	Increased ACC deaminase activity	Cotton (*Gossypium hirsutum* L.)	Wu et al. (2012)
Salinity	*Pseudomonas syringae*, Mk1; *Pseudomonas fluorescens*, Mk20 and *Pseudomonas fluorescens* Biotype G, (Mk25)	ACC deaminase activity and increased water use efficiency	Mung bean (*Vigna radiata* L.)	Ahmad et al. (2012)
Salinity field	*Rhizobium* and *Pseudomonas*	Phytohormones (IAA) production and ACC deaminase activity	Mung bean (*Vigna radiata* L.)	Ahmad et al. (2013)
Salinity field	*Acinetobacter* spp. and *Pseudomonas* sp.	Production of ACC deaminase and IAA	Barley (*Hordeum vulgare* L.) and oats (*Avena sativa* L.)	Chang et al. (2014)
Salinity	*Streptomyces* sp. strain PGPA39	ACC deaminase activity and IAA production and phosphate solubilization	Tomato (*lycopersicon esculentum* L.)	Palaniyandi et al. (2014)
Drought	*Paenibacillus polymyma* and *Rhizobium tropici*	Altered hormonal balance and stomatal conductance	Bean (*Phaseolus vulgaris* L.)	Figueiredo et al. (2008)
Flooding	*Pseudomonas* sp. and *Enterobacter* sp.	ACC deaminase activity	Tomato (*Solanum lycopersicumm* L.)	Grichko and Glick (2001)
Cd	*Variovorax paradoxus*, *Rhodococcus* sp. and *Flavobacterium* sp.	ACC deaminase activity stimulated the plant growth	Indian mustard (*Brassica juncea* L.)	Belimov et al. (2015)

3.5.2.2 Nutrients Uptake

PGPR possess the potential to enhance the bioavailability of essential micro and macro-nutrients to plants under nutrient deficient environment. It is well known that atmosphere contains 78% of nitrogen, which is not available to plants. However, nitrogen fixing rhizobacteria (diazotrophs) can fix atmospheric nitrogen in degraded soils and offers advantage to the host plants (Dobbelaere et al. 2003; Nonnoi et al.

2012). Phosphorous is also considered major essential macronutrient after nitrogen and its deficiency can inhibit the plant growth (Ullah et al. 2015a). Soils contain abundant phosphorous which is not available to plants because it is present in insoluble and mineralized form. It is only taken up in monobasic ($H2PO^{4-)}$ or dibasic ($HPO4^{2-}$) soluble forms (Glass 1989). There are some PGPRs which are reported to solubilize the insoluble phosphates by the release of organic acids, acidification of soil, exchange reactions and chelation (Chung et al. 2005) and mineralization of organic phosphates by secreting phosphatases an extracellular enzyme (Gyaneshwar et al. 2002). Certain important phosphorous solubilizing PGPRS are *Pseudomonas*, *Bacillus* and *Rhizobium* (Chen et al. 2006; Rodríguez et al. 2006).

Usually, Iron is an essential nutrient and important cofactor for many enzyme catalyzed reactions. Soil contains iron in the ferric state ($Fe^{3+)}$ react to form insoluble form (oxyhydroxides and hydroxides) and unavailable to both microorganisms and plant roots. Siderophores act as solubilizing agents for iron from minerals or organic compounds under conditions of iron limitation (Indiragandhi et al. 2008; Ahemad and Kibret 2014). Most of the siderophores are water-soluble and can be divided into extracellular siderophores and intracellular siderophores (Ahemad and Kibret 2014). However, certain strains of PGPR produce siderophore which chelated the insoluble form of iron and successively make it available to microorganisms and plant roots (Ma et al. 2011; Rajkumar et al. 2010). PGPR inoculations have been reported to improve the bioavailabilty of Fe in plants with simultaneously stimulated of plant growth (Burd et al. 2000; Safronova et al. 2006; Sharma et al. 2003). In addition to this, siderophores produce by PGPR alleviate the stresses posed by metal degraded soil (Braud et al. 2009). Siderophores also known to forms firm complexes with heavy metals (Neubauer et al. 2000), thereby reducing its toxicity. Some of the examples of PGPR in nutrient uptake are listed in Table 3.2.

3.5.2.3 Remediation of Metal Contaminated Soils

PGPR may use more than one of these mechanisms to boost the plant growth in heavy metals contaminated soils; (1) Production of phytohormones (2) Siderophore production (3) Synthesis of ACC deaminase enzymes and (4) Nitrogen fixation and solubilization of phosphate. In addition to this, functional group such as sulfhydryl, carboxyl, hydroxyl, sulfonate, amine and amide groups (negatively charge) present on the cell surface that can bind efficiently with heavy metals (Pulsawat et al. 2003). Number of researcher reported that PGPR facilitate the plant based reclamation i.e., phytoremediation of metal contaminated soils (Table 3.3). Some strain of PGPR are known to produce biosurfactants (Sheng et al. 2008a; Nie et al. 2010) and forms complexes with heavy metals stimulating desorption and solubility of metals from the soils and thereby enhance metal uptake (Rajkumar et al. 2012). For instance, biosurfactants produced by *Pseudomonas aeruginosa* BS2 increased the mobility and solubility of Cd and Pb (Ullah et al. 2015a). Gadd (2010) reported that under heavy metal stress certain plants and fungi carry out enzymatic synthesis of phytochelatins. Phytochelatins are important cysteine rich metal binding peptides which

Table 3.2 Some examples of PGPR in synthesizing siderophores and nutrients uptake

Stress	PGPRS	Mechanisms	Host plant	reference
Salinity	*Klebsiella oxytoca*	Plant height and dry weight, also enhanced the absorption of N, P, K and Ca, and decreased the absorption of Na	Cotton (*Gossypium hirsutum* L.)	Yue et al. (2007)
Salinity	*Azospirillum* Sp.	Nitrate reductase and nitrogenase activity	Maize (*Zea mays* L.)	Hamdia et al. (2004)
Salinity	*Pseudomonas putida* Rs-198	Increase the absorption of the Mg^{2+}, K^{+} and Ca^{2+} and decrease the uptake of the Na^{2+} from the soil	Cotton (*Gossypium hirsutum* L.)	Yao et al. (2010)
Salinity	*Streptomyces* sp. strain PGPA39	Phosphate solubilization	Tomato (*lycopersicon esculentum* L.)	Palaniyandi et al. (2014)
Drought	*Pseudomonas mendonica*	Phosphate solubilization	Lettuce (*Lactuca sativa* L.)	Kohler et al. (2008)
Heavy metals	*Streptomyces tendae* F4	Promoted plant growth, facilitated soil metal solubilization; enhanced Cd and Fe uptake	Sunflower (*Helianthus annuus* L.)	Dimkpa et al. (2009)
Ni	*Streptomyces acidiscabies* E13	Promoted plant growth by binding of Fe and Ni	Cowpea (*Vigna unguiculata* L.)	Dimkpa et al. (2008)

is synthesized by glutathione as its immediate precursor. Genetically engineered the rhizobacteria strains which have the ability produce phytochelatin in response different heavy metals (Kang et al. 2007). Ullah et al. (2015a) reported the presence of another heavy metal binding polyppeptide called metallothioneins which is rich in cystein. Low molecular weight metallothioneins has high affinity for Cd, Cu, Ag and Hg compared to other metals. Gene encoding metallothionein has been found in a diverse group of organisms (Sriprang et al. 2002) including PGPR. For instance, gene encoding metallothionein produced by *Mesorhizobium huakuii* subsp. *rengi* that upon inoculation to *Astragalus sinicus* L. possibly helping to make Cd more bioavailable (Ullah et al. 2015a).

3.6 Application of PGPR and Its Constraint

PGPR is very operative for stimulating plant growth and development under salinity, drought and heavy metals degraded soils (Mayak et al. 2004b; Nadeem et al., 2014). For instance, *Pseudomonas fluorescens* MSP-393 could serve as the perfect

Table 3.3 Some examples of PGPR assisted phytoremediation of heavy metals

Heavy metals	PGPR	Mechanisms	Host plant	reference
Ni, Pb, Zn and Cr	*Kluyvera ascorbata* SUD165	Provide tolerance and ACC deaminase activity	Canola (*Brassica napus* L.)	Burd et al. (1998)
Cd, Pb and As	*Ochrobactrum* sp. (CdSP9), *Bacillus* sp. (PbSP6) and *Bacillus* sp.(AsSP9)	Phytoremediation	Rice (*Oryza sativa* L.)	Pandey et al. (2013)
Cd	*Pseudomonas putida*	Phytoremediation	Arugula (*Eruca sativa* L).	Kamran et al. (2015)
Cu	*Pseudomonas asplenii*	Promote plant growth	Canola (*Brassica napus* L.) and Reed (*Phragmites australis*)	Reed and Glick (2005), Reed et al. (2005)
Cu	*Pseudomonas fluorescens* Avm and *Rhizobium leguminosarum* bv *phaseoli* CPMex46	Improved Cu and Fe translocation from root to shoot	Alfalfa (*Medicago sativa* L.)	Carrillo-Castaneda et al. (2003)
Cd	*Pseudomonas aeruginosa*	Stimulated plant growth; reduced Cd uptake	Indian mustard (*Brassica Juncea* L.) and pumpkin (*Cucurbita* maxima L.)	Sinha and Mukherjee (2008)
Ni	*Pseudomonas putida*	Phytoremediation	Canola (*Brassica napus* L.)	Farwell et al. (2006)

bioinoculant for crops in salt affected soils (Paul and Nair 2008). According to Sandhya et al. (2009) PGPR inoculated seedlings improved the soil aggregation and root adhering soil and improve the water use efficiency in plants. According to Huang et al. (2004a, b), application of PGPR increased the removal of organic pollutant probably by enhancing plants propagation and endurance in soils that were severely polluted. Likewise, Braud et al. (2009) reported that inoculation of *Pseudomonas aeruginosa* significantly increased the bioavailability of Cr and Pb compared with uninoculated one (controls). Moreover, they also observed that *P. aeruginosa* significantly improve the phytoextraction potential of maize (*Zea mays* L.) plants. Inoculation of *Klebsiella* sp. promoted tall fescue growth and enhanced reclamation efficiency in petroleum-contaminated saline-alkaline soil (Liu et al. 2014). Interestingly, Naveed et al. (2014) inoculated PGPR strain (*Burkholderia phytofirmans* PsJN) in wheat (*Triticum aestivum* L.) under field conditions. They found that, *B. phytofirmans* PsJN improved the photosynthetic efficiency, water use efficiency, ionic balance and antioxidant levels in plants. *B. phytofirmans* inoculation also increased the N, P, K and protein concentration in the grains of wheat in *Triticum aestivum* L. in comparision to uninoculated control. Thus, it has been established that the PGPR inoculation is an effective measure for growth and development of plant in reclamation programme. However, the usefulness of PGPR also depends on soil mineral content, climatic variations, host plant and indigenous rhizosphere population. It has also been observed that under definite environmental conditions the results found in a field were different from the laboratory conditions (Zhender et al. 1999; Smyth et al. 2011). The variation in results might be due to the reason that inoculum has less ability to compete with an indigenous population. Strigul and Kravchenko (2006) observed that inoculum potential was greatly affected by the antagonism with native inhabitants for micro and macronutrient and niches. Certain strains of PGPR were variable concerning their potential (Dewey et al. 1999; Nadeem et al. 2009). Therefore, application of particular strains under a specific environmental conditions possibly operative for gaining maximum profits to the host plants (Nadeem et al. 2014).

3.7 Conclusion

Land degradation is a serious environmental issue, as it limits the plant growth and development that ultimately reduces productivity. In order to reclaim the degraded soil physical and chemical approaches are being employed. Of these approaches, biological approach is an ecofriendly and cost effective. In this context, Plant growth promoting rhizobacteria, exhibit various characteristics, including nitrogen fixation and phosphate solubilization, detoxifying heavy metals, synthesis of biosurfactants, phytochelatin and siderophores and organic acids. Apart from this, PGPR also produce ACC deaminase enzymes, phytohormones like IAA and cytokinin, that appeared to be responsible in improving the plant growth which indirectly increases the efficiency of reclamation. In addition, application of PGPR is developing area of awareness and has revealed a considerable improvement in plant growth in situ, which needs to be further consolidated through field trials under ecologically distinct conditions.

Acknowledgements The authors gratefully acknowledge the Head, Department of Botany, University of Allahabad, Allahabad, for providing the necessary facilities and also grateful to acknowledge the University Grant Commission and CSIR for providing financial assistance to Ifra Zoomi and Pragya Srivastava and Ovaid Akhtar.

References

Ahmad M, Zahir ZA, Asghar HN, Arshad M (2012) The combined application of rhizobial strains and plant growth promoting rhizobacteria improves growth and productivity of mung bean (*Vigna radiata* L.) under salt-stressed conditions. Ann Microbial 62(3):1321–1330

Ahmad M, Zahir ZA, Asghar HN, Asghar M (2011) Inducing salt tolerance in mung bean through coinoculation with rhizobia and plant-growth-promoting rhizobacteria containing 1-aminocyclopropane-1-carboxylate deaminase. Can J Microbiol 57(7):578–589

Ahmad M, Zahir ZA, Khalid M, Nazli F, Arshad M (2013) Efficacy of *Rhizobium* and *Pseudomonas* strains to improve physiology, ionic balance and quality of mung bean under salt-affected conditions on farmer's fields. Plant Physiol Biochem 63:170–176

Ahmed A, Hasnain S (2014) Auxins as one of the factors of plant growth improvement by plant growth promoting rhizobacteria. Pol J Microbiol 63(3):261–266

Ahemad M, Kibret M (2014) Mechanisms and applications of plant growth promoting rhizobacteria: current perspective. J King Saud Univ Sci 26(1):1–20

Alexander M (1991) Introduction to soil microbiology. Krieger, Malabar

Arora NK, Tewari S, Singh S, Lal N, Maheshwari DK (2012) PGPR for protection of plant health under saline conditions. In: Maheshwari DK (ed) Bacteria in agrobiology: stress management. Springer, Berlin, pp 239–258

Arshad M, Saleem M, Hussain S (2007) Perspectives of bacterial ACC deaminase in phytoremediation. Trends Biotechnol 25(8):356–362

Arshad M, Shaharoona B, Mahmood T (2008) Inoculation with *Pseudomonas* spp. containing ACC-deaminase partially eliminates the effects of drought stress on growth, yield, and ripening of pea (*Pisum sativum* L.) Pedosphere 18(5):611–620

Ashraf MA, Hussain I, Rasheed R, Iqbal M, Riaz M, Arif MS (2017) Advances in microbe-assisted reclamation of heavy metal contaminated soils over the last decade: a review. J Environ Manag 198:132–143

Belimov AA, Dodd IC, Safronova VI, Shaposhnikov AI, Azarova TS, Makarova NM, Davies WJ, Tikhonovich IA (2015) Rhizobacteria that produce auxins and contain 1-amino-cyclopropane-1-carboxylic acid deaminase decrease amino acid concentrations in the rhizosphere and improve growth and yield of well-watered and water-limited potato (*Solanum tuberosum*). Ann Appl Biol 167(1):11–25

Belimov AA, Safronova VI, Sergeyeva TA, Egorova TN, Matveyeva VA, Tsyganov VE, Borisov AY, Tikhonovich IA, Kluge C, Preisfeld A, Dietz KJ (2001) Characterization of plant growth promoting rhizobacteria isolated from polluted soils and containing 1-aminocyclopropane-1-carboxylate deaminase. Can J Microbiol 47(7):642–652

Belimov AA, Wenzel WW (2009) The role of rhizosphere microorganisms in heavy metal tolerance of higher plants. Asp Appl Biol 98:81–90

Bhattacharyya PN, Jha DK (2012) Plant growth-promoting rhizobacteria (PGPR): emergence in agriculture. World J Microbiol Biotechnol 28(4):1327–1350

Bianco C, Defez R (2009) *Medicago truncatula* improves salt tolerance when nodulated by an indole-3-acetic acid-overproducing *Sinorhizobium meliloti* strain. J Exp Bot 60(11):3097–3107

Braud A, Jézéquel K, Bazot S, Lebeau T (2009) Enhanced phytoextraction of an agricultural Cr-and Pb-contaminated soil by bioaugmentation with siderophore-producing bacteria. Chemosphere 74(2):280–286

Burd GI, Dixon DG, Glick BR (1998) A plant growth-promoting bacterium that decreases nickel toxicity in seedlings. Appl Environ Microbiol 64(10):3663–3668

Burd GI, Dixon DG, Glick BR (2000) Plant growth-promoting bacteria that decrease heavy metal toxicity in plants. Can J Microbiol 46(3):237–245

Carrillo-Castaneda G, Munoz JJ, Peralta-Videa JR, Gomez E, Gardea-Torresdey JL (2003) Plant growth-promoting bacteria promote copper and iron translocation from root to shoot in alfalfa seedlings. J Plant Nutr 26:1801–1814

Chang P, Gerhardt KE, Huang XD, XM Y, Glick BR, Gerwing PD, Greenberg BM (2014) Plant growth-promoting bacteria facilitate the growth of barley and oats in salt-impacted soil: implications for phytoremediation of saline soils. Int J Phytoremediation 16(11):1133–1147

Chen HM, Zheng CR, Tu C, Shen ZG (2000) Chemical methods and phytoremediation of soil contaminated with heavy metals. Chemosphere 41(1):229–234

Chen YP, Rekha PD, Arun AB, Shen FT, Lai WA, Young CC (2006) Phosphate solubilizing bacteria from subtropical soil and their tricalcium phosphate solubilizing abilities. Appl Soil Ecol 34(1):33–41

Cheng Z, Park E, Glick BR (2007) 1-Aminocyclopropane-1-carboxylate deaminase from *Pseudomonas putida* UW4 facilitates the growth of canola in the presence of salt. Can J Microbiol 53:912–918

Chung H, Park M, Madhaiyan M, Seshadri S, Song J, Cho H, Sa T (2005) Isolation and characterization of phosphate solubilizing bacteria from the rhizosphere of crop plants of Korea. Soil Biol Biochem 37(10):1970–1974

Compant S, Duffy B, Nowak J, Clément C, Barka EA (2005) Use of plant growth-promoting bacteria for biocontrol of plant diseases: principles, mechanisms of action, and future prospects. Appl Environ Microbiol 71(9):4951–4959

Costa JM, Loper JE (1994) Characterization of siderophore production by the biological control agent Enterobacter cloacae. MPMI-Mol Plant Microbe Interact 7(4):440–448

Dary M, Chamber-Pérez MA, Palomares AJ, Pajuelo E (2010) "In situ" phytostabilisation of heavy metal polluted soils using *Lupinus luteus* inoculated with metal resistant plant-growth promoting rhizobacteria. J Hazard Mater 177(1):323–330

Dash S, Gupta N (2011) Microbial bioinoculants and their role in plant growth and development. Int J Biotechnol Mol Biol Res 13:232–251

Dewey FM, Wong YL, Seery R, Hollins TW, Gurr SJ (1999) Bacteria associated with *Stagonospora* (Septoria) *nodorum* increase pathogenicity of the fungus. New Phytol 144(3):489–497

Dimkpa CO, Merten D, Svatos A, Büchel G, Kothe E (2009) Siderophores mediate reduced and increased uptake of cadmium by *Streptomyces tendae* F4 and sunflower (*Helianthus annuus*), respectively. J Appl Microbiol 107:1687–1696

Dimkpa CO, Svatos A, Dabrowska P, Schmidt A, Boland W, Kothe E (2008) Involvement of siderophores in the reduction of metal-induced inhibition of auxin synthesis in *Streptomyces* spp. Chemosphere 74:19–25

Dobbelaere S, Vanderleyden J, Okon Y (2003) Plant growth-promoting effects of diazotrophs in the rhizosphere. Crit Rev Plant Sci 22(2):107–149

Dregne HE (2002) Land degradation in the drylands. Arid Land Res Manag 16:99–132

Egamberdieva D (2009) Alleviation of salt stress by plant growth regulators and IAA producing bacteria in wheat. Acta Physiol Plant 31(4):861–864

Egamberdiyeva D (2007) The effect of plant growth promoting bacteria on growth and nutrient uptake of maize in two different soils. Appl Soil Ecol 36(2):184–189

El-Howeity MA, Asfour MM (2012) Response of some varieties of canola plant (*Brassica napus* L.) cultivated in a newly reclaimed desert to plant growth promoting rhizobacteria and mineral nitrogen fertilizer. Ann Agric Sci 57(2):129–136

El-Iklil MK, Benichou M (2000) Salt stress effect on epinasty in relation to ethylene production and water relations in tomato. Agronomy 20:399–406

Farwell AJ, Vesely S, Nero V, Rodriguez H, McCormack K, Shah S, Dixon DG, Glick BR (2007) Tolerance of transgenic canola plants (*Brassica napus*) amended with plant growth-promoting bacteria to flooding stress at a metal-contaminated field site. Environ Pollut 147:540–545

Farwell AJ, Vesely S, Nero V, Rodriguez H, Shah S, Dixon DG, Glick BR (2006) The use of transgenic canola (*Brassica napus*) and plant growth-promoting bacteria to enhance plant biomass at a nickel-contaminated field site. Plant Soil 288(1–2):309–318

Figueiredo MV, Burity HA, Martínez CR, Chanway CP (2008) Alleviation of drought stress in the common bean (*Phaseolus vulgaris* L.) by co-inoculation with *Paenibacillus polymyxa* and *Rhizobium tropici*. Appl Soil Ecol 40(1):182–188

Frankenberger JWT, Arshad M (1995) Microbial synthesis of auxins. In: Frankenberger WT, Arshad M (eds) Phytohormones in soils. Marcel Dekker, New York, pp 35–71

Gadd GM (2010) Metals, minerals and microbes: geomicrobiology and bioremediation. Microbiology 156(3):609–643

García de Salamone IE, Hynes RK, Nelson LM (2001) Cytokinin production by plant growth promoting rhizobacteria and selected mutants. Can J Microbiol 47(5):404–411

Giri B, Kapoor R, Mukerji KG (2003) Influence of arbuscular mycorrhizal fungi and salinity on growth, biomass, and mineral nutrition of *Acacia auriculiformis*. Biol Fertil Soils 38(3):170–175

Gisladottir G, Stocking M (2005) Land degradation control and its global environmental benefits. Land Degrad Dev 16(2):99–112

Glass ADM (1989) Plant nutrition: an introduction to current concepts. Jones and Bartlett Publishers, Boston, p 234

Glick BR (1995) The enhancement of plant growth by free-living bacteria. Can J Microbiol 41(2):109–117

Glick BR (2010) Using soil bacteria to facilitate phytoremediation. Biotechnol Adv 28(3):367–374

Glick BR (2014) Bacteria with ACC deaminase can promote plant growth and help to feed the world. Microbiol Res 169(1):30–39

Glick BR, Bashan Y (1997) Genetic manipulation of plant growth-promoting bacteria to enhance biocontrol of phytopathogens. Biotechnol Adv 15(2):353–378

Glick BR, Karaturovic DM, Newell PC (1995) A novel procedure for rapid isolation of plant growth promoting Pseudomonas. Can J Microbiol 41:533–536

Glick BR, Todorovic B, Czarny J, Cheng Z, Duan J, McConkey B (2007) Promotion of plant growth by bacterial ACC deaminase. Crit Rev Plant Sci 26(5–6):227–242

Gray EJ, Smith DL (2005) Intracellular and extracellular PGPR: commonalities and distinctions in the plant–bacterium signaling processes. Soil Biol Biochem 37(3):395–412

Grayston SJ, Prescott CE (2005) Microbial communities in forest floors under four tree species in coastal British Columbia. Soil Biol Biochem 37(6):1157–1167

Grichko VP, Glick BR (2001) Amelioration of flooding stress by ACC deaminase-containing plant growth-promoting bacteria. Plant Physiol Biochem 39:11–17

Gurska J, Wang W, Gerhardt KE, Khalid AM, Isherwood DM, Huang XD, Glick BR, Greenberg BM (2009) Three year field test of a plant growth promoting rhizobacteria enhanced phytoremediation system at a land farm for treatment of hydrocarbon waste. Environ Sci Technol 43(12):4472–4479

Gyaneshwar P, Kumar GN, Parekh LJ, Poole PS (2002) Role of soil microorganisms in improving P nutrition of plants. Plant Soil 245:83–93

Hamdia MA, Shaddad MA, Doaa MM (2004) Mechanisms of salt tolerance and interactive effects of *Azospirillum brasilense* inoculation on maize cultivars grown under salt stress conditions. Plant Growth Regul 44(2):165–174

Huang XD, El-Alawi Y, Penrose DM, Glick BR, Greenberg BM (2004a) A multi-process phytoremediation system for removal of polycyclic aromatic hydrocarbons from contaminated soils. Environ Pollut 130(3):465–476

Huang XD, El-Alawi Y, Penrose DM, Glick BR, Greenberg BM (2004b) Responses of three grass species to creosote during phytoremediation. Environ Pollut 130(3):453–463

Indiragandhi P, Anandham R, Kim K, Yim W, Madhaiyan M, Sa T (2008) Induction of defense responses in tomato against *Pseudomonas syringae* pv. tomato by regulating the stress ethylene level with *Methylobacterium oryzae* CBMB20 containing 1-aminocyclopropane-1-carboxylate deaminase. World J Microbiol Biotechnol 24(7):1037–1045

Jackson MB (1985) Ethylene and responses of plants to soil waterlogging and submergence. Annu Rev Plant Physiol 36(1):145–174

Jegatheesan V, Ravishankar H, Shu L, Wang J (2016) Application of green and physico-chemical technologies in treating water polluted by heavy metals. In: Green Technologies for Sustainable Water Management. American Society of Civil Engineers, Reston, pp 579–614

Kamran MA, Syed JH, Eqani SA, Munis MF, Chaudhary HJ (2015) Effect of plant growth-promoting rhizobacteria inoculation on cadmium (Cd) uptake by *Eruca sativa*. Environ Sci Pollut Res 22(12):9275–9283

Kang CH, Kwon YJ, So JS (2016) Bioremediation of heavy metals by using bacterial mixtures. Ecol Eng 89:64–69

Kang SH, Singh S, Kim JY, Lee W, Mulchandani A, Chen W (2007) Bacteria metabolically engineered for enhanced phytochelatin production and cadmium accumulation. Appl Environ Microbiol 73(19):6317–6320

Kaymak HC (2010) Potential of PGPR in agricultural innovations. In: Maheshwari DK (ed) Plant growth and health promoting bacteria. Springer, Berlin/Heidelberg, pp 45–79

Kloepper JW, Lifshitz R, Zablotowicz RM (1989) Free-living bacterial inocula for enhancing crop productivity. Trends Biotechnol 7(2):39–44

Kloepper JW, Schroth MN (1981) Relationship of in vitro antibiosis of plant growth-promoting rhizobacteria to plant growth and the displacement of root microflora. Phytopathology 71(10):1020–1024

Kloepper-Sams PJ, Park SS, Gelboin HV, Stegeman JJ (1987) Specificity and cross-reactivity of monoclonal and polyclonal antibodies against cytochrome P-450E of the marine fish scup. Arch Biochem Biophy 253(1):268–278

Kohler J, Hernández JA, Caravaca F, Roldán A (2008) Plant-growth-promoting rhizobacteria and arbuscular mycorrhizal fungi modify alleviation biochemical mechanisms in water-stressed plants. Funct Plant Biol 35(2):141–151

Lim KT, Shukor MY, Wasoh H (2014) Physical, chemical, and biological methods for the removal of arsenic compounds. Biomed Res Int 2014:503784

Liu W, Hou J, Wang Q, Ding L, Luo Y (2014) Isolation and characterization of plant growth-promoting rhizobacteria and their effects on phytoremediation of petroleum-contaminated saline-alkali soil. Chemosphere 117:303–308

Ludwig AA, Saitoh H, Felix G, Freymark G, Miersch O, Wasternack C, Boller T, Jones JD, Romeis T (2005) Ethylene-mediated cross-talk between calcium-dependent protein kinase and MAPK signaling controls stress responses in plants. Proc Natl Acad Sci U S A 102(30):10736–10741

Ma W, Guinel FC, Glick BR (2003) Rhizobium leguminosarum biovar viciae 1-aminocyclopr opane-1-carboxylate deaminase promotes nodulation of pea plants. Appl Environ Microbiol 69:4396–4402

Ma Y, Prasad MN, Rajkumar M, Freitas H (2011) Plant growth promoting rhizobacteria and endophytes accelerate phytoremediation of metalliferous soils. Biotechnol Adv 29(2):248–258

Macdonald SE, Quideau SA, Landhäusser SM (2012) Rebuilding boreal forest ecosystems after industrial disturbance. In: Vitt D, Bhatti J (eds) Restoration and reclamation of boreal ecosystems, attaining sustainable development, vol 20. Cambridge University Press, pp 123–161

Madhaiyan M, Poonguzhali S, Ryu J, Sa T (2006) Regulation of ethylene levels in canola (*Brassica campestris*) by 1-aminocyclopropane-1-carboxylate deaminase-containing *Methylobacterium fujisawaense*. Planta 224(2):268–278

Mahar A, Wang P, Ali A, Awasthi MK, Lahori AH, Wang Q, Li R, Zhang Z (2016) Challenges and opportunities in the phytoremediation of heavy metals contaminated soils: a review. Ecotoxicol Environ Safe 126:111–121

Mani D, Kumar C (2014) Biotechnological advances in bioremediation of heavy metals contaminated ecosystems: an overview with special reference to phytoremediation. Int J Environ Sci Technol 11:843–872

Martínez-Viveros O, Jorquera MA, Crowley DE, Gajardo GM, Mora ML (2010) Mechanisms and practical considerations involved in plant growth promotion by rhizobacteria. J Soil Sci Plant Nutri 10(3):293–319

Mayak S, Tirosh T, Glick BR (2004a) Plant growth-promoting bacteria confer resistance in tomato plants to salt stress. Plant Physiol Biochem 42:565–572

Mayak S, Tirosh T, Glick BR (2004b) Plant growth-promoting bacteria that confer resistance to water stress in tomato and pepper. Plant Sci 166(2):525–530

Nadeem SM, Ahmad M, Zahir ZA, Javaid A, Ashraf M (2014) The role of mycorrhizae and plant growth promoting rhizobacteria (PGPR) in improving crop productivity under stressful environments. Biotechnol Adv 32(2):429–448

Nadeem SM, Zahir ZA, Naveed M, Arshad M (2007) Preliminary investigations on inducing salt tolerance in maize through inoculation with rhizobacteria containing ACC deaminase activity. Can J Microbiol 53(10):1141–1149

Nadeem SM, Zahir ZA, Naveed M, Arshad M (2009) Rhizobacteria containing ACC-deaminase confer salt tolerance in maize grown on salt-affected fields. Can J Microbiol 55(11):1302–1309

Naveed M, Mitter B, Reichenauer TG, Wieczorek K, Sessitsch A (2014) Increased drought stress resilience of maize through endophytic colonization by *Burkholderia phytofirmans* PsJN and *Enterobacter* sp. FD17. Environ Exp Bot 97:30–39

Neubauer U, Furrer G, Kayser A, Schulin R (2000) Siderophores, NTA, and citrate: potential soil amendments to enhance heavy metal mobility in phytoremediation. Int J Phytoremediation 2(4):353–368

Nie M, Yin X, Ren C, Wang Y, Xu F, Shen Q (2010) Novel rhamnolipid biosurfactants produced by a polycyclic aromatic hydrocarbon-degrading bacterium *Pseudomonas aeruginosa* strain NY3. Biotechnol Adv 28(5):635–643

Nonnoi F, Chinnaswamy A, de la Torre VS, de la Pena TC, Lucas MM, Pueyo JJ (2012) Metal tolerance of rhizobial strains isolated from nodules of herbaceous legumes (*Medicago* spp. and *Trifolium* spp.) growing in mercury-contaminated soils. Appl Soil Ecol 61:49–59

Ortíz-Castro R, Contreras-Cornejo HA, Macías-Rodríguez L, López-Bucio J (2009) The role of microbial signals in plant growth and development. Plant Signal Behav 4(8):701–712

Oves M, Khan MS, Zaidi A, Ahmad E (2012) Soil contamination, nutritive value, and human health risk assessment of heavy metals: an overview. In: Zaidi A, Wani PA, Khan MS (eds) Toxicity of heavy metals to legumes and bioremediation. Springer, New York, pp 1–27

Padilla FM, Pugnaire FI (2006) The role of nurse plants in the restoration of degraded environments. Front Ecol Environ 4:196–202

Palaniyandi SA, Damodharan K, Yang SH, Suh JW (2014) *Streptomyces* sp. strain PGPA39 alleviates salt stress and promotes growth of 'Micro Tom'tomato plants. J Appl Microbiol 117(3):766–773

Pandey S, Ghosh PK, Ghosh S, De TK, Maiti TK (2013) Role of heavy metal resistant *Ochrobactrum* sp. and *Bacillus* spp. strains in bioremediation of a rice cultivar and their PGPR like activities. J Microbiol 51(1):11

Patten CL, Glick BR (2002) Role of *Pseudomonas putida* indoleacetic acid in development of the host plant root system. Appl Environ Microbiol 68(8):3795–3801

Paul D, Nair S (2008) Stress adaptations in a plant growth promoting rhizobacterium (PGPR) with increasing salinity in the coastal agricultural soils. J Basic Microbiol 48(5):378–384

Pulsawat W, Leksawasdi N, Rogers PL, Foster LJ (2003) Anions effects on biosorption of Mn (II) by extracellular polymeric substance (EPS) from *Rhizobium etli*. Biotechnol Lett 25(15):1267–1270

Rajkumar M, Ae N, Prasad MN, Freitas H (2010) Potential of siderophore-producing bacteria for improving heavy metal phytoextraction. Trends Biotechnol 28(3):142–149

Rajkumar M, Sandhya S, Prasad MN, Freitas H (2012) Perspectives of plant-associated microbes in heavy metal phytoremediation. Biotechnol Adv 30(6):1562–1574

Reed ML, Glick BR (2005) Growth of canola (*Brassica napus*) in the presence of plant growth-promoting bacteria and either copper or polycyclic aromatic hydrocarbons. Can J Microbiol 51(12):1061–1069

Reed ML, Warner BG, Glick BR (2005) Plant growth–promoting bacteria facilitate the growth of the common reed *Phragmites australisin* the presence of copper or polycyclic aromatic hydrocarbons. Curr Microbiol 51(6):425–429

Remans R, Beebe S, Blair M, Manrique G, Tovar E, Rao I, Croonenborghs A, Torres-Gutierrez R, El-Howeity M, Michiels J, Vanderleyden J (2008) Physiological and genetic analysis of root responsiveness to auxin-producing plant growth-promoting bacteria in common bean (*Phaseolus vulgaris* L.) Plant Soil 302(1–2):149–161

Rodríguez H, Fraga R (1999) Phosphate solubilizing bacteria and their role in plant growth promotion. Biotechnol Adv 17(4):319–339

Rodríguez H, Fraga R, Gonzalez T, Bashan Y (2006) Genetics of phosphate solubilization and its potential applications for improving plant growth-promoting bacteria. Plant Soil 287(1):15–21

Rodríguez-Díaz M, Rodelas-Gonzalés P-CC, Martínez-Toledo MV, González-López J (2008) A review on the taxonomy and possible screening traits of plant growth promoting rhizobacteria. In: Ahmad I, Pichtel J, Hayat S (eds) Plant–bacteria interactions: strategies and techniques to promote plant growth. Wiley, Weinheim, pp 55–80

Safronova VI, Stepanok VV, Engqvist GL, Alekseyev YV, Belimov AA (2006) Root-associated bacteria containing 1-aminocyclopropane-1-carboxylate deaminase improve growth and nutrient uptake by pea genotypes cultivated in cadmium supplemented soil. Biol Fertili Soils 42(3):267–272

Sandhya V, Ali SKZ, Grover M, Reddy G, Venkateswarlu B (2009) Alleviation of drought stress effects in sunflower seedlings by the exopolysaccharides producing *Pseudomonas putida* strain GAP-P45. Biol Fertili Soils 46:17–26

Saravanakumar D, Samiyappan R (2007) ACC deaminase from *Pseudomonas fluorescens* mediated saline resistance in groundnut (*Arachis hypogea*) plants. J Appl Microbiol 102(5):1283–1292

Shaharoona B, Arshad M, Zahir ZA, Khalid A (2006) Performance of Pseudomonas spp. containing ACC-deaminase for improving growth and yield of maize (*Zea mays* L.) in the presence of nitrogenous fertilizer. Soil Biol Biochem 38(9):2971–2975

Sharma A, Johri BN, Sharma AK, Glick BR (2003) Plant growth-promoting bacterium Pseudomonas sp. strain GRP 3 influences iron acquisition in mung bean (*Vigna radiata* L. Wilzeck). Soil Biol Biochem 35(7):887–894

Sheng X, He L, Wang Q, Ye H, Jiang C (2008) Effects of inoculation of biosurfactant-producing Bacillus sp. J119 on plant growth and cadmium uptake in a cadmium-amended soil. J Hazard Mater 155(1):17–22

Sinha S, Mukherjee SK (2008) Cadmium-induced siderophore production by a high Cd-resistant bacterial strain relieved Cd toxicity in plants through root colonisation. Curr Microbiol 56:55–60

Smyth EM, McCarthy J, Nevin R, Khan MR, Dow JM, O'Gara F, Doohan FM (2011) In vitro analyses are not reliable predictors of the plant growth promotion capability of bacteria; a *Pseudomonas fluorescens* strain that promotes the growth and yield of wheat. J Appl Microbiol 111(3):683–692

Sriprang R, Hayashi M, Yamashita M, Ono H, Saeki K, Murooka Y (2002) A novel bioremediation system for heavy metals using the symbiosis between leguminous plant and genetically engineered rhizobia. J Biotechnol 99(3):279–293

Strigul NS, Kravchenko LV (2006) Mathematical modeling of PGPR inoculation into the rhizosphere. Environ Monit Assess 21(8):1158–1171

Tank N, Saraf M (2010) Salinity-resistant plant growth promoting rhizobacteria ameliorates sodium chloride stress on tomato plants. J Plant Interact 5(1):51–58

Tilak KV, Ranganayaki N, Pal KK, De R, Saxena AK, Nautiyal CS, Mittal S, Tripathi AK, Johri BN (2005) Diversity of plant growth and soil health supporting bacteria. Curr Sci 10:136–150

Ullah A, Heng S, Munis MFH, Fahad S, Yang X (2015a) Phytoremediation of heavy metals assisted by plant growth promoting (PGP) bacteria: a review. Environ Exp Bot 117:28–40

Ullah A, Mushtaq H, Ali H, Munis MFH, Javed MT, Chaudhary HJ (2015b) Diazotrophs-assisted phytoremediation of heavy metals: a novel approach. Environ Sci Pol 22:2505–2514

Upadhyay SK, Singh DP, Saikia R (2009) Genetic diversity of plant growth promoting rhizobacteria isolated from rhizospheric soil of wheat under saline condition. Curr Microbiol 59(5):489–496

Verma JP, Yadav J, Tiwari KN, Lavakush SV (2010) Impact of plant growth promoting rhizobacteria on crop production. Int J Agric Res 5(11):954–983

Vessey JK (2003) Plant growth promoting rhizobacteria as biofertilizers. Plant Soil 255(2):571–586

Wang KL, Li H, Ecker JR (2002) Ethylene biosynthesis and signaling networks. Plant Cell 14(1):131–151

Wang L, Coles N, Wu C, Wu J (2014) Effect of long-term reclamation on soil properties on a coastal plain, southeast China. J Coast Res 30(4):661–669

Wilhite DA (2000) Drought as a natural hazard. In: Wilhite DA (ed) Drought: a global assessment. Routledge, London, pp 3–18

Wu Z, Yue H, Lu J, Li C (2012) Characterization of rhizobacterial strain Rs-2 with ACC deaminase activity and its performance in promoting cotton growth under salinity stress. World J Microbiol Biotechnol 28(6):2383–2393

Yao L, Wu Z, Zheng Y, Kaleem I, Li C (2010) Growth promotion and protection against salt stress by *Pseudomonas putida* Rs-198 on cotton. Eur J Soil Biol 46(1):49–54

Yue H, Mo W, Li C, Zheng Y, Li H (2007) The salt stress relief and growth promotion effect of Rs-5 on cotton. Plant Soil 297(1–2):139–145

Zahir ZA, Ghani U, Naveed M, Nadeem SM, Asghar HN (2009) Comparative effectiveness of Pseudomonas and Serratia sp. containing ACC-deaminase for improving growth and yield of wheat (*Triticum aestivum* L.) under salt-stressed conditions. Arch Microbiol 191(5):415–424

Zahir ZA, Munir A, Asghar HN, Shaharoona B, Arshad M (2008) Effectiveness of rhizobacteria containing ACC deaminase for growth promotion of peas (*Pisum sativum*) under drought conditions. J Microbiol Biotechnol 18(5):958–963

Zehnder GW, Yao C, Murphy JF, Sikora ER, Kloepper JW, Schuster DJ, Polston JE (1999) Microbe-induced resistance against pathogens and herbivores: evidence of effectiveness in agriculture. In: Induced plant defenses against pathogens and herbivores: biochemistry, ecology and agriculture. APS Press, St. Paul, p 33

Zhuang X, Chen J, Shim H, Bai Z (2007) New advances in plant growth-promoting rhizobacteria for bioremediation. Environ Int 33(3):406–413

Chapter 4
Promising Applications for the Production of Biofuels Through Algae

Nafe Aziz, Ram Prasad, Amr I.M. Ibrahim, and Ahmed I.S. Ahmed

4.1 Introduction

Production of fuel from biomass is gaining importance nowadays due to scarcity and increase in the cost of fossil fuels, it also causes pollution. According to previous studies algae are considered having the fuel properties (Demirbas 2008). Algae contain protein carbohydrates, lipid in varying amount. The amount of lipid content can vary according to the types of algae and its cultivated conditions, there are some algae which contain or can accumulate more than 40% of the fatty acid content (Becker 1994). Algae are having the potential to produce oil per acre more efficiently than any other feedstocks used to make biodiesel (Demirbas 2008). Biodiesel are mono alkyl ester, which are the long chain of fatty acids which can be formed into transesterification of oil, either from vegetable or animal fats. Biodiesel has been used in many parts of the world such as United States, Europe and Brazil with an annual production of 57 million L to 1 billion L (Gerpan 2005). The main problem facing the commercialization of biodiesel is its high of oil feedstock. There are various research going on in order to minimize the cost of oil feedstock. The

N. Aziz
Amity Institute of Microbial Technology, Amity University Uttar Pradesh, Noida, UP, India

Cyanobacterial Lab, Department of Bioscience, Jamia Millia Islamia, New Delhi, India

R. Prasad (✉)
Amity Institute of Microbial Technology, Amity University Uttar Pradesh, Noida, UP, India

Department of Mechanical Engineering, Johns Hopkins University, Baltimore, MD, USA
e-mail: rprasad@amity.edu; rpjnu2001@gmail.com

A.I.M. Ibrahim
Plant Pathology Unit, Plant Protection Department, Desert Research Center, Cairo, Egypt

A.I.S. Ahmed
Amity Institute of Microbial Technology, Amity University Uttar Pradesh, Noida, UP, India

Plant Pathology Unit, Plant Protection Department, Desert Research Center, Cairo, Egypt

J.K. Patra et al. (eds.), *Microbial Biotechnology*,
https://doi.org/10.1007/978-981-10-6847-8_4

cereal crops, trees, grasses, algae and cyanobacteria utilize solar energy, water and carbon dioxide by the process of photosynthesis, and stores their energy (in the form of starch, lipids and sugars) which can be used for the production of bio-fuel.

Bio-fuels based on plants (such as sugarcane, rapeseed, soybean, corn, Jatropa and Pongamia are extensively used for the production of bio-fuels around the world, mostly in Brazil, USA, Europe, South-East Asia. In these years, most of the industries are using oil (triglycerides) from the raw materials (Jatropa, Pongamia, sunflower, soybean, etc) with having intention to use bio-fuel as petroleum based diesel, in this the triglycerides are transesterified into fatty acid alkyl esters, which are utilized in diesel engine without modification of engine (Knothe 2006; Knothe 2010). As the oleaginous plants reduce the emission of greenhouse gases and the biodiesel produced from it also reduces the emissions of pollutants (SO_2, CO, CO_2, etc). As these plants are grown on arable and non-arable land and forest area and are the important sources of raw materials for humans and animals so these plants cannot be used as only source to produce biodiesel. As Algae fuel is an alternative fossil fuel various companies and government are giving funds in order to reduce the cost on algae oil.

In order to overcome these problems the researchers are utilizing another way for producing biodiesel using microalgae. Microalgae also having ability to absorb phosphates and nitrates (Cadoret and Bernard 2008), can accumulate high amount of fatty acids and having high yielding capacity than other oleaginous plants (Chisti 2007).

4.2 History

Algae as a bio-energy feedstock were known from mid-twentieth century. During 1950s the use of algae for producing energy started in 1955; Meier, Oswald and Golueke (1960) suggested that methane gas can be produce by anaerobic digestion. Actually various species of algae were discovered during 1940s which can produce oil droplets. During 1950s–1960s various key nutrient (nitrogen or silicon) were discovering which upon starvation can lead to more production of lipid. In 1978–1996, "The aquatic species program on fuel was started at DOE'S National renewable Energy Laboratory (NREL) and funded many academic institutes. In around 1980s the work were mostly done on Hydrogen production. In 2007, various species were discovered which can produce lipids, more than 60% of their dry weight (Chisti 2007).

During 2006 the world were mainly used biomass and waste by 58%, hydropower by 31% and other were by 12% it also include solar, wind and geothermal. It is also estimated that, up to 2030 the production of bio-fuel will achieve up to 153 billion of 12KWatt-H (US Energy Information Administration 2011). Bio-fuel is the way for first, second and third generation bio-fuels.

4.2.1 *First Generation Bio-fuels*

First generation bio-fuels are basically produced from the crop plants such as sugarcane (*Saccharum* sp.), oilpalm (*Elaeis oleifera*), sugarbeet (*Beta vulgaris),* rapeseed (*Brassica napus*), soyabeans (*Glycine max*), wheat (*Triticum* sp.) and corn (*Zea mays*) etc., and also from the fermentation of sugars to produced ethanol for fuel production (Knothe 2010; Natural Resources Canada 2011). As plant utilizes arable land which causes decrease in the availability of land and its fertility can also be lost.

4.2.2 *Second Generation Bio-fuels*

Second generation bio-fuels generally based on cellulose, can be found in crops and are not used as food by humans. Cellulose can be extracted from wood, leaves, straw, etc., which are utilized to produced alcohol, bio-oil, bio-hydrogen, etc. (Demirbas 2009; Roman-Leshkov et al. 2007; Demirbas 2010).

The most important of these include lignocellulosic processes which convert cellulose-based products of plants into liquid fuels. Sorghum, Myscanthus, Camelina, switchgrass (*Panicum virgatum*) and poplar trees (*Populus* sp.) are currently the most prominent'non-food' plant candidates for these approaches (Carpita and McCann 2008; Li et al. 2007; Schmer et al. 2008).

4.2.3 *Third Generation Bio-fuels*

Third generation bio-fuels are produced from microorganism. For examples algae, yeast, fungi, etc. Most of the researches are going on third generation bio-fuels and the bio-fuels produced from it and have similar characteristics as that of other sources.

Second and third generation bio-fuels are much better than first generation, because it can reduce the CO_2 emission by 150%. There are several applications of algae for different products (Fig. 4.1).

4.3 Algal Biology

Biodiesel produced from agricultural crops using current technology cannot sustainably replace fossil-based fuels in terms of its cost and environment impact. However, biodiesel from algae seems to have the potential as the alternative renewable bio-fuel, replacing fossil-based fuels. Producing renewable bio-fuels from algae will have no conflict with food supply and also contribute to reducing greenhouse gas emissions.

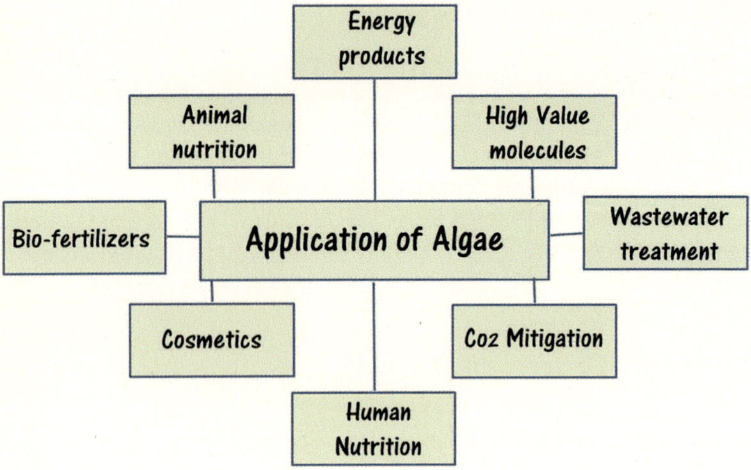

Fig. 4.1 Applications of algae for different products

Microalgae can be preferred as a valuable source of bio-fuels. It can produce different kinds of bio-fuels such as, bio-methane, by anaerobic digestion of algae (Sialve et al. 2009; Spolaore et al. 2006), Bio-hydrogen, by photo-biological process (Fedorov et al. 2005), Bio-ethanol by fermentation (Choi et al. 2010), liquid oil by thermal liquefaction (Banerjee et al. 2002) and Biodiesel (Johnson and Wen 2009). Bio-fuel or bio-oil are produced from biomass (microalgae) under high temperature and in absence of oxygen, the bio-fuels which are formed of three phases, the vapour phases, liquid phases and solid phases. The liquid phase is called bio-oil and its characteristics vary from feed-stocks used and condition for its process. Microalgae are the diverse group of organisms which have the ability to carry out photosynthesis and are generally found in the aquatic region. Microalgae can be single cellular or multi cellular, prokaryotic or eukaryotic. Microalgae may contains more than 75% of lipids and can produce 770 times more lipids as compared to any oleaginous plants and can be culturing more than once in a year (Chisti 2007) (Table 4.1).

4.3.1 Characteristics of Microalgae

Microalgae are present in diverse area where light and water are present. It can be found in ocean, ice, rivers, soils, etc. Microalgae are distinguished according to their pigmentation, structure and metabolism. Microalgae vary according to their sizes. Microalgae are generally of small size due to which it can perform effective photosynthesis. According to their sizes microalgae are divided into four groups such as, Micro-plankton (20–1000 μm), Nano-planktons (2–100 μm), Ultra-plankton (0.5–15 μm) and the Pico-plankton (0.2–2 μm), (Callieri and Stockner 2002).

Table 4.1 Comparison of biodiesel which obtained from different sources

Crop	Oil yield (L/ha)	Land area	Percent of existing
Corn	172	1540	846
Soybean	446	594	326
Canola	1190	223	122
Jatropa	1892	140	77
Coconut	2689	99	54
Oil palm	5950	45	24
Microalgae[a]	136,900	2	1.1
Microalgae[b]	58,700	4.5	2.5

Needed (M ha) US cropping area
[a]70% oil (by wt) in biomass
[b]30% oil (by wt) in biomass

4.3.2 *Taxonomic Classification*

According to the taxonomy, microalgae are divided into four main groups such as, Diatoms (Bacillariophyceae), green algae (Chlorophyceae), Cyanobacteria or blue green algae (Cyanophyceae), and golden algae (Chrysophyceae).

There are other groups also which come under this taxonomy such as yellow green algae (Xanthophyceae), golden algae (Chrysophyceae*)* red algae (Rhodophyceae), brown algae (Phaeophyceae), dinoflagellates (Dinophyceae) (Gerken et al. 2013; Ryckebosch et al. 2014) (Table 4.2). These groups are not generally used for biodiesel production. The most group of algae used for biodiesel production are green algae which includes *Botrycoccus* sp., *Dunaliella* sp., and *Chlorella* sp. (Garofalo 2010) for biodiesel.

4.3.3 *Lipids in Microalgae*

Lipid in microalgae can be increased to biodiesel production by growing microalgae in photo-bioreactor under nutrient (Nitrogen, Silicon, etc.) limited condition. This condition is suitable for most of the algae. For example, microalgae such as *Navicula* has lipid which can be increased from 22% to 49% (g lipid/g dry weight) in absence of silicon and increased up to 58% under nitrogen limitation. There are some algae which have no effect on lipid content under limited condition, such as *Amphora* and *Cyclotella* (Sheehan et al. 1998). The oil levels of 20–50% are quite common in microalgae (Chisti 2007), for examples, *Chlorella* sp. (28–32); *Botryococcus braunii* (29–75); *Nannochloris* sp. (20–35); *Nannochloropsis* sp. (31–68); *Neochloris oleoabundans* (35–54); *Dunaliella primolecta* (23); *Dunaliella tertiolecta* (36–42); *Scenedesmus* sp. (45–50); *Thalassiosira pseudonama* (20–30). *Isochrysis* sp. (25–33).

Table 4.2 Properties of major algae taxonomic groups

S. no	Taxonomic group	Chlorophyll	Carotenoids	Bilo protein	Storage products	Flagellation & cell structure
1	Bacillariophyta	a,. c	β- carotene ± -carotene Rarely fucoxanthin	–	Chrysolaminarin oils	1 apical flagellum in male gametes
2	Chloro phycophyta (green algae)	a, b	β- carotene ± -carotene Rarely carotene and lycopene lutein	–	Starch, oils	1,2,4 to many, equal, apical or subapical flagella
3	Chrysophycophyta (golden algae)	a, c	β-carotene, fucoxanthin	–	Chrysolaminarin oils	1 or 2 unequal, apical flagella in some cell surface covered by characteristic scales
4	Cyanobacteria (blue green algae)	a, c	β carotene, phycobilins	–		
5	Phaeco phycophyta (brown algae)	a, c	β-carotene, fucoxanthin	–	Laminarin, soluble carbohydrates, oils	2 lateral flagella
6	Dinophyta	a, c	β-carotene, peridinin, neoperididnin dinoxanthin, neodinoxanthin	–	Starch, oils	2 lateral, 1 trailing, 1 girdling flagellum
7	Rhodo Phycophyta (red algae)	a, rarely d	β-carotene, zeaxanthin,±β carotene	Phycoerythrin, Phycocyanin	Floridean starch oils	Flagella absent

There is other nutrient such as phosphate which has impact on lipid content of the microalgae. For example, upon increasing the concentration of the phosphate there is an increase in the concentration of microalgae under limited condition. Upon increase in concentration from 0.14 to 0.37 mg/L (Xin et al. 2010) there is an increase in concentration of microalgae from 0.14 to 0.37 g/L while there is decreased in lipid concentration in microalgae from 5.3% to 23.5% (g lipid/g dry weight).

Lastly the osmotic shock can also help in the production of lipid. For example, an increase in the NaCl concentration from 3.5 to 7 g/L, there is an increase in the lipid production from 60% to 67% (g lipid/g dry weight).

4.4 Metabolism Classification

Microalgae are classified into four types i.e. photoautotrophs, heterotrophs, mixotrophs and photoheterotrophs (Chen et al. 2011).

4.4.1 Photoautotrophic Microalgae

It converts light energy into inorganic carbon (CO_2) and water to produce biomass during the process of photosynthesis. As photoautotrophic microalgae contains high level of lipids but their biomass produced by these are low, as compared to the heterotrophic microalgae, between 0.117 and 1.54 kg/m3/day (Chisti 2007).

4.4.2 Heterotrophic Microalgae

It requires organic carbon as a source of carbon and energy. It can be produced in closed container and are more effective than photo-autotrophic species for biodiesel production. For example, the biomass produced by *Chlorella* sp., under heterotrophic conditions is 7.4 kg/m3/day, as compared to photo-autotrophic species and lipid present up to 58% (g lipid/g dry weight).

4.4.3 Mixotrophic Microalgae

It grown in both light as well as dark condition, inorganic and organic carbon sources. For example, when *Chlorella vulgaris* is grown in the dark, its biomass productivity increases to 151 mg/L/day and, in presence of light and glucose its biomass productivity increases to 254 mg/L/day.

4.4.4 Photo-Heterotrophic Microalgae

This type of metabolism required light and organic carbon for growth (Chen et al. 2011). For example, in *Chlorella minutissima,* glycerol as carbon source and having light intensity 35µe/m^2/s, used to produce microalgae at a biomass concentration up to 8.2 g/L after 15 days of culture.

4.5 Criteria for Stain Selection

Microalgae for biodiesel production should contain following criteria: It should be carbon dioxide tolerant and its uptake, temperature tolerant, stable in specific bioreactor and can produce valuable secondary products, require specific growth condition & resistant to infection, execute auto inhibitor and easy to harvest & processing, and also have potential for genetic manipulation (Giostri et al. 2016).

4.6 Advantages of Algal Feed-Stocks

Algae are favoured feedstock due to its high energy density and fuel nature it is used in biodiesel production by researchers and entrepreneurs around the world. Various advantages of algal bio-fuel are as follows: Algae can be produced in large amount, having high yield per acre of cultivation, It does not require arable land for cultivation & fewer amounts of nutrients which minimize the competition with other organisms; It can use waste water, saline water, & produced water due to which it can reduce the competition against fresh water; recycle carbon (CO_2) emitted from the industrial waste and power plants (Hu et al. 2008).

Algae can produce valuable co-products and various kinds of bio-fuels; high growth rate of algae makes it possible to full fill the demands of bio-fuels with the use of limited land and without causing any potential biomass deficit; Due to high tolerance of microalgae under high CO_2 content in gas stream allows high- efficiency CO_2 mitigation; Nitrous oxide release could be minimized when microalgae are used for biodiesel production and micro algal farming could be potentially more cost effective than conventional farming (Hu et al. 2008).

4.7 The Major Disadvantages of Microalgae for Bio-fuel Production Are as Follows

When low biomass concentration in the micro algal culture due to the limit of light penetration, which in combination with the small size of algal cells makes the harvested biomass more costly.

The large water content of harvested algal biomass also means its drying would be an energy-consuming process. The high capital costs of and the rather intensive care required by a microalgal farming facility compared to a conventional agricultural farm is another factor that impedes the commercial implementation of the biofuels from microalgae strategy (NAABB 2014).

4.8 Biodiesel Production

4.8.1 Culturing

Culturing of microalgae is to be done in large scale in order to produce biodiesel. Large scale production can be done in the open ponds as well as in closed systems (fermentors).

4.8.2 Open Ponds

In the open ponds the microalgae are grown as photoautotrophs in presence of light. It is not suitable for the growth of microalgae because the ponds are open, contaminants are present in the form of protozoa, bacteria, or other microalgae which can affect the microalgae.

4.8.3 Photo-Bioreactors

It is a closed and continuous culture systems in which culturing can be done up to 6.7 g/L (Bai et al. 2011; Ranjbar et al. 2008) by using fresh or sea water. Different types of photo-bioreactor are available which can use in indoor or outdoor, these bioreactors are tubular, flat plate, airlift, bubble column and stirred tank (Xu et al. 2009). Under control condition of temperature, pressure, CO_2 concentration, etc., microalgae can be produced in large quantity. But its production cost is higher than the open ponds (Carvalho et al. 2006). Culturing of microalgae can also be done by using both open ponds and bio reactor. In this technique algae is first grown in photo-bioreactor under controlled temperature and then it is transferred into open ponds for few days for growing (Huntley and Redalja 2007).

4.8.4 Fermentors

It is generally the costly processes and are used for the production of heterotrophic microalgae in presence of organic sources of carbon (glucose, fructose, galactose, acetate, glycerol and acetic acid (Fig. 4.2). By this process high concentration of biomass can be achieved up to 150 g/L (Wu and Shi 2008; Cantin 2010).

4.8.5 Microalgae Harvesting

Microalgae can be harvested by different techniques such as, centrifugation, flocculation, gravity, sedimentation, filtration, screening, flotation or electrophoresis techniques (Chen et al. 2011).

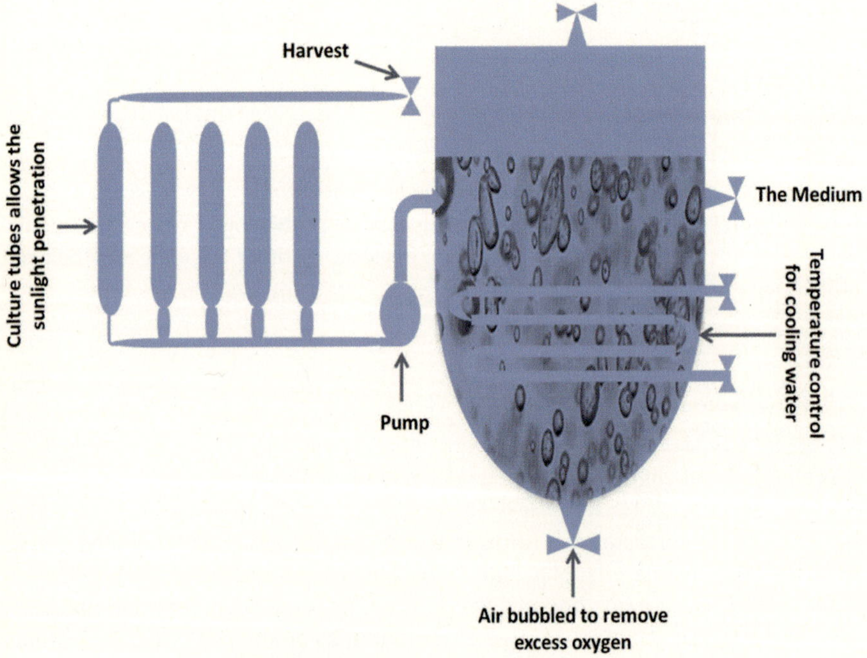

Fig. 4.2 Fermentation process for the production of heterotrophic microalga

4.9 Techniques Involved in Extraction of Lipids

There are different techniques involved in order to extract lipid from microalgae to produce biodiesel. The techniques which mainly involve are.

4.9.1 Solvent Extraction

Through this method, firstly the algae are dried by different method such as freeze drying, spray drying, oven drying or vacuum evaporation. After drying of algae various chemical solvents are used. In small scale generally chloroform-methanol mixture are used with having high extraction yield up to 83% (g lipid/g dry weight) (Yaguchi et al. 1997). Hexane is used as a less polar solvent and low toxicity and having less affinity toward non-lipid contaminants (Halim et al. 2010), from this it can obtain lipid up to 55% (g lipid/g dry weight). For example, *Chlorella protothecoides*. Other low toxic chemical solvent used are alcohol, n-hexane extraction.

4.9.2 Supercritical Carbon Dioxide Extraction

Supercritical CO_2 is generally non-toxic, reusable, and can be utilized at low temperatures (less than 40 °C) (Andrich et al. 2005). This system requires sophisticated instrument and large amount of energy. As this technique are utilized to extract lipids from algae to transform lipid to biodiesel. For example, by using supercritical CO_2 extraction technique at 60 °C and pressure 30 MPa in *Chlorococcum* sp. Lipid can be extracted higher than hexane Soxhlet extraction method.

4.9.3 Physiochemical Extraction

This technique includes various processes such as microwave, autoclave, osmotic shock, freeze drying, sonication (Cooney et al. 2009). Through these techniques algae can be disrupted physically to extract lipids from it. Microwave and bead beating techniques are mostly used. For example, in *Botryococcus* sp., the lipid content increases upon pre-treatment of microwave (Lee et al. 2010).

4.9.4 Biochemical Extraction

In these techniques lipids are extracted from algae by using cellulose hydrolysis before extraction. For example there is an increase in lipid concentration from 52% to 54% (g Lipid/g dry weight) through hydrolysis of sugars (Kumar et al. 2015).

4.9.4.1 Direct *(in situ)* Transesterification

In these method algae is directly treated with alcohol and a catalyst without extraction of lipids from it. The catalysts used are acid catalyst and alkali catalyst. The acid catalyst which could be sulphuric acid, hydrochloric acid as well as acetyl chloride (CH_3COCl) which can produced higher FAME up to 56% (g FAME/g dry weight) (Cooney et al. 2009). There are other polar solvent such as hexane or chloroform could be utilized to produce biodiesel in high amount (Johnson and Wen 2009). Heterogeneous catalyst (SrO) could also be used and it is also being used with microwave which is more effective in direct transesterification. For example, in Nanochloropsis FAME yield could be increased from 7% to 37% (g FAME/g dry weight) in direct transesterification by using microwave coupled with heterogeneous catalyst (Koberg et al. 2011).

4.9.4.2 Transesterification

It is a process of making biodiesel in which the microalgal lipid (triglycerides) reacted with methanol in the presence of catalyst. During the process methyl ester of fatty acids called biodiesel and glycerol are formed. During the reaction triglycerides are converted to di-glycerides and then to mono-glycerides and lastly to glycerol. Actually the transesterification depends upon certain things which are as follows (Nelson et al. 1996; Canakci and Van Gerpen 1999).

4.9.4.2.1 Lipids

In transesterification the amount of biodiesel yields depend upon the quality of lipids. The quality of lipid is very important characteristics in production of biodiesel because the nature and amount of lipid vary from species to species. Some microalgae contain 93% (g/g lipid) of phospholipids and glycolipids (Williams and Laurens 2010).

4.9.4.2.2 Alcohols

During the transesterification process alcohol used are Methanol, ethanol, butanol, etc. In this process methanol are generally used (Chisti 2007). Alcohol used during the production of biodiesel from vegetable oils in ratio of 6:1 of methanol to oil molar ratio for transesterification but for the production of biodiesel from microalgae the methanol to oil ratio is higher as compared. For example, during direct transesterification for 8 h at 25 °C (Ehimen et al. 2010) there is a decrease in the specific gravity (SG) of the biodiesel from 0.8887 to 0.8849 in which the molar ratio of methanol to oil was increased from 105:1 to 524:1.

4.9.4.2.3 Catalyst

Catalysts used in transesterification of microalgae are generally in homogeneous state (Tan and Lee 2011). Homogeneous alkaline catalysts used in transesterification of vegetable oils include Sodium or Potassium hydroxide and Sodium or Potassium methoxide. In large scale production of biodiesel it is generally used as catalyst which can produce 4000 times (Chisti 2007) faster rate as compared to acid catalyst (H_2SO_4, HCl and H_2SO_3) and are much cheaper as well (Helwani et al. 2009). Homogeneous acidic catalysts are used generally in transesterification of microalgae lipids. The biodiesel production can be increased more than 50 times by using HCl than NaOH (Nagle and Lemke 1990) at same conditions of transesterification (0.1 h, 70 °C). The acid catalyst which is used is CH_3COCl (Cooney et al. 2009), HCl (Tran et al. 2009) or H_2SO_4 (Miao and Wu 2006). Catalyst can also be used in combination of homogeneous acid-alkaline to transesterify lipids from microalgae. For example H_2SO_4-CH_3OK, KOH-HCl. During this process H_2SO_4 is first used for 2 h at 50 °C and then CH_3OH is used for 2 h at 55 °C (Halim et al. 2010) through this process biodiesel can be obtaining up to 44% (g FAME/ g lipid). Transesterification can also be done using heterogeneous catalysts. It could be acidic, alkaline or enzymatic in nature. The alkaline catalyst generally used is SrO (Koberg et al. 2011), calcium oxide (CaO) or magnesium oxide (MgO) (Umdu et al. 2009). In enzymatic catalyst generally lipase is used.

4.9.4.2.4 Reaction Time, Temperature and Stirring

During the transesterification processes the reaction time has positive effect on the specific gravity of the biodiesel produced. For example, in *Chlorella* at 30 °C with H_2SO_4 as catalyst (Ehimen et al. 2010) in reaction time from 0.25 to 12 h the Specific Gravity (SG) decreases from 0.914 to 0.884. The temperature has less effect on biodiesel production as compared to reaction time. For example, in presence of catalyst H_2SO_4 (2.25 mol/L) the biodiesel yield are similar at both 30 and 50 °C, respectively. Stirring has positive effect on biodiesel yield and quality (Ehimen et al. 2010).

4.10 Purification of Biodiesel and by Products

Purification of biodiesel is done in order to increase the biodiesel yield. During this process, hot water (50 °C), organic solvents such as hexane (Halim et al. 2010; Wiltshire et al. 2000) and water are used for a liquid-liquid separation (Couto et al. 2010; Lewis et al. 2000). For transesterification of lipid using non polar co-solvent, only water is used to separate biodiesel from the by-products (Johnson and Wen 2009). Based on first generation biodiesel (Leung et al. 2010) there are three steps which are followed during purification, water-washing, dry washing and then membrane extraction. According to the micro-algal lipid and vegetable oil composition, the by-product which could be present as unreacted lipid, water, alcohol, chlorophyll, metal and glycerol. Glycerol is the most important by-product having the consumption rate of 600 k ton/year. Glycerol is used in pharmaceutical, cosmetic and soap industries (Bondioli 2003). Glycerol can be transformed into other value added products by different methods such as chemical, thermo-chemical or biological conversion. By chemical method glycerol can be converted into propylene glycol, acrylic acid, propanol, propionic acid, allyl alcohol by the oxidation and reduction of glycerol. At low temperature through catalytic process glycerol can be converted into hydrogen (H_2) using nickel, platinum as catalyst. Through biological conversion, glycerol can be converted into alcohol (ethanol, butanol) through fermentation and other by-product like H_2, succinic acid (Yazdani and Gonzalez 2007). Anaerobic digestion of by-products could be another method for biodiesel production from microalgae by which it can be made very cost effective.

4.11 Opportunities and Challenges

Today the algae are the abundant, affordable, and sustainable sources of feed-stocks in the bio-fuel industries. Algae are considered as a part and parcel for producing advanced bio-fuels. The cellulosic bio-fuels can be benefited by agricultural and engineering process. On cultivating algae parallel with agriculture on similar scale there is increased in the production of algae with increase enterprises. Algae bio-fuels process can be commercialized by achieving the public-private partnerships and maintaining the regulation and standards. By reviewing the technology gaps and cross-cutting needs, the roadmap aims to guide researchers and engineers, policymakers, federal agencies, and the private sector in implementing a nationally coordinated effort toward developing a viable and sustainable algal bio-fuel industry (Hannon et al. 2010; Borowitzka 1999) (Table 4.3).

4.12 Regulation and Standards

The primary objective of the roadmap is to highlight the technical challenges and opportunities related to algal bio-fuels for commercialization. The R&D activities are carried out under a framework of standards, regulation, and policy. The

Table 4.3 Various process steps and its Research and Development (R&D) Challenges

Process steps	R&D challenges
Algal biology	Strain should be maximum diversity Develop small scale, high-throughput screening technology Open-access database having detailed characterization of collected strains Genetics and biochemical pathway should be investigated for production of precursors of fuel Gene manipulation techniques should be used to improve the strains
Algal cultivation	Algal cultivation multiple methods of culturing should be investigated such as open, closed, phototrophic, heterotrophic and mixotrophic growth Stable and robust cultures should be achieved at commercial scale Productivity system of fuel precursors (e.g. lipids) should be optimized Land, water and nutrients should be cost effective and sustainable Impacts and environmental risk should be identify and addressed
Harvesting and dewatering	Multiple harvesting methods should be used such as filtration, flocculation, floatation, centrifugation, sedimentation and mechanized seaweed harvesting Energy intensity during the process should be minimized Capital and operating cost should be lowered.
Extraction and fractionation	Multiple methods should be investigated in order to get high yield of oil such approaches like sonication, microwave, solvent systems, sub critical water, selective extraction and secretion Intermediates product achieved in high quantity Energy should be minimize during process Recycling mechanisms investigate to minimize waste
Fuel conversion	Multiple method of liquid transportation of fuels should be investigate by different means such as direct fuel production, thermo-chemical/catalytic conversion, anaerobic digestion and biochemical conversion Catalyst are improved by maintaining its specificity, activity and durability of fuels Contaminants and reaction inhibitor should be reduced High conversion rate can be achieved by scale up condition
Co-products	Value added co-product should be identify from algal remnants such as biogas, fertilizers, bio-plastics and surfactants Extraction and recovery of co-products should be optimize Market analysis are conducted in order to meet the applicable standards by checking its quality and safety
Distribution and utilization	In order to distribute the bio-fuels firstly energy and cost should be optimize For utilization of algal bio-fuels its regulation and customer requirement should be completed like engine performance and material compatibility

algae bio-fuels developer has to understand and ensure that algae are legally and safely developed, and its end product should have attained consumption standards. As there in no existing standards for process of bio fuel production, and the R&D can develop some laws and standards (Hannon et al. 2010; Bagnoud-Velásquez et al. 2015).

4.13 Characteristics of Biodiesel

The physiochemical properties of biodiesel are almost similar to that of diesel. There are various important characteristics of the biodiesel which are as follows:

4.13.1 Cetane Number

It tells about the quality of ignition of biodiesel which will increase with the number of carbon and it decreases as the number of unsaturated carbon bonds increases. The cetane number of many species based on their FAME content are in the range of 39–54, and the cetane number for petro-diesel are in the range from 47 to 51(Stansell et al. 2012).

4.13.2 Heat of Combustion

Biodiesel are suitable to use in diesel engine. Heat of combustion increases as the carbon chain length increases (Knothe 2005). By using heterotrophic microalgae lipid can be extracted in presence of H_2SO_4 in methanol, (Miao and Wu, 2006) biodiesel with a heat of combustion of 35.4 MJ/L can be obtain which is in the range of diesel fuel (36–38 MJ/L).

4.13.2.1 Viscosity

It increases as the number of carbon increases and decreases with the degree of unsaturation (Knothe 2005). Transesterification of oils helps the biodiesel to decrease the viscosity of oil between 4 and 6 mm2/s (40 °C) (National Renewable Energy Laboratory 2009).

4.13.2.2 Oxidation of Biodiesel

It will happen when the FAME comes in contact with oxygen it transformed into hydrogen peroxides, aldehydes, acids and other oxygenates, and can form deposits (Knothe 2005). Upon increase on unsaturation there is an increase in oxidation of biodiesel. Biodiesel can be stored for few months by adding antioxidants (Knothe and Steidley 2005; National Renewable Energy Laboratory 2009).

4.13.2.3 Cold Flow Properties of Biodiesel

It is also important properties of biodiesel, it can be measured by knowing its cloud and pour points. Cloud points are the formation of visible crystals on decreasing the temperature in the biodiesel (Knothe 2005). Pour points defined as the temperature at which the biodiesel does not flow. Cloud and pour points can be maintained upon unsaturation and function of molar ratio of biodiesel in diesel.

4.13.2.4 Lubricity

Lubricity of biodiesel is lower as compared to that of petro-diesel. Lubricity is the ability to reduce the friction between the solid surfaces in relative motion (Chevron Corporation 2007).

4.14 Other Importance of Microalgae: Bio-Hydrogen

Hydrogen is an important fuel with wide applications in fuel cells, liquefaction of coal, and upgrading of heavy oils (e.g., Bitumen). Hydrogen can be produced biologically by a variety of means, including the steam reformation of bio-oils (Wang et al. 2007), dark and photo fermentation of organic materials and photolysis of water catalyzed by special micro algal species (Kapdan and Kargi 2006).

Microalgae have the capacity of producing a vast array of high-value bioactive compounds that can be used as pharmaceutical compounds, health foods, and natural pigments (Oh et al. 2003; Jiang 2000). Some well-studied examples include acetylic acids, b-carotene (Huang et al. 2006; Del Campo et al. 2007), vitamin B (Wen and Chen 2003), ketocarotenoid, astaxanthin and polyunsaturated fatty acids (Ip and Chen 2005) and lutein (Blanco et al. 2007) (Shi et al. 2002) (Table 4.4). The economic feasibility of micro algal bio-fuel production should be significantly enhanced by a high-value co-product strategy, which would conceptually to involve sequentially the cultivation of microalgae in a micro algal farming facility (CO_2 mitigation), extracting bio-reactive products from harvested algal biomass, thermal processing (pyrolysis, liquefaction, or gasification), extracting high-value chemicals from the resulting liquid, vapour, and/or solid phases, and reforming/upgrading bio-fuels for different applications.

Table 4.4 Some high-value bioproducts extracted from microalgae

Product group	Applications	Examples	Reference
Phycobiliproteins, carotenoids	Pigments, cosmetics, pro-vitamins, pigmentation	Phycocyanin (*Spirulina platensis)*	Furuki et al. (2003)
		β-carotene (*Dunaliella salina*)	Borowitzka (1991)
		Astaxanthin and Leutin (*Haematococcus pluvialis*)	Olaizola (2003)
Polyunsaturated fatty acids (PUFAs)	Food additive, nutraceutics	Eicosapentaenoic acid (*Chlorella minutissima*)	Cardozo et al. (2007)
		Docosahexaenoic acid (DHA) (*Schizochytrium* sp.)	Valencia et al. (2007)
		Arachidonic acid (AA) (*Parietochlorisincise*)	Bigogno et al. (2002)
Vitamins	Nutrition	Biotin (*Euglena gracilis*)	Baker et al. (1981)
		α-Tocopherol (Vitamin E) (*Euglena gracilisa*)	Survase et al. (2006)
		Ascorbic acid (Vitamin C) (*Prototheca moriformis, Chlorella* sp.)	Running et al. (2002)

4.15 Companies for Algae Bio-fuel Production

There are many companies involved in producing bio-fuel from algae worldwide. Significant advances have already been made in the research of manipulating the metabolism of algae and engineering of algae systems to produce large amounts of oil while absorbing equally impressive quantities of carbon dioxide. Some countries have achievements related to bio-fuel from algae such as Greon (Bulgaria), Alvigor AG (Germany), Alpha Biotech (France); Alga fuel, S.A (Portugal); Algae Link (Cadiz, Spain); Alvigor AG (Switzerland); algaeLink N.V, ingrepro B.V, LGem B.V (Netherland), Oil Fox (Argentina), Algae Fuel System, Centurion Biofuels, Pond Biofuels Inc. (Canada); Recursos Renovables Alternativos (Maxico); Algae Fuel, Algae Fuel System, Alginol, Algae wheel, Aquatic Energy, Algoil Energy, PetroSun and Algae BioFuels Inc., Solazyme (USA); Bio Fuels Pty Ltd. (A Victor Smorgon Group Company), Algae Tec. (Australia) and Aquaflow Bionomic Corporation-ABC (New Zealand).

4.16 Future Prospects of Algal Biofuels

Algal bio-fuels systems, which produce fuels from algae such as eukaryotic algae or cyanobacteria, and have many advantage that their productivity is not dependent on soil fertility and contribute substantially to the environment without increasing the pressure on arable land and forest ecosystems. Algae can also be grown in saline water and can produce large amount of feed stocks for the production of bio-fuels. The bio-fuels which can be produced are biodiesel, methane, ethanol, butanol and hydrogen based on their efficient production of starch, sugar and oils. During the growth of microalgae it absorbs CO_2 from both atmosphere and industrial sources and help in capturing of carbon. The waste biomass which remains after production of fuels can be utilized for the production of charcoal like products through pyrolysis which is suitable for long term storage. Bio-char can also be used as a fuel there by reducing the dependency on coal, or marketed as a soil additive. Microalgae bio-fuels system eliminates both the food versus fuel and potential forest versus fuel problems. Now algae bio-fuel system has achieved economic viability. There are various companies which actively developing these technology for commercial operation. Based on the recent economic studies the by-products produced during the production of bio-fuels have much importance, it is because these high-value products can effects in order to lowering the construction costs, biomass productivity and the price of the dominated products.

There are various efforts which can be applied in order to lower the cost of the bio fuels commercially, which are as follows: Improving solar energy conversion efficiency; Lowering the bioreactor capital cost; Co-production of high value products for generating additional income; Temperature, O_2 and CO_2 regulation; Nutrients recycling; Optimizing media for biomass and bio-fuel production- carbon, nutrients and pH; and by avoiding contamination.

4.17 Conclusion

Micro algal biodiesel is technically feasible. It is the only renewable biodiesel that can potentially completely displace liquid fuels derived from petroleum. Producing low-cost Micro algal biodiesel requires primarily improvements to algal biology through genetic and metabolic engineering. Use of the bio-refinery concept and advances in photo-bioreactor engineering will further lower the cost of production. In view of their much greater productivity, photo-bioreactors are used in producing much of the Micro algal biomass required for making biodiesel. Photo-bioreactors provide a controlled environment that can be tailored to the specific demands of highly productive microalgae to attain a consistently good annual yield of oil. Microalgae can produce a large variety of novel bio-products with wide applications in medicine, food, and cosmetic industries.

References

Andrich G, Nesti U, Venturi F, Zinnai A, Fiorentini R (2005) Supercritical fluid extraction of bioactive lipids from the microalga *Nannochloropsis* sp. Eur J Lipid Sci Technol 107(6):381–386

Bagnoud-Velásquez M, Refardt D, Vuille F, Ludwig C (2015) Opportunities for Switzerland to contribute to the production of algal biofuels: the hydrothermal pathway to bio-methane. Chimia (Aarau) 69(10):614–621. https://doi.org/10.2533/chimia.2015.614

Bai M, Cheng C, Wan H, Lin Y (2011) Microalgal pigments potential as by products in lipid production. J Taiwan Inst Chem Eng 42:783–786

Baker ER, McLaughlin JJA, Hutner SH (1981) Water-soluble vitamins in cells and spent culture supernatants of *Poteriochromonas stipitata*, *Euglena gracilis*, and *Tetrahymena thermophila*. Arch Microbiol 129:310–313

Banerjee A, Sharma R, Chisti Y, Banerjee UC (2002) *Botryococcus braunii*- a renewable source of hydrocarbons and other chemicals. Crit Rev Biotechnol 22(3):245–279

Becker EW (1994) In: Baddiley J et al (eds) Microalgae: biotechnology and microbiology. Cambridge University Press, Cambridge/New York

Bigogno C, Khozin-Goldberg I, Boussiba S, Vonshak A, Cohen Z (2002) Lipid and fatty acid composition of the green oleaginous alga *Parietochloris incisa*, the richest plant source of arachidonic acid. Phytochemistry 60:497–503

Blanco AM, Moreno J, Del Campo JA, Rivas J, Guerrero MG (2007) Outdoor cultivation of lutein-rich cells of *Muriellopsis* sp. in open ponds. Appl Microbiol Biotechnol 73:1259–1266

Bondioli P (2003) From oil seeds to industrial products: present and near future of oleochemistry. Ital J Agron 7(2):129–135

Borowitzka LJ (1991) Development of western biotechnology's algal β-carotene plant. Bioresour Technol 38:251–252

Borowitzka MA (1999) Commercial production of microalgae: ponds, tanks, tubes and fermenters. J Biotechnol 70:313–321

Cadoret J, Bernard O (2008) La production de biocarburant lipidique avec des microalgues: promesses et defis. J de la Societe de Biologie 202(3):201–211

Callieri C, Stockner JG (2002) Freshwater autotrophic picoplankton: a review. J Limnol 61(1):1–14

Canakci M, Van Gerpen JH (1999) Biodiesel production via acid catalysis. Trans ASAE 42(5):1203–1210

Cantin I (2010) La production de biodiesel à partir des microalgues ayant un métabolisme hétérotrophe. Maîtrise en environnement (M.Env), Université de Sherbrooke, Sherbrooke, pp 1–82

Cardozo KHM, Guaratini T, Barros PM, Falcao VR, Tonon AP, Lopes NP, Campos S, Torres MA, Souza AO, Colepicolo P, Pinto E (2007) Metabolites from algae with economical impact. Comp Biochem Physiol 146:60–78

Carpita NC, McCann MC (2008) Maize and sorghum: genetic resources for bio-energy grasses. Trends Plant Sci 13:415–420

Carvalho AP, Meireles LA, Malcata FX (2006) Microalgal reactors: a review of enclosed system designs and performances. Biotechnol Prog 22:1490–1506

Chen C, Yeh K, Aisyah R, Lee D, Chang J (2011) Cultivation, photobioreactor design and harvesting of microalgae for biodiesel production: a critical review. Bioresour Technol 102(1):71–81

Chisti Y (2007) Biodiesel from microalgae. Biotechnol Adv 25(3):294–306

Choi SP, Nguyen MT, Sim SJ (2010) Enzymatic pretreatment of *Chlamydomonas reinhardtii* biomass for ethanol production. Bioresour Technol 101:5330–5336

Cooney M, Young G, Nagle N (2009) Extraction of bio-oils from microalgae. Sep Purif Rev 38:291–325

Corporation C (2007) Diesel Fuels Technical Review. Chevron Products Company, San Ramon, pp 1–116. Available from http://www.chevron.com/products/prodserv/fuels/documents/Diesel_Fuel_Tech_Review.pdf

Couto RM, Simões PC, Reis A, Da Silva TL, Martins VH, Sánchez Vicente Y (2010) Supercritical fluid extraction of lipids from the heterotrophic microalga *Crypthecodinium cohnii*. Eng Life Sci 10(2):158–164

Del Campo JA, Garcia-Gonzalez M, Guerrero MG (2007) Outdoor cultivation of microalgae for carotenoid production: current state and perspectives. Appl Microbiol Biotechnol 74:1163–1174

Demirbas A (2008) Relationships derived from physical properties of vegetable oil and bio-diesel fuels. Fuel 87:1743–1748

Demirbas AH (2009) Inexpensive oil and fats feedstocks for production of biodiesel. Energy Edu. Sci Technol 23:1–13

Demirbas MF (2010) Microalgae as a feedstock for biodiesel, *Energy Edu.* Sci Technol 25:31–43

Ehimen EA, Sun ZF, Carrington CG (2010) Variables affecting the in situ transesterification of microalgae lipids. Fuel 89(3):677–684

Fedorov AS, Kosourov S, Ghirardi ML, Seibert M (2005) Continuous hydrogen photoproduction by *Chlamydomonas reinhardtii*: using a novel two-stage, sulfate limited chemostat system. Applied biochemistry and biotechnology - part a enzyme engineering and. Biotechnology 121(1–3):403–412

Furuki T, Maeda S, Imajo S, Hiroi T, Amaya T, Hirokawa T, Ito K, Nozawa H (2003) Rapid and selective extraction of phycocyanin from *Spirulina platensis* with ultrasonic cell disruption. J Appl Phycol 15:319–324

Garofalo R (2010) Algae and aquatic biomass for a sustainable production of 2nd Generation biofuels, http://www.aquaculture.ugent.be

Gerken HG, Donohoe B, Knoshaug EP (2013) Enzymatic cell wall degradation of *Chlorella vulgaris* and other microalgae for biofuels production. Planta 237:239–253

Gerpan JV (2005) Biodiesel processing and production. Fuel Process Technol 86:1097–1107

Giostri A, Binotti M, Macchi E (2016) Microalgae coffering in coal power plants: innovative system layout and energy analysis. Renew Energy 95:449–464

Halim R, Gladman B, Danquah MK, Webley PA (2010) Oil extraction from microalgae for biodiesel production. Bioresour Technol 102(1):178–185

Hannon M, Gimpel J, Tran M, Rasala M, Mayfield S (2010) Biofuels from algae: challenges and potential. Biofuels 1(5):763–784

Helwani Z, Othman MR, Aziz N, Kim J, Fernando WJN (2009) Solid heterogeneous catalysts for transesterification of triglycerides with methanol: a review. Appl Catal 363:1–10

Hu Q, Sommerfeld M, Jarvis E, Ghirardi M, Posewitz M, Seibert M, Darzins A (2008) Microalgal triacylglycerols as feedstocks for biofuel production: perspectives and advances. Plant J 54:621–639

Huang JC, Chen F, Sandmann G (2006) Stress-related differential expression of multiple b-carotene ketolase genes in the unicellular green alga *Haematococcus pluvialis*. J Biotechnol 122:176–185

Huntley M, Redalje D (2007) CO2 mitigation and renewable oil from photosynthetic microbes: a new appraisal. Mitigat Adapt Strat Global Change 12:573–608

Ip PF, Chen F (2005) Employment of reactive oxygen species to enhance astaxanthin formation in *Chlorella zofingiensis* in heterotrophic culture. Process Biochem 40:3491–3496

Jiang FC (2000) Algae and their biotechnological potential. Kluwer Academic Publishers, Dordrecht/Boston/ London

Johnson MB, Wen Z (2009) Production of biodiesel fuel from the micro alga *Schizochytrium limacinum* by direct transesterification of algal biomass. Energy Fuel 23:5179–5183

Kapdan IK, Kargi F (2006) Bio-hydrogen production from waste materials. Enzym Microb Technol 38:569–582

Knothe G (2005) Cetane numbers–heat of combustion–why vegetable oils and their derivatives are suitable as viscosity of biodiesel. In: Knothe G, Krahl J, Gerpen JV (eds) The biodiesel handbook. AOCS Press, Campaign, pp 81–82

Knothe G (2006) Analyzing biodiesel: standards and other methods. J Am Oil Chem Soc 83(10):823–833

Knothe G (2010) Biodiesel and renewable diesel: a comparison. Prog Energy Combust Sci 36(3):364–373

Knothe G, Steidley KR (2005) Lubricity of components of biodiesel and petrodiesel, the origin of biodiesel lubricity. Energy Fuels 19(3):1192–1200

Koberg M, Cohen M, Ben-Amotz A, Gedanken A (2011) Bio-diesel production directly from the microalgae biomass of *Nannochloropsis* by microwave and ultrasound radiation. Bioresour Technol 102:4265–4269

Kumar RR, Raoand PH, Arumugam M (2015) Lipid extraction methods from microalgae: a comprehensive review. Front Energy Res. https://doi.org/10.3389/fenrg.2014.00061

Lee J, Yoo C, Jun S, Ahn C, Oh H (2010) Comparison of several methods for effective lipid extraction from microalgae. Bioresour Technol 101:575–577

Leung DYC, Wu X, Leung MKH (2010) A review on biodiesel production using catalyzed transesterification. Appl Energy 87:1083–1095

Lewis T, Nichols PD, McMeekina TA (2000) Evaluation of extraction methods for recovery of fatty acids from lipid- producing micro-heterotrophs. J Microbiol Methods 42:107–116

Li X, Xu H, Wu Q (2007) Large-scale biodiesel production from microalga *Chlorella protothecoides* through heterotrophic cultivation in bioreactors. Biotechnol Bioeng 98(4):764–771

Miao X, Wu Q (2006) Biodiesel production from heterotrophic microalgal oil. Bioresour Technol 97(6):841–846

NAABB (National Alliance for Advanced Biofuels and Bioproducts) (2014) National Alliance for Advanced Biofuels and Bio-products Full Final Report. Donald Danforth Plant Science Center. http://energy.gov/eere/bioenergy/downloads/national-alliance-advanced-biofuels-and-bioproducts-synopsis-naabb-final

Nagle N, Lemke P (1990) Production of methyl-ester fuel from microalgae. Appl Biochem Biotechnol 24:355–361

National Renewable Energy Laboratory (2009) Biodiesel handling and use guide. pp. 1–56

Natural Resources Canada (2011) Government of Canada calls on industry to participate in new bio-fuels initiative, In: Natural Resources Canada, 25.05.2011, Available from

Nelson LA, Foglia TA, Marmer WN (1996) Lipase catalyzed production of biodiesel. JAOCS 73:1191–1195

Oh HM, Choi A, Mheen TI (2003) High-value materials from microalgae. Korean J Microbiol Biotechnol 31:95–102

Olaizola M (2003) Commercial development of microalgal biotechnology: from the test tube to the marketplace. Biomol Eng 20:459–466

Oswald WJ, Golueke CG (1960) Biological transformation of solar energy. Adv Appl Microbiol 11:223–242

Ranjbar R, Inoue R, Shiraishi H, Katsuda T, Katoh S (2008) High efficiency production of astaxanthin by autotrophic cultivation of Haematococcus pluvialis in a bubble column photobioreactor. Biochem Eng J 39(3):575–580

Roman-Leshkov Y, Barrett CJ, Liu ZY, Dumesic JA (2007) Production of dimethylfuran for liquid fuels from biomass-derived carbohydrates. Nature 447:982–985

Running JA, Severson DK, Schneider KJ (2002) Extracellular production of Lascorbic acid by *Chlorella protothecoides*, *Prototheca* species, and mutants of *Prototheca moriformis* during aerobic culturing at low pH. J Ind Microbiol Biotechnol 29:93–98

Ryckebosch E, Bruneel C, Termote-Verhalle R, Muylaert K, Foubert I (2014) Influence of extraction solvent system on extractability of lipid components from different microalgae species. Algal Res 3:36–43

Schmer MR, Vogel KP, Mitchell RB, Perrin RK (2008) Net energy of cellulosic ethanol from switchgrass. Proc Natl Acad Sci U S A 105:464–469

Sheehan J, Combreco V, Duffield J, Graboski M, Shapouri H (1998) An overview of biodiesel and petroleum diesel life cycles. National Renewable Energy Laboratory, Golden, pp 1–47

Shi XM, Jiang Y, Chen F (2002) High-yield production of lutein by the green microalga *Chlorella protothecoides* in heterotrophic fedbatch culture. Biotechnol Prog 18:723–727

Sialve B, Bernet N, Bernard O (2009) Anaerobic digestion of microalgae as a necessary step to make microalgal biodiesel sustainable. Biotechnol Adv 27:409–416

Spolaore P, Joannis-Cassan C, Duran E, Isambert A (2006) Commercial applications of microalgae. J Biosci Bioeng 101(2):87–96

Stansell GR, Gray VM, Sym SD (2012) Microalgal fatty acid composition: implications for biodiesel quality. J Appl Phycol 24:791–801

Survase SA, Bajaj IB, Singhal RS (2006) Biotechnological production of vitamins. Food Technol Biotechnol 44:381–396

Tan KT, Lee KT (2011) A review on supercritical fluids (SCF) technology in unstainable biodiesel production: potential and challenges. Renew Sust Energ Rev 15(5):2452–2456

Tran H, Hong S, Lee C (2009) Evaluation of extraction methods for recovery of fatty acids from *Botryococcus braunii* LB 572 and *Synechocystis* sp. PCC 6803. Biotechnol Bioprocess Eng 14(2):187–192

U.S. Energy Information Administration (2011) Annual energy outlook, pp 1–246, Available from http://www.eia.doe.gov/forecasts/aeo/pdf/0383 (2011.pdf)

Umdu ES, Tuncer M, Seker E (2009) Transesterification of *Nannochloropsis oculata* microalga's lipid to biodiesel on Al_2O_3 supported CaO and MgO catalysts. Bioresour Technol 100:2828–2831

Valencia I, Ansorena D, Astiasaran I (2007) Development of dry fermented sausages rich in docosahexaenoic acid with oil from the microalgae *Schizochytrium* sp.: influence on nutritional properties, sensorial quality and oxidation stability. Food Chem 104:1087–1096

Wang Z, Pan Y, Dong T, Zhu X, Kan T, Yuan L, Torimoto Y, Sadakata M, Li Q (2007) Production of hydrogen from catalytic steam reforming of bio-oil using $C1_2A7$-O-based catalysts. Appl Catal A 320:24–34

Wen ZY, Chen F (2003) Heterotrophic production of eicosapentaenoic acid by microalgae. Biotechnol Adv 21:273–294

Williams PJlB, Laurens LML (2010) Microalgae as biodiesel & biomass feedstocks: Review & analysis of the biochemistry, energetics & economics. Energy Environ Sci 3(5):554–590

Wiltshire KH, Boersma M, Möller A, Buhtz H (2000) Extraction of pigments and fatty acids from the green alga *Scenedesmus obliquus* (Chlorophyceae). Aquat Ecol 34(2):119–126

Wu Z, Shi X (2008) Rheological properties of *Chlorella pyrenoidosa* culture grown heterotrophically in a fermentor. J Appl Phycol 20(3):279–282

Xin L, Hong-ying H, Ke G, Ying-xue S (2010) Effects of different nitrogen and phosphorus concentrations on the growth, nutrient uptake, and lipid accumulation of a freshwater microalga *Scenedesmus* sp. Bioresour Technol 101(14):5494–5500

Xu L, Weathers PJ, Xiong X, Liu C (2009) Microalgal bioreactors: challenges and opportunities. Eng Life Sci 9(3):1787189

Yaguchi T, Tanaka S, Yokochi T, Nakahara T, Higashihara T (1997) Production of high yields of docosahexaenoic acid by *Schizochytrium* sp. strain SR21. J Am Oil Chem Soc 74:1431–1434

Yazdani SS, Gonzalez R (2007) Anaerobic fermentation of glycerol- a path to economic viability for the bio fuels industry. Curr Opin Biotechnol 18(3):213–219

Chapter 5
Advances in Microbial Keratinase and Its Potential Applications

Dipak K. Sahoo, H.N. Thatoi, Bhabatosh Mitra, Keshab C. Mondal, and Pradeep K. Das Mohapatra

5.1 Introduction

Enzymes are known to be very useful biocatalysts for various industrial processes and chemical reactions. Their applicability as technical, food and feed enzymes revolutionized the market scenario of industries. The applications of different enzymes are focused on various industrial markets including leather, detergents and textiles, pulp and paper, pharmaceuticals, chemical, food and beverages, biofuels, animal feed, personal care, and others. It was estimated that the global market for industrial enzymes is expected to reach to nearly $7.1 billion by 2018, registering a five-year compound annual growth rate (CAGR) of 8.2%. Europe, USA, Japan and Denmark are the major enzyme producing countries in the world. The industrial enzyme market is principally occupied by hydrolytic enzymes which include proteases, amylases, lipases, esterases, cellulases, xylanases, phytases etc. More than 75% of these industrial enzymes are hydrolyses (Paranthaman et al. 2009) out of which 85% is proteases.

Keratins are valuable fibrous animal proteins with high mechanical stability. A significant amount of keratinous waste in the form of feathers, wool, nails, and horn are available as byproducts of agroindustrial processing and from other sources (Onifade et al. 1998; Brandelli et al. 2010). These keratinaceous materials are insoluble, rigid and resistant to degradation by common proteolytic enzymes like pepsin, trypsin and papain due to presence of high degree of cross-linking by disulphide

D.K. Sahoo • K.C. Mondal • P.K. Das Mohapatra (✉)
Department of Microbiology, Vidyasagar University, Midnapore, West Bengal, India
e-mail: pkdmvu@gmail.com

H.N. Thatoi
Department of Biotechnology, North Orissa University, Baripada, Odisha, India

B. Mitra
Department of Bio-Sciences and Bio-Technology, Fakir Mohan University, Balasore, Odisha, India

J.K. Patra et al. (eds.), *Microbial Biotechnology*,
https://doi.org/10.1007/978-981-10-6847-8_5

bonds, hydrogen bonds and hydrophobic interactions (Bradbury 1973; Sahni et al. 2015). Keratin, by virtue of its insolubility and resistance to common proteolytic enzymes, is not attacked by most living organisms. Nevertheless, keratin does not accumulate in nature. The biological agencies from eukaryotes like clothes moth larvae, carpet beetles and chewing lice are known to digest keratin partially (Waterhouse 1957). The remaining of the undigested keratins is degraded by keratinolytic microorganisms (Noval and Nickerson 1959). These microorganisms are more efficient to degrade keratinous substrates like feather, hair, wool, nail, horn etc. upto a significant level. These microorganisms have been isolated from diverse environments like poultry waste, soil from feather dumping site, tannery wastes, hot springs, soil from limestone quarry and even from Antarctic soils (Brandelli et al. 2010; Kshetri and Ningthoujam 2016). The ability of microbes to degrade keratin and the level of keratinase production vary according to species, type of substrates and also culture conditions.

Keratinases are a special class of protease that displays the capability of degrading insoluble keratin containing substrates. This enzymes draw attention because of their several potential applications in different sectors of the society like agroindustrial, biomedical and pharmaceutical fields. Keratinase producing microorganisms are largely isolated from keratin-enriched environments. Hydrolysis of keratinous wastes by keratinases brings up an eco-friendly approach about disposal and recycling of waste to obtain value-added by-products. The investigations reported that keratinase production primarily occur through submerged fermentation (SmF) (Gupta and Ramnani 2006; Brandelli et al. 2010) though solid state fermentation (SSF) process (Moreira et al. 2007; Rai et al. 2009; Sahoo et al. 2015) and immobilized condition was also reported (Prakash et al. 2010; Chitturi and Lakshmi 2016). The keratinases and their hydrolytic products have many important biotechnological applications as in leather industry (Prakash et al. 2010; Paul et al. 2014; Bouacem et al. 2016), detergent formulation (Rai and Mukherjee 2011; Sivakumar et al. 2013a), production of feedstuffs, fertilizers and films (Gupta and Ramnani 2006; Kornillowicz-Kowalska and Bohacz 2011), used for pharmaceutical and cosmetic purposes (Brandelli 2008; Villa et al. 2013), in bioremediation (Gupta and Ramnani 2006), prion protein degradation (Langveld et al. 2003; Okoroma et al. 2013) etc. The commercial products of keratinases are available in the market today under different trade names. Versazyme from *B. licheniformis* PWD-1 has led to significant improvements in broiler performance (Odetallah et al. 2005), and also helpful for growth and feed utilization of broilers (Stark et al. 2009). Another commercial keratinase named Cibenza DP100™ of *B. licheniformis* PWD-1 is used for successful and sustainable growth of piglets (Wang et al. 2011). There are some other keratinases named as Valkerase (BRI), Prionzyme (Genencor), PURE100 (Proteos Biotech), Kernail-Soft PB (Proteos Biotech), Cibenza DP100™ (Novus International) are available in market. Considering the versatile and potential applications, the enzyme keratinase has provided strong force to study this group of proteases. Diverse groups of microorganisms are being discovered every year which have able to produce keratinases that hasten valorization of keratin waste and thereby protect our environment and life.

5.2 Keratin – Structure and Functions

The word "keratin" comes from the Greek word "*kera*" meaning horn, first appears in the literature around 1850 to describe the material that made up of hard tissues like animal horns and hooves. Keratins were defined as all intermediate filament-forming proteins with specific physicochemical properties and produced in vertebrate epithelia (Bragulla and Homberger 2009). They are insoluble fibrous structural proteins with high mechanical stability and low degradation rate (Bradbury 1973). The recalcitrant keratin substrates are degraded by microbial keratinases through attacking the disulfide bonds in it and convert them from complex to simple forms (Table 5.1).

The principal role of keratins is to protect cells from mechanical and non-mechanical stress and others like defense, aggression, cell signaling etc. The major functions are: (i) Keratins persuade the structural design and mitotic division of the epithelial cells, (ii) It also helps to maintain structural integrity of epithelial cells, sustain mechanical stress to protect the cell from variation in hydrostatic pressure, (iii) Keratin filaments are involved in cell signaling, cell transport, cell compartmentalization and cell differentiation, (iv) These proteins also influence the cell metabolic process, (v) It involve in the transport of membrane bound vesicles of epithelial cell.

Keratins can be classified as α-keratin and β-keratin according to the presence of alpha-helix and beta-pleated sheet in their secondary structure. Alpha-keratins are generally found in hair, wool, quills, horns, hooves, nails, stratum corneum, whale baleen and hagfish slime etc. The alpha helix is the secondary structure of keratin protein in which every backbone N-H group donates a hydrogen bond to the backbone C = O group of the amino acid located four residues earlier along the protein sequence (hydrogen bonding). X-ray diffraction study shows that two right-handed α-helical chains form a left-handed dimer (45 nm long) by disulfide bonds. Then dimmers aggregate end-to-end and stagger side-by-side via disulfide linkages to form protofilament (2 nm in diameter). Two protofilaments laterally associate to form a superhelix known as protofibril; four protofibrils coalesce into a circular/helical structure called intermediate filament (IF) with a diameter of 7 nm. Then, the IFs bunches into a supercoiled conformation, and associate with the matrix proteins of keratin (Kornillowicz-Kowalska and Bohacz 2011; Chou and Buehler 2012; Wang et al. 2016). The molecular weight of alpha-keratin comprises 40–70 kDa.

Table 5.1 Sources of keratin

Structure containing keratin	Source
Hoof, horn, fur, wool, skin	Cow, sheep, goat, pig, horse, tapir and rhinoceros
Hair, nail, skin	Primate
Stratum corneum	Human
Feather, claw and beak	Bird
Osteoderm	Turtle, tortoise, crocodile and alligator

The beta keratins are found in feathers, claws and beak of birds, and scales and claws of reptiles. They have right handed beta helical structure that consists of beta strands which are a stretch of polypeptide chain (3–10 amino acids). One polypeptide chain folds to form four β-strands (parallel) which twist to form the distorted β-sheet; two of them assemble and run in opposite directions to form a beta-keratin intermediate filament. The diameter of the filament is 4 nm and pitch length 9.5 nm (Wang et al. 2016). The structure of β-pleated sheet is stabilized mainly by hydrogen bonds and peptide bonds. The hydrogen bonds between the beta strands help to form a sheet like structure and the planarity forces a β-sheet to be pleated. The molecular weight of beta-keratin comprises 10–22 kDa.

According to sulfur content, keratins are grouped into hard and soft keratin. Hard keratins are strong and less flexible due to presence of high amount of disulfide bond and found in feathers, hairs, hoofs and nails. Whereas, soft keratins are more flexible due to low content of disulfide bonds and present in skin and callus (Schrooyen et al. 2001; Tork et al. 2010). Intermediate filaments of hard keratins is embedded in a matrix of cystine rich proteins and give toughness of epidermal appendages, whereas soft keratins preferentially form loosely-packed bundles of IFs (Coulombe et al. 2000) with less sulphur content (Zoccola et al. 2009). The soft keratin contains about 2% cystine whereas the hard keratin has about ~ 22% (Kornillowicz-Kowalska and Bohacz 2011).

Mammalian epithelial Keratins are also classified into Type I (acidic) and Type II (basic to neutral) keratin. Type I keratins (Type I cytokeratins) are cytokeratins that constitute the Type I intermediate filaments (IFs) which is present in all mammalian epithelial cells. Their molecular weight ranges from 40–64 kDa. Type II keratins (Type II cytokeratins) constitutes the Type II intermediate filaments (IFs) of the intracytoplasmatic cytoskeleton, which is present in all mammalian epithelial cells. Their molecular weight ranges from 52–67 kDa.

5.3 Source of Microbial Keratinases

Keratinases produced by microorganisms like bacteria, fungi and actinomycetes are frequently isolated from keratin-rich environments.The sources of some remarkable keratinase producers are listed in Table 5.2.

In bacteria, keratinase production and keratin degradation is mostly confined to Gram-positives, including *Bacillus*, *Lysobacter*, *Nesternokia*, *Kocurica* and *Microbacterium*. Among them, *Bacillus* sp. appears as the major keratinase producing organisms. Different strains of *Bacillus licheniformis* and *Bacillus subtilis* are predominantly found as keratinolytic (Lin et al. 1999; Cai et al. 2008a, b; Rai et al. 2009; Tork et al. 2013), but other species such as *Bacillus pumilus*, *B. cereus*, *B. weihenstephanensis*, *B. thuringiensis* also produce keratinases (Kim et al. 2001; Werlang and Brandelli 2005; Kumar et al. 2008; Sahoo et al. 2012; Sivakumar et al. 2013a). Some strains of Gram-negative bacteria, viz. *Vibrio*, *Xanthomonas*, *Stenotrophomonas*, *Chryseobacterium*, *Fervidobacterium*, *Thermoanaerobacter*,

Table 5.2 Source of keratinolytic microorganisms

Source	Microorganism	References
Soil	*Bacillus* sp. FK 28	Pissuwan and Suntornsuk (2001)
	Bacillus sp. FK 46	Suntornsuk and Suntornsuk (2003)
	Bacillus sp. H62	Kazzaz et al. (2015)
	Kocuria rosea LBP-3	Bernal et al. (2003)
	Bacillus licheniformis, Bacillus subtilis ATCC 6633	Zerdani et al. (2004)
	Bacillus licheniformis FK14	Suntornsuk et al. (2005)
	Bacillus cereus DCUW	Ghosh et al. (2008)
	Brevibacillus brevis US575	Jaouadi et al. (2013)
Soil containing deer fur	*Stenotrophomonas* sp. D1	Yamamura et al. (2002)
Soil from feather dumping site	*Bacillus weihenstephanensis* PKD5	Sahoo et al. (2012)
Alkaline mud and soil samples	*Bacillus pseudofirmis* AL-89	Gassessse et al. (2003)
Soil from limestone quarry	*Bacillus* sp. MBRL 575	Kshetri and Ningthoujam (2016)
Antarctic soil	*Streptomyces flavis* 2BG	Gousterova et al. (2005)
Poultry waste	*Bacillus subtilis* KS1	Kim et al. (2001)
	Xanthomonas maltophila POA-1	De Toni et al. (2002)
	Pseudomonas aeruginosa	Raju and Divakar (2013)
	Klebsiella sp. BTSUK	Gurav et al. (2016)
Poultry waste digester	*Bacillus licheniformis* PWD1	Williams et al. (1990)
Industrial poultry waste	*Chryseobacterium* sp. kr6	Riffel et al. (2003)
	Microbacterium arborescens kr10	Thys et al. (2004)
Slaughter house waste sample	*Streptomyces* sp. S7	Tatineni et al. (2008)
Bovine hair, skin wastes, oil sample	*Bacillus subtilis* S 14	Macedo et al. (2005)
Rotted feather	*Bacillus licheniformis* K-508	Manczinger et al. (2003)
Hot spring	*Fervidobacterium pennavorans*	Friedrich and Antranikian (1996)
Geothermal hot spring	*Thermoanaerobacter keratinophilus* sp. nov.	Riessen and Antranikian (2001)
	Fervidobacterium islandicum AW-1	Nam et al. (2002)
Hydrothermal hot spring	*Caldicoprobacter algeriensis* strain TH7C1^{T}	Bouacem et al. (2016)
Compost	*Bacillus licheniformis* RG1	Ramnani and Gupta (2004)
Solfataric muds	*Clostridium sporogenes*	Ionata et al. (2008)
Hornmeal	*Bacillus subtilis* MTCC (9102)	Balaji et al. (2008)
Water	*Bacillus licheniformis RP1*	Haddar et al. (2011)

Psuedomonas and *Nesterenkonia* have also been reported as keratinolytic organism (Gupta and Ramnani 2006; Gopinath et al. 2015).

The most keratinolytic fungi belong to the following genera like *Chrysosporium, Microsporum, Aspergillus, Trichophyton, Penicillum, Scopulariopsis.* Among them some are dermatophytic (e.g., *Microsporum* species) and other are nondermatophytic (e.g., *Crysosporium* species), though dermatophytes have least commercial value (Gradisar et al. 2000). Besides these, actinomycetes from the *Streptomyces* group, viz. *Streptomyces pactum, S. albidoflavus* K1–02, *S. thermoviolaceus* SD8, *S. thermonitrificans, S. graminofaciens, S. flavis* 2BG and thermophilic *Streptomyces gulbarguensis* have been isolated from various sites (Syed et al. 2009; Brandelli et al. 2010). In addition to that, a new species of thermophilic keratinolytic anaerobic bacterium *Keratinibaculum paraultunense* Nov. KD-1 was isolated from grassy marshland (Huang et al. 2013).

5.4 Structural Characteristics of Keratinase

Keratinase from *Bacillus licheniformis* PWD-1 is a serine-type protease (Lin et al. 1995). Comparative analysis of nucleotide sequence of the serine protease genes from *B. licheniformis* PWD-1, Carlsberg NCIMB 6816, ATCC 12759, and NCIMB 10689 revealed that the *kerA*-encoded protease of PWD-1 differs from the others only by having V222 rather than A222 near the active site serine S220 (Evans et al. 2000). Keratinases from *B. licheniformis* MKU3 and MSK103, showed 99% similarity with both kerA and subtilisin Carlsberg (Radha and Gunasekaran 2007) and 87% with kerA (Yoshioka et al. 2007) respectively. Keratinase KerRP from *Bacillus licheniformis* RPk differs from KerA of *B. licheniformis* PWD-1, subtilisin Carlsberg, and keratinase of *B. licheniformis* by 2, 4 and 62 amino acids, respectively (Fakhfakh et al. 2009; Brandelli et al. 2010). The cloned gene *aprA* from *B. subtilis* showed significant similarity and homology with subtilisins (Zaghloul 2001). Purified keratinase from *B. subtilis* KS-1 showed an N-terminal sequence which is similar to other serine proteases of *B. subtilis* (Suh and Lee 2001). Keratinase from *B. subtilis* S14, shows an identical N-terminal sequence to subtilisin E (Macedo et al. 2005). The 1.7A° resolution crystal structure (Kim et al. 2004) revealed that the configuration and N-terminal sequence of a keratinase named fervidolysin, from *Fervidobacterium pennivorans* showed high homology with the subtilisin-like proteases (Kluskens et al. 2002). Jayalakshmi et al. (2011) was predicted the binding pockets for keratinase of *Bacillus licheniformis*. Three pockets were predicted and the ligand was docked with GLY residue of the protein in the second predicted binding pocket.

The *In silico* study of bacterial keratisaes by Banerjee et al. (2014) revealed that, among the fifteen keratinase sequences ten belongs to genus *Bacillus* with high level of sequence homology. Their amino acid number and molecular weight ranges from 349 to 381 and 35715.9 to 39560.5 dalton, respectively. The superfamily search indicate that all the ten keratinase producing *Bacillus* genus have two domain;

the large parts of those sequences were related to Subtilisin-like superfamily, subtilases family and the small parts were related to protease propeptides/inhibitors superfamily, subtilase propeptides/inhibitors family. The hydrophobicity and surface accessibility analysis of the model structure of *Bacillus* genus, revealed the extracellular nature of keratinase.

5.5 Mechanism of Keratinolysis

Over the years, different hypotheses have been proposed in relation to the mechanism of keratin degradation by microbial keratinases. Kunert (1976) proposed that the keratinolysis occurs in two steps, one is sulfitolysis i.e. reduction in the disulfide bonds and another is proteolysis i.e. breakdown of peptide bonds. In sulfitolysis, the inorganic sulphite produced by keratinolytic organisms cleaved the disulphide bonds that cause denaturation of keratin structure.

In prokaryotes, the enzyme disulphide reductase reduced the disulphide bonds of keratin to produce denature keratin that decomposed by keratinolytic protease to produce peptides and amino acids. The breakdown of disulfide bonds causes the conformational changes of keratins and thereby exposes more sites for keratinase action (Vignardet et al. 2001). The growth of keratinophilic microorganisms on keratinous substrates, release the thiol groups which supports the reduction of disulfide bonds that helps for efficient keratin degradation (Daroit et al. 2009). So, it can be say that sulfitolysis and proteolysis occurs synergistically to achieve complete degradation of keratin by keratinolytic microorganisms, its enzymes and metabolites (Brandelli et al. 2010).

In eukaryotes, the keratin degradation was reported through ubiquitin-proteasome pathway (Rogel et al. 2010; Sahni et al. 2015). It was also reported that the process of native keratin degradation by fungi is an enzymatic lysis along with mechanical destruction by eroding mycelium complex (Kornillowicz-Kowalska and Bohacz 2011). Recently Lange et al. (2016) hypothesized Lytic Polysaccharide Mono-Oxygenases (LPMOs) mediated degradation process in keratin-degrading fungi. They deduced a model about the degradation of keratin by LPMOs in keratin-degrading fungi; LPMOs break the glycosylation bond leading to change of steric formation and charge that further lead to disassembly of the keratin filaments. The de-assembled keratins are degraded into smaller peptides and amino acids.

Purified enzyme alone is not able to decompose fully the recalcitrant keratin structure (Ramnani and Gupta 2007; Inada and Watanabe 2013). It was reported that the degradation of keratin by purified keratinase of *Bacillus* sp. MTS was enhanced by the purified disulfide reductase, compared to activity of individual enzyme (Rahayu et al. 2012). Fang et al. (2013) isolated the three kertinolytic enzymes (serine protease, serine-metalloprotease and disulfide reductase) from *Stenotrophomonas maltophilia* BBE11–1 and it was found that any of them could not showed keratinolytic activity independently. Purified keratinases are generally less effective on native keratin, probably due to the removal of disulphide bond reduction components during the purification process (Nam et al. 2002; Brandelli et al. 2010).

5.6 Production of Keratinase

Production of keratinases is an important step to fulfill the biotechnological applications in industrial level. The major investigations reported that keratinase production occur through submerged cultivation process though solid-state processes has also been demonstrated. The potential of keratinase production by immobilized microorganisms was also reported.

5.6.1 Submerged Fermentation (SmF) for Keratinase Production

Submerged fermentation involves the production of biomolecules (enzymes), by microorganisms in a vessel with a broth of nutrients. When the fermentation is over, the contents are emptied for downstream processes. The major investigations revealed that the keratinase production has been achieved through submerged cultivation processes (Table 5.3). In some of the instance, thermostable keratinase was also produced through submerged fermentation from thermophilic bacteria of *Caldicoprobacter algeriensis* isolated from hot spring and the maximum keratinase activity recorded after 24-h of incubation at 50 °C as 21,000 U/ml (Bouacem et al. 2016).

5.6.2 Solid State Fermentation (SSF) for Keratinase Production

From past few years, the production of keratinolytic enzymes through solid-state processes has also been demonstrated (Hölker and Lenz 2005; Rai et al. 2009; El-Gendy 2010; Sahoo et al. 2015) (Table 5.3). Use of solid-state fermentation (SSF) has gained importance because of its unique advantages over SmF, like low production cost, saving of water, requirement of minimal energy and stability of the product (Hölker and Lenz 2005). Gioppo et al. (2009) reported that the fungus *Myrothecium verrucaria* able to grow and produce keratinase in submerged (93.0 ± 19 U/ml) and solid state (98.8 ± 7.9 U/ml) fermentation in which poultry feather powder (PFP) used as substrate. The highest levels of keratinase activity was found as 168.0 ± 28 U/ml in submerged and 189.0 ± 26 U/ml in solid state culture, when the PFP cultures supplemented with cassava bagasse.

Table 5.3 Environmental parameters for the production of keratinase by microorganisms through SmF

Microorganisms	Production of keratinase				Degradation time	References
Bacteria	Substrate	pH	Temp. (°C)	Optimum keratinase production		
Bacillus licheniformis PWD1	Chicken feather	7.5	50	30 h	10 days	Williams et al. (1990)
Fervidobacterium pennavorans	Chicken feather	6.3	70	–	48 h	Friedrich and Antranikian (1996)
Kocuria rosea LBP-3	Chicken feather	7.5	40	36 h	96 h (55%)	Bernal et al. (2003)
Bacillus subtilis KS1	Chicken feather	5–9	40	84 h	72 h	Kim et al. (2001) and Suh and Lee (2001)
Bacillus sp. FK 28	Chicken feather and feather meal	7.5	37	3 days	–	Pissuwan and Suntornsuk (2001)
[a]*Thermoanaerobacter keratinophilus* sp. nov.	Chicken feather and wool	6.8	70	96 h	10 days (70%)	Riessen and Antranikian (2001)
Bacillus licheniformis K-508	Chicken feather	7	45	–	4 days	Manczinger et al. (2003)
Xanthomonas maltophila POA-1	Feather meal	7	30	72 h	–	De Toni et al. (2002)
[a]*Fervidobacterium islandicum* AW-1	Chicken feather	7	70	48 h	48 h	Nam et al. (2002)
Stenotrophomonas sp. D1	Keratin powder, chickenfeather	7	20	4 days	2.5 days	Yamamura et al. (2002)
Bacillus pseudofirmis AL-89 and *Nesternokia* sp. AL-20	Chicken feather	–	37	–	48–60 h	Gassessse et al. (2003)
Chryseobacterium sp. kr6	Chicken feather	8	25–30	48 h	72 h	Riffel et al. (2003)
Bacillus sp. FK 46	Chicken feather	9	37	5 days	5 days (85%)	Suntornsuk and Suntornsuk (2003)

(continued)

Table 5.3 (continued)

Microorganisms	Production of keratinase				Degradation time	References
Bacillus licheniformis RG1	Chicken feather	7	37	72 h	24 h	Ramnani and Gupta (2004)
Microbacterium arborescens kr 10	Chicken feather	7	30	36 h	72 h	Thys et al. (2004)
Bacillus subtilis S 14	Bovine hair	7.5	–	–	–	Macedo et al. (2005)
Bacillus licheniformis FK14	Feather meal	7.5	50	3 days	5 days	Suntornsuk et al. (2005)
Bacillus cereus DCUW	Feather	7.5	30–37	–	–	Ghosh et al. (2008)
Clostridium sporogenes	Chicken feathers	7.0	42	–	7 days	Ionata et al. (2008)
Bacillus subtilis MTCC (9102)	Hornmeal	7.0	37	120 h	–	Balaji et al. (2008)
B. licheniformis H62	Feather-meal	7.0	40	45 h	–	Kazzaz et al. (2015)
Bacillus sp. khayat (immobilised)	Chicken feathers	8.0	27	18 h	18 h	Yusuf et al. (2015)
Fungi						
Aspergillus flavus K-03	Chicken feather	8.0	28	16 days	–	Kim (2007)
Aspergillus terreus	Feather powder	7.0–8.0	40	25 days	–	Koutb et al. (2012)
Aspergillus flavus S125	Keratin substrate	9.0	55	6 days	–	Mini et al. (2015)
Onygena corvina	Duck feathers	8.0	25	–	–	Huang et al. (2015)

[a]*SSF* Solid state fermentation

5.6.3 Production of Keratinase from Immobilized Microorganisms

The immobilized microorganisms have also been employed for keratinase production and keratin degradation (Table 5.4) (Devi and Lakshmi 2015; Saieb et al. 2016). Whole-cell immobilization was successful in sodium alginate beads for continuous production of keratinase and feather degradation by *B. halodurans* PPKS-2 (Prakash et al. 2010). Chitturi and Lakshmi (2016) reported that sodium alginate immobilized *Bacilli* sp. (MBF11, MBF20 and MBF45) were efficient in the production of keratinase bringing 100% degradation of feather in 4–5 days as free cells. Further, cells could be recycled up to 3 batches over 21 days in sodium alginate beads making the process economical and feasible.

5.6.4 Keratinase Production Using Recombinant Strains

Recombinant DNA technology has been employed in isolation and cloning of keratinase genes for improvement of enzyme production and to allow its commercialization. Different recombinant microbial strains were developed to enhance keratinase production and keratin degradation (Brandelli et al. 2010; Yang et al. 2016).

The *kerA* gene which encodes keratinase of *B. licheniformis*, was also cloned for extracellular expression in *P. pastoris*, resulting in a recombinant enzyme that was glycosylated and even though active on azokeratin (Porres et al. 2002). Increased enzyme stability has been achieved by recombinant keratinases when *B. licheniformis* MKU3 keratinase expressed in *Pichia pastoris* X33 (Radha and Gunasekaran 2009). The gene *kerA* (1047 bp) encoding the keratinase from *B. licheniformis* was cloned into two conventional vectors, pET30a and pET32a, and expressed in *Escherichia coli* that showed same optimum temperature of about 50 °C but different optimum pH of 7.5 and 8.5 (Hu et al. 2013). The *ker* gene from *Bacillus licheniformis* BBE11-1 was also expressed in *Escherichia coli*, *Bacillus subtilis*, *and Pichia pastoris* to increase the production of keratinase (Liu et al. 2014). A keratinolytic gene from thermophilic bacterium like *Geobacillus stearothermophilus* AD-11 was cloned and expressed in *Escherichia coli* BL21

Table 5.4 Keratinase production by immobilized microorganisms

Immobilized organism	Material for immobilization	References
Bacillus sp. PPKS-2	Sodium alginate beads	Prakash et al. (2010)
Bacillus sp. MBF11, MBF20, MBF 21 and MBF45	Chotosan matrix	Devi and Lakshmi (2015)
Bacillus licheniformis (St. 24)	2% Ca-alginate	Saieb et al. (2016)
Bacillus sp. MBF11, MBF20 and MBF45	Sodium alginate, polyacrylamide, agar-agar and gelatin matrices	Chitturi and Lakshmi (2016)

(DE3) and it was found that recombinant keratinase (RecGEOker) showed optimal activity at pH 9 and 60 °C (Gegeckas et al. 2015). A keratinolytic protease gene (*kerD*, 1251 bp) from *Aspergillus niger* was cloned and expressed in *Escherichia coli* (Chen et al. 2015). The recombinant enzyme has optimum pH and temperature 8.0 and 70 °C respectively and hydrolyzed a broad range of substrates. The recombinant plasmids harboring *kerK* were extracted from *B. subtilis* SCK6 and transformed into *B. amyloliquefaciens* K11. The recombinant *B. amyloliquefaciens* K11 exhibited enhanced feather-degrading capacity with shortened reaction time (12 h) with increased keratinolytic activity (1500 U/mL) by 6-fold (Yang et al. 2016).

5.6.5 Statistical Approach for Higher Keratinase Production

Optimization of the selected physical and chemical components/factors can be successfully brought about the involvement of statistical approaches like Plackett-Burman design (PBD) and response surface methodology (RSM) for significant amount of keratinase production (Bernal et al. 2006; Rai and Mukherjee 2011; Sahoo et al. 2015). On the other hand, the concentration of optimal factors can be deduced using central composite design (CCD), followed by analysis using the RSM (Daroit et al. 2011; Bach et al. 2012).

5.6.6 Factors Effecting Keratinase Production

Composition of fermentation medium and cultural conditions is the two important factors that affect on the enzyme production in a fermentation process. Different chemical and physical factors effect on keratinase production by various organisms are listed in Table 5.3.

5.6.6.1 Substrate for Keratinase Production

Biotechnological applicability of keratinase demands higher production of keratinase. The enzyme production is usually induced by keratin substrates as sole source of carbon and nitrogen (Williams et al. 1990; Gousterova et al. 2005; Anbu et al. 2008), though non keratinous substrates (soybean meal, casein, skim milk, soy flour, gelatine) also act as inducers of keratinase production (Brandelli et al. 2010).

5.6.6.2 Effect of Carbon Supplement on Enzyme Production

All organisms require energy and the carbon source for their growth. In any fermentation the amount of carbon compounds utilized by microorganism depends on type, source and nature of carbon. The degree of carbon utilization by bacteria varies from organism to organism; hence study on different carbon sources is essential. The addition of glucose (Ramnani and Gupta 2004; Son et al. 2008), sucrose (Cai and Zheng 2009), starch (Syed et al. 2009), molasses (Cheng et al. 1995), and bagasses (Gioppo et al. 2009) enhance the enzyme production. On the other hand, some reports are also present as enzyme production was suppressed by carbon source like glucose (Sahoo et al. 2012), sucrose or lactose through catabolic repression (Brandelli et al. 2010).

5.6.6.3 Nitrogen Supplementation for Enzyme Production

The nitrogen sources have most profound influences on enzyme production by keratinolytic microorganisms next to carbon. A wide range of inorganic and organic nitrogen compounds can be utilized for structural and functional purpose of the cell. Supplementary nitrogen sources, like peptone, tryptone, urea, yeast extract, chloride, ammonium sulphate, sodium nitrate have been reported to enhance enzyme production (Brandelli et al. 2010; Sahoo et al. 2012). Therefore the supplementation of nitrogen for enzyme production is variable and it depends on the concentration, medium composition and type of microorganisms (Cai and Zheng 2009).

5.6.6.4 Fermentation Time

Fermentation time is one of the most important factors for keratinase production. It was reported that, incubation period varied from hours to days depending on type of species. Gurav and Jadhav (2013) reported the keratinase production by *Chryseobacterium* sp. RBT for 6–48 h, where 30 h was found as optimum incubation period. The keratinase production by *B. cereus* TS1 for 24–120 h incubation period was found as optimum at 72 h (Sivakumar et al. 2013b). The incubation period 2 days was found for optimum keratinase production by *Stenotrophomonas maltophila* R-13 (Jeong et al. 2010) and *B. cereus* LAU08 (Lateef et al. 2010). The enzyme production by *B. pumilus* A1 (Fakhfakh et al. 2011), *Streptomyces exfoliatus* CFS1068 (Jain et al. 2012) and *B. weihenstephaensis* PKD5 (Sahoo et al. 2012) have the optimum incubation period of 4, 6 and 7 days respectively. Anbu et al. (2008) reported keratinase production by a fungus, *Trichophyton* sp. HA-2 for 1–7 weeks fermentation, where 5 weeks was found as optimum for enzyme production. Gioppo et al. (2009) reported keratinase production by another fungal strain, *Myrothecium verrucaria* for 0–9 days incubation period, where 6 days found as optimum.

5.6.6.5 Fermentation pH

The pH has a strong influence on the microbial metabolic pathways and products generation. Each microorganism has an optimum pH for its growth, activity and high product yields. The pH optima also depend on factors like temperature, type and concentration of substrate, chemical properties of the medium, type and source of the enzyme and the purity of the enzyme (Pandey et al. 2001). It has been observed that a pH range from 6 to 9 supports keratinase production and feather degradation in most microorganisms. Though alkaline pH possibly favors keratin degradation as it modifies cystine residues to lathionine (Friedrich and Antranikian 1996), making it accessible for keratinase action.

5.6.6.6 Temperature

Temperature is a very critical parameter in determining keratinase production (Gupta and Ramnani 2006). Temperature for keratinase production ranges from 28 °C to 50 °C for most keratinophilic microorganisms. Thermophillic organisms like *Thermoanaerobacter* and *Fervidobacterium* spp. has high production temperature of about 70 °C (Friedrich and Antranikian 1996; Riessen and Antranikian 2001; Nam et al. 2002). Production of keratinase at 20 °C (Psychrotrophic) has also been reported for *Stenotrophomonas* sp. D1 (Yamamura et al. 2002).

5.6.6.7 Aeration/Agitation Rate on Enzyme Production

The majorities of fermentation process is aerobic and require oxygen. The oxygen demand of the fermentation process (both in SmF and SSF) is satisfied by aeration and agitation. Failure to supply adequate oxygen may lead to adverse changes in enzymatic makeup (Rolinson 1952) or death of the organisms (Hromatka et al. 1951) and less product yield. Laboratory-scale cultures may be aerated by shaking the conical flask (with culture) on a platform contained in a controlled environmental chamber. Aeration has been found between 100–150 rpm for keratinase production by *Bacillus* sp. FK28 (Pissuwan and Suntornsuk 2001), *B. cereus* LAU08 (Lateef et al. 2010), *Streptomyces exfoliatus* CFS1068 (Jain et al. 2012), *B. weihenstephaensis* PKD5 (Sahoo et al. 2012), *B. cereus* TS1 (Sivakumar et al. 2013b). Aeration at 180 rpm for keratinase production by *Vibrio* sp. Kr2 and *Chryseobacterium* sp. kr6 respectively has been reported (Riffel et al. 2003; Sangali and Brandelli 2000). Jeong et al. (2010) have maintained aeration at 200 rpm for keratiase production by *B. megaterium* F7–1 and *B. pumilus* FH-9 respectively. Whereas Fakhfakh et al. (2011) has kept aeration at 250 rpm while studying keratinase production by *B. pumilus* A1.

5.6.6.8 Inoculum Size

An inoculum is the population of microorganisms or cells that is introduced in the fermentation medium. Inoculum size is an important factor affecting cell growth and product formation. Several researchers used different inoculum size from 1–5% (v/v) for keratinase production (Table 5.5) (El-Refai et al. 2005; Park and Son 2009; Ningthoujam et al. 2016). High inoculum size of 10% (v/v) was optimal for keratinase production by *Bacillus subtilis* KD-N2 with hair substrate (Cai and Zheng 2009). The inoculum size of 50% (optical density at 600 nm 0.59 ± 0.05) was found to be suitable for optimum keratinase production by *B. subtilis* MTCC9102 strain using horn meal as a substrate in solid-state fermentation (Kumar et al. 2010).

5.7 Applications of Keratinase

In modern decade, microbial keratinases have attracted enormous attention due to their huge biotechnological applications in feed formulation, fertilizer, detergent, leather and textile industries, and also for pharmaceutical and biomedical sectors.

5.7.1 Feather Meal

Feather meal is produced conventionally by hydrolyzing poultry feathers with high temperature and pressure, and then grinding and drying. It is generally used in formulation of animal feed and organic fertilizer. Protein quality inconsistency is an important factor regarding its use in livestock food. Variable nutrient composition and nutrient bioavailability was found in feather meal (Grazziotin et al. 2008). This conventional processing destroys different amino acids like methionine, lysine and tryptophan with poor digestibility (Wang and Parson 1997). In this concern, the use

Table 5.5 Various inoculum size used for keratinase production

Inoculum size	Organism	Reference
1%	*Bacillus* sp. FK28	Pissuwan and Suntornsuk (2001)
	B. pumilus FH-9	El-Refai et al. (2005)
2%	*B. megaterium* F7–1	Park and Son (2009)
	B. weihenstephaensis PKD5	Sahoo et al. (2012)
4%	*B. cereus* TS1	Sivakumar et al. (2013a, b)
5%	*Stenotrophomonas maltophila* R-13	Jeong et al. (2010)
	B. cereus LAU08	Lateef et al. (2010)
	Pseudomonas aeruginosa	Raju and Divakar (2013)
	Amycolatopsis sp. strain MBRL 40	Ningthoujam et al. (2016)

of microbial keratinases could overcome this disadvantage by hydrolyzing feather into nutritionally wealthy animal feed (Onifade et al. 1998). The keratinase from *B. licheniformis* PWD-1 named as *versazyme* (trade name) supplemented with diets improved the growth of broiler chickens (Odetallah et al. 2005). Feather meal produced by keratinase from *Bacillus licheniformis* ER-15 contained 14% nitrogen, 44% carbon with all essential amino acids and showed 73% in-vitro digestibility (Tiwary and Gupta 2012). Additional utilization of keratinase supplemented diets has led to reduction in feed requirement (Wang et al. 2011). The nutritive value of feather meal was evaluated by use of keratinases (K6 and K82) from *B. licheniformis* KUB-K0006 and *B. pumilus* KUB-K0082 respectively and it was found that both the keratinase had a positive effect on the protein quality of the feather meal (Eaksuree et al. 2016).

5.7.2 Organic Fertilizer

High concentration of nitrogen and amino acids in keratin hydrolysates facilitates their use as bio-fertilizers and in foliar fertilization (Gupta and Ramnani 2006). The hydrolysate resulting from the bioconversion of keratinous wastes could also be useful for preparation of nitrogen fertilizers or soil amendments (Gousterova et al. 2005; Vasileva-Tonkova et al. 2009). In this concern, feather digests showed comparable results to a reference fertilizer in a 27-day plant growth assay using carrots and Chinese cabbage (Kim et al. 2005). Keratin-based organic fertilizer produced by *Aspergillus niger* used at the rate of 3 kg/10 kg of soil played a significant role in growth of organism and yield, and least susceptibility to diseases of cowpea-*Vigna unguiculata* (Adetunji et al. 2012). Paul et al. (2013) reported the role of feather-hydrolysate in enhancing the number of root hair. Chicken feathers along with keratinolytic strain of *Bacillus subtilis* PF1 in soil promote the growth of *Vigna radiata* plant (Bhange et al. 2016).

5.7.3 Production of Flim, Coating, Glue

The partially hydrolyzed keratin waste by the keratinolytic microorganisms or their keratinase enzyme utilized for the production of biodegradable films, coatings and glues (Gupta and Ramnani 2006; Brandelli et al. 2010).

5.7.4 *Production of Biofuel, Metahane, Bio Hydrogen*

Keratin hydrolysates can serve as nitrogen rich source that could be employed in production of methane gas and fuel pellets (Ichida et al. 2001), and also in biohydrogen generation with the help of *B licheniformis* KK1 and *Thermococcus litoralis* (Bálint et al. 2005; Sahni et al. 2015).

5.7.5 *Use in Cosmetics*

Keratin hydrolysates are used in hair care products including shampoo, coloring spray, toner, etc. to provide strength and protection to hair fibers against chemical and environmental damage and also improve the texture (Masato et al. 2011; Villa et al. 2013). The hydrolysates have been applied on the surface of heels, knees or elbows as finishing agents to get smooth surface through preventing damage to the skin (Shigeru 2012). Keratin hydrolysates also protect nail enamels against damage by chemicals.

5.7.6 *Mosquitocidal Toxin*

The keratin-rich substrate was employed for cultivation of entomopathogenic bacteria and production of mosquitocidal toxins (Poopathi and Abidha 2008; Brandelli et al. 2010).

5.7.7 *In Leather Industry*

Leather processing involves a series of steps like pre-tainning, taininng and post tanning. For dehairing of skin/hide in the pre taining process, sodium sulphide and lime are conventionally used which creates immense pollution and wastage of water. Approximately 35–40 liter of water is used per kg of hide processed (Sahni et al. 2015). In leather industry, keratinases from most of the *Bacillus* sp. exhibits significant dehairing efficiency without the degradation of collagen layer of skins due to selective breakdown of keratin tissue at the base of follicle; and pulling out intact hair without affecting the tensile strength of leather (Macedo et al. 2005). Keratinases from *B. subtilis* KD-N2 (Cai et al. 2008a, b), *Bacillus* sp. PPKS-2 (Prakash et al. 2010), *Trichoderma harzianum* MH-20 (Ismail et al. 2012), *Paenibacillus woosongensis* TKB2 (Paul et al. 2014) exhibits remarkable dehairing capabilities without the degradation of collagen. Keratinolytic enzyme from *B. weihenstephanensis* PKD5 alone could dehaired the cow, goat and sheep skin within 10 h and produce better leather quality (Sahoo et al. 2016).

5.7.8 Formulation of Detergents

Proteolytic enzymes have dominated the detergent market; especially alkaline protease which is captured 89% approximately. Keratinase have the property of detergent enzymes as they act on protein substrates like feathers; making a smart additive for hard-surface cleaning. They hydrolyzed the fixed proteins on the surface, and can remove stains including keratinous soiled collars, cuffs, as well as blood stains (Gupta and Ramnani 2006). Keratinases also used as additives for cleaning up of drains clogged with keratinous wastes (Farag and Hassan 2004). Keratinases from different organism like *B. pumilus*, *B. thuringiensis*, *Paecilomyces lilacinus* and *Paenibacillus woosongensis* TKB2 have been explored as efficient detergent agent.

5.7.9 Textile Industry

Conventionally, organic chlorides were employed in wool processing to control the felting shrinkage of wool fiber. Use of protease enzymes (Novozymes form *B. licheniformis*) is regarded as an ecologically safe alternative compared to use of organic chlorides. But the proteases penetrate deep inside the wool fiber and damage it, thereby lead to loss of weight and tensile strength of the fiber (Shen et al. 2007). In this concern, use of keratinase enzymes would better as they selectively target the keratinous layer of wool without damaging other parts of the fiber. Keratinase enzyme from different sources like *Stenotrophomonas maltophilia* DHHJ (Cai et al. 2008a, b), *B. thuringiensis* L11 (Infante et al. 2010), *B. licheniformis* (Liu et al. 2013), have been demonstrated to improve felt-shrink resistance and dyeing property without the loss of fiber weight.

5.7.10 Milk Clotting Agent

Milk clotting activity was seen by crude keratinase from *B. weihenstephanensis* PKD5 and also in the presence of $CaCl_2$ and $MnSO_4$, in single or in combination (Sahoo et al. 2015). The highest clotting activity (43.6 SU/ml) was achieved in presence of $CaCl_2$ and $MnSO_4$ compare to crude enzyme alone (30.7 SU/ml). This milk clotting property may helpful for cheese making in dairy industry as substitute of calf rennet. Moreover, the crude keratinase will be ideal milk-coagulant due to unlimited production through fermentation, cost effectiveness and acceptable by lacto-vegetarians.

5.7.11 Waste Water Treatment/Waste Management

Keratinolytic microorganisms and their enzymes could be used in wastewater treatment; in removing clogs in bathroom drainpipes; and also to clear obstructions in sewage systems (Takami et al. 1992; Tapia and Simões 2008).

5.7.12 Pathogenic Prion (PrP) Protein Degradation

Prions are infectous proteinaceous particles that are responsible for different diseases like mad cow disease, scrapie, kuru (Gupta and Ramnani 2006). It has two forms- one is harmless PrP^{C} and another is infectous PrP^{Sc}. The infectous PrP^{Sc} is higly resistant to common proteolytic enzymes due to presence of more beta sheet in PrP^{Sc} (43%) than PrP^{C} (3%). Keratinase enzymes degraded the PrP^{Sc} infected brain tissues by KerA of *B. licheniformis* PWD-1 (Langveld et al. 2003). Keratinases from a diverse group of bacteria and actinomycetes have been shown to degrade prion proteins in laboratory experimental conditions (Gupta et al. 2012). Some of them work in high temperatue and pH (Mitsuiki et al. 2002; Gupta et al. 2013), other works without pre-treatments. It would also useful for decontamination of medical and lab instruments, contact lenses and dental tools (Langveld et al. 2003).

5.7.13 Treatment of Acne and Psoriasis

Acne is a common skin problem that occurs due to blocking at opening of the sebaceous gland by the presence of excessive keratin (Selvam and Vishnupriya 2012). Keratinases from *B. licheniformis* used for acne treatment as they dissolve dead cells and keratin that obstructs the sebaceous glands (Saha and Dhanasekaran 2010). Psoriasis is a condition that creates scaly appearance of skin. The U.S food and drug administration (FDA) has approved keratinase for treating the disease.

5.7.14 Transungual Drug Delivery

Nail disorders, like dystrophies, infections and deformities causes painful complications (Kumar and Raju 2013). Worldwide, nearly 700 million people suffer from a common fungal infection of the nail known as onychomycosis. They require a long time treatment and face a major problem of reoccurrence. Currently oral and topical medications available for treatment of nail disorders though they face several complications like pain, skin rashes, and liver damage to poor drug permeability across the nail plate (Rajendra et al. 2012). Various problems regarding these methods can be overcome by use of keratinases that cleave the keratin protein of nail plate and thereby loosing the plate and enhancing drug permeability (Gradisar et al. 2005; Mohorcic et al. 2007). Keratinase from *Paecilomyces marquandii* was shown to partially disrupt the nail plates and increase the drug permeability (Mohorcic et al. 2007). The keratinase KerN (subtilisin-γ-glutamyl transpeptidase) has also been reported to enhance the drug delivery through nails (Tiwary and Gupta 2010). Besides, a few keratinase-based commercial products like Fixa Fungus TM, Kernail-Soft PB and Pure100 Keratinase, are available in the market for treating nail disorders. Keratinases was utilized for skin tissue permeabilization to enhance drug delivery by hyperkeratotic lesions. A porous sheet immobololized with keratinase was developed that loosens the skin and improves drug permeability (Susumu et al. 1998).

5.7.15 Dermatophytosis Therapy

Dermatophytosis is a clinical condition caused by fungal infection of the skin in humans, pets, sheep, goats and cattle. It is commonly known as "ringworm". The fungi that cause parasitic infection, collectively known as dermatophytes, feed on keratin that found in the outer layer of skin, hair and nails. The thermostable keratinase of *Candida albicans* AP3 (survived at 80 °C) have high preventive and therapeutic property against dermatophytes in the tropics (Selvam and Vishnupriya 2012).

5.7.16 Pearl Bleaching

During pearl formation, organic impurities like free cells of mother of pearl oysters, mucilage and necrotic part of mantle tissue pieces are trapped in the pearls and they need to be processed to enhance their jewel quality (Gupta et al. 2013). Hydrogen peroxide are used in mild concentration to bleach for lightening of the pearl nacre without damaging its quality since the pearl surface is contaminated with organic impurities like cells, tissues, and mucilage. Zhang et al. (2010) have explored the use of keratinases during pearl bleaching. Keratinases can be used during an initial treatment to remove keratin impurities from the outside of pearl followed by the conventional processing/bleaching methods.

5.7.17 Removal of Ear Wax

Earwax (cerumen) primarily consists of shed layers of skin, with 60% of keratin (Gupta et al. 2013). The high amount of keratin present in cerumen makes the keratinolytic protease as possible cerumenolytic agent. Cagle et al. (2001) have formulated a composition for removing human cerumen that includes bicarbonate along with combination of enzymes like lipase, amylase and protease/keratinase.

5.7.18 *Processing of Edible bird's Nest*

Nests of swiftlets are consumed by people as a medicinal food. Before consumption, various impurities like feather and plumage are cleaned by several processes like sieving, hand picking, and warm water treatment that are time consuming and less effective (Marcone 2005). The enzyme keratinases can be employed to degrade keratin protein present in feather and plumage and thereby improve the processing of nests for consumption (Gupta et al. 2013).

5.7.19 *Recovery of Silver from Photographic Film*

In X-ray films, metallic silver is impregnated in gelatin layer. After use, they are dumped near many hospitals and pathological laboratories. In this comcern, keratinolytic enzymes offer an eco-friendly and cost effective method of silver recovery from photographic films. Gelatin present in X-ray films used as substrate of keratinase, as a result metallic silver is released into solution (Han et al. 2012; Verma et al. 2017).

5.8 Conclusion

Keratinase is an important enzyme that has capability to degrade the recalcitrant keratinous materials. A number of microorganisms from different genera are capable of producing keratinases which display huge diversity in their biochemical and biophysical characteristics. In recent decade, the keratinase enzymes have paying attention due to their various applications in feed, detergent, pharmaceuticals and leather industry. The enzyme and their hydrolytic products help to produce bioenergy to meet fuel scarcity, degradation of infectious prion protein, enhanced drug delivery and formulation of personal care products. Keratinases also have other established applications like production of biotoxin with mosquito larvicidal activities, recovery of silver from photographic flims, clear obstructions in sewage systems. Efforts have been made to isolate new potent keratinolytic microorganisms capable of higher enzyme production. The ability of microorganisms to grow and produce substantial levels of keratinase using keratinous substrate could open up new opportunities for the valorization of these wastes and reduction of harmful effects on the environment. Besides, manipulation of culture conditions along with application of genetic engineering technology are being employed that may also help to extend enzyme production for various applications of this group of enzyme.

References

Adetunji CO, Makanjuola OR, Arowora KA, Afolayan SS, Adetunji JB (2012) Production and application of keratin-based organic fertilizer from microbially hydrolyzed feathers to cowpea (*Vigna unguiculata*). Int J Sci Eng Res 3(12):22–29

Anbu P, Hilda A, Sur HW, Hur BK, Jayanthi S (2008) Extracellular keratinase from *Trichophyton* sp. HA-2 isolated from feather dumping soil. Int Biodeterior Biodegrad 62(3):287–292

Bach E, Sant'Anna V, Daroit DJ, Correa APF, Segalin J, Brandelli A (2012) Production one-step purification and characterization of a keratinolytic protease from *Serratia marcescens* P3. Process Biochem 47(12):2455–2462

Balaji S, Kumar MS, Karthikeyan R, Kumar R, Kirubanandan S, Sridhar R, Sehgal PK (2008) Purification and characterization of an extracellular keratinase from a hornmeal-degrading *Bacillus subtilis* MTCC (9102). World J Microbiol Biotechnol 24(11):2741–2745

Bálint B, Bagi Z, Tóth A, Rákhely G, Perei K, Kovács KL (2005) Utilization of keratin-containing biowaste to produce biohydrogen. Appl Microbiol Biotechnol 69(4):404–410

Banerjee A, Sahoo DK, Thatoi H, Pati BR, Mondal KC, Sen A, Mohapatra PKD (2014) Structural characterization and active site prediction of bacterial keratinase through molecular docking. J Bioinform 1(4):67–82

Bernal C, Cairó J, Coello N (2006) Purification and characterization of a novel exocellular keratinase from *Kocuria rosea*. Enzym Microb Technol 38:49–54

Bernal C, Vidal L, Valdivieso E, Coello N (2003) Keratinolytic activity of *Kocuria rosea*. World J Microbiol Biotechnol 19(3):255–261

Bhange K, Chaturvedi V, Bhatt R (2016) Ameliorating effects of chicken feathers in plant growth promotion activity by a keratinolytic strain of *Bacillus subtilis* PF1. Bioresour Bioprocess 3(1):1–10

Bouacem K, Bouanane-Darenfed A, Jaouadi NZ, Joseph M, Hacene H, Ollivier B, Fardeau ML, Bejar S, Jaouadi B (2016) Novel serine keratinase from *Caldicoprobacter algeriensis* exhibiting outstanding hide dehairing abilities. Int J Biol Macromol 86: 321–328

Bradbury JH (1973) The structure and chemistry of keratin fibers. Adv Protein Chem 27:111–211

Bragulla HH, Homberger DG (2009) Structure and functions of keratin proteins in simple, stratified, keratinized and cornified epithelia. J Anat 214(4):516–559

Brandelli A (2008) Bacterial keratinases: useful enzymes for bio-processing agroindustrial wastes and beyond. Food Bioprocess Tech 1(2):105–116

Brandelli A, Daroit DJ, Riffel A (2010) Biochemical features of microbial keratinases and their production and applications. Appl Microbiol Biotechnol 85(6):1735–1750

Cagle GD, Owen GR, Ridruejo NJ, Wall GM (2001) Compositions for removing human cerumen Patent: EP1337228

Cai C, Zheng X (2009) Medium optimization for keratinase production in hair substrate by a new *Bacillus subtilis* KD-N2 using response surface methodology. J Ind Microbiol Biotechnol 36(7):875–883

Cai CG, Chen JS, Qi JJ, Yin Y, Zheng XD (2008a) Purification and characterization of keratinase from a new *Bacillus subtilis* strain. J Zhejiang Univ Sci B 9(9):713–720

Cai CG, Lou BG, Zheng XD (2008b) Keratinase production and keratin degradation by a mutant strain of *Bacillus subtilis*. J Zhejiang Univ Sci B 9(1):60–67

Chen X, Zhou B, Xu M, Jia G, Huang Z, Zhao H, Liu G (2015) Prokaryotic expression and characterization of a keratinolytic protease from *Aspergillus niger*. Biologia 70(2):157–164

Cheng SW, HM H, Shen SW, Takagi H, Asano M, Tsai YC (1995) Production and characterization of keratinase of a feather degrading *Bacillus licheniformis* PWD-1. Biosci Biotechnol Biochem 59(12):2239–2243

Chitturi CMK, Lakshmi VV (2016) Immobilization of bacilli in arious matrices to enhance the utilization of keratinase and its comparison. Int J Curr Microbiol Appl Sci 5(2):389–396

Chou CC, Buehler MJ (2012) Structure and mechanical properties of human trichocyte keratin intermediate filament protein. Biomacromolecules 13(11):3522–3532

Coulombe PA, Bousquet O, Ma L, Yamada S, Wirtz D (2000) The 'ins' and 'outs' of intermediate filament organization. Trends Cell Biol 10(10):420–428

Daroit DJ, Corrêa APF, Brandelli A (2009) Keratinolytic potential of a novel *Bacillus* sp. P45 isolated from the Amazon basin fish *Piaractus mesopotamicus*. Int Biodeterior Biodegrad 63(3):358–363

Daroit DJ, Corrêa APF, Brandelli A (2011) Production of keratinolytic proteases through bioconversion of feather meal by the Amazonian bacterium *Bacillus* sp. P45. Int Biodeterior Biodegrad 65(1):45–51

DeToni CH, Richter MF, Chagas JR, Henriques JAP, Termignoni C (2002) Purification and characterization of an alkaline serine endopeptidase from a feather-degrading *Xanthomonas maltophila* strain. Can J Microbiol 48(4):342–348

Devi DA, Lakshmi VV (2015) Enhancing the utilisation of keratinases by using immobilised chitosan beads. IOSR J Pharm Biol Sci 10(4):1–4

Eaksuree W, Prachayakitti A, Upathanpreecha T, Taharnklaew R, Nitisinprasert S, Keawsompong S (2016) In vitro and in vivo evaluation of protein quality of enzymatic treated feather meals. Springer Plus 5(1):1–6

El-Gendy MMA (2010) Keratinase production by endophytic *Penicillium* spp. Morsy1 under solid-state fermentation using rice straw. Appl Biochem Biotechnol 162(3):780–794

El-Refai HA, AbdelNaby MA, Gaballa A, El-Araby MH, Abdel Fattah AF (2005) Improvement of the newly isolated *Bacillus pumilus* FH9 keratinolytic activity. Process Biochem 40(7):2325–2332

Evans KL, Crowder J, Miller ES (2000) Subtilisins of *Bacillus* spp. hydrolyze keratin and allow growth on feathers. Can J Microbiol 46(11):1004–1011

Fakhfakh N, Kanoun S, Manni L, Nasri M (2009) Production and biochemical and molecular characterization of a keratinolytic serine protease from chicken feather-degrading *Bacillus licheniformis* RPk. Can J Microbiol 55(4):427–436

Fakhfakh N, Ktari N, Haddar A, Mnif IH, Dahmen I, Nasri M (2011) Total solubilisation of the chicken feathers by fermentation with a keratinolytic bacterium *Bacillus pumilus* A1 and the production of protein hydrolysate with high antioxidative activity. Process Biochem 46(9):1731–1737

Fang Z, Zhang J, Liu B, Du G, Chen J (2013) Biochemical characterization of three keratinolytic enzymes from *Stenotrophomonas maltophilia* BBE11-1 for biodegrading keratin wastes. Int Biodeterior Biodegrad 82:166–172

Farag AM, Hassan MA (2004) Purification, characterization and immobilization of a keratinase from *Aspergillus orizae*. Enzym Microb Technol 34(2):85–93

Friedrich AB, Antranikian G (1996) Keratin degradation by *Fervidobacterium pennavorans,* a novel thermophilic anaerobic species of the order thermatogales. Appl Environ Microbiol 62(8):2875–2882

Gassessse A, Kaul RH, Gashe BA, Mattiasson BO (2003) Novel alkaline proteases from alkalophilic bacteria grown on chicken feather. Enzym Microb Technol 32(5):519–524

Gegeckas A, Gudiukaitė R, Debski J, Citavicius D (2015) Keratinous waste decomposition and peptide production by keratinase from *Geobacillus stearothermophilus* AD-11. Int J Biol Macromolec 75:158–165

Ghosh A, Chakrabarti K, Chattopadhyay D (2008) Degradation of raw feather by a novel high molecular weight extracellular protease from newly isolated *Bacillus cereus* DCUW. J Ind Microbiol Biotechnol 35(8):825–834

Gioppo NMR, Moreira-Gasparin FG, Costa AM, Alexandrino AM, Souza CGM, Peralta RM (2009) Influence of the carbon and nitrogen sources on keratinase production by *Myrothecium verrucaria* in submerged and solid state cultures. J Ind Microbiol Biotechnol 36(5):705–711

Gopinath SCB, Anbu P, Lakshmipriya T, Tang TH, Chen Y, Hashim U, Ruslinda AR, MKMd A (2015) Biotechnological aspects and perspective of microbial Keratinase production. Biomed Res Int 2015:1–10

Gousterova A, Braikova D, Goshev I, Christov P, Tishinov K, Vasileva-Tonkova E, Haertlé T, Nedkov P (2005) Degradation of keratin and collagen containing wastes by newly isolated thermoactinomycetes or by alkaline hydrolysis. Lett Appl Microbiol 40(5):335–340

Gradisar H, Friedrich J, Krizaj I, Jerala R (2005) Similarities and specificities of fungal keratinolytic proteases: comparison of keratinases of *Paecilomyces marquandii* and *Doratomyces microspores* to some known proteases. Appl Environ Microbiol 71(7):3420–3426

Gradisar H, Kern S, Friedrich J (2000) Keratinase of *Doratomyces microspores*. Appl Microbiol Biotechnol 53:196–200

Grazziotin A, Pimentel FA, de Jong EV, Brandelli A (2008) Poultry feather hydrolysate as a protein source for growing rats. Braz J Vet Res Anim Sci 45:61–67

Gupta R, Rajput R, Sharma R, Gupta N (2013) Biotechnological applications and prospective market of microbial keratinases. Appl Microbiol Biotechnol 97(23):9931–9940

Gupta R, Ramnani P (2006) Microbial keratinases and their prospective applications: an overview. Appl Microbiol Biotechnol 70(1):21–33

Gupta R, Sharma R, Beg QK (2012) Revisiting microbial keratinases: next generation proteases for sustainable biotechnology. Crit Rev Biotechnol 33(2):216–228

Gurav RG, Jadhav JP (2013) Biodegradation of keratinous waste by *Chryseobacterium* sp. RBT isolated from soil contaminated with poultry waste. J Basic Microbiol 53(2):128–135

Gurav RG, Mirajkar DB, Savardekar AV, Pisal SM (2016) Microbial degradation of poultry feather biomass by *Klebsiella* sp. BTSUK isolated from poultry waste disposal site. Res J life Sci bioinform pharm Chem Sci 1(6):279–288

Haddar A, Hmidet N, Ghorbel-Bellaaj O, Fakhfakh-Zouari N, Sellami-Kamoun A, Nasri M (2011) Alkaline proteases produced by *Bacillus licheniformis* RP1 grown on shrimp wastes: application in chitin extraction chicken feather-degradation and as a dehairing agent. Biotechnol Bioprocess Eng 16(4):669–678

Han M, Luo W, Gu Q, Yu X (2012) Isolation and characterization of a keratinolytic protease from a feather-degrading bacterium *Pseudomonas aeruginosa* C11. Afr J Microbiol Res 6(9):2211–2221

Hölker U, Lenz J (2005) Solid-state fermentation—are there any biotechnological advantages? Curr Opin Microbiol 8(3):301–306

Hromatka O, Ebner H, Csoklich C (1951) Investigation of vinegar fermentation. IV: about the influence of a total interruption of the aeration. Enzymologia 15(3):134–153

Hu H, He J, Yu B, Zheng P, Huang Z, Mao X, Yu J, Han G, Chen D (2013) Expression of a keratinase (*kerA*) gene from *Bacillus licheniformis* in *Escherichia coli* and characterization of the recombinant enzymes. Biotechnol Lett 35(2):239–244

Huang Y, Busk PK, Lange L (2015) Production and characterization of Keratinolytic proteases produced by *Onygena corvina*. Fungal Genom Biol 5:119. https://doi.org/10.4172/2165-8056.1000119

Huang Y, Sun Y, Ma S, Chen L, Zhang H, Deng Y (2013) Isolation and characterization of *Keratinibaculum paraultunense* gen. Nov., sp. Nov., a novel thermophilic, anaerobic bacterium with keratinolytic activity. FEMS Microbiol Lett 345(1):56–63

Ichida JM, Krizova L, LeFevre CA, Keener HM, Elwell DL, Burtt EH (2001) Bacterial inoculum enhances keratin degradation and biofilm formation in poultry compost. J Microbiol Methods 47(2):199–208

Inada S, Watanabe K (2013) Draft genome sequence of *Meiothermus ruber* H328 which degrades chicken feathers and identification of proteases and peptidases responsible for degradation. Genome Announc 1(3):1–2

Infante I, Morel MA, Ubalde MC, Martinez-Rosales C, Belvisi S, Castro- Sowninski S (2010) Wool degrading *Bacillus* isolates: extracellular protease production for microbial processing of fabrics. World J Microbiol Biotechnol 26(6):1047–1052

Ionata E, Canganella F, Bianconi G, Benno Y, Sakamoto M, Capasso A, Rossi M, La CF (2008) A novel keratinase from *Clostridium sporogenes* bv Pennavorans bv. Nov., a thermotolerant organism isolated from solfataric muds. Microbiol Res 163(1):105–112

Ismail AMS, Housseiny MM, Abo-Elmagd HI, El-Sayed NH, Habib M (2012) Novel keratinase from *Trichoderma harzianum* MH-20 exhibiting remarkable dehairing capabilities. Int Biodeterior Biodegrad 70:14–19

Jain R, Jain PC, Agrawal SC (2012) Feather degradation by *Streptomyces exfoliatus* CFS 1068. Ann Microbiol 62(3):973–978

Jaouadi NZ, Rekik H, Badis A, Trabelsi S, Belhoul M, Yahiaoui AB, Aicha HB, Toumi A, Bejar S, Jaouadi B (2013) Biochemical and molecular characterization of a serine keratinase from *Brevibacillus brevis* US575 with promising keratin-biodegradation and hide-dehairing activities. PLoS One 8(10):1–17

Jayalakshmi T, Krishnamoorthy P, Kumar GR, Sivamani P (2011) Docking conformation of *Bacillus licheniformis* keratinase enzyme and PMSF ligand. Drug Invent Today 3(8):200–202

Jeong JH, Lee OM, Jeon YD, Kim JD, Lee NR, Lee CY, Son HJ (2010) Production of keratinolytic enzyme by a newly isolated feather-degrading *Stenotrophomonas maltophilia* that produces plant growth-promoting activity. Process Biochem 45(10):1738–1745

Kazzaz AE, Feizi ZH, Guvenmez HK (2015) Keratinolytic protease production and characterization from *Bacillus* sp. isolated from poultry wastes. Int J Appl Biol Pharm 6(4):63–73

KD M, Paul M, Mathew J (2015) Optimization of conditions for production of keratinase by *Aspergillus flavus* by submerged fermentation. Int J Pharma Bio Sci 6(4):377–385

Kim JD (2007) Purification and characterization of a keratinase from a feather-degrading fungus *Aspergillus flavus* Strain K-03. Mycobiology 35(4):219–225

Kim JM, Choi YM, Suh HJ (2005) Preparation of feather digests as fertilizer with *Bacillus pumilis* KHS-1. J Microbiol Biotechnol 15(3):472–476

Kim JM, Lim WJ, Suh HJ (2001) Feather-degrading *Bacillus* species from poultry waste. Process Biochem 37(3):287–291

Kim JS, Kluskens LD, de Vos WM, Huber R, vand der Oost J (2004) Crystal structure of fervidolysin from *Fervidobacterium pennivorans* a keratinolytic enzyme related to subtilisin. J Mol Biol 335(3):787–797

Kluskens LD, Voorhorst WGB, Siezen RJ, Schwerdtfeger RM, Antranikian G, van der OJ, de Vos WM (2002) Molecular characterization of fervidolysin a subtilisin-like serine protease from the thermophilic bacterium *Fervidobacterium pennivorans*. Extremophiles 6(3):185–194

Kornillowicz-Kowalska T, Bohacz J (2011) Biodegradation of keratin waste: theory and practical aspects. Waste Manag 31(8):1689–1701

Koutb M, Morsy FM, Bagy MMK, Hassan EA (2012) Optimization of extracellular keratinase production by *Aspergillus terreus* isolated from chicken's litter. J Adv Lab Res Biol 3(3):210–216

Kshetri P, Ningthoujam DS (2016) Keratinolytic activities of alkaliphilic *Bacillus* sp. MBRL 575 from a novel habitat limestone deposit site in Manipur, India. SpringerPlus 5(1):1–16

Kumar AG, Swarnalatha S, Gayathri S, Nagesh N, Sekaran G (2008) Characterization of an alkaline active—thiol forming extracellular serine keratinase by the newly isolated Bacillus pumilus. J Appl Microbiol 104(2):411–419

Kumar R, Balaji S, Uma TS, Mandal AB, Sehgal PK (2010) Optimization of influential parameters for extracellular keratinase production by *Bacillus subtilis* (MTCC9102) in solid state fermentation using horn meal—a biowaste management. Appl Biochem Biotechnol 160(1):30–39

Kumar TP, Raju PN (2013) Transungual drug delivery: a promising route to treat nail disorders. Int J Pharm Sci Rev Res 2(4):22–33

Kunert J (1976) Keratin decomposition by dermatophytes II: presence of S-sulfocysteine and cysteic acid in soluble decomposition products. Z Allg Mikroboil 16(2):97–105

Lange L, Huang Y, Busk PK (2016) Microbial decomposition of keratin in nature a new hypothesis of industrial relevance. Appl Microbiol Biotechnol 100:2083–2096

Langveld JP, Wang JJ, Van de Wiel DF, Shih GC, Garssen GJ, Bossers A, Shih JC (2003) Enzymatic degradation of prion protein in brain stem from infected cattle and sheep. J Infect Dis 188(11):1782–1789

Lateef A, Oloke JK, Kana EG, Sobowale BO, Ajao SO, Bello BY (2010) Keratinolytic activities of a new feather-degrading isolate of *Bacillus cereus* LAU 08 isolated from Nigerian soil. Int Biodeterior Biodegrad 64(2):162–165

Lin X, Inglis GD, Yanke LJ, Cheng KJ (1999) Selection and characterization of feather-degrading bacteria from canola meal compost. J Ind Microbiol Biotechnol 23(2):149–153

Lin X, Kelemen DW, Miller ES, Shih JC (1995) Nucleotide sequence and expression of *kerA*, the gene encoding a keratinolytic protease of *Bacillus licheniformis* PWD-1. Appl Environ Microbiol 61(4):1469–1474

Liu B, Zhang J, Xiangru-Liao BL, Du G, Chen J (2013) Expression and characterization of extreme alkaline oxidation-resistant keratinase from *Bacillus licheniformis* in recombinant *Bacillus subtilis* WB600 expression system and its application in wool fiber processing. World J Microbiol Biotechnol 29(5):825–832

Liu B, Zhang J, Gu L, Du G, Chen J, Liao X (2014) Comparative analysis of bacterial expression systems for keratinase production. Appl Biochem Biotechnol 173(5):1222–1235

Macedo AJ, da Silva WOB, Gava R, Driemeier D, Henriques JAP, Termignoni C (2005) Novel keratinase from *Bacillus subtilis* S14 exhibiting remarkable dehairing capabilities. Appl Environ Microbiol 71(1):594–596

Manczinger L, Rozs M, Vagvolgyi C, Kevei F (2003) Isolation and characterization of a new keratinolytic *Bacillus licheniformis* strain. World J Microbiol Biotechnol 19(1):35–39

Marcone MF (2005) Characterization of the edible bird's Nest the "caviar of the east". Food Res Int 38(10):1125–1134

Masato Y, Shinji Y, Yasuko M (2011) Thermoprotecting agent for hair. Patent: JP2011046632

Mitsuiki S, Sakai M, Moriyama Y, Goto M, Furukawa K (2002) Purification and some properties of keratinolytic enzyme from analkaliphilic *Nocardiopsis* sp. TOA-1. Biosci Biotechnol Biochem 66(1):64–167

Mohorcic M, Torkar A, Friedrich J, Kristl J, Murdan S (2007) An investigation into keratinolytic enzymes to enhance ungual drug delivery. Int J Pharm 332(1):196–201

Moreira FG, de Souza CG, Costa MA, Reis S, Peralta RM (2007) Degradation of keratinous materials by the plant pathogenic fungus *Myrothecium verrucaria*. Mycopathologia 163(3):153–160

Nam GW, Lee DW, Lee HS, Lee NJ, Kim BC, Choe EA, Hwang JK, Suhartono MT, Pyun YR (2002) Native-feather degradation by *Fervidobacterium islandicum* AW-1 a newly isolated keratinase producing thermophilic anaerobe. Arch Microbiol 178(6):538–547

Ningthoujam DS, Devi LJ, Devi PJ, Kshetri P, Tamreihao K, Mukherjee S, Devi S (2016) Optimization of Keratinase production by *Amycolatopsis* sp. strain MBRL 40 from a limestone habitat. J Bioprocess Biotech 6:282. https://doi.org/10.4172/2155-9821.1000282

Noval JJ, Nickerson WJ (1959) Decomposition of native keratin by *Streptomyces fradiae*. J Bacteriol 77(3):251–263

Odetallah NH, Wang JJ, Garlich JD, Shih JCH (2005) Versazyme supplementation of broiler diets improves market growth performance. Poult Sci 84(6):858–864

Okoroma EA, Purchase D, Garelick H, Morris R, Neale MH, Windl O, Abiola OO (2013) Enzymatic formulation capable of degrading scrapie prion under mild digestion conditions. PLoS One 8(7):1–7

Onifade AA, Al-Sane NA, Al-Musallam AA, Al-Zarban S (1998) A review: potentials for biotechnological applications of keratin degrading microorganisms and their enzymes for nutritional improvement of feathers and other keratins as livestock feed resources. Bioresour Technol 66(1):1–11

Pandey A, Szakacs G, Soccol CR, Rodriguez-Leon JA, Soccol VT (2001) Production purification and properties of microbial phytases. Bioresour Technol 77(3):203–214

Paranthaman R, Alagusundaram K, Indhumathi J (2009) Production of protease from rice mill wastes by *Aspergillus niger* in solid state fermentation. World J Agric Sci 5(3):308–312

Park GT, Son HJ (2009) Keratinolytic activity of *Bacillus megaterium* F7-1, a feather-degrading mesophilic bacterium. Microbiol Res 164(4):478–485

Paul T, Das A, Mandal A, Jana A, Maity C, Adak A, Halder SK, Das Mohapatra PK, Pati BR, Mondal KC (2014) Effective dehairing properties of keratinase from *Paenibacillus woosongensis* TKB2 obtained under solid state fermentation. Waste Biomass Valor 5(1):97–107

Paul T, Haldar SK, Das A, Bera S, Maity C, Mandal A, Das PS, Das Mohapatra PKD, Pati BR, Mondal KC (2013) Exploitation of chicken feather waste as a plant growth promoting agent using keratinase producing novel isolate *Paenibacillus woosongensis* TKB2. Biocat Agri Biotechnol 2(1):50–57

Pissuwan D, Suntornsuk W (2001) Production of keratinase by *Bacillus* sp. FK 28 isolated in Thailand. Kasetsart J 35:171–178

Poopathi S, Abidha S (2008) Biodegradation of poultry waste for the production of mosquitocidal toxins. Int Biodeterior Biodegrad 62(4):479–482

Porres JM, Benito MJ, Lei XG (2002) Functional expression of keratinase (kerA) gene from *Bacillus licheniformis* in *Pichia pastoris*. Biotechnol Lett 24(8):631–636

Prakash P, Jayalakshmi SK, Sreeramulu K (2010) Production of keratinase by free and immobilized cells of *Bacillus halodurans* strain PPKS-2: partial characterization and its application in feather degradation and dehairing of the goat skin. Appl Biochem Biotechnol 160(7):1909–1920

Radha S, Gunasekaran P (2007) Cloning and expression of keratinase gene in *Bacillus megaterium* and optimization of fermentation conditions for the production of keratinase by recombinant strain. J Appl Microbiol 103(4):1301–1310

Radha S, Gunasekaran P (2009) Purification and characterization of keratinase from recombinant *Pichia* and *Bacillus* strains. Protein Expr Purif 64(1):24–31

Rahayu S, Syah D, Suhartono MT (2012) Degradation of keratin by keratinase and disulfide reductase from *Bacillus* sp. MTS of Indonesian origin. Biocatal Agric Biotechnol 1(2):152–158

Rai SK, Konwarh R, Mukherjee AK (2009) Purification characterization and biotechnological application of an alkaline β-keratinase produced by *Bacillus subtilis* RM-01 in solid-state fermentation using chicken-feather as substrate. Biochem Eng J 45(3):218–225

Rai SK, Mukherjee AK (2011) Optimization of production of an oxidant and detergent-stable alkaline β-keratinase from *Brevibacillus* sp. strain AS-S10-II: application of enzyme in laundry detergent formulations and in leather industry. Biochem Eng J 54(1):47–56

Rajendra VK, Baro A, Kumari A, Dhamecha DL, Lahoti SR, Shelke SD (2012) Transungual drug delivery: an overview. J Appl Pharm Sci 2(1):203–209

Raju EVN, Divakar G (2013) Production of keratinase by using *Pseudomonas aeruginosa* isolated from poultry waste. Int J pharm chem. Boil sci 3(1):79–86

Ramnani P, Gupta R (2004) Optimization of medium composition for keratinase production by *Bacillus licheniformis* RG1 using statistical methods involving response surface methodology. Biotechnol Appl Biochem 40(2):191–196

Ramnani P, Gupta R (2007) Keratinases vis-a-vis conventional proteases and feather degradation. World J Microbiol Biotechnol 23(11):1537–1540

Riessen S, Antranikian G (2001) Isolation of *Thermoanaerobacter keratinophilus* sp. Nov., a novel thermophilic anaerobic bacterium with keratinolytic activity. Extremophiles 5(6):399–408

Riffel A, Lucas F, Heeb P, Brandelli A (2003) Characterization of a new keratinolytic bacterium that completely degrades native feather keratin. Arch Microbiol 179(4):258–265

Rogel MR, Jaitovich A, Ridge KM (2010) The role of the ubiquitin proteasome pathway in keratin intermediate protein filament protein degradation. Proc Am Thoracic Soc 7(1):71–76

Rolinson GN (1952) Respiration of *Penicillium chrysogenuin* in penicillin fermentations. J Gen Microbiol 6:336–343

Saha S, Dhanasekaran D (2010) Isolation and screening of keratinolytic actinobacteria from keratin waste dumped soil in Tiruchirapalli and Nammakal, Tamilnadu, India. Curr Res J Biol Sci 2(2):124–131

Sahni N, Sahota PP, Phutela UG (2015) Bacterial keratinases and their prospective applications: a review. Int J Curr Microbiol Appl Sci 4(6):768–783

Sahoo DK, Das A, Thatoi HN, Mondal KC, Mohapatra PK (2012) Keratinase production and biodegradation of whole chicken feather keratin by a newly isolated bacterium under submerged fermentation. Appl Biochem Biotechnol 167(5):1040–1051

Sahoo DK, Halder SK, Das A, Jana A, Paul T, Thatoi H, Mondal KC, Mohapatra PKD (2015) Keratinase production by *Bacillus weihenstephanensis* PKD5 in solid-state fermentation and its milk clotting potential. Ind J Biotechnol 14:200–207

Sahoo DK, Halder SK, Thatoi H, Mohapatra PKD (2016) Potentiality of *Bacillus weihenstephanensis* PKD5 Keratinase for eco-friendly dehairing of skins and hide. Biotechnol An Ind J 12(12):1–8

Saieb FM, Boaker SA, El-komy HM, Issa A (2016) Production and estimation of keratinase by immobilized and free *Bacillus licheniformis* (St 24). J Appl Environ Microbiol 3(5):119–122

Sangali S, Brandelli A (2000) Feather keratin hydrolysis by a *Vibrio* sp. strain kr2. J Appl Microbiol 89(5):735–743

Schrooyen PM, Dijkstra PJ, Oberthür RC, Bantjes A, Feijen J (2001) Partially carboxymethylated feather keratins 2 thermal and mechanical properties of films. J Agric Food Chem 49(1):221–230

Selvam K, Vishnupriya B (2012) Biochemical and molecular characterization of microbial keratinase and its remarkable applications. Int J Pharm Biol Sci Arch 3(2):267–275

Shen JS, Cavaco-Paulo A, Guubitz GM, Rushforth M, Lenting HBM (2007) Development and industrialisation of enzymatic shrink-resist process based on modified proteases for wool machine washability. Enzym Microb Technol 40(7):1656–1661

Shigeru F (2012) Skin cosmetic kneaded composition and method for producing same and method for using skin cosmetic kneaded composition. Patent: US20120305017

Sivakumar T, Balamurugan P, Ramasubramanian V (2013a) Characterization and applications of keratinase enzyme by *Bacillus thuringiensis* TS2. Int J Future Biotechnol 2(1):1–8

Sivakumar T, Shankar T, Thangapand V, Ramasubram V (2013b) Optimization of cultural condition for keratinase production using *Bacillus cereus* TS1. Insights Microbiol 3(1):1–8

Son HJ, Park HC, Kim HS, Lee CY (2008) Nutritional regulation of keratinolytic activity in *Bacillus pumilis*. Biotechnol Lett 30(3):461–465

Stark CR, Spencer BE, Shih JCH, Chewing CG, Wang JJ (2009) Evaluation of keratinase stability in pelleted broiler diets. J Appl Poult Res 18(1):30–33

Suh HJ, Lee HK (2001) Characterization of a keratinolytic serine protease from *Bacillus subtilis* KS-1. J Protein Chem 20(2):165–169

Suntornsuk W, Suntornsuk L (2003) Feather degradation by *Bacillus* sp. FK 46 in submerged cultivation. Bioresour Technol 86(3):239–243

Suntornsuk W, Tongjun J, Onnim P, Oyama H, Ratanakanokchai K, Kusamran T, Oda K (2005) Purification and characterisation of keratinase from a thermotolerant feather-degrading bacterium. World J Microbiol Biotechnol 21(6):1111–1117

Susumu S, Yasuo K, Toshiyuki Y (1998) Skin agent. Patent: JPS63115813

Syed GD, Lee JC, Li WJ, Kim CJ, Agasar D (2009) Production, characterization and application of keratinase from *Streptomyces gulbargensis*. Bioresour Technol 100(5):1868–1871

Takami H, Nakamura S, Aono R, Horikoshi K (1992) Degradation of human hair by a thermostable alkaline protease from alkaliphilic *Bacillus* sp. no AH-101. Biosci Biotechnol Biochem 56(10):1667–1669

Tapia DMT, Simões MLG (2008) Production and partial characterization of keratinase produced by a microorganism isolated from poultry processing plant waste water. Afr J Biotechnol 7(3):296–300

Tatineni R, Doddapaneni KK, Potumarthi RC, Vellanki RN, Kandathil MT, Kolli N, Mangamoori LN (2008) Purification and characterization of an alkaline keratinase from *Streptomyces* sp. Bioresour Technol 99(6):1596–1602

Thys RCS, Lucas FS, Riffel A, Heeb P, Brandelli A (2004) Characterization of a protease of a feather-degrading *Microbacterium* species. Lett Appl Microbiol 39(2):181–186

Tiwary E, Gupta R (2010) Medium optimization for a novel 58 kDa dimeric keratinase from *Bacillus licheniformis* ER-15; biochemical characterization and application in feather degradation and dehairing of hides. Bioresour Technol 101(15):6103–6110

Tiwary E, Gupta R (2012) Rapid conversion of chicken feather to feather meal using dimeric keratinase from *Bacillus licheniformis* ER-15. J Bioprocess Biotech 2:123. https://doi.org/10.4172/2155-9821.1000123

Tork S, Aly MM, Nawar L (2010) Biochemical and molecular characterization of a new local keratinase producing *Pseudomomanas* sp. MS21. Asian J Biotechnol 2(1):1–13

Tork SE, Shahein YE, El-Hakim AE, Abdel-Aty AM, Aly MM (2013) Production and characterization of thermostable metallo-keratinase from newly isolated *Bacillus subtilis* NRC3. Int J Biol Macromol 55:169–175

Vasileva-Tonkova E, Gousterova A, Neshev G (2009) Ecologically safe method for improved feather wastes biodegradation. Int Biodeterior Biodegrad 63(8):1008–1012

Verma A, Singh H, Anwar S, Chattopadhyay A, Tiwari KK, Kaur S, Dhilon GS (2017) Microbial keratinases: industrial enzymes with waste management potential. Crit Rev Biotechnol 37(4):476–491

Vignardet C, Guillaume YC, Michel L, Friedrich J, Millet J (2001) Comparison of two hard keratinous substrates submitted to the action of a keratinase using an experimental design. Int J Pharm 224(1):115–122

Villa ALV, Aragao MRS, Santos EP, Mazotto AM, Zingali RB, Souza EP, Vermelho AB (2013) Feather keratin hydrolysates obtained from microbial keratinases: effect on hair fiber. BMC Biotechnol 13(1):1–11

Wang B, Yang W, McKittrick J, Meyers MA (2016) Keratin: structure mechanical properties occurrence in biological organisms and efforts at bioinspiration. Prog Mater Sci 76:229–318

Wang D, Piao XS, Zeng ZK, Lu T, Zhang Q, Li PF, Xue LF, Kim SW (2011) Effects of keratinase on performance nutrient utilization intestinal morphology intestinal ecology and inflammatory response of weaned piglets fed diets with different levels of crude protein. Asian-Aust J Anim Sci 24(12):1718–1728

Wang X, Parson CM (1997) Effect on processing systems on protein quality of feather meals and hog hair meals. Poult Sci 76(3):491–496

Waterhouse DF (1957) Digestion in insects. Annu Rev Entomol 2:1–18

Werlang PO, Brandelli A (2005) Characterization of a novel feather-degrading *Bacillus* sp. strain. Appl Biochem Biotechnol 120(1):71–79

Williams CM, Richter CS, Mackenzie JM, Shih JC (1990) Isolation, identification and characterization of a feather-degrading bacterium. Appl Environ Microbiol 56(6):1509–1515

Yamamura S, Morita Y, Hasan Q, Rao SR, Murakami Y, Yokoyama K, Tamiya E (2002) Characterization of a new keratin-degrading bacterium isolated from deer fur. J Biosci Bioeng 93(6):595–600

Yang L, Wang H, Lv Y, Bai Y, Luo H, Shi P, Huang H, Yao B (2016) Construction of a rapid feather-degrading bacterium by over expression of a highly efficient alkaline keratinase in its parent strain *Bacillus amyloliquefaciens* K11. J Agric Food Chem 64(1):78–84

Yoshioka M, Miwa T, Horii H, Takata M, Yokoyama T, Nishizawa K, Watanabe M, Shinagawa M, Murayama Y (2007) Characterization of a proteolytic enzyme derived from a *Bacillus* strain that effectively degrades prion protein. J Appl Microbiol 102(2):509–515

Yusuf I, Shukor MY, Syed MA, Yee PL, Siti NAS, Ahmad A (2015) Investigation of keratinase activity and feather degradation ability of immobilised *Bacillus* sp. Khayat in the presence of heavy metals in a semi continuous fermentation. J Chem Pharm Sci 8(2):342–347

Zaghloul TI (2001) Cloned *Bacillus subtilis* alkaline protease (*aprA*) gene showing high level of keratinolytic activity. Appl Biochem Biotechnol 70(72):199–205

Zerdani I, Faid M, Malki A (2004) Feather wastes digestion by new isolated strains *Bacillus* sp. in Morocco. Afr J Biotechnol 3(1):67–70

Zhang S, Long L, Yin H, Xiao Z, Li Q, Zhang S, Tian X, Li C, Chen S, Yin T, Chen Y (2010) Pearl albefaction method mediated by keratinase and combined with redox. Patent: CN100579412

Zoccola M, Aluigi A, Tonin C (2009) Characterisation of keratin biomass from butchery and wool industry wastes. J Mol Struct 938(1):35–40

Chapter 6
Exploitation of Fungi and Actinobacteria for Sustainable Agriculture

Reshma R. Anilkumar, Lekshmi K. Edison, and N.S. Pradeep

6.1 Introduction

Soil microorganisms presents a connecting link with plants and mineral nutrients in soil. Thus, they are amassing accruing interest as biofertilizers. They form symbiotic association with approximately 80% of crop species (Araujo et al. 2009). They provide minerals, nutrients and water to the host plant in return for photosynthetic products. These microorganisms can acquire nutrients from soil volumes that are inaccessible to roots of the crop plants. Thus, these organisms can mitigate the limitation in plant growth caused by an inadequate nutrient supply. Far more than nutritional supply, these association also benefits the plants with add on benefits like drought resistance, tolerance to salinity and disease suppression. Metals like Zn, Cu and Fe play prominent roles in the subcellular compartments of the plants, but they are toxic at higher concentrations. The microorganisms associated with the plant rhizosphere play a role in alleviating toxicity caused by heavy metal in the host plants and those plants are also capable in tolerating high metal concentrations in the soil (Nihorimbere et al. 2011) Fungi and actinomycetes play an unequivocal role on the ecosystem, as they augment the structure of soil and its aggregation and drive the structure of plant communities and productivity. Thus, soil microorganisms are prime biotic soil components lacking of which, can lead to an inefficient functioning of the ecosystem.

Restoring the existing level of microorganism luxuriance can prove as a substitute to ordinary fertilization procedures for tenable agriculture, a crucial objective for farmers meeting the global recession. The crucial approach endorsed to accomplish this target is the open reenrichment of propagules into a particular soil. Nonetheless, the discovery of these actinomycetes and fungi in the level of

R.R. Anilkumar • L.K. Edison • N.S. Pradeep (✉)
Microbiology Division, Jawaharlal Nehru Tropical Botanic Garden and Research Institute, Palode, Trivandrum, Kerala, India
e-mail: drnspradeep@gmail.com

J.K. Patra et al. (eds.), *Microbial Biotechnology*,
https://doi.org/10.1007/978-981-10-6847-8_6

application needs the understanding of how they suit and proceed to the respective ecosystem and soil management and of the processes that pave to the formulation of a functional symbiosis, including the mechanisms associated with nutrient transfer. In this chapter, various applications of fungi and actinomycetes in agriculture are discussed with main emphasis on their use in sustainable agriculture.

6.2 Vesicular Arbuscular Mycorrhiza (VAM)

The symbiotic VAM fungi that colonize and settle the roots and soils of crop and weed plants form an element of the ecology of agricultural systems since they have a considerable impact on the functioning, strength and stability of any ecosystem. They include significant soil biota that commit essentially to the fertility and durability of the man- made ecosystems (Mobasser and Tavassoli 2013). Fungi form a close symbiotic association with most of the terrestrial greens. Most of the deciduous or evergreen trees have ectomycorrhizas- in this case the roots are externally surrounded by the hyphae of some fungi such as Boletus Phallus, Scleroderma, Amantia, Tricholoma etc. Those fungi decompose soil organic matter and the leaf litter in the soil. Ectomycorrhiza stimulates the growth of the used-seedlings, easily absorb nutrients like phosphorous, calcium, nitrogen, potassium etc. and then passed to the tissues of roots (Ramanankierana et al. 2007). Being saprophytes they decompose the organic matter and augment the fertility of the soil. In this way a symbiotic relationship is established between the fungi and roots. In a sustainable agricultural system characterized by low levels of disturbance, the significance of mycorrhizae will be similar to that in natural ecosystems at advanced successional stages. In such systems, the role of mycorrhizae may be expressed not only by the procurement of nutrients by one host plant, but also by a redistribution of nutrients between many host plants and between host plants and soil.

Today's agriculture highlight on maximum production of commodities for consumption. This approach has induced the germplasm selection favouring translocation of Carbon compounds to different portions of crop plants. The proper allocation of Carbon to the roots is desired to maintain soil structure and safeguard subsequent harvests. The general loss of Carbon from the plant to its sorroundings is mediated by exudation. As it is lost, the population density of soil microbiota fumble steepily with increasing distance from root. The mycelia of VAM fungi, in particular, not only control the composition and rate of flow of root exudates, but contain a massive portion of Carbon that are derived from roots. These fungal biomass is also accessible as a substrate to microbial metabolism (Harrier 2001).

6.2.1 Impact of Cultural Practices and Agro Chemicals on VAM

The level of cultural stress in soil is a magnitude of the sustainability of agriculture in the soil. Just as it is, the degree of impoverishment of VAM microflora is a symbol of decreased stability of plant-soil system. Soil disturbances alter root and soil colonization by VAM fungi. The disruption of mycorrhiza formation conclude in decreased phosphorous uptake by plants. Tillage effects upset the density of infective propagules required to re-establish VAM colonization differently. Hence, the roots become more naive to pathogen attack by infectious organisms. Crop rotation demonstrate to be an effective solution to this problem, but the exact reason is not well known. Enhanced VAM colonization and spore production were established in different crops, as a result of rotation (Abbasi et al. 2015). It also influence the species composition of VAM microflora growing with different crop plants. A better perspective of rotation effects would help in selecting more effective VAM fungi for use in agriculture. In intercrop systems, VAM fungi colonize and link the roots of adjacent plants and thereby mutual enhancement of productivity by both plants is observed. They are also engaged in transfer and distribution of nutrients in such plant communities. They indirectly affect the nitrogen inputs in a legume intercropping system by increasing the phosphorous availability of plants. Mycorrhiza formation may also build upon grazing intensity. This depends on the capability of the host plant to provide the symbionts with photosynthetic products.

Agro-chemicals are profitable in controlling pests and pathogens, but their application also results in aimless killing of beneficial microorganisms, such as VAM fungi. These fungi are integral parts of both the plant and population of soil microbes. Biocides, may therefore influence VAM fungi either directly or in an indirect manner through their effects on host plant or through the soil organisms associated with the plant. These complex interactions are of precise interest in sustainable agriculture, because the retardation induced by biocide in VAM hyphae, the subsequent stoppage of root colonization by them, and an alteration in species combinations in soil are significant for soil stability and production by plants. The use of agrochemicals have varied effects on VAM. They embellish or hinder VAM colonization and sporulation, improve mycorrhiza formation by controlling hyperparasites and VAM in turn revamp host plant resistance to pesticide stress. It is important to have a thorough understanding of using biocides based on its effects on VAM fungi and on pathogens and pests since it affects the potentiality of VAM's biocontrol ability (Sukarno et al. 1993). The fungi in the mycorrhizal symbiosis are major elements affecting plant and soil health. They have a dominant role in agriculture as agents of plant productivity and soil conservation. As of now, we do not have much knowledge on the effects of specific VAM-fungal isolates on specific plant or soil problems. This information have to be gathered before these fungi can be fully integrated into agricultural practice as a management tool. Research on the VAM plant-soil system is vital on many fronts to achieve this goal: (a) Elucidation of the conditions for large scale production of host-free inocula (b) Choice of effective isolates from naturally occurring populations or by artificial, directional selection methods; and (c) Identification of specific cultural and environmental conditions which can be pacified by mycorrhizae (d) provide the field user with specific product-use recommendations.

6.3 Myco-bio Control of Insect Pests

Biopesticides are biological substances that manage pests without posing a toxic threat to environment. They include natural enemies of pests, phytochemicals produced by them and their byproducts which effectively regulate the pests that are harmful to plant crops (Gupta and Dikshit 2010). Biopesticides have a significant role in protecting crops although most commonly used in conjunction with other aids like chemical pesticides as part of concerted Integrated Pest Management. These microbial pesticides prove to be a substitute to chemical insecticides for enhancing target specificity as well as ecological safety lest they are used uniquely or merging with other pest management programmes. For instance there are fungi that can control weeds as well as insects.

Approximately about US $ 10,000 million has estimated to be a loss in the agricultural production in India in relation to the pest associated damages in the field and during storage. The application of fungi to minimize the frequency of insects and reducing its pathogenesis thereby resulting in a reduced crop loss is referred to as Mycobiocontrol (Chet et al. 1993). About 700 species of fungi are listed as pathogens. Few of these fungi are having confined host ranges, for instance, *Aschersonia aleyrodes* affects scale insects and whiteflies, whereas other species of fungi have an expanded host range, with particular isolates being extra specific to distinct pests. Species like Aspegillus and Penicillium are facultative pathogens whereas species like Cordyceps are obligate pathogens.

Pathogenic fungi that infects the fungi are found as spores in the environment. When an insect comes in contact with a spore either on the surface of a plant or from soil or from air or from the dead insects, spores stick to the body of the insect. These spores penetrate the insect cuticle (often at joints or creases where its protective lining is thin), and grows throughout its body. Most of the insects also produce toxins in the host body which reduces its time of killing and also prevent entry of other microbes (Faria and Wraight 2001). After the insect death, fungus exits through the exoskeleton of insect, usually from the above mentioned thinner areas and once again begins spore production. The spores are liberated and scattered by the action of wind, rain or contact in the surroundings. The insect victims often have a "fuzzy" appearance, because the fungi protrude out of the exoskeleton to produce spores. Most commercial fungal strains produce green or white spores, although the colour of fungi can vary overtime.

Most of the fungi which is used to manage pets belong to hyphomycetes. Some of the species can be mass multiplied, hence used for commercial production. Some of the fungi in this group cause natural outbreaks in the soil when conditions are favourable. Their host range are broad. Commercial fungal strains often target aphids, ants, caterpillars, grasshoppers, thrips, whiteflies, weevils, Colorado potato beetle, and mealybugs. One of the most widely used fungal biopesticide include the hyphomycete fungus, *Beauveria bassiana.*

6.3.1 Entomopathogenic Fungi

Entomopathogenic fungi are natural managers of insect populations and are potent mycoinsecticide agents against various insect pests in agriculture. These fungi penetrate host cuticle, facilitating its entry into the hemolymph, produce toxins, and utilize the maximum available nutrients in the haemocoel therby avoiding immune response of the insects. The conidia of these fungi is applied since they sporulate after application. Fungal entomopathogens serves as an alternative to insecticide and the combined application of insecticide with fungal entomopathogens can turn out to be very useful for management of insecticide resistance. Table 6.1 showing some bioactive products derived from Entomopathogenic fungi commercially used for agricultural field applications.

Fungi belonging to this group cause outbreaks in natural population but are difficult to produce by mass multiplication, hence not used commercially. They are host specific, certain species affects aphids. Inspite of the adversities in commercial production, they have a large impact on the infecting pest populations. Using fungi as biocontrol agents have an added advantage of controlling even sucking insects since they infect the host tissues by invading directly through the cuticle rather than the hosts ingesting them. Insects in the orders Coleoptera Hemiptera, Orthoptera, Thysanoptera, and Lepidoptera, comprise most of the targets. Chinese caterpillar fungus controls insect pests of crop plants, the spores being applied to (Florez 2002). The other widely used fungi include those that control Colorado potato beetles, Citrus rust mites, leaf hoppers and Spittlebugs. This method is cheap and less detrimental than chemical pesticides.

6.3.2 Bio-Management of Insect-Pests by Different Entomopathogenic Fungi

6.3.2.1 *Beauveria sp. Beauveria bassiana*

A filamentous fungus, also named as imperfect fungus belonging to a class of insect pathogenic deuteromycete. Particular strains of Beauveria act only against specific hosts. A wide variety of *B. bassiana* spp. have been obtained from a variety of medicinal or agriculturally important insects worldwide. These fungi is found in soil naturally all around the world and is pathogenic to many insects causing physiological disease such as white muscardine disease. *Beauveria* sp. is the highly host

Table 6.1 Bioactive products derived from Entomopathogenic fungi commercially used for field application

Product	Fungus	Biological action (Sandhu et al. 2000)
Mycotal	*Verticillum lecanii*	Fungal pesticide
Pfr 21	*Paecilomyces fumosoroscus*	Fungal pesticide
Verelac	*Verticillum lecanii*	Sucking pests
Beevicide	*Beauveria bassiana*	Borer type pests
Grubkill	Selected fungus and bacteria	Borers and sucking pests
Pelicide	*Paecilomyces lilacinus*	Effective against nematode
Biologic Bio 1020	*Metarhizium anizopliae*	Mycelium granules as pesticide
Bioter	*Verticillum lecanii*	Effective against termites
Brocaril	*Beauveria bassiana*	Wetteble powder used as pesticide
Ostrinil	*Beauveria bassiana*	Microgranules of mycelium used as pesticide
Boverol	*Beauveria bassiana*	Dry pellets as pesticide
Naturalis	*Beauveria bassiana*	Liquid formation as pesticide
Mycontrol- WP	*Beauveria bassiana*	Wetteble powder used as pesticide
Betel	*Beauveria brongniartii*	Microgranules of mycelium used as pesticide
Engerlingspilz	*Beauveria brongniartii*	Barley kernels colonized with fungus used as pesticide
Biopath	*Metarhizium anizopliae*	Conidia on a medium used as pesticide
Biomite	*Verticillum lecanii* and other entomopathogenic organisms	Effective against mites
Biogreen	*Metarhizium anizopliae*	Conidia produced on grain used as pesticide
Naturalis-O and BotaniGard	*Beauveria bassiana*	Effective against whiteflies
Trypae Mix	*Trichoderma* and *Paecilomyces*	Effective against fungal pathogens and nematodes in soil

specific (Sandhu et al. 1993). The hosts of medicinal significance include vectors for agents of tropical infectious diseases such as *Glossina morsitans,* tsetse fly, and sand fly *Phlebotomus* that transmits *Leishmania* and bugs belonging to genera *Triatoma* and *Rhodnius*, that transmits Chagas disease. The most important agricultural hosts include Coloradopotato beetle, American bollworm *Helicoverpa armigera*, *Hyblaeapara* and *Eutectona machaeralis* and other termites.

The long term effects of entomopathogenic fungi on pest suppression in case of an epizootic is attributed to the high level of its persistence in the host population. It is a good bioinsecticide to manage a number of pests such as whitefly, termites, and in malaria-transmitting mosquitoes. It is the asexually reproducing form (anamorph) of *Cordyceps bassiana* (Sandhu et al. 2001). The teleomorphic form has been found only in eastern Asia. This fungus, ubiquitously found is the most frequent causative agent of disease associated with dead and obsolescent insects in nature and has been scrutinized worldwide as a control agent of hypogeous species. The sub-terranean level of Curculionidae weevils are highly vulnerable to white muscardine disease.

B. bassiana comprise of geographically and genetically distinct variants as hosts which differ in their ability to cause pathogenesis, the spores of which are sprayed on infected crops as an emulsified suspension or wettable powder. *B. Bassiana* targets a large number of arthropod hosts and hence regarded as a nonselective biological insecticide. *B. bassiana* is also applied against the, pine caterpillars *Dendrolimus* spp., European corn borer *Ostrinia Mubilalis* and green leafhoppers *Nephotettix* spp (Thakur et al. 2005).

6.3.2.2 *Verticillium lecanii*

Verticillium lecanii is widely seen fungi which cause large epizootic in tropical and subtropical regions, as well as in warm and humid environments. In south Korean greenhouses, *V. Lecanii* is used as effective bio agent against *Trialeurodes vaporariorum*. This fungus invade nymphs and adults and get cemented to the leaf underside by means of a filamentous mycelium (Nunez et al. 2008). *Verticillium lecanii* has been instrumental in manging whitefly, many aphids, including *Myzus persicae*. The cereal cyst nematodes showed an enormous drop in population when treated with *V.lecanii*. *Verticillium chlamydosporium* has a broad host range among cyst and root-knot nematodes but is highly variable and only some isolates proved to be potent commercial biological control agents (Kim et al. 2002).

6.3.2.3 *Metarhizium spp. Metarhizium anisopliae*

A major pathogen affecting pests and is probed for mycobiocontrol of injurious insect pests. A thorough bioactivity of *M. anisopliae* has been investigated on teak skeletonizer *Eutectona machaeralis* and it has been reported to be a potential mycobiocontrol agent of teak pest. The spore production of *M. anisopliae* by solid state fermentation makes its production easier (Sandhu et al. 2000).

6.3.2.4 *Nomuraea sp. Nomuraea rileyi*

It is a dimorphic hyphomycete that leads to death in insect pests. *N. rileyi* kills Lepidoptera class insects like Spodoptera litura and some insects belonging to Coleoptera. The fungi being environmental friendly and host specific, are used in insect pest management (Mathew et al. 1998). The mode of infection and development of *N.rileyi* have been reported for various insect hosts such as *Heliothis zea*, *Trichoplusia ni*, *Plathypena scabra*, *Pseudoplusia includes*, *Bombyx mori*, and *Anticarsia gemmatalis*, now *Spilosoma* was found to be severely attacked by *Nomuraea rileyi*, hence studied in detail for its mycobiocontrol. Similarly an epizootic of *Nomuraea rileyi* on the hedge plant eater *Junonia orithya* proved to be the best alternative to manage the same.

6.3.2.5 *Paecilomyces sp. Paecilomyces*

It belongs to a genus of nematophagous fungus which is detrimental to harmful nematodes. Hence this is used as a bionematicide to control nematodes by its application onto soil. *Paecilomyces lilacinus* attacks root- knot nematodes and assimilates eggs of cyst nematodes. After its discovery in 1979, this fungus turned out to be the subject of appreciable biological control research *Paecilomyces fumosoroseus* (Hyphomycetes) is one among the crucial natural enemies of whiteflies worldwide, and causes the sickness "Yellow Muscardine". The capability of this fungus to grow largely over the leaf surface under humid conditions is a feature that enhances its ability to spread briskly through whitefly populations (Wraight et al. 2000).

The fungi suppress and kills *Bemisia tabaci* multitudes. Epizootics by *Paecilomyces fumosoroseus* also result in considerable reductions in *B. tabaci* populations during or shortly following rainy seasons and in prolonged periods of humid conditions in the field or greenhouse (Faria and Wraight 2001). But, epizootics of naturally occurring fungi cannot be confided upon for control. Many fungi have the capacity to cause remarkable mortality, and advancement of natural epizootics is not only dependent on the climatic conditions, but also determined by many crop production practices. Epizootics often occur after acute injury has previously been inflicted by whiteflies. *P. fumosoroseus* is perfect for controlling the nymphs of whitefly. The nymphs exhibit "feathery" form and are enclosed by mycelia and conidia. *P. furiosus* is also used to control mosquito sp. *Culex pipiens*.

Advantages of using fungi as biopesticide include: (1) Non-toxic and non-pathogenic to wildlife, humans, and other organisms not closely related to target (2) Mass production of spores of these fungi is relatively easy, so comparably priced with other biocontrol agents (3) Application can be done with spray rigs, hence easily adapted to existing application technology (4) broad host range, so can achieve control of multiple pests with the same product (5) persistence of seasonal infection tends to be low for most fungi (6) most microbial insecticides can be used in combination with artificial chemical insecticides/ pesticides since the microbial product is not deactivated by residues of conventional insecticides (7) Enhance the root and plant growth by increasing the beneficial soil microflora. By this way they take a part in the increase of the crop yield.

Disadvantages of fungal biopesticide includes: (1) High concentration of spores needed to get adequate control of pests (2) Time required by fungi to get rid of the pests is too long (3) Non-target mortality of beneficial insects (4) Environmental factors also can affect the fungal activity which limit their effectiveness as biocontrol agents of pests (5) Each application of the microbial insecticide is specific to only a certain class of insects (6) Special formulation and storage procedures are necessary for some microbial pesticides. Even though these procedures may obstruct the production and distribution of few products, storage requirements need not vigorously limit the management of microbial insecticides that are broadly available. (Store all pesticides, inclusive of microbial insecticides, in accordance with label directions).

6.4 Integrated Pest Management

Integrated Pest Management is a method used to control pests in an environmentally culpable manner. Biological control is a doctrine of cultural control of plant pathogens that chiefly involves the change of biotic and abiotic environments from one that devours disease/pathogen to one that dispirits the accumulation of infective or parasitic material and curtails the activity of the pathogen (Kalra 2007). These potential biocontrol fungi are mostly saprophytic in nature and proliferate abundantly in various natural soils. The most important fungi used as biocontrol agents against plant pathogens are – *Trichoderma*, *Gliocladium*, *Aspergillus*, *Penicillium*, *Neurospora*, *Chaetomium*, *Dactylella*, *Arthrobotrys*, *Catenaria*, *Paecilomyces*, *Glomus* etc.

Isolates of *Beauveria bassiana* (Balsamo) Vuillemin has been used in conjunction with conventional insecticides for biological control of rice stink bug, Oebalus pugnax (F.) in the laboratory and in small-plot field experiments. Entomopathogens (like Verticillium lecani) have shown promise for augmentative biological control of Scirtothrips dorsalis, and there was scope for identifying more adapted and virulent strains of the entomopathogens. The commercial mycoinsecticide 'Boverin' formulated on *B. bassiana* with low doses of trichlorophon have been employed to weaken the second-generation outbreaks of *Cydia pomonella.* Higher insect mortality was also observed when *B. bassiana* and sublethal doses of insecticides were tested to control Colorado potato beetle (*Leptinotarsa decemlineata*), resulting in higher rates of synergism among two agents. The combination treatment of fungi *Beauveria bassiana* and *Metarrhizium* along with new generation pesticides have showed higher dose mortality response of disease causing insects than their sole treatment (Hajeck 1994).

6.5 Fungi in the Conversion of Agricultural Wastes to Compost

The residues from crop plants are produced abundantly butis an underutilized renewable resource in agriculture. The approximate amount of residues is estimated to be 620 million tons. Half of the quantity is used for roofing purposes, animal feeds, fuel and packing stuffs. Burning of these residues is an easy way of disposing them but tends to air pollution, cause soil erosion and reduces the efficiency of herbicides in soil. It also causes respiratory problems and fog issues. The application of agro residues in soil, eventhough increase soil health, decrease subsequent crop yields due to the production of phytotoxins, allelochemicals etc. (Singh and Nain 2014).

Lignocellulose is composed of polymers like cellulose, hemicelluloses and lignin. Hence the microorganisms which can breakdown these polymers with their enzymes are efficiently used in their breakdown. The more complicated is the polymer, the more elaborate network of enzymes required for its breakdown. There are mainly three types of fungi which lives on dead wood that preferentially degrade one or more wood components *viz.* soft rot fungi, brown rot fungi and white rot fungi. Soft rot fungi decompose cellulose but degrade lignin slowly and incompletely. The brown rot fungi exhibit preference for lignin, hence largely focussed on demethylation. White rot fungi are capable of degrading both lignin and cellulose. In majority of soils, 80% of the fungal population belongs to the genera *Aspergillus* and *Penicillium.*

6.6 Fungi in Humus Formation

Once the plants and animals die, there is a generation of large amount of organic wastes. Agricultural wastage, forest litter, etc. also plays a vital role in organic or bio-waste formation. The fungi and bacteria play the key roles for degradation of these. When fungi degrades such organic wastes, these generates a kind of organic nutrient for plants called humus. Humus is none other than degenerated plant and animal bodies. During the formation of humus, Carbondioxide gas (CO2) is formed, which is utilised by green plants during photosynthesis. Humus is hence a degenerative product of cellulose, hemicellulose, lignin, proteins, nucleic acid, etc. The major part of the humus consists of Humic acid, Humins, Fulvic acid, etc. It maintains physical and chemical properties of soils supporting various biological activities. During humus formation, all those complex organic molecules are degraded in steps. Mentioned below are some of the complex organic molecules along with the fungi degrading them:

- *Cellulose: Aspergillus, Penicillium, Chaetomium, Fusarium, Trichoderma, Cladosporium, Alternaria, Humicola, Phoma*, etc.
- Hemicellulose: *Aspergillus, Penicillium, Fusarium, Chaetomium, Glomerella*, etc.
- Pectin: *Aspergillus, Penicillium, Fusarium, Rhizopus, Monilia*, etc.
- Lignin: Many white rod fungi of Basidiomycotina and many *Agaricus* spp.

And thus fungi upgrades minerals and other nutrients in soil, increasing fertility.

6.7 Fungal Enzymes in Agriculture

Soil is an important component of all terrestrial ecosystems as well as a main source of production in agriculture. The functioning of an ecosystem and its biochemical functions are influenced by soil. The overall enzyme activity in soil consists of various intracellular and extracellular enzymes that originate from microorganisms and from plants and animals.

Soil enzymes are significant in soil functioning due to the following features: (1) they play a critical role in the decomposition of organic materials and the transformation of organic matter, (2) they release available nutrients to plants, (3) they participate in N2 fixation, nitrification and denitrification processes, and (4) they take part in the detoxification of xenobiotics, such as pesticides, industrial wastes, etc.

Amylase: An alpha amylase enzyme produced by *Aspergillus niger* and *Aspergillus oryzae.* Rapid acting hydrolase enzyme particularly active in the mildly acidic pH range and degrades a variety of starch containing substrates. It can also be used as feed additives to increase the utilization of feedstuffs by hydrolyzing the starch contained in feeds to dextrins and sugars.

Cellulase: Cellulases are inducible enzymes produced by a wide variety of microorganisms which includes both fungi and bacteria. *Trichoderma*, and *Aspergillus* are the most widely employed cellulose producers. Many enzyme preparations comprising of various combinations of cellulases, hemicellulases, and pectinases have immense applications in the field of agriculture for stimulating growth of crops and managing plant diseases.

Pectinase: Pectin is a polymer of carbohydrate group esterifying with methanol. It is an important component of plant cell wall. The maximum amount of pectin is present in middle lamella of cells. Plant pathogens attack target cells by producing number of cell degrading enzyme which facilitates the entry and expansion of pathogen in the host tissue. *Aspergillus niger, Aspergillus japonicus, Chaetomium globosum* and *Aspergillus flavus* are potent pectinase producers.

Invertase: Invertase acts on 1, 4 glycoside linkage of sucrose and splits it into D-glucose and D-fructose. It is intracellular as well as extracellular enzyme. Invertase is also referred as β- fructofuranosidase as it catalyses hydrolysis of the terminal non-reducing residue of β –fructofuranoside. *Thermomyces lanuginosus, Candida utilis, Penicillium chrisogenum, Saccharomyces cerevisiae* and *S. Carlsbergensis* are examples of invertase producing fungi.

6.8 Phytohormone Production by Fungi

A phytohormone is an organic substance manufactured in defined organs of the plant that can be transported to other sites, where it brings about specific morphological, physiological and biochemical responses. Nevertheless, phytohormones are also effective in tissues where they are created. In addition, various soil bacteria and fungi are also phytohormone producers. The most commonly accepted classes of phytohormones, known as the "classical five", are: the auxins, cytokinins, gibberellins, ethylene and abscisic acid,. The capacity to synthesize cytokinins, Gas and IAA, is common among soil and plant-associated bacteria and fungi responsible for plant growth promotion, symbiotic associations and also pathogenesis. *Aspergillus niger, Penicillium citrinum, Trichoderma harzianum* are major exogenous phytohormone producers in the plants.

6.8.1 Auxins From Fungi Play a Positive Role in Plant–Fungus Interactions

The hormones derived from indole capable of plant development is termed as auxin. The processes such as cell division, differentiation and organ formation needs auxins (Barker and Tagu 2000). Auxins also control biotic and abiotic stress responses in plants. Auxins are involved in symbiotic interactions between plants and bacteria or fungi. They are required for the initiation of nodule formation in the nitrogen-fixative bacterial symbiosis and for the invasion of mycorrhizal fungi. They are also involved in plant–pathogen interactions (Benjamins and Scheres 2008).

6.8.2 Cytokinins

Cytokinins are plant hormones derived from ATP/ADP/AMP or from the tRNA degradation pathway. They have a decisive role in plant developmental processes, such as root and shoot formation, through the regulation of cell cycle and cell differentiation. They are also involved in the delay of senescence and in source–sink nutrient distribution. CKs are probably involved in 'green island' formation, a photosynthetically active zone often found around lesions caused by biotrophic fungi (Chanclud et al. 2016).

6.8.3 Gibberelic Acid

GAs are terpenoid hormonal compounds identified for the first time as being produced by Gibberella fujikuroi. GAs are involved in the control of germination, flowering, cell division and internode elongation. In mycorrhizal interaction, the GA content is increased in plants (Brian and Elson 1954).

6.8.4 Abscissic Acid (ABA)

ABA is the key hormone for plant abiotic stress responses (Peleg and Blumwald 2011) and it is also involved in seed dormancy by acting antagonistically with the GA pathway. In plants, ABA is well known to induce stomatal closure and thus to contribute to plant drought tolerance (Crocoli et al. 1991).

6.8.5 *Ethylene*

A gaseous hormone involved in plant physiology and defence which also affects fungal development. Ethylene is a gaseous compound first discovered for its role in fruit maturation. ET was later shown to be involved in senescence, germination, flowering and the inhibition of root and shoot growth (Bleecker and Kende 2000).

6.9 Actinobacteria in Agriculture

One of the most fascinating group of organisms' actinobacteria or actinomycetes comes under largest taxonomic units within the domain bacteria. The word "Actinomycetes" are originated from "atkis" (a ray) and "mykes" (fungus), Greek words, because they having the features of both bacteria and fungi. But with satisfactory unique features delineate them into 'Kingdom bacteria'. These are aerobic, sporulaing, gram positive bacteria comes under the order actinomycetales, especially with guanine and cytosine rich DNA. Although they are unicellular, they do not possess cell wall and characterized with slender, nonseptate distinct substrate and aerial mycelium. Most actinobmycetes produce powdery colonies and firmly sticking to agar surface and producing fungi like hyphae and conidia/sporangia in culture media. They are the potential producers of several secondary metabolites, include antibiotics, immunosuppressive agents, antitumor agents, and enzymes (Chaudhary et al. 2013).

Actinomycetes displays a range of unique prokaryotic life cycle and play an important role in organic matter recycling of soil ecosystem (Veiga et al. 1983). They are the most abundant soil organisms, produces a characteristic "earthy smell" because the existence of metabolite "geosmin" and grows as thread-like filaments in the soil (Sprusansky et al. 2005). The Actinomycetes are ubiquitous group of microbes extensively distributed in nature all around the world (Srinivasan et al. 1991). They are largely soil occupants (Kuster 1968) but widely distributed in diverse habitats including sediments collected from deep sea vents (Colquhoun et al. 1998), from the deepest depth of Mariana Trench (Pathom-aree et al. 2006), cryophilic soil taken from Antarctica (Moncheva et al. 2002) and also has been reported from desert soil (Diraviyam et al. 2011). A comparative survey on Actinomycete population has been demonstrated that it is greatly found in surface layer of soils and decreases gradually when the depth increased (Takahashi and Omura 2003) (Fig. 6.1).

Actinobacteria are characterized by the development of branching filaments or rods with nonseptate hyphae (Fig. 6.2). Several special conditions, septa may be visualized in different forms. The sporulating mycelium may be straight, branching or nonbranching or spiral shaped. The spores are cylindrical, spherical or oval. The cell wall has a rigid structure that helps for maintaining the shape of the cell also prevents from breaking of the cells under high osmotic pressure. The wall contains

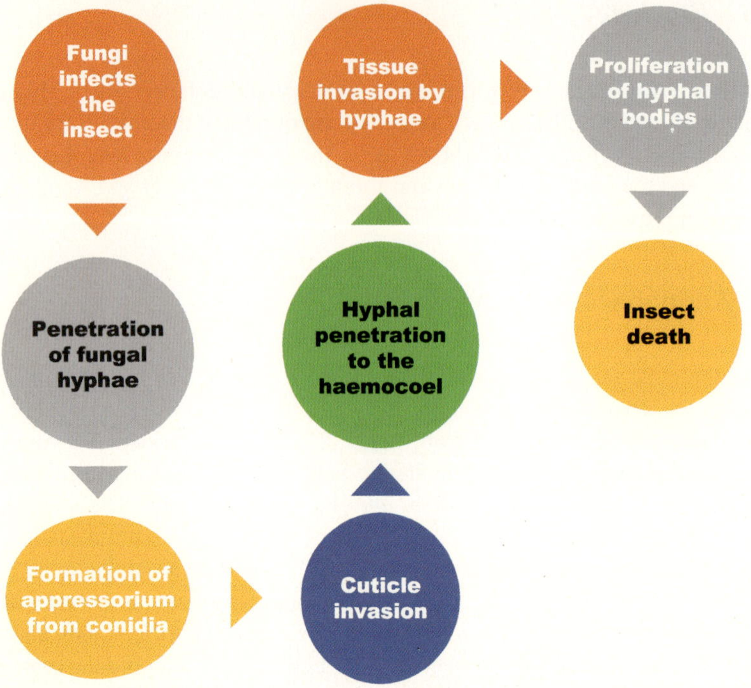

Fig. 6.1 Depiction of infection process by Entomopathogenic fungi

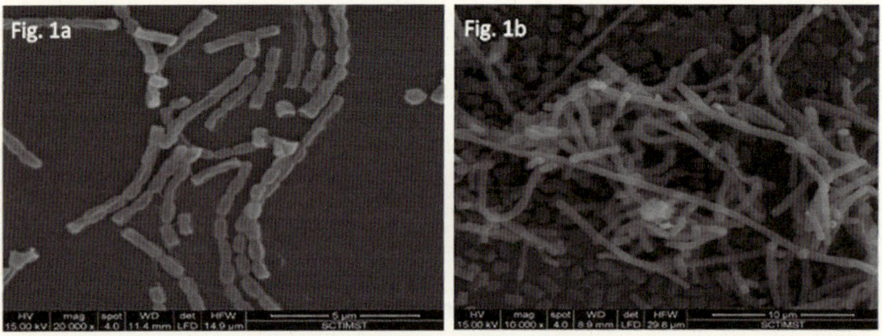

Fig. 6.2 Scanning electron microscopy images of Actinobacteria showing branching rods (Fig. 1a: Streptomyces sp. TBG-AL19) or filaments (Fig. 1b: Streptomyces sp. TBG-AL13)

variety of composite compounds like peptidoglycan [glycan chains with alternating N-acetyl-d-muramic acid (NAM), N-acetyl-d-glucosamine (NAG) and diaminopimelic acid (DAP)], teichoic and teichuronic acids and some polysaccharides (Manuselis and Mahon 2007). They shows similar cell wall chemical composition like in gram positive bacteria but their well-developed cultural and morphological characteristics, finely separated Actinomycetes from all other common bacteria

(Das et al. 2008). They are included volume four of Bergey's Manual of Determinative Bacteriology, comes under the order Actinomycetales. It is separated in to four families- Streptomycetaceae, Actinomycetaceae, Mycobacteriaceae, and Actinoplanaceae (Williams et al. 1989). Streptomyces and Micromonospora are the two generally defined actinobacterial genus. Although the Streptomyces genus is recognized as the largest reservoir of natural bioactive products (Terkina et al. 2006). About 75% of available natural antibiotics are produced in the members of genus Streptomyces (Jimenez-Esquilin and Roane 2005).

6.10 Plant Growth Promotion by Actinobacteria

The popular soil actinobacteria indicated their optimal development in neutral or alkaline conditions. The filamentous sporulating actinobacteria have fascinated superior interest because of their ability to flourish in extremely diverse soil circumstances and also due to their significant ecological role in nutrient cycling. Furthermore, these are existent widely in the plant rhizosphere and secrete innumerable agro active compounds. In the last few years, Actinobacteria gained much attention are also included in the category of plant growth promoting rhizobacteria (PGPR), free living agriculturally important bacteria, due to its robust antimicrobial potential, and dominant soil saprophytic nature (Franco-Correa et al. 2010). These bacteria have voluminous beneficial properties on agricultural production by overwhelming microbial plant pathogens, improving nutrient availability and increasing assimilation. Hence, the use of plant growth promoting Actinobacteria (PGPA) diminishes the negative impact of inorganic fertilizers, thus by improving crop quality, fertility and yield. Actinomycetes are actively involved agricultural productivity by production of plant growth promoting substances such as plant hormones, siderophores etc. and actively involved in increasing soil fertility and in stress alleviation. Figure 6.3 representing the roles of actinobacteria that helps maintaining sustainable agriculture.

6.10.1 Actinobacterial "Geosmin" as Soil Fertility Indicator

Actinobacteria are well-known producers of organic compound "geosmin", is responsible for the earthy odour of soil mostly after the rain. Among the actinomycetes, the *Streptomyces* strains are the most common producers of this volatile compound and released into the soil after the death of these microorganisms. The geosmin biosynthesis in *Streptomyces coelicolor* was has been showed (Jiang et al. 2006, 2007). During geosmin biosynthesis, the substrate farnesyl diphosphate is converts to geosmin by a single enzyme, geosmin synthase, in a two-step reaction. It is a bicyclic alcohol ($C_{12}H_{22}O$) and a derivative of decalin, frequently used for soil biological fertility. The soil intense "geosmin in soil are the major indicator of its

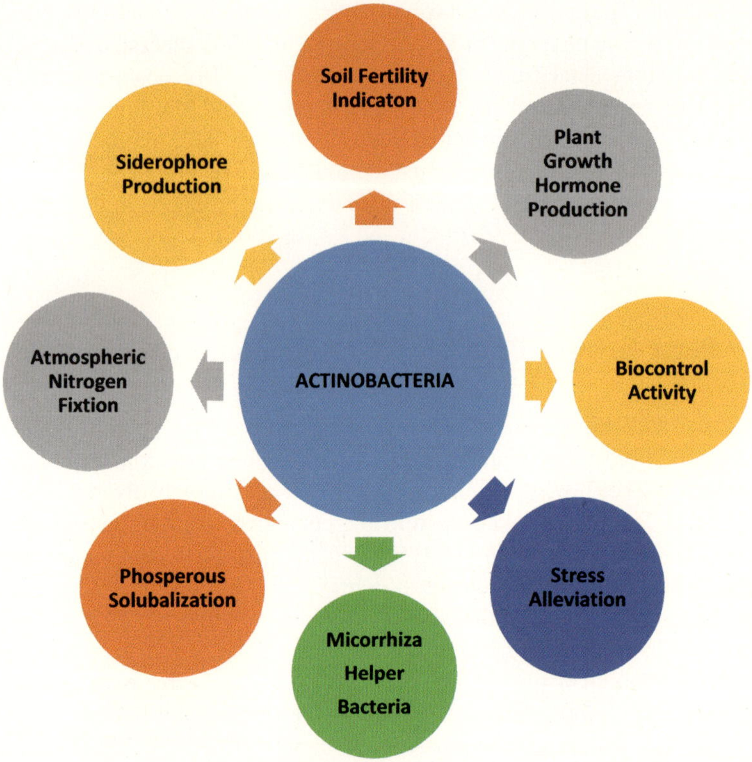

Fig. 6.3 Actinobacteria in sustainable agriculture

fertility status. Soils with this earthy smell are considered as more fertile. The geosmin smell can be detected by human nose up to five parts per trillion. Besides to execute the role as soil fertility indicator, actinobacteria can also be involved in the nutrients biogeocycling particularly of phosphorous, nitrogen and iron.

6.10.2 Plant Growth Hormone Production by Actinobacteria

A known plant growth regulator, Indole acetic acid (IAA), an active form of auxins, plays a vibrant role in plant development. It promotes lateral roots and apical meristem development along with roots elongation (López et al. 2004). IAA are produced by several Actinobacteria in substantial quantities (Ghosh et al. 2011). In Streptomyces IAA production is tryptophan dependent (Lin and Xu 2013). A significant quantity of IAA (52.3 $\mu g.ml^{-1}$) was made by *Streptomyces* sp. obtained from the medicinal plants rhizosphere region. *Streptomyces filipinensis* no. 26 isolate produced IAA stimulated the growth of tomato under greenhouse environments (Khamna et al. 2009). 29 Actinobacterial isolates studied in soil from yam rhizosphere region found that 28 isolates produced IAA and there 11 were stimulated in vitro growth of Arabidopsis (Palaniyandi et al. 2013a, b). Similarly, many IAA produced Actinobacteria has been reported to increase plant shoot and root lengths. *Streptomyces* genus (Da Silva Sousa et al. 2008; Shrivastava et al. 2008), *Frankia* sp. (Sivasithamparam et al. 2003), *Nocardia* sp. (El-Tarabily and Sivasithamparam 2006), *Kitasatospora* sp. (Shrivastava et al. 2008) have been broadly considered as IAA producers. IAA helps for the germination of *Streptomyces atroolivacezlz* spore by acting as an endogen regulator and also involved in the actinobacteria differentiation. Three actinobacterial species, *Streptomyces olivaceoviridis*, *S. rochei* and *S. rimosus* cultures, not only produces IAA, but the studies indicated that they are excellent producers of cytokinin and gibberellins -like substances, also showed enhanced the growths in wheat plants (Aldesuquy et al. 1998).

The reports that supporting cytokinin producion by actinobacteria are very few when compared with IAA production. According to Joshi and Loria 2007, cytokinin producing Actinomycetes like *Streptomyces turgidiscabies* and *Rhodococcus fascians* are pathogenic to plants and causes tobacco leafy galls. *Streptomyces hygroscopicus*, an endophytic actinobacterium synthesised pteridic acids A and B along with auxin-like activity, promoted the hypocotyls of kidney beans adventitious roots formation (Ortı´z-Castro et al. 2008). *Streptomyces turgidiscabies* Car8 have a gene cluster for cytokinin biosynthesis and produces the leafy galls (Joshi and Loria 2007). An actinobacterium, *Arthrobacter globiformis*, was found to produce small amounts of a gibberellin-like substance.

6.10.3 Actinobacterial Siderophores in Crop Protection

Siderophores are low molecular weight, organic molecules chelating ferric ion produced by various microorganisms surviving under iron-limiting conditions, increases the uptake of iron in microorganism's compounds. The function of siderophore is to scavenge environmental ferric iron, which are inaccessible to microorganisms at physiological pH (7.35–7.40) and converted as mineral, which is virtually always essential and accessible to the microbial cell (Saha et al. 2015). In agriculture, siderophores promotes plant growth and can used equally as an eco-friendly biocontrol agent that substitute to harmful chemical pesticides. Iron is a micronutrient, in plants it is essential for redox reactions, chlorophyll biosynthesis and some other physiological activities. Iron limitation in soil significantly decreases the yield in agriculture sector. Microbial siderophores can be used as an efficient and easily available iron source in plants. Siderophore can also use as a prospective biocontrol representative against phyto-pathogens. Siderophores strongly bind with the iron, thus limits the availability of iron plant pathogens and enabling the killing of phytopathogens due to iron limitation (Ahmed and Holmstrom 2014). Numerous studies have been conducted to demonstrate the biocontrol role of siderophores in crop protection.

Actinobacteria is one the supreme group of microorganism involved in siderophores production. An endophytic Streptomyces sp obtained from the rhizosphere of a Thai jasmine rice plant produced considerable amount of Siderophore prompted plant growth and evidently raised root- shoot lengths and biomass (Rungin et al. 2012). Actinobacterial strains such as Thermobifida and Streptomyces MCR3 synthesis a great amount of hydroxamate-type siderophores using the glucose as the sole carbon source. The plants also have a mechanism to increasing the structure of microbial community around the root soil areas. They synthesize certain phenolic exudates from roots that enhances the development of additional siderophore-secreting microbes. This improves the iron solubility and moreover enhances iron uptake in plants (Jin et al. 2010).

6.10.4 Atmospheric Nitrogen Fixation by Actinobacteria

Nitrogen fixation in plants is referred as the assimilation of gaseous N into amino acids. Nitrogen fixation by actinobacteria have been broadly reported (Clawson and Benson 1999; Tjepkema et al. 2002). The heterotrophic actinobacteria necessitate carbon sources to acquire the energy essential for nitrogen fixation. Each bacteria varies in the ability of nitrogen fixation and carbon metabolism, showing dissimilar units in acetylene reduction assay (ARA). It is centred on distinguishing the existence nitrogenase enzyme, which reduces nitrogen (N2) to ammonium. This essay also estimates the enzymatic reduction of acetylene to ethylene. Most extensively studied actinobacterial nitrogen fixation is in *Frankia*, mostly it lives in symbiotic relationship with dicotyledons. It has an exceptional feature, the vesicles specialized for nitrogen fixation. The *Frankia* infected with plants through symbiosis are known as actinorhizal plants and produces root nitrogen-fixing nodules (Yamaura et al. 2010). Apart from Frankia, *Streptomyces thermoautotrophicus*, a thermophilic Actinobacteria can fix atmospheric nitrogen. The nitrogenase enzyme in *S. thermoautotrophicus* is O_2 insensitive, and also utilizes N_2 as nitrogen source, is exclusive in biological nitrogen fixation (Gadkari et al. 1992).

Several PGPA creates a symbiosis relationship between other nitrogen-fixing microorganisms. For example, *Streptomyces lydicus* WYEC108 induces root nodulation only after inoculating with *Rhizobium* sp in cow pea plant. It works as a nodule colonizes on the cell surface layers of the nodules, which increases nodule size and bacteriods vigour by enhancing iron and other nutrients assimilation. *Streptomyces kanamyceticus* revealed a negative effect in the nodule formation when they are inoculated with *Bradyrhizobium japonicum*. Though, the co-inoculation of *Streptomyces kanamyceticus* with *Bradyrhizobium japonicum,* an antibiotic-resistant strain, showed a positive effect in increasing root nodule vigour and number. This a very important observation specifies that Streptomyces producing antimicrobial substances masked the facilitate nodulation. The symbiosis with *Frankia* sp. actinobacteria such as *Micromonospora, Streptomyces* and *Actinoplanes* were capable to stimulating root nodule formation in *Discaria trinervis*. However, studies indicated that co-inoculation of *Frankia* mycelia did not encourage root nodulation. But, *Frankia* culture filtrates promoted root nodulation, suggesting the occurrence of nodule-inducing substances in culture filtrate (Solans et al. 2009). In addition, actinobacteria were described to stimulate symbiosis between plants and mycorrhizal fungi (Frey-Klett et al. 2007).

6.10.5 Phosphorus Solubilization by Actinobacteria

Phosphorus (P) is a key vital macronutrients for plants. Phosphorus deficiencies are wide spectrum because major phosphorus portion in soil is unavailable to plants, and mainly it is applied as phosphatic manure in soil. However, a major percentage of the used phosphorus is promptly immobilized and becoming inaccessible for plants due to the formation of metal complexes with Al, Fe and Si. Many soil microorganisms involved phosphorus transformation processes of soil, solubilizing soil phosphorus and accessible it for plants growth. Phosphate solubilization is widely exhibited in actinobacteria such as *Streptomyces, Micromonospora, Micrococcus, Thermobifida* and *Kitasatospora*. Although the mechanism behind actinobacterial phosphorus solubilisation is not fully understood. Actinobacteria with ability to solubilizing rock phosphate were reported that it promotes the wheat plants growth in vitro and in vivo conditions (Hamdali et al. 2008). Actinobacterial P-solubilizing strains have a dual benefit, they are revealed to suppress damping off affected by Pythium ultimum and also increased wheat growth in P-deficient soil. This dual assistance is advantageous in cumulative agricultural production (Oliveira et al. 2009). PGPA solubilize P by the producing organic acid and by acidification of rhizosphere. Furthermore, phosphorus availability is recognised by the chelating cations such as Fe^{+2}, Al^{+3} or Ca^{+2} helps in the solubilization of phosphate. Actinobacteria secretes phosphatases such as phytases and acidic/alkaline phosphatases which can hydrolyse phytate, constitutes up to 60% of soil organic phosphorus (Palaniyandi et al. 2013a, b). Actinobacteria such as *Nocardia* sp., *Micromonospora* sp., *Actinomadura* sp., *Actinoplanes* sp., *Rhodococcus* sp., *Microbispora sp. and Streptosporangium sp.* produce alkaline or acidic phosphatase enzymes, depending on reaction conditions.

6.10.6 Actinobacteria in Plant Biotic Stress Alleviation

Abiotic stresses like salinity, drought, heavy metal contamination and nutrient stress reduced agricultural productivity at a significant level. These stresses often cause the production of gaseous hormone ethylene in plants which negatively affects plant growth. Some PGPA have the ability to produce stress alleviating compound thus by enhancing plant growth by several mechanisms. One well studied mechanism is ACC deaminase production by actinobacteria. Enzyme ACC deaminase converts

ACC, an ethylene precursor in plants, to α- ketobutyrate and ammonia, in that way dropping stress at ethylene level and enlightening plant growth (Glick 2005). ACC deaminase activities were reported from some halotolerant actinobacteria such as *Corynebacterium variabile, Micrococcus yunnanensis,* and *Arthrobacter nicotianae*, promots canola plants growth under salt stress conditions (Siddiqui et al. 2010). ACC deaminase activity reported in Arthrobacter sp. EZB4 from pepper plants significantly reduced some osmotic stress-inducible gene expressions. ACC activity was also detected in Streptomyces filipinensis no. 15 strain. When this strain was co-inoculated with tomato plants, it significantly reduces ACC deaminase levels in roots and shoots and promotes plant growth (Sziderics et al. 2007). Recent studies on 29 actinobacterial strains from yam rhizosphere revealed only 6 were showed ACC deaminase activity, belonged to the genus Streptomyces (Palaniyandi et al. 2013a, b). Recently a novel type Actinobacterial drought stress tolerance was also reported. Inoculation of mountain laurel tissue-cultured seedlings with Streptomyces padanus AOK-30, an endophyte, showed callose accumulation in cell wall, which enhanced drought tolerance in seedlings.

6.10.7 Antagonistic Activity against Plant Pathogens

Actinobacteria have been recognised as one of the chief antagonistic microbe against some plant pathogens based on their ability to secrete metabolic compounds, which inhibit the pathogens growth by competing for nutrients. Antibiotics produces actinobacteria in rhizosphere region thus helps for inhibiting the growth of fungal pathogens, which in turn promots effective rhizosphere colonization. For example, antibiotics methyl vinyl ketone produced by actinobacteria alters pathogenic fungal morphology and finally kill them. Streptomyces genus have widely exploited for antibiotic production and have revealed antagonistic activity against *Pythium aphanidermatum, Alternaria* sp., *Colletotrichum higginsianum, Fusarium oxysporum and Acremonium lactucum* (Hong et al. 2002). According to Molano et al. an antibiotic actinomycin, synthesized by *Nocardia* sp. showed in vitro inhibition contrary to Fusarium oxysporum isolated from rhizosphere soil sample. Antifungal agents produced by Actinomycetes shows wide spectrum activity against plant fungal pathogens. An antifungal metabolite Mildiomycin, isolated from *Streptoverticillium rimofaciens* inhibits fungal protein biosynthesis and is intensely active against powdery mildews on various crops. The main site of action of these fungicide are the location of chitin synthesis in fungal cell walls. Examples of some actinobacterial antifungal agents and their functions are showing in Table 6.2. *Streptomyces lydicus* WYEC 108 producing water soluble biofungicide was permitted by Natural Industries Inc., TX, USA and registered as Actinovate soluble in 2004 and it effectively controls some common soilborne and foliar diseases.

Actinomycetes frequently produces hydrolytic enzymes against enormous spectrum of pathogenic fungi. Production of certain cell wall-degrading enzymes like glucanase and chitinase also causes fungal cell wall degradation and hinder the

Table 6.2 Actinobacterial antifungal agents and mode of actions

Antifungal agent	Producing Actinomycetes	Mode of action	References
Actinomycins	*Streptomyces anulatus*	Protein synthesis inhibition	Bister et al. (2004)
Validamycin	*Streptomyces hygroscopicus*	It inhibits enzyme trehalase	Iwasa et al. (1970)
Tetracenomycin	*Streptomyces canus*	Inhibits DNA replication	Zhang et al. (2013)
Resistomycin	*Streptomyces canus*	Inhibits DNA and RNA synthesis	Zhang et al. (2013)
Polyoxin B	*Streptomyces cacaoi*	Inhibits chitin synthesis	Isono et al. (1965)
Nikkomycin	*Streptomyces tendae*	Inhibits chitin synthesis	Bormann et al. (1985)
Streptothricin	*Streptomyces lavendulae*	Protein synthesis inhibition	Waksman and Woodruff (1942).
Natamycin	*Streptomyces natalensis*	Targets ergosterol in fungal membrane	Struyk et al. (1958)
Oligomycin	*Streptomyces diastatochromogenes*	Inhibitor for ATP synthase	Smith et al. (1954)
Fungichromin	*Streptomyces padanus*	Prevents oospore induction	Shih et al. (2003)
Kasugamycin	*Streptomyces kasugaensis*	Inhibits protein synthesis	Umezawa et al. (1965)
Amphotericin B	*Streptomyces nodosus*	Targets ergosterol in fungal membrane	Linke et al. (1974)
Transvalencin	*Nocardia transvalensis*	Squalene epoxidase inhibition	Hoshino et al. (2004)
Blasticidin	*Streptomyces griseochromogenes*	Inhibits protein synthesis	Takeuchi et al. (1958)
Galbonolides	*Streptomyces galbus*	Inhibits sphingolipid biosynthesis	Fauth et al. (1986)
Chloramphenicol	*Streptomyces venezuelae*	Prevents protein chain elongation	Matsuoka et al. (1953)
Candicidin	*Streptomyces griseus*	Targets ergosterol in fungal membrane	Acker and Lechevalier (1954)

growth. Chitin is an important structural component of fungal cell wall, composed by residues of N-acetyl-D-Glucosamine, which can hydrolysed by chitinase enzymes. Actinobacteria have been reported as the dominant organisms involved in the production of chitinase enzyme. *Streptomyces* species are the chief chitinolytic microbial group and promising fungal antagonist (Asha poorna and Pradeep 2016). Numerous chitinolytic enzymes have been recognised in some actinobacterial species such as *Streptomyces aureofaciens*, *Streptomyces antibioticus*, *Streptomyces lividens*, *Streptomyces halsteii* AJ-7, *Streptomyces plicatus*, and *Streptomyces lydicus* WYEC108.

Actinomycetes fungus antagonism is also related with the production of β-glucanase enzymess. A best effective fungal antagonist against *Phytophthora* spp is *Streptomyces* sp. EF-14 secreted β-1,3 glucanase and β-1,6 glucanase. Soil added *Streptomyces nigellus* strain NRC 10 reduced damping off diseases affected by *Pythium ultimum* in tomato plants. Studies designated that these strains are tremendous producers of β-1,4 glucanases and β-1,3 glucanases (Helmy et al. 2010). β-1,4, β-1,3, and β-1,6 glucanases produced from *Actinoplanes philippinesis, Micromonospora chalcea* and *Microbispora rosea* caused *Pythium aphanidermatum* hyphae lysis and thereby reduces cucumber damping- off disease (El-Tarabily 2006). Some *Streptomyces* strains produced β-1,3-glucanase such as CAI-24, CAI-127, CAI-121, KAI-32 and KAI-90 demonstrated significant *Fusarium oxysporum* f. sp. Cicero biocontrol activity causes Fusarium wilt of chickpea (Gopalakrishnan et al. 2013). According to Lekshmi et al. (2017) exo- β-1,4- glucanase activities in Streptomyces sp. are also considered as an indicator of environmental and soil quality changes.

6.10.8 Actinobacteria as Mycorrhiza (MA) Helper Bacteria

Bacteria also present inside mycorrhizas as colonies, hence the plants takes these strains as beneficial for the symbiosis. Microorganisms encourage the formation mycorrhiza through several activities, like fungal propagules stimulation in pre-symbiotic infective stages, enable the formation of inputs points in the roots and also increasing growth rate. Actinomycetes have the ability to promote mycelial growth that is correlated with their influence on mycorrhizal formation predominantly in hyphal growth promotion. Streptomyces also evolved in mechanisms to facilitate mycorrhiza formation by stimulating fungal growth and by reducing plant defence responses. At the time of rhizobacterial infection, plants attain a high resistance against plant pathogen attack. Later, the investigations revealed that such disease resistance have been induced by some endophytic *Streptomyces* sp. Based on the studies of Carpenter-Boggs et al. 1995, an actinobacteria, *Streptomyces orientalis*, have the ability to secrete volatile compounds, have an advantageous effect on *Gigaspora margarita* spore germination. An Auxofuran compound released by mycorrhiza helper *Streptomyces* spp. AcH 505 influences fungal metabolism and helps mycorrhizal formation by improving root colonization and also prompts a systemic defense response against mycorrhizal fungus (Schrey et al. 2007).

6.11 Conclusion

The current farming interest is predominantly placed in eco-friendly and sustainable agricultural practices. Efficient microorganisms and their products may improve plant growth in many ways compared to synthetic fertilizers, pesticide and insecticides and help in sustainability of environment and crop productivity. Nowadays sustainable agriculture is vital as it compromises the prospective to meet our agricultural necessities. This kind of agriculture fully utilized environmental resources through special farming technique and at the same time it is environment friendly and warrants healthy and safe agricultural foodstuffs. The practice of plant growth-promoting microorganisms, for improving fertility of soils, increasing crop yield and reducing the worst deleterious impact of chemical fertilizers, has developed as a most attractive strategies for emerging sustainable agriculture. The use of fungi and actinobacteria in agriculture offers an environmentally sustainable approach for agricultural production and overall global health. Hence, agro active natural compounds, effective for sustainable farming practices from fungi and actinobacteria are not fully illustrated and that are currently considered as a foremost research area in the field of agriculture, biotechnology and microbiology. Current and future advances in our knowledge about of diversity, mechanisms, applications and formulations of plant growth promoting microorganisms facilitating reliable development and management of sustainable agricultural systems.

References

Abbasi H, Akthar A, Sharf R (2015) Vesicular Arbuscular Mycorrhizal (VAM) Fungi: a tool for sustainable agriculture. Am J Plant Nutr Fert Technol 5(2):40–49

Acker RF, Lechevalier H (1954) Some nutritional requirements of Streptomyces griseus 3570 for growth and candicidin production. Appl Microbiol 2:152–157

Ahmed E, Holmstrom SJM (2014) Siderophores in environmental research: roles and applications. J Microbial Biotechnol 7:196–208

Aldesuquy HS, Mansour FA, Abo-Hamed SA (1998) Effect of the culture filtrates of Streptomyces on growth and productivity of wheat plants. Folia Microbiol 43:465–470

Araujo ASF, Leite LFC, Santos VB, Carneiro RFV (2009) Soil microbial activity in conventional and organic agricultural systems. Sustainability 1:268–276

Asha Poorna C, Pradeep NS (2016) Identification of novel Chitinolytic Streptomyces. spp from a sacred grove and it's in vitro antagonistic activity analysis. Int J Curr Microbiol App Sci 5(8):916–928

Barker S, Tagu D (2000) The roles of auxins and cytokinins in mycorrhizal symbioses. J Plant Growth Regul 19:144–154

Benjamins R, Scheres B (2008) Auxin: the looping star in plant development. Annu Rev Plant Biol 59:443–465

Bister B, Bischoff D, Strobele M, Riedlinger J, Reicke A, Wolter F, Bull AT, Zahner H, Fiedler HP, Sussmuth RD (2004) Abyssomicin C: a polycyclic antibiotic from a marine Verrucosispora strain as an inhibitor of the p-aminobenzoic acid/tetrahydrofolate biosynthesis pathway. Angew Chem Int Ed Engl 43:2574–2576

Bleecker AB, Kende H (2000) Ethylene: a gaseous signal molecule in plants. Annu Rev Cell Dev Biol 16:1–18

Bormann C, Huhn W, Zahner H, Rathmann R, Hahn H, Konig WA (1985) Metabolic products of microorganisms. 228. New nikkomycins produced by mutants of Streptomyces tendae. J Antibiot 38:9–16

Brian P, Elson G (1954) The plant-growth-promoting properties of gibberellic acid, a metabolic product of the fungus Gibberella fujikuroi. J Sci Food Agric 5:602–612

Carpenter-Boggs L, Loynachan TE, Stahl PD (1995) Spore germination of Gigaspora margarita stimulated by volatiles of soil-isolated actinomycetes. Soil Biol Biochem 27:1445–1451

Chanclud E, Kisiala A, Emery NRJ, Chalvon V, Ducasse A, RomitiMichel C, Gravot A, Kroj T, Morel JB (2016) Cytokinin production by the rice blast fungus is a pivotal requirement for full virulence. PLoS Pathog 12:e1005457

Chaudhary HS, Bhavana S, Anju RS, Saurabh S (2013) Diversity and versatility of Actinomycetes and its role in antibiotic production. J App Pharm Sci 3:S83–S94

Chet T, Schichler H, Haran S, Appenheim AB (1993) Cloned chitinase and their role in biological control of plant pathogenic fungi. In: Proceedings of the international symposium on chitin enzymology, pp 47–48

Clawson ML, Benson DR (1999) Natural diversity of Frankia strains in actinorhizal root nodules from promiscuous hosts in the family Myricaceae. Appl Environ Microbiol 65(10):4521–4527

Colquhoun JA, Mexson J, Goodfellow M, Ward AC, Horikoshi K, Bull AT (1998) Novel rhodococci and other mycolate actinomycetes from the deep sea. Antonie Van Leeuwenhoek 74:27–40

Crocoli C, Kettner J, Dorffling K (1991) Abscisic acid in saprophytic and parasitic species of fungi. Phytochemistry 30:1059–1060

Da Silva Sousa C, Fermino Soares A, da Silva Garrido M (2008) Characterization of Streptomyceswith potential to promote plant growth and biocontrol. Sci Agric 65:50–55

Das S, Lyla PS, Khan SA (2008) Distribution and generic composition of culturable marine actinomycetes from the sediments of Indian continental slope of Bay of Bengal. Chinese J Oceanol Limnol 26:166–177

Diraviyam T, Radhakrishnan M, Balagurunathan R (2011) Antioxidant activity of melanin pigment from Streptomyces species D5 isolated from Desert soil, Rajasthan, India. Drug Invent Today 3:12–13

El-Tarabily KA (2006) Rhizosphere-competent isolates of streptomycete and non-streptomycete actinomycetes capable of producing cell-wall-degrading enzymes to control Pythium aphanidermatum damping-off disease of cucumber. Can J Bot 84:211–222

El-Tarabily K, Sivasithamparam K (2006) Non-streptomycete actinomycetes as biocontrol agents of soil-borne fungal plant pathogens and as plant growth promoters. Soil Biol Biochem 38:1505–1520

Faria M, Wraight SP (2001) Biological control of *Bemisia tabaci* with fungi. Crop Prot 20(9):767–778

Fauth U, Zahner H, Muhlenfeld A, Achenbach H (1986) Galbonolides A and B: two non-glycosidic antifungal macrolides. J Antibiot 39:1760–1764

Florez FJP (2002) Fungi for coffee berry borer control- Colombia proceedings of the 35th annual meeting of the Society of Invertebrate Pathology. Sci Technol 11(2):245–250

Franco-Correa M, Quintana A, Duque C, Suarez C, Rodríguez MX, Barea JM (2010) Evaluation of actinomycete strains for key traits related with plant growth promotion and mycorrhiza helping activities. Appl Soil Ecol 45:209–217

Frey-Klett P, Garbaye J, Tarkka M (2007) The mycorrhiza helper bacteria revisited. New Phytol 176:22–36

Gadkari D, Mo¨rsdorf G, Meyer O (1992) Chemolithoautotrophic assimilation of dinitrogen by Streptomyces thermoautotrophicus UBT1: identification of an unusual N2-fixing system. J Bacteriol 174:6840–6843

Ghosh S, Ghosh P, Maiti T (2011) Production and metabolism of indole acetic acid (IAA) by root nodule bacteria (Rhizobium): a review. J Pure Appl Microbiol 5:523–540

Glick B (2005) Modulation of plant ethylene levels by the bacterial enzyme ACC deaminase. FEMS Microbiol Lett 251:1–7

Gopalakrishnan S, Srinivas V, Sree Vidya M, Rathore A (2013) Plant growth-promoting activities of Streptomyces spp. in sorghum and rice. SpringerPlus 2:574

Gupta S, Dikshit AK (2010) Biopesticides: an ecofriendly approach for pest control. J Biopest 3(1):186–188

Hajeck AE, Leger RJ (1994) Interactions between fungal pathogens and insect hosts. Annu Rev Entomol 39:293–322

Hamdali H, Hafidi M, Virolle M, Ouhdouch Y (2008) Rock phosphate-solubilizing actinomycetes: screening for plant growth-promoting activities. World J Microbiol Biotechnol 24:2565–2575

Harrier LA (2001) The arbuscular mycorrhizal symbiosis: a molecular review of the fungal dimension. J Exp Bot 52(1):469–478

Helmy SM, Zeinat K, Mahmoud SA, Moataza S, Nagwa M, Amany H (2010) Streptomyces nigellus as a biocontrol agent of tomato damping-off disease caused by pythium ultimum. Electron J Pol Agric Univ 13:4

Hong T, Cheng C, Huang J, Meng M (2002) Isolation and biochemical characterization of an endo-1, 3-β-glucanase from Streptomyces sioyaensis containing a C-terminal family 6 carbohydrate-binding module that binds to1, 3-β-glucan. Microbiology 148:1151–1159

Hoshino Y, Mukai A, Yazawa K, Uno J, Ando A, Mikami Y, Fukai T, Ishikawa J, Yamaguchi K (2004) Transvalencin A, a thiazolidine zinc complex antibiotic produced by a clinical isolate of Nocardia transvalensis. II. Structure elucidation. J Antibiot 57:803–807

Isono K, Nagatsu J, Kawashima Y, Suzuki S (1965) Studies on polyoxins, antifungal antibiotics. Part I. Isolation and characterization of polyoxins A and B. Agric Biol Chem (Tokyo) 29:848–854

Iwasa T, Yamamoto H, Shibata M (1970) Studies on validamycins, new antibiotics. I. Streptomyces hygroscopicus var. limoneus nov. var., validamycin- producing organism. J Antibiot 23:595–602

Jiang J, He X, Cane DE (2006) Geosmin biosynthesis. Streptomyces coelicolor germacradienol/germacrene D synthase converts farnesyl diphosphate to geosmin. J Am Chem Soc 128:8128–8129

Jiang J, He X, David E, Cane DE (2007) Biosynthesis of the earthy odorant geosmin by a bifunctional Streptomyces coelicolor enzyme. Nat Chem Biol 3:711–715

Jimenez-Esquilin AE, Roane ETM (2005) Antifungal activities of actinomycete strains associated with high-altitude sagebrush rhizosphere. J Ind Microbiol Biotechnol 32(8):378–381

Jin CW, Li GX, Yu XH, Zheng SJ (2010) Plant Fe status affects the composition of siderophore-secreting microbes in the rhizosphere. Annals of botany, mcq 071

Joshi MV, Loria R (2007) Streptomyces turgidiscabies possesses a functional cytokinin biosynthetic pathway and produces leafy galls. Mol Plant Microbe Interact 20:751–758

Kalra A, SPS K (2007, 2007) Research and development priorities for biopesticide and biofertiliser products for sustainable agriculture in India. In: Teng PS (ed) Business potential for agricultural biotechnology. Asian Productivity Organisation, pp 96–102

Khamna S, Yokota A, Lumyong S (2009) Actinomycetes isolated from medicinal plant rhizosphere soils: diversity and screening of antifungal compounds, indole-3- acetic acid and siderophore production. World J Microbiol Biotechnol 25:649–655

Kim JJ, Lee MH, Yoon CS, Kim HS, Yoo JK, Kim KC (2002) Control of cotton aphid and greenhouse whitefly with a fungal pathogen. J Natl Inst Agric Sci Technol:7–14

Kuster E (1968) Taxonomy of soil actinomycetes and related organisms. In: Gray S, Parkinson T (eds) Ecology of soil bacteria. Liverpool University Press, Liverpool, pp 322–336

Lekshmi KE, Reshma RA, Gayathri V, Pradeep NS (2017) Enzyme activity profiling of exo -β-1,4-glucanase, endo-β-1,3- glucanase and protease in streptomyces species from highland, midland and lowland areas of Kerala, India. Int J Adv Res 5(2):2431–2444

Lin L, Xu X (2013) Indole-3-acetic acid production by endophytic Streptomyces sp. En-1 isolated from medicinal plants. Curr Microbiol 67:209–217

Linke HA, Mechlinski W, Schaffner CP (1974) Production of amphotericin B-14C by Streptomyces nodosus fermentation, and preparation of the amphotericin B-14C-methyl-ester. J Antibiot 27:155–160

López J, Acosta M, Sánchez-Bravo J (2004) Role of basipetal auxin transport and lateral auxin movement in rooting and growth of etiolated lupin hypocotyl. Physiol Plant 121:294–304

Manuselis G, Mahon CR (2007) In: Manon CR, Lehman DC, Mauselis G (eds) Textbook of diagnostic microbiology. Saunders, pp 3–13

Mathew SO, Sandhu SS, Rajak RC (1998) Bioactivity of *Nomuraea rileyi* against *Spilosoma obliqua*: effect of dosage, temperature and relative humidity. J Indian Bot Soc 77:23–25

Matsuoka M, Yagishita K, Umezawa H (1953) Studies on the intermediate metabolism of chloramphenicol production. II. On the carbohydrate metabolism of Streptomyces venezuelae. Jpn J Med Sci Biol 6:161–169

Mobasser H, Tavassoli A (2013) Study of vesicular arbuscular mycorrhizal (VAM) fungi symbiosis with maize root and it effect on yield components, yield and protein content of maize in water deficit condition. J Nov Appl Sci 2(10):456–460

Moncheva P, Tishkov S, Dimitrova N, Chipeva V, Bogatzevska N (2002) Characteristics of soil actinomycetes from Antartica. J Cult Coll 3:3–14

Nihorimbere V, Ongena M, Smargiassi M, Thonart P (2011) Beneficial effect of the rhizosphere microbial community for plant growth and health. Biotechnol Agron Soc Environ 15(2):327–337

Nunez E, Iannacone J, Gomez H (2008) Effect of two entomopathogenic fungi in controlling *aleurodicus cocois* (Curtis, 1846) (Hemiptera: Aleyrodidae). Chilean J Agric Res 68(1):21–30

Oliveira CA, Alves VMC, Marriel IE, Gomes EA, Scotti MR, Carneiro NP, CT G˜e, Schaffert RE, Sa´ NMH (2009) Phosphate solubilizing microorganisms isolated from rhizosphere of maize cultivated in an oxisol of the Brazilian Cerrado Biome. Soil Biol Biochem 41:1782–1787

Ortı´z-Castro R, Valencia-Cantero E, Lopez-Bucio J (2008) Plant growth-promotion by Bacillus megaterium involves cytokinin signaling. Plant Signal Behav 3:263–265

Palaniyandi SA, Yang SH, Zhang L, Suh JW (2013a) Effects of actinobacteria on plant disease suppression and growth promotion. Appl Microbiol Biotechnol 97:9621–9636

Palaniyandi SA, Yang SH, Suh JW (2013b) Extracellular proteases from Streptomyces phaeopurpureus ExPro138 inhibit spore adhesion, germination and appressorium formation in Colletotrichum coccodes. J Appl Microbiol 115:207–217

Pathom-aree W, Stach JE, Ward AC, Horikoshi K, Bull AT, Goodfellow M (2006) Diversity of actinomycetes isolated from challenger deep sediment (10,898 m) from the Mariana Trench. Extremophiles 10:181–189

Peleg Z, Blumwald E (2011) Hormone balance and abiotic stress tolerance in crop plants. Curr Opin Plant Biol 14:290–295

Ramanankierana N, Ducousso M, Prin Y, Thioulouse J, Randrianjohany E, Ramaroson L, Kisa M, Galiana A, Duponnois R (2007) Arbuscular mycorrhizas and ectomycorrhizas of *Uapaca bojeri* L. (Euphorbiaceae): sporophore diversity, patterns of root colonization, and effects on seedling growth and soil microbial catabolic diversity. Mycorrhiza 17(3):195–208

Rungin S, Indananda C, Suttiviriya P, Kruasuwan W, Jaemsaeng R, Thamchaipenet A (2012) Plant growth enhancing effects by a siderophore-producing endophytic streptomycete isolated from a Thai jasmine rice plant (Oryza sativa L. cv. KDML105). Antonie Van Leeuwenhoek 102(3):463–472

Saha M, Sarkar S, Sarkar SBK, Bhattacharjee S, Tribedi B (2015) Microbial siderophores and their potential applications: a review. Environ Sci Pollut Res:1–16

Sandhu SS, Rajak RC, Agarwal GP (1993) Studies on prolonged storage of *Beauveria bassiana* conidia: effects of temperature and relative humidity on conidial viability and virulence against chickpea borer *Helicoverpa armigera*. Biocontrol Sci Tech 3:47–53

Sandhu SS, Rajak RC, Hasija SK (2000) Potential of entomopathogens for the biological management of medically important pest progress and prospect glimpses in plant. Sciences:110–117

Sandhu SS, Unkles SE, Rajak RC, Kinghorn JR (2001) Generation of benomyl resistant Beauveria bassiana strains and their infectivity against *Helicoverpa armigera*. Biocontrol Sci Technol 11(2):245–250

Schrey SD, Salo V, Raudaskoski M, Hampp R, Nehls U, Tarkka MT (2007) Interaction with mycorrhiza helper bacterium Streptomyces sp. AcH 505 modifies organisation of actin cytoskeleton in the ectomycorrhizal fungus Amanita muscaria (fly agaric). Curr Genet 52:77–85

Shih HD, Liu YC, Hsu FL, Mulabagal V, Dodda R, Huang JW (2003) Fungichromin: a substance from Streptomyces padanus with inhibitory effects on Rhizoctonia solani. J Agric Food Chem 51:95–99

Shrivastava S, D'Souza FD, Desai PD (2008) Production of índole-3-acetic acid bye immobilized actinomycete (Kitasatospora sp.) for soil applications. Curr Sci 94(12):1595–1604

Siddiqui MH, Mohammad F, Khan MN, Al-Whaibi MH, Bahkali AH (2010) Nitrogen in relation to photosynthetic capacity and accumulation of osmoprotectant and nutrients in Brassica genotypes grown under salt stress. Agric Sci China 9:671–680

Singh S, Nain L (2014) Microorganisms in the conversion of agricultural wastes to compost. Proc Indian Natn Sci Acad 80(2):473–481. June 2014 Spl. Sec

Sivasithamparam FJ, Rincón J, Martín JF (2003) Characterization of the iron regulated desA promoter of Streptomyces pilosus as a system for controlled gene expression in Actinomycetes. Microb Cell Fact 2:5

Smith RM, Peterson WH, McCoy E (1954) Oligomycin, a new antifungal antibiotic. Antibiot Chemother (Northfield) 4:962–970

Solans M, Vobis G, Wall LG (2009) Saprophytic Actinomycetes promote nodulation in Medicago sativa-Sinorhizobium meliloti symbiosis in the presence of high nitrogen. J Plant Growth Regul 28:106–114

Sprusansky O, Stirrett K, Skinner D, Denoya C, Westpheling J (2005) The bkdR gene of Streptomyces coelicolor is required for morphogenesis and antibiotic production and encodes a transcriptional regulator of a branched-chain amino acid dehydrogenase complex. J Bacteriol 187:664–671

Srinivasan MC, Laxman RS, Deshpande MV (1991) Physiology and nutrition aspects of actinomycetes – an overview. World J Microbial Biotechnol 7:171–184

Struyk AP, Hoette I, Drost G, Waisvisz JM, van Eek T, Hoogerheide JC (1958) Pimaricin, a new antifungal antibiotic. In: Welch H, Marti-Ibanez F (eds) Antibiotics annual 1957–1958. Medical Encylopedia, Inc., New York, pp 878–885

Sukarno N, Smith SE, Scott ES (1993) The effect of fungicides on vesicular- arbuscular mycorrhizal symbiosis. New Phytol 25:139–147

Sziderics AH, Rasche F, Trognitz F, Sessitsch A, Wilhelm E (2007) Bacterial endophytes contribute to abiotic stress adaptation in pepper plants (Capsicum annuum L.) Can J Microbiol 53:1195–1202

Takahashi Y, Omura S (2003) Isolation of new actinomycete strains for the screening of new bioactive compounds. J Gen Appl Microbiol 49:141–154

Takeuchi S, Hirayama K, Ueda K, Sakai H, Yonehara H (1958) Blasticidin S, a new antibiotic. J Antibiot 11:1–5

Terkina IA, Parfenova VV, Ahn TS (2006) Antagonistic activity of Actinomycetes of Lake Baikal. Appl Biochem Microbiol 42(2):173–176

Thakur R, Rajak RC, Sandhu SS (2005) Biochemical and molecular characteristics of indigenous strains of the entomopathogenic fungus *Beauveria bassiana* of Central India. Biocontrol Sci Tech 15(7):733–744

Tjepkema JD, Cashon RE, Beckwith J, Schwintzer CR (2002) Hemoglobin in Frankia, a nitrogen-fixing actinomycete. Appl Environ Microbiol 68(5):2629–2631

Umezawa H, Okami Y, Hashimoto T, Suhara Y, Hamada M, Takeuchi T (1965) A new antibiotic, kasugsmycin. J Antibiot 18:101–103

Veiga M, Esparis A, Fabregas J (1983) Isolation of cellulolytic Actinomycetes from marine sediments. Appl Environ Microbiol 46:286–287

Waksman SA, Woodruff HB (1942) Selective antibiotic action of various substances of microbial origin. J Bacteriol 44:373–384

Williams ST, Goodfellow M, Alderson G (1989) Genus Streptomyces Waksman and Henrici 1943, 339AL. In: Williams ST, Sharpe ME, Holt JG (eds) Bergey's manual of systematic bacteriology, vol 4. Williams and Wilkins Co, Baltimore, pp 2452–2492

Wraight SP, Carruthers RI, Jaronski ST, Bradley CA, Garza CJ, Wraight SG (2000) Evaluation of the entomopathogenic fungi *Beauveria bassiana* and *Paecilomyces fumosoroseus* for microbial control of the silverleaf whitefly, *Bemisia argentifolii*. Biol Control 17(3):203–217

Yamaura M, Uchiumi T, Higashi S, Abe M, Kucho K (2010) Identification by suppression subtractive hybridization of Frankia genes induced under nitrogen-fixing conditions. Appl Environ Microbiol 76:1692–1694

Zhang YL, Li S, Jiang DH, Kong LC, Zhang PH, JD X (2013) Antifungal activities of metabolites produced by a termite-associated Streptomyces canus BYB02. J Agric Food Chem 61:1521–1524

Chapter 7
Bacterial Siderophore as a Plant Growth Promoter

A. Pahari, A. Pradhan, S.K. Nayak, and B.B. Mishra

7.1 Introduction

Agriculture accords a major share of capital revenue in both developing and developed countries with establishing employment and food security. Sustainable agriculture is extremely imperative in present changing world due to its possibility for meeting the upcoming agricultural needs. To reach the goal of 321 million tonnes of food grain production by 2020, the nutrient requirement will be 28.8 million tonnes and availability will be 21.6 million tonnes only, with a deficit of about 7.2 million tonnes (Arun 2007). To meet this requirement, a huge amount of chemical fertilizers is applied to replenish soil Nitrogen and Phosphorus in agricultural fields. For this purpose, in 1960s the Green Revolution-I was started with objectives of application of chemical fertilizers like pesticides, herbicide, and weedicide. But excessive use of these chemical fertilizers adversely affects the soil fertility, soil microbial diversity, surface and ground water etc. Because these chemical fertilizers contain acids, including sulfuric and hydrochloric acids and due to the excessive use, they tend to destroy the beneficial microbes present in the soil. Chemical fertilizers and pesticides also accumulated in the environment causing pollution, bioaccumulation &n biomagnifications and spread disease. Moreover, application of fertilizer can have a detrimental effect not only on soil health but also facilitates the growth of soil-borne pathogens. James (1981) reported that annual loss of 13–20% in production of economic crops is solely due to soil borne diseases. The picture is worsening in India, *viz.* more than 50% loss in crops are due to soil pathogens (Rajash 2005). In India,

A. Pahari • A. Pradhan • B.B. Mishra (✉)
Department of Microbiology, College of Basic Science & Humanities, Orissa University of Agriculture and Technology, Bhubaneswar, Odisha, India
e-mail: bb_mishra58@yahoo.com

S.K. Nayak
Department of Biotechnology, College of Engineering and Technology, Biju Patnaik University of Technology, Bhubaneswar, Odisha, India

J.K. Patra et al. (eds.), *Microbial Biotechnology*,
https://doi.org/10.1007/978-981-10-6847-8_7

compelling the search for alternatives of chemical fertilizers and pesticides is the primary focus of current agricultural trends.

In the era of sustainable agriculture, Organic farming with application of biofertilizer has emerged worldwide by targeting the demand rise in safe & healthy food and long-term sustainability. It not only ensures food safety but also adds to the soil biodiversity (Megali et al. 2013). Biofertilizers have longer shelf life and doesn't adversely effects to the ecosystem which is the main advantage of using biofertilizer. Organic farming includes the application of compost, biofertilizers, biopesticides etc. which may not serve as a complete substitute but can be an effective supplement to decrease application of agrochemicals. Organic farming is mainly depends upon the soil normal microbial flora including all types of eubacteria, archaebacteria and eukarya (fungi) including the arbuscular mycorrhiza fungi (AMF) and plant growth promoting rhizobacteria (PGPR) (Sahoo et al. 2014).

Now a days much attention has been given to the use of biofertilizers in the field of agriculture because it is an important alternative source of plant nutrition. Biofertilizers are living formulation or latent cells of efficient strains of microorganisms such as bacteria, algae or fungi that can be directly applied to seed, soil or composting areas. So the increasing number of such organisms accelarate the decomposition process of the complex nutrients and convert it into the available form which can be directly uptake by plants. They are biologically active products, avails essential nutrients to plants and may be nitrogen fixers, phosphate solubilizers, sulphuroxidisers or decomposers. Simply, they are known as bio-inoculants and on application increases yield of crop plants (Vessey 2003).

7.2 Microbial Diversity of Soil

Soil quality may be summerised as the ability of a specific soil to function within ecosystem boundaries with plant and animal growth and productivity, improving air and water quality and supporting human health. Soil organisms are directly responsible for soil ecosystem processes like soil organic matter decomposition and production of different antibiotic substances (Nayak et al. 2012), cycling of nutrients. So the soil (micro) biological parameters may possess as sensitive indicators for soil ecological disturbances (Dick 1992). In case of soil enzyme activities and exopolysaccharides, soil micro flora is used as potential biochemical or ecological indicators of soil quality (Dick 1994). According to Islam and Weil (2000), soil quality can be determined by the total microbial biomass, based on soil samples of contrasting management systems. It has been already proved that ten billion microorganisms can be present in 1 g of soil, may harbor up to of possibly thousands of different species and among them, 10% of the microorganisms observed under the microscope which can be cultivated and characterized (Roselló-Mora and Amann 2001).

Plants usually exude a carbon-rich compound that nourishes the microbes (Farrar et al. 2014). Plants also exude various chemicals in response to environmental factors. Soil bacteria able to sense these chemosignals and also secrete chemicals that can activate complex plant defenses within the plants (Glick 2012). Some of these bacteria also play an important role in agriculture crops and termed as Agriculturally important microorganism (AIMs). AIMs have diverse applications in agriculture and allied sciences (Arora et al. 2005) and are designated as a large group of frequently unknown or ill-defined group of microorganisms (Higa and Parr 1994). They interact in soil and with plants to render beneficial effects which are sometimes unpredicted. *Azotobacter chroococcum, Azospirillum basilensis, Bacillus weihenstephanensis, Bradyrhizobium* sp., *Paenibacillus* sp., *Pseudomonas corrugate, Rhizobium* sp., etc. have established plant growth promotion efficiency. The soil residing microbes protects crops through enhancing the plant resistance capacity and against different biotic and environmental factors as biotic elicitor's. According to Singh et al. (2011), microorganisms are valuable asset in managing various weeds, pests and other diseases. For example, fungi can colonize at above ground parts of the plants and provide its benefits to drought as well as heat tolerance and simultaneously insect resistance and other plant disorders.

7.3 Biofertilizer- A Substitute to Chemical Fertilizers

Application of fertilizers has become an essential and vital option to amplify productivity in agricultural activities. A non-organic fertilizer primarily comprises of salts of phosphate, nitrate, ammonium and potassium. According to Reddy et al. (2002), the deficiency of soil phosphate results due to the shrinkage of nutrients reservoirs in soil after the harvest session, which is than later replenished with the instant application of chemical fertilizers. However in recent years, fertilizer consumption has increased exponentially throughout the world causing serious environmental threats including accumulation of heavy metals by plant that enter into the food chain. Serpil (2012) reported that fertilizer application leads to water, soil and air pollution ultimately decreasing the soil health. Due to increasing cost and negative impact of chemical fertilizers on soil health, sustainable approach for application of organic fertilizers persuaded search for an effective alternative with organic manure including application of plant growth promoting rhizobacteria (PGPR) in agriculture practices. For example, *Pseudomonas corrugata*, a soil bacterium showing antagonistic activities against phytopathogens like *Alternaria alternate* and *Fusarium oxysporum* can be used as a bio-pesticides (Trivedi et al. 2008).

Biofertilizers, which involves microbial inoculants, exhibit numerous advantages such as nutrient recycling, both for current intake as well as residual and synthetically proliferated microbial formulations that enhance soil fertility and crop productivity, thus, can be an ideal supplement or substitute to chemical fertilizers. Though useful effect of legumes in nitrogen fixation, improvement in soil texture as well as fertility has been known to us since century or more than that but currently commercial use and application of such potent organism becomes new interest and custom.

7.4 Microorganisms as Plant Growth Promoters

Thin region of soil is considered as rhizosphere which is remain in contact with different minearls, vitamins & amino acids which is collectively called as root exudates and due to which it contain several microorganisms. Rhizosphere provides lively surroundings and environment which shelters different group of microorganisms. Some of the bacteria can grow at rhizosphere as well as root region which directly and indirectly augments growth of plant and hence considered as Plant Growth Promoting Rhizobacteria (PGPR). Although all parts of the plant are colonized by microorganisms but rhizosphere represents the main source of bacteria due to its plant-beneficial activities. Use of PGPR is considered as alternative source which lessens the use of chemical fertilizer in agricultural field and avoids the growth of plant pathogens by broad mechanisms (De Weger et al. 1995; Glick 1995; Tilak et al. 2005).

PGPR can be differentiated into two separate classes, one is extracellular plant growth promoting rhizobacteria (ePGPR) and another one is intracellular plant growth promoting rhizobacteria (iPGPR) (Viveros et al. 2010). Extracllular PGPR are chiefly available on the outer part of the root surface which is termed as rhizoplane or else gap between the cells of root cortex and intracellular PGPR usually present within the nodular arrangements of root cells. Genus belongs to extracellular PGPR like *Agrobacterium*, *Bacillus*, *Burkholderia*, *Azotobacter*, *Azospirillum*, *Arthrobacter*, *Flavobacterium*, *Chromobacterium*, *Caulobacter*, *Erwinia*, *Micrococcus*, *Pseudomonas* and *Serratia* (Ahemad and Kibret 2014). *Rhizobiaceae* family belongs to intracellular PGPR which includes species of *Allorhizobium*, *Bradyrhizobium*, *Mesorhizobium* and *Rhizobium*. Different species of Endophytes and *Frankia* can symbiotically fix atmospheric nitrogen with the higher plants (Bhattacharyya and Jha 2012).

PGPR supports the direct & indirect growth of plants through different mechanism. In the direct mechanism, it contributes towards utilization of nutrients, N_2 fixation, Solubilization of phosphorus, Production of siderophore, IAA and other growth phytohormones (Glick 1995; Bhardwaj et al. 2014; Pahari et al. 2016). Plant pathogenic microorganisms those are major trouble to maintain sustainability in agriculture and environmental steadiness are controlled by indirect mechanism through production of antibiotics, siderophore, lytic enzymes (Table 7.1). Application of PGPR is a capable ecological and green alternative to achieve sustainable soil fertility as well as growth of plants.

7.5 Effects of Iron on Plants

Iron is recognized as an essential micronutrient as early as 1845 and fourth most lavish constituent in the earth. In huge amount it occurs in various types of soil. Iron is an abundant element but it is mostly lacking micronutrients due to its insolubility

Table 7.1 Growth promoting substances released By PGPR strains

PGPR	Plant growth promoting traits	References
Pseudomonas fluorescens	IAA, siderophores, antifungal activity	Dey et al. (2004)
Bacillus sp.	P-solubilization, Ammonia	Canbolat et al. (2006)
Baciilus subtilis	IAA, phosphate solubilization	Zaidi et al. (2006)
Azotobacter sp.,	IAA, ammonia production	Joseph et al. (2007)
Rhizobium sp. (pea)	IAA, siderophores, HCN, ammonia	Wani et al. (2007)
Enterobacter sp.	ACC deaminase, IAA, siderophore, phosphate solubilization	Kumar et al. (2008)
Pseudomonasaeruginosa	IAA, siderophores, HCN, ammonia, phosphate solubilization	Ahemad and Khan (2012)
Bacillus sp. JQ408711	IAA, siderophores, antifungal activity, ammonia, phosphate solubilization	Pradhan et al. (2014)

in nature of certain variety of (Fe^{+3}). These compounds are drop down in extremely tough soils and the chief component of the red soils of tropical regions. Some of the earliest soils, Fossil remnants hold an adequate amount of iron to supply iron ore which is insoluble to meet plant needs. Quick oxidation from Fe^{2+} to Fe^{3+} takes place in oxygen and neutral pH which finally undergoes into insoluble ferric oxyhydroxide [$Fe(OH)_2$] and this form is nearly not accessible by microbes. Iron deficiencies may occur if the ferrous (Fe^{+2}) iron is not gradually released from the soil minerals (Thompson and Troeh 1973). Two oxidizing agents namely Manganese and Copper can convert Fe^{2+} to the more insoluble Fe^{3+}. Plant absorbs (Fe^{+2}) form which is necessary for chlorophyll development and also play an important role in a number of enzymes which take part in respiratory mechanism of the plants (Schneider et al. 1968). It plays vital role in various metabolic procedures which includes TCA, ETC, oxidative phosphorylation & photosynthesis process (Messenger and Barclay 1983; Fardeau et al. 2011).

Soil and plant genotype interaction is important for iron nutrition (SSSA 1991). Efroymson et al. (1997) reviewed several studies which checked sensitivity of plant to iron from different solutions of soil. Iron paucity results in difference between green and less green (or yellow) tissue which occurs in plants and chlorophyll shortage in plants called as chlorosis (Wallihan 1966). Insoluble Fe^{+2}oxides is the major cause of chlorosis in plants. Iron accessibility in the soil to plants decreases with increase in redox potential as well as increase in pH of soil.

7.6 Bacterial Siderophore as Iron Chelator

Due to less availability of Fe^{3+} in the soil, nature has gifted explicit uptake approach in microbes like development of siderophores. Siderophores (Greek: "iron carriers") are defined low molecular weight (500–100 dt), ferric ion specific chelating agents synthesized by many bacteria like *Pseudomonas, Azotobacter, Bacillus,*

Enterobacter, Serratia, Azospirillum and *Rhizobium* (Glick et al. 1999; Loper and Henkels 1999; Pahari and Mishra 2017), under iron-limited conditions (Neilands 1981). Not only siderophore forage iron from the surroundings to create mineral which is very important and accessible to the microbes but also they form complexes with other metals like Molybdenum, Manganese, Cobalt, and Nickel in the environment and enhance availibilty to microbial cells (Bellenger et al. 2008; Braud et al. 2009a, b). On cellular level biosynthesis of siderophores is takes place which is carried out by a set of enzymes and these enzymes are particular for its respective siderophore. Bacterial chromosome or Plasmid contains the corresponding genes. There are over 500 bio-molecules that are classified as siderophores; hence, different multiple genes and regulators are involved in their biosynthesis, transport, and re-import into the cell (Challis 2005; Visca et al. 2007). However many of the siderophores are peptides which are synthesized by members of the multi-enzyme family. The non-ribosomal peptide synthetase (NRPS) also directs the synthesis of microbial peptide antibiotics (Barona-Gómez et al. 2004). Non-ribosomal peptide synthetase (NRPS) dependent mechanism shows that there was no involvement of mRNAs in the biosynthesis procedure. Still, other siderophores, including many of the hydroxamate-type ones, are synthesized based on mechanisms that are NRPS-independent (Challis 2005).

From five decades, it is assumed that, the most aerobic and facultative anaerobic microorganisms synthesize at least one siderophore. But according to Essen et al. (2007), in both aerobic and anaerobic situation *Pseudomonas stutzeri* CCUG 36651, a facultative aerobic bacterium can make siderophores but with difference in type of siderophore which varies as per the aerobic and anaerobic state. It has already been reported that *Pseudomonas stutzeri* CCUG 36651 produce four ferrioxamine siderophores in presence of oxygen whereas no ferroxamine siderophores were developed in absence of oxygen or in anaerobic situation. In microorganisms siderophores have been associated with kind of virulence mechanisms which was pathogenic to both animals and plants (Winkelman and Drechsel 1997).

7.7 Types of Siderophore

According to the oxygen ligands for Fe^{3+} organization, siderophores can be differentiated to three main categories, namely, hydroxamates, catecholates, and carboxylates (Table 7.2).

7.7.1 Hydroxymate Type of Siderophore

Hydroxymate type of siderophore is mostly produced by the bacteria and fungi. Mostly groups of hydroxamate are belongs to C (=O) N-(OH) R and R is an amino acid. Every individual group of hydroxamate provides two molecules of oxygen and thereby with iron form a bidentate ligand. Therefore with Fe^{3+}, every single

Table 7.2 List of bacteria which can produce different types of siderophore

Types of siderophore	Name of siderophore	Siderophore-producing bacteria	References
Hydroxymate	Ferribactin	*Pseudomonas fluorescens*	Maurer and Keller-Schierlein (1968)
	Unknown	*Escherichia coli*	Kannahi and Senbagam (2014)
	Unknown	*Pseudomonas putida*	Sayyed et al. (2005)
	Unknown	*Micrococcus luteus*	Cabaj and Kosakowska (2009)
	Unknown	*Methylobacterium radiotolerans*	Lacava et al. (2008)
	Unknown	*Methylobacterium zatmanii*	Lacava et al. (2008)
	Desferrioxamine B, desferrioxamine	*Streptomyces coelicolor*	Saharan and Nehra (2011)
	Unknown	*Halorubrum saccharovorum*	Dave et al. (2006)
Catecholate	Enterobactin	*Escherichia coli*	Saharan and Nehra (2011)
	Pyoverdine	*Pseudomonas aeruginosa*	Peek et al. (2012)
	Salmochelins	*Salmonella enterica*	Hantke et al. (2003)
	Bacillibactin	*Bacillus anthracis*	Saharan and Nehra (2011)
	Bacillibactin	*Bacillus subtilis*	Saharan and Nehra (2011), and May et al. (2001)
	Petrobactin, Bacillibactin	*Bacillus cereus, Bacillus anthracis*	Wilson et al. (2006)
	Bacillibactin	*Bacillus thuringiensis*	Wilson et al. (2006)
	Vibriobactin	*Vibrio cholera*	Saharan and Nehra (2011), and Griffiths et al. (1984)
	Agrobactin	*Agrobacterium tumefaciens*	Dave et al. (2006)
	Parabactin	*Paracoccus denitrificans*	Dave et al. (2006)
Carboxylate	Rhizobactin	*Rhizobium meloti*	Drechsel et al. (1995)
	Staphyloferrin A	*Staphylococcus hyicus*	Meiwes et al. (1990)
	Staphyloferrin A, Staphyloferrin B	*Staphylococcus aureus*	Beasley et al. (2011)
	Unknown	*Halococcuss accharolyticus*	Dave et al. (2006)
	Unknown	*Halorubrum saccharovorum*	Dave et al. (2006)
	Unknown	*Haloterrigena turkmenica*	Dave et al. (2006)
	Unknown	*Halogeometricum* sp.	Dave et al. (2006)

siderophore forms a complex of hexadentate octahedral structure. Hydroxamates forms 1:1 complexes with ferric iron. Winkelmann 2007 opined that against hydrolysis ferric hydroxamate complexes remain constant and in the natural environment enzymatic degradation takes place when pH is more than 1. Generally Hydroxamate siderophores when bound to iron shows strong absorption between 425 and 500 nm. Fungi produced hydroxamates e.g. ferrichromes, coprogens & fusigenes and bacteria produced ferrioxamines (Winkelmann 2007). *Erwinia*, *Nocardia*, *Streptomyces*,

Arthrobacter, *Chromobacterium* and *Pseudomonas* species were soil bacteria that produce Ferrioxamine type siderophores. *Ustilago sphaerogena* is the example of fungal species that produce ferrichrome type siderophores (Emery 1971) whereas *Aspergillus fumigates* produce ferricrocin (Wallner et al. 2009). Coprogens are produced by fungal species such as *Trichoderma* sp. and Fusigen by the fungi *Fusarium* sp. (Diekmann and Zahner 1967).

7.7.2 Catecholate Type of Siderophore

Catecholate is another important group of siderophore, mainly produced by the bacteria. The structure of the backbone can be polyamine, a peptide or a macrocyclic lactone. Siderophores with one, two or three catecholate or phenolate chelating groups attached to a peptide backbone are reported (Dave et al. 2006). In 1970, isolation was done from culture fluids of *E. coli*, *Aerobacter aerogenes*, and *Salmonella typhimurium* which produced enterobactin (also termed as enterochelin) and it was the first tricatechol siderophore. Exceptionally high complex formation constant, low redox potential for physiological reductants and extremely strong pH-dependence is the most important feature of enterobactin (cyclic trimester of 2.3- dihydroxy benzoylserine). This exceptional properties pertinent to its physiological reactions and due to which enterobactin is generally the most intensively analyzed siderophores.

7.7.3 Carboxylate Type of Siderophore

This type of siderophores is mainly produced by bacteria like *Rhizobium*, *Staphylococcus* and fungi like *Mucorales*. Carboxylate type of siderophores mainly exhibits hydroxy and carboxylate donor groups and many of them belongs to the mixed ligand group. It is already reported that very few numbers of carboxylate siderophores having an iron chelating moiety composed of only α–hydroxy donor groups and carboxylate. Bacteria like *Staphylococci*, *Rhizobium melilot*, *Mucorals* produce Staphyloferrin A &B, rhizobactin and rhizoferrin carboxylate siderophore respectively. The carboxylate type of siderophore can be detected by Vogels chemical test (Dave and Dube 2000).

7.8 Phytosiderophore

Reports available indicate that plants can also produce siderophore. Under iron limiting conditions, graminaceous plants like barley and wheat have the capability to chelate iron from insoluble form (Kraemer et al. 2006). Phytosiderophores are the Fe^{3+}-chelating compounds secreted by graminaceous plant can form specific strong complexes with Fe^{3+}. This type of siderophores have hexadentate ligands that coordinate $Fe^{3+}$with their amino and carboxyl groups (Ma 2005; Singh et al. 2011). When phytosiderophore is released in the rhizosphere region, it chelates the iron from the soil by forming a complex of Fe^{3+}–with phytosiderophore which can be directly transported across the root plasma membrane (Römheld and Marschner 1986; Dell'mour et al. 2012). Among all the phytosiderophores, the most common and first identified phytosiderophore is Mugineic acid (MA) (Takemoto et al. 1978). The molecular weight of phytosiderophores are ranged between 500 and 1000 Da, where the molecular massof microbial siderophores ranged between 200 and 2000 Da (Neilands 1981). Some of the phytosiderophores such as avenic acid and distichonic acid have also been isolated from the graminaceous oats plants (*Avenasativa*) and from beer barley (*Hordeum vulgate*) respectively (Nomoto et al. 1981). Several studies also showed that some plant species like wheat, rye and barley produce a high concentration of phytosiderophores and more resistant to iron deficiency than other species like maize, rice and sorghum, which produce a lower concentration of phytosiderophores (Masuda et al. 2009; Kobayashi et al. 2010).

7.9 Mechanism of Siderophore Mediate Iron Transport

Iron is the key element of many enzymes like catalases, superoxide dismutases (SOD) and peroxidases. So insufficient amount iron not only hamper the cellular metabolism and growth but also lower the oxidative defenses and lead to oxidative injury. So, all organisms require a well organized controlled systems for the iron uptake and storage. Siderophoreis a part of a multi-component system for transporting ferric iron into a cell. When the siderophore is released in the environment, it first binds with iron (Fe^{+3}) tightly and then the siderophore-iron complex is transported to the cell interior through the cell membrane using the specific siderophore receptors. Different types of siderophore specific receptors are present in different bacteria. In outer membrane, receptor protein Fec A & Fep A are present and Ton B-Exb B-Exb D protein complex present in the inner membrane. Some other proteins like ATP-dependent Fec CDE- Fep CDE, an inner membrane protein and periplasmic binding protein also participate in their on transport mechanism. In iron limiting condition, bacteria synthesize siderophore and number of receptor molecules increases. When the siderophore is released from the cell, the membrane receptors protein bind with iron and form iron-siderophore complex and this complex is transported into the cell via Fec A and Fep A, an outer membrane receptor (OM). After that it is transported to ABC-Transporter systems i.e. Fec C,D,E and Fep C,D,E (from ATP-binding cassette) which is assembled of two proteins, one acts as permease and the other protein hydrolyze ATP to provide energy for transport (Boos and Eppler 2001). Finally siderophore iron complex is released in the cytosol with the help of membrane protein Ton B. In the cytoplasm, the iron released from the complex by the help of hydrolytic destruction of the siderophore molecule or by the help of Ent A, B, C, D protein or the reduction of Fe^{3+} by NADPH linked

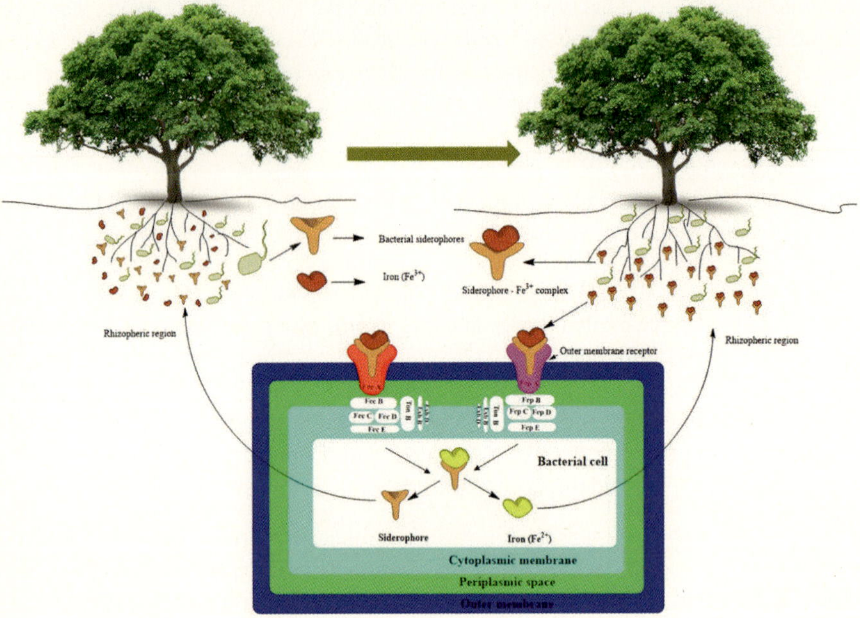

Fig. 7.1 Mechanism of siderophore mediated iron transport

siderophore reductase. The final Fe^{2+} does not have a high affinity for siderophore and therefore separated from the siderophore-iron complex and the siderophores either get degraded or recycled by excretion through efflux pump system (Fig. 7.1).

7.10 Application of Siderophore-Producing Bacteria

7.10.1 Siderophore as a Plant Growth Promoter

Siderophores are eco-friendly alternative to the hazardous chemical pesticide and can be used in the different agricultural sectors. Iron is an essential micronutrient and it is required for some important physiological activities, chlorophyll biosynthesis and redox reactions in plant (Briat et al. 1995; Schenk et al. 2012). So iron deficient condition significantly reduces the quality and quantity of Crop production. However, microbe-plant and siderophore-plant interactions under iron deficiency have been investigated by several workers.

From past few decades, it has been considered that different species of *Pseudomonas*can enhance plant growth by producing pyoverdine siderophores (Kloepper et al. 1980; Gamalero and Glick 2011). Mahmoud and Abd-Alla (2001) reported that hydroxymate type of siderophore producing *Pseudomonas* sp. enhanced the nodulation and nitrogen fixation of mung bean plant as compared to plant infected with *Bradyrhizobium* strain alone.

In addition to pseudomonads, other bacteria found in the rhizosphere region of *Azadirachta indica* produces ferrioxamines siderophore which transfer the iron to the plant for the growth and development of shoot and root (Verma et al. 2011; Crowley 2006). Powell et al. (1980) reported that hydroxamate siderophores exists in different soils as well as in the aquatic environments.

It has been reported that excessive accumulation of heavy metals is toxic for most of the plants and responsible for the contamination of soil which decreases the soil fertility and soil microbial activity (McGrath et al. 1995). In this concern hydroxamate type of siderophore present in soil play an important role to immobilize the metals. Masalha et al. (2000) concluded that microbial siderophore can be considered as an efficient iron source for the plant because they found that microbial siderophores production was totally suppressed when the plants were grown under sterile conditions. Similarly some of the other bacteria like *Escherichia coli* from rye grass (*Loliumperenne* sp.) and endo-rhizosphere of sugarcane (*Saccharum* sp.) & an endophytic *Streptomyces* sp. isolated from the roots of a Thai jasmine rice plant enhance plant growth and significantly elevated root and shoot biomass and lengths (Gangwar and Kaur 2009; Rungin et al. 2012).

7.10.2 Siderophore as Potential Biocontrol Agent

Siderophore producing bacteria plays an important role in the biological control against certain phytopathogens. Bacteria produce siderophore and it is bind with the iron strongly and make it unavailable for the plant pathogens, therefore inhibiting the growth of phytopathogens (Beneduzi et al. 2012; Ahmed and Holmstrom 2014). This is especially true for *Pseudomonas* and *Bacillus* sp. which are popular biocontrol agents (Beneduzi et al. 2012). Reddy and Battu (2009) proved that the antagonistic activity of *Pseudomonas fluorescens* against rice fungal pathogens i.e. *P. Oryzae* and *R. Solani* occurred both in presence and absence of $FeCl_3$ which indicated the siderophore mediation along with antifungal. Moreover, siderophore producing rhizobacteria also inhibit the other phytopathogenic fungi such as *Phytophthora parasitica* (Seuk et al. 1988), *Fusarium oxysporumveridianthi* (Buysens et al. 1996), *Phythiumultimum* (Hamdan et al. 1991) and *Sclerotinia sclerotiorum* (Mc Loughlin et al. 1992). For the first time, Kloepper et al. (1980), illustrated that siderophore producing bacteria like *Pseudomonas fluorescens* strains A1, BK1, TL3B1can be used as a biological control agent against *Erwinia carotovora*. Similarly different species Pseudomonads are also involved in the control of wilt diseases of potato caused by *Fusarium oxysporum* by production of Pyoverdine siderophores (Schippers et al. 1987). It is also suppress the growth of *Gaeumannomyces graminis* in wheat, barley, peanuts and maize (Voisard et al. 1989; Pal et al. 2001). Yu et al. (2011) reported that *Bacillus subtilis* also produce different types siderophore which have a significant role for the bio-control of *F. oxysporum* in pepper.

7.10.3 Application of Siderophore as Bioremediator

Over the time, much attention has been given to investigate the potential use of siderophores in metal bioremediation due to rapid accumulation of heavy metals and metalloids in soil from petrochemical industry (Zhang et al. 2010; Wuana and Okieimen 2011). Apart from binding with ferric iron, siderophores also plays a significant role in chelation of other toxic metals e.g., Cr^{3+}, Al^{3+}, Cu^{2+}, Eu^{3+} and Pb^{2+} (Nair et al. 2007; Rajkumar et al. 2010; O'Brien et al. 2014). Depending upon the concentration of metals in the growth medium, siderophore production can be regulated the production of siderophores (Schalk et al. 2011; Braud et al. 2010). For example, in presence of Al^{3+}, Cu^{2+}, Cr^{2+}, Ga^{3+}, Mn^{2+} and Ni^{2+},pyoverdine production was upregulated in *P. aeruginosa* (Braud et al. 2009a, b), azotochelin biosynthesis was stimulated by molybdenum in *Azotobacter vinelandii* (Duhme et al. 1998), schizokinen and *N*-di-oxyschizokinen production was enhanced by high concentrations of aluminum in *B. Megaterium* (Hu and Boyer 1996). Similarly production of Desferrioxamines B, E and Cch were stimulated by Cd and Ni even in the presence of ferric iron (Dimkpa et al. 2008). Thus, siderophores can be considered as agriculturally important tool for heavy metal remediation (Rajkumar et al. 2010). Siderophores have a strong affinity for a particular type of metal other than iron which is totally depends upon the ligand functionalities and siderophore-metal complex is formed (Hernlem et al. 1999). Neubauer et al. (2000) reported that, in high pH conditions siderophores such as desferrioxamine B can chelate Co^{3+}better than Fe^{3+}. Likewise, *Azotobacter vinelandii*can produce siderophores like azotochelin and azotobactin which are very useful for Mo and V acquisition (Wichard et al. 2009); *P. fluorescens* can mobilize the metals like Ni and Co from waste material (acid-leached ore) of a former uranium mine (Edberg et al. 2010). It is already proved that siderophores produced by *Agrobacterium radiobacter* can remove approximately 54% pollutant from a metal-contaminated soil and pyoverdine siderophore can mobilize U^{6+}, Np^{5+} &other metals from uranium mine waste (Behrends et al. 2012; Wang et al. 2011).

7.11 Conclusion

In a view of focus on organic farming, microbial diversity and soil health has gained considerable attention in recent years. Applications of organic manures directly and/or indirectly increased soil microbiome that plays a pivotal role in maintaining soil fertility and increasing productivity. Plant Growth Promoting Rhizobacteria (PGPR) are microbes that colonize in the rhizosphere region of the crop plant. With their innate capacity through direct and indirect mechanisms, these organisms felicitate growth of the crop plant and subsequently productivity. Siderophore producing bacteria present in the rhizosphere region is of significant importance in the field of agriculture. Siderophore are low molecular weight phenolic compounds have the capacity to chelate Fe^{3+} irons and reduce to Fe^{2+} by siderophore and supplement to the crop plant. In addition to supplementing iron to the plant, siderophore also prevent the growth of the soil borne phytopathogens which are mostly iron dependent. Hence it is envisaged to application of siderophore producing bacteria in crop field to increase growth and productivity of plant.

References

Ahemad M, Khan MS (2012) Effect of fungicides on plant growth promoting activities of phosphate solubilizing *Pseudomonas putida* isolated from mustard (Brassica compestris) rhizosphere. Chemosphere 86:945–950

Ahemad M, Kibret M (2014) Mechanisms and applications of plant growth promoting rhizobacteria: current perspective. J King Saud Univ Sci 26:1–20

Ahmed E, Holmstrom SJM (2014) Siderophores in environmental research: roles and applications. J Microbial Biotechnol 7:196–208

Arora DK, Saikia R, Dwievdi R, Smith D (2005) Current status, strategy and future prospects of microbial resource collections. Curr Sci 89:488–495

Arun KS (2007) Bio-fertilizers for sustainable agriculture. Mechanism of solubilization, 6th edn. Agribios publishers, Jodhpur, pp 196–197

Barona-Gómez F et al (2004) Identification of a cluster of genes that directs Desferrioxamine biosynthesis in Streptomyces coelicolor M145. J Am Chem Soc 126:16282–16283

Beasley FC, Marolda CL, Cheung J, Buac S, Heinrichs DE (2011) *Staphylococcus aureus*transporters Hts, Sir, and Sst capture iron liberated from human transferrin by Staphyloferrin A, Staphyloferrin B, and catecholamine stress hormones, respectively, and contribute to virulence. Infect Immun 79:2345–2355

Behrends T, Krawczyk-Bärsch E, Arnold T (2012) Implementation of microbial processes in the performance assessment of spent nuclear fuel repositories. Appl Geochem 27:453–462

Bellenger JP, Wichard T, Kustka AB, Kraepiel AML (2008) Uptake of molybdenum and vanadium by a nitrogen-fixing soil bacterium using siderophores. Nat Geosci 1:243–246

Beneduzi A, Ambrosini A, Passaglia LM (2012) Plant growth-promoting Rhizobacteria (PGPR): their potential as antagonists and biocontrol agents. Genet Mol Biol 35:1044–1051

Bhardwaj D, Ansari MW, Sahoo RK, Tuteja N (2014) Biofertilizers function as key player in sustainable agriculture by improving soil fertility, plant tolerance and crop productivity. Microb Cell Fact 13:66

Bhattacharyya PN, Jha DK (2012) Plant growth-promoting Rhizobacteria PGPR: emergence in agriculture. J World Microbiol Biotechnol 28:1327–1350

Boos W, Eppler T (2001) Prokaryotic binding protein dependent ABC transporters. In: Winkelmann G (ed) Microbial transport systems. Wiley-VCH, Weinheim, pp 77–114

Braud A, Hannauer M, Mislin GLA, Schalk IJ (2009a) The *Pseudomonas aeruginosa* pyochelin-iron uptake pathway and its metal specificity. J Bacteriol 191:3517–3525

Braud A, Jézéquel K, Bazot S, Lebeau T (2009b) Enhanced phytoextraction of an agricultural Cr- and Pb-contaminated soil by bioaugmentation with siderophore-producing bacteria. Chemosphere 74:280–286

Braud A, Geoffroy V, Hoegy F, Mislin GLA, Schalk IJ (2010) Presence of the siderophorespy-overdine and pyochelin in the extracellular medium reduces toxic metal accumulation in *Pseudomonas aeruginosa* and increases bacterial metal tolerance. Environ Microbiol R 2:419–425

Briat JF, FobisLoisy I, Grignon N, Lobreaux S, Pascal N, Savino G, Thoiron S, Wiren N, Wuytswinkel O (1995) Cellular and molecular aspects of iron metabolism in plants. Biol Cell 84:69–81

Buysens S, Heungens K, Poppe J, Hofte M (1996) Involvement of Pyochelin and pioverdin in suppression of *Pseudomonas aeruginosa* 7NSK2. Appl Environ Microbiol 62(3):865–871

Cabaj A, Kosakowska A (2009) Iron-dependent growth of and siderophore production by two heterotrophic bacteria isolated from brackish water of the southern Baltic Sea. Microbiol Res 164:570–577

Canbolat MY, Bilen S, Cakmakc R, Sahin F, Aydın A (2006) Effect of plant growth-promoting bacteria and soil compaction on barley seedling growth, nutrient uptake, soil properties and rhizospheremicroflora. Biol Fertil Soils 42:350–357

Challis GL (2005) A widely distributed bacterial pathway for siderophore biosynthesis independent of nonribosomal peptide synthetases. Chem Bio Chem 6:601–611

Crowley DA (2006) Microbial siderophores in the plant rhizosphere. In: Barton LL, Abadia J (eds) Iron nutrition in plants and rhizospheric microorganisms. Springer, Netherlands, pp 169–189

Dave BP, Dube HC (2000) Chemical characterization of fungal siderophores. Indian J Exp Biol 38:56–62

Dave BP, Anshuman K, Hajela P (2006) Siderophores of halophilic archaea and their chemical characterization. Indian J Exp Biol 44:340–344

De Weger LA, van der Bij AJ, Dekkers LC, Simons M, Wijffelman CA, Lugtenberg BJJ (1995) Colonization of the rhizosphere of crop plants by plant-beneficial pseudomonads. FEMS Microbiol Ecol 17:221–228

Dell'mour M, Schenkeveld W, Oburger E, Fischer L, Kraemer S, Puschenreiter M et al (2012) Analysis of iron-phytosiderophore complexes in soil related samples: LC-ESI-MS/MS versus CE-MS. Electrophoresis 33:726–733

Dey R, Pal KK, Bhatt DM, Chauhan SM (2004) Growth promotion and yield enhancement of peanut (*Arachishypogaea* L.) by application of plant growth-promoting rhizobacteria. Microbiol Res 159:371–394

Dick RP (1992) A review: long-term effects of agricultural systems on soil biochemical and microbial parameters. Agric Ecosyst Environ 40:25–26

Dick RP (1994) Soil enzyme activities as indicators of soil quality. In: Doran JW, Coleman DC, Bezdicek DF, Stewart BA (eds) Defining soil quality for a sustainable environment: Minneapolis, SSSA Special Publication, 35. Soil Science Society of America, New York, pp 107–124

Diekmann H, Zahner H (1967) Konstitution von Fusigen and dessen Abbauzu Δ2-Anhydromevalonsaurelacton. Eur J Biochem 3(2):213–218

Dimkpa CO, Svatoš A, Dabrowska P, Schmidt A et al (2008) Involvement of siderophores in the reduction of metal-induced inhibition of auxin synthesis in *Streptomyces* spp. Chemosphere 74:19–25

Drechsel H, Tschierske M, Thieken A, Jung G, Zahner H, Winkelmann G (1995) The carboxylate type siderophorerhizoferrin and its analogs produced by directed fermentation. J Ind Microbiol 14:105–112

Duhme AK, Hider RC, Naldrett MJand Pau RN (1998) The ability of the molybdnem-azotochelin complex and its effect on siderophore production in Azotobacter vnelandii. J Biol Inorg Chem 3(5):520–526

Edberg F, Kalinowski BE, Holmström SJM, Holm K (2010) Mobilization of metals from uranium mine waste: the role of pyoverdines produced by *Pseudomonas fluorescens*. Geobiology 8:278–292

Efroymson RA, Will M E, Suter GW II (1997) Toxicological benchmarks for screening contaminants of potential concern for effects on terrestrial plants. 1997 Revision ES/ER/TM-85/R3

Emery T (1971) Role of ferrochrome as a ferric ionophore in *Ustilago sphaerogena*. Biochemistry 10:1483–1488

Essen SA, Johnsson A, Bylund D, Pedersen K, Lundstrom US (2007) Siderophore production by Pseudomonas stutzeri under aerobic and anaerobic conditions. Appl Environ Microbiol 73(18):5857–5864

Fardeau S, Mullie C, Dassonville-Klimpt A, Audic N, Sonnet P (2011) Bacterial iron uptake: a promising solution against multidrug resistant bacteria. In: Science against microbial pathogens: communicating current research and technological advances. Formatex Research Center, Badajoz, pp 695–705

Farrar K, Bryant D, Cope-Selby N (2014) Understanding and engineering beneficial plant–microbe interactions: plant growth promotion in energy crops. Plant Biotechnol J 12:1193–1206

Gamalero E, Glick BR (2011) Mechanisms used by plant growth promoting bacteria. In: Bacteria in agrobiology: plant nutrient management. Springer, Berlin, Heidelberg, pp 17–46

Gangwar M, Kaur G (2009) Isolation and characterization of endophytic bacteria from endorhizosphere of sugarcane and ryegrass. Internet J Microbiol 7:139–144

Glick BR (1995) The enhancement of plant growth by free-living bacteria. Can J Microbiol 41:109–117

Glick BR (2012) Plant growth-promoting bacteria: mechanisms and applications. Scientifica 2012:15

Glick BR, Patten CL, Holguin G, Penrose DM (1999) Biochemical and genetic mechanisms used by plant growth promoting bacteria. Imperial College Press, London

Griffiths GL, Sigel SP, Payne SM, Neilands JB (1984) Vibriobactin, a siderophore from Vibrio cholerae. J Biol Chem 259(1):383–385

Hamdan H, Weller D, Thomashow L (1991) Relative importance of fluorescens siderophores and other factors in biological control of *Gaeumannomyces graminis* var. Tritici by *Pseudomonas fluorescens* 2-79 and M4-80R. Appl Environ Microbiol 57(11):3270–3277

Hantke K, Nicholson G, Rabsch W, Winkelmann G (2003) Salmochelins, siderophores of *Salmonella enterica* and uropathogenic *Escherichia coli* strains, are recognized by the outer membrane receptor. Iron. Proc Natl Acad Sci 100:3677–3682

Hernlem BJ, Vane LM, Sayles GD (1999) The application of siderophores for metal recovery and waste remediation: examination of correlations for prediction of metal affinities. Water Res 33:951–960

Higa T, Parr JF (1994) Beneficial and effective microorganisms for a sustainable agriculture and environment. International Nature Farming Research Center, Atami

Hu X, Boyer GL (1996) Siderophore-mediated aluminium uptake by *Bacillus megaterium* ATCC 19213. Appl Environ Microbiol 62:4044–4048

Islam KR, Weil RR (2000) Soil quality indicator properties in mid-Atlantic soils as influenced by conservation management. J Soil Water Conserv 55:69–78

James WC (1981) Estimated losses of crops from plant pathogens. In: Pimentel D (ed) Handbook of pest management in agriculture. CRC Press, Boca Raton, pp 79–94

Joseph B, Patra RR, Lawrence R (2007) Characterization of plant growth promoting rhizobacteria associated with chickpea (*Cicerarietinum* L.) Int J Plant Prod 2:141–152

Kannahi M, Senbagam N (2014) Studies on siderophore production by microbial isolates obtained from rhizosphere soil and its antibacterial activity. J Chem Pharm Res 6:1142–1145

Kloepper JW, Leong J, Teintze M, Schiroth MN (1980) Enhanced plant growth by siderophores produced by plant growth promoting Rhizobacteria. Nature 286:885–886

Kobayashi T, Nakanishi H, Nishizawa NK (2010) Recent insights into iron homeostasis and their application in graminaceous crops. Proc Jpn Acad Ser B Phys Biol Sci 86:900–913

Kraemer SM, Crowley DE, Kretzschmar R (2006) Geochemical aspects of phytosiderophore promoted iron acquisition by plants. Adv Agron 91:1–46

Kumar KV, Singh N, Behl HM, Srivastava S (2008) Influence of plant growth promoting bacteria and its mutant on heavy metal toxicity in Brassica juncea grown in fly ash amended soil. Chemosphere 72:678–683

Lacava PT, Silva-Stenico ME, Araujo WL, Simionato AVC, Carrilho E, Tsai SM, Azevedo JL (2008) Detection of siderophores in endophytic bacteria Methylobacterium spp. associated with *Xylellafastidiosa* subsp. pauca. Pesq Agrop Brasileira 43(4):521–528

Loper JE, Henkels MD (1999) Utilization of heterologous siderophore enhances levels of iron available to *Pseudomonasputida*in the rhizosphere. Appl Environ Microbiol 65:5357–5363

Ma JF (2005) Plant root responses to three abundant soil minerals: silicon, aluminum and iron. Crit Rev Plant Sci 24:267–281

Mahmoud ALE, Abd-Alla MH (2001) Siderophore production by some microorganisms and their effect on Bradyrhizobium-Mung Bean symbiosis. Int J Agric Biol 03(2):157–162

Masalha J, Kosegarten H, Elmaci O, Mengel K (2000) The central role of microbial activity for iron acquisition in maize and sunflower. Biol Fertil Soils 30:433–439

Masuda H, Usuda K, Kobayashi T, Ishimaru Y, Kakei Y, Takahashi M et al (2009) Overexpression of the barley nicotianamine synthase gene HvNAS1 increases iron and zinc concentrations in rice grains. Rice 2:155–166

Maurer B, Keller-Schierlein W (1968) Ferribactin, a Siderochrome from Pseudomonas fluorescens Migula: 61. Mitteilung Ferribactin, einSiderochromaus *Pseudomonas fluorescens* Migula. Arch Microbiol 60:326–339

May JJ, Wendrich TM, Marahiel MA (2001) The dhb Operon of *Bacillus subtilis* encodes the biosynthetic template for the catecholicsiderophore 2, 3-dihydroxybenzoate-glycine-threonine trimeric ester bacillibactin. J Biol Chem 276:7209–7217

Mc Loughlin T, Quinn J, Bettermann A, Bookland R (1992) *Pseudomonas cepacia*suppression of sunflower. *Pseudomonas cepacia*. Wilt fungus and role of antifungal compounds in controlling the disease. Appl Environ Microbiol 58(3):1760–1763

McGrath SP, Chaudri AM, Giller KE (1995) Long-term effects of metals in sewage sluge on soils, microorganisms and plants. J Ind Microbiol 14(2):94–104

Megali L, Glauser G, Rasmann S (2013) Fertilization with beneficial microorganisms decreases tomato defences against insect pests. Agron Sustain Dev 34:649. https://doi.org/10.1007/s13593-013-0187-0

Meiwes J, Fiedler HP, Haag H, Zahner H, Konetschny-Rapp S, Jung G (1990) Isolation and characterization of Staphyloferrin A, a compound with siderophore activity from Staphylococcus hyicus DSM 20459. FEMS Microbiol Lett 67:201–206

Messenger AJ, Barclay R (1983) Bacteria, iron and pathogenicity. Biochem Educ 11(2):54–63

Nair A, Juwarkar AA, Singh SK (2007) Production and characterization of siderophores and its application in arsenic removal from contaminated soil. Water Air Soil Pollut 180:199–212

Nayak SK, Santra GH, Mishra BB (2012) Antimycotic potential of bacterial ethanolic extract isolated from acid soil of Mahishapat, Odisha. J Pure Appl Microbiol 6(4):1853–1858

Neilands JB (1981) Microbial iron compounds. Annu Rev Biochem 50:715–731

Neubauer U, Nowak B, Furrer G, Schulin R (2000) Heavy metal sorption on clay minerals affected by the siderophore desferroixamine B. Environ Sci Technol 34:2749–2755

Nomoto K, Mino Y, Ishida T, Yoshioka H, Ota N, Inoue M et al (1981) X-ray crystal structure of the copper (II) complex of mugineic acid, a naturally occurring metal chelator of graminaceous plants. J Chem Soc Chem Commun 7:338–339

O'Brien S, Hodgson DJ, Buckling A (2014) Social evolution of toxic metal bioremediation in *Pseudomonas aeruginosa*. Proc R Soc B Biol Sci 281(1787):20140858

Pahari A, Mishra BB (2017) Characterization of Siderophore producing Rhizobacteria and its effect on growth performance of different vegetables. Int J Curr Microbiol App Sci 6(5):1398–1405. doi: https://doi.org/10.20546/ijcmas.2017.605.152

Pahari A, Dangar TK, Mishra BB (2016) Siderophore quantification of bacteria from Sundarban and its effect on growth of Brinjal (*Solanum melongena.* L). The Bioscan 11(4):2147–2151

Pal KK, Tilak KV, Saxena AK, Dey R, Singh CS (2001) Suppression of maize root diseases caused by *Macrophomina Phaseolina*, *Fusarium moniliforme* and *Fusarium graminearum* by plant growth promoting Rhizobacteria. Microbiol Res 156:209–223

Peek ME, Bhatnagar A, McCarty NA, Zughaier SM (2012) Pyoverdine, the major siderophore in *Pseudomonas aeruginosa*, evades NGAL recognition. Inter disciplinary perspectives on infectious diseases. 2012

Powell PE, Cline GR, Reid CPP, Szaniszlo PJ (1980) Occurrence of hydroxamate siderophore iron chelators in soils. Nature 287:833–834

Pradhan A, Baisakh B, Mishra BB (2014) Plant growth characteristics of bacteria isolated from rhizosphere region of *Santalum album*. J Pure Appl Microbiol 8(6):4245–5054

Rajash P (2005) Effect of plant growth promoting Rhizobacteria on Canola (*Brassica napus* L) and Lentil (*Lens culinaris*) Plants ETD Project

Rajkumar M, Ae N, Prasad MNV, Freitas H (2010) Potential of siderophore-producing bacteria for improving heavy metal phytoextraction. Trends Biotechnol 28:142–149

Reddy P, Battu RMS (2009) Siderophore-mediated antibiosis of rhizobacterial fluorescent *Pseudomonads* against Rice fungal pathogens. Int J Pharm Tech Res 1:227–229

Reddy MS, Kumar S, Khosla B (2002) Biosolubilization of poorly soluble rock phosphates by *Aspergillustubingensis* and *Aspergillusniger*. Bioresour Technol 84:187–189

Römheld V, Marschner H (1986) Evidence for a specific uptake system for iron phytosiderophores in roots of grasses. Plant Physiol 80:175–180

Rosselló-Mora R, Amann R (2001) The species concept for prokaryotes. FEMS Microbiol Rev 25:39–67

Rungin S, Indananda C, Suttiviriya P, Kruasuwan W, Jaemsaeng R, Thamchaipenet A (2012) Plant growth enhancing effects by a siderophore-producing endophytic *streptomycete* isolated from a Thai jasmine rice plant (*Oryza sativa* L. cv. KDML105). Antonie Van Leeuwenhoek 102(3):463–472

Saharan BS, Nehra V (2011) Plant growth promoting rhizobacteria: a critical review. Life Sci Med Res 21:1–30

Sahoo RK, Ansari MW, Pradhan M, Dangar TK, Mohanty S, Tuteja N (2014) Phenotypic and molecular characterization of efficient native *Azospirillum* strains from rice fields for crop improvement. Protoplasma 251:943. https://doi.org/10.1007/s00709-013-0607-7

Sayyed RZ, Badgujar MD, Sonawane HM, Mhaske MM, Chincholkar SB (2005) Production of microbial iron chelators (siderophores) by fluorescent Pseudomonads. Indian J Biotechnol 4:484–490

Schalk IJ, Hannauer M, Braud A (2011) New roles for bacterial siderophores in metal transport and tolerance. Environ Microbiol 13:2844–2854

Schenk PM, Carvalhais LC, Kazan K (2012) Unraveling plant–microbe interactions: can multi-species transcriptomics help? Trends Biotechnol 30:177–184

Schippers B, Bakker AW, Bakker PA (1987) Interactions of deleterious and beneficial rhizosphere microorganisms and the effect of cropping practices. Annu Rev Phytopathol 25:339–358

Schneider EL, Loke SP, Hopkins DT (1968) GLC analysis of cyclopropenoid acids. JAOCS 45:585–590

Serpil S (2012) An agricultural pollutant: chemical fertilizer. Int J Environ Sci 3(1):77–80

Seuk C, Paulita T, Baker R (1988) Attributes associate with increased biocontrol activity of fluorescent *Pseudomonads*. J Plant Pathol 4(3):218–225

Singh JS, Pandey VC, Singh DP (2011) Efficient soil microorganisms: a new dimension for sustainable agriculture and environmental development. Agric Ecosyst Environ 140:339–353

SSSA (1991) In: Mortvedt JJ, Cox FR, Shuman LM, Welch RM (eds) Micronutrients in agriculture, 2nd edn. Soil Science Society of America, Inc, Madison

Takemoto T, Nomoto K, Fushiya S, Ouchi R, Kusano G, Hikino H et al (1978) Structure of mugineic acid, a new amino acid possessing an iron-chelating activity from roots washings of water-cultured *Hordeumvulgare*. L. Proc Jpn Acad Ser B 54:469–473

Thompson LM, Troeh FR (1973) Soils and soil fertility, 3rd edn. McGraw-Hill Book Company, New York

Tilak KVBR, Ranganayaki NL, Pal KK, De R, Saxena AK, Nautiyal CS, Mittal S, Tripathi AK, Johri BN (2005) Diversity of plant growth and soil health supporting bacteria. Curr Sci 89:136–149

Trivedi P, Pandey A, Palni LS (2008) In vitro evalution of antagonistic properties of *Psudomonascoruugata*. Microbiol Res 163:329–336

Verma VC, Singh SK, Prakash S (2011) Bio-controland plant growth promotion potential of siderophore producing endophytic *Streptomyces* from *Azadirachta indica* Juss. J Basic Microbiol 51:550–555

Vessey JK (2003) Plant growth promoting rhizobacterianas biofertilizers. Plant and Soil 255:571–586

Visca P, Imperi F, Lamont IL (2007) Pyoverdine siderophores: from biogenesis to biosignificance. Trends Microbiol 15:22–30

Viveros OM, Jorquera MA, Crowley DE, Gajardo G, Mora ML (2010) Mechanisms and practical considerations involved in plant growth promotion by rhizobacteria. J Soil Sci Plant Nutr 10:293–319

Voisard C, Keel C, Haas D, Defago G (1989) Cyanide production by Pseudomonas fluorescens helps suppress black root rot of tobacco under gnotobiotic conditions. EMBO J 8(2):351–358

Wallihan EF (1966) Iron. In: Chapman HD (ed) Diagnostic criteria for plants and soils. University of California, Div Agric Sci, Riverside, pp 203–212

Wallner A, Blatzer M, Schrettl M, Sarg B, Lindner H, Haas H (2009) Ferricrocin, a siderophore involved in intra- and transcellular iron distribution in *Aspergillus fumigatus*. Appl Environ Microbiol 75:4194–4196

Wang Q, Xiong D, Zhao P, Yu X, Tu B, Wang G (2011) Effect of applying an arsenic-resistant and plant growth-promoting rhizobacterium to enhance soil arsenic phytoremediation by *Populusdeltoides* LH05–17. J Appl Microbiol 111:1065–1074

Wani PA, Khan MS, Zaidi A (2007) Effect of metal tolerant plant growth promoting Bradyrhizobium sp. (vigna) on growth, symbiosis, seed yield and metal uptake by greengram plants. Chemosphere 70:36–45

Wichard T, Bellenger JP, Morel FM, Kraepiel AM (2009) Role of the siderophore azotobactin in the bacterial acquisition of nitrogenase metal cofactors. Environ Sci Technol 43:7218–7224

Wilson MK, Abergel RJ, Raymond KN, Arceneaux JE, Byers BR (2006) Siderophores of *Bacillus anthracis*, *Bacillus cereus* and *Bacillus thuringiensis*. Biochem Biophys Res Commun 348:320–325

Winkelman G, Drechsel H (1997) Microbial siderophores. In: Kleinkauf H, von Dohren H (eds) Products of secondary metabolism, vol 7. Wiley VCH, Weinheim, pp 200–246

Winkelmann G (2007) Ecology of siderophores with special reference to the fungi. Biometals 20:379–392

Wuana RA, Okieimen FE (2011) Heavy metals in contaminated soils: a review of sources, chemistry, risks and best available strategies for remediation. ISRN Ecol 2011:20–31

Yu X, Ai C, Xin L, Zhou G (2011) The siderophore producing bacterium, Bacillus subtilis CAS15, has a biocontrol effect on Fusarium wilt and promotes the growth of pepper. Eur J Soil Biol 47:138–145

Zaidi A, Khan MS, Aamil M (2006) Bioassociative effect of rhizospheric microorganisms on growth, yield, and nutrient uptake of green gram. J Plant Nutr 27:601–612

Zhang MK, Liu ZY, Wang H (2010) Use of single extraction methods to predict bioavailability of heavy metals in polluted soils to rice. Commun Soil Sci Plant Anal 41(7):820–831

Chapter 8
Role of Microbial Technology in Agricultural Sustainability

Sushanto Gouda, Suman Nayak, Shristy Bishwakarma, Rout George Kerry, Gitishree Das, and Jayanta Kumar Patra

8.1 Introduction

Green revolution has been one of the greatest achievements of mankind that changed the course of agriculture production and practices all over the world. It has doubled the crop yield, helped in eradication of poverty from several South Asian countries and developed the field of agriculture technology. The success of green revolution for food crop productivity growth was explicitly based on the premise that all round participation of appropriate institutional mechanisms, technology spill over across political and agro-climatic boundaries could be formed (Pingali 2012). However, neither private firms nor national governments had sufficient incentive to invest in the research and development of such international public goods, soil fertility, climate change or human health thereby leading to excessive use of inorganic fertilizers and plant protection chemicals for maximizing crop yield. The indiscriminate uses of inorganic or synthetic fertilizers have resulted in deterioration of physical, chemical and biological health of cultivated land (Pingali 2012; Bagyaraj 2014;

S. Gouda
Amity Institute of Wildlife Science, Amity University, Noida, Uttar Pradesh, India

S. Nayak
Department of Biotechnology and Biomedical Engineering, National Institute of Technology, Rourkela, Odisha, India

S. Bishwakarma
MITS School of Biotechnology, Bhubaneswar, Odisha, India

R.G. Kerry
P.G. Department of Biotechnology, Academy of Management & Information Technology, Khurda, Odisha, India

G. Das • J.K. Patra (✉)
Research Institute of Biotechnology & Medical Converged Science, Dongguk University-Seoul, Ilsandong-gu, Gyeonggi-do, South Korea
e-mail: jkpatra@dongguk.edu

J.K. Patra et al. (eds.), *Microbial Biotechnology*,
https://doi.org/10.1007/978-981-10-6847-8_8

Singh et al. 2017). Consequently, we are now concern about sustainable agriculture and the issue of feeding the ever increasing world population that is estimated to be around ten billion by the next 30 years.

As rapid development continues around the globe, peoples are changing their livelihood and lifestyle. More and more people are moving towards cities and are attracted towards urban life. With the ever increasing population, and limited number of persons available for agricultural practices, the need for food and resources is mounting tremendous pressure on agriculture system for sustainability of mankind (Guihéneuf et al. 2016). Taking into account the different forms of challenges such as soil fertility, increase of agriculture productivity, minimizing the usage of inorganic fertilizers, restoration of agriculture or cultivable land and the demand to feed the growing population; the need for development in agriculture system and its technology is of paramount concern.

Conventional techniques or measures taken by various governmental agencies across the globe is mostly restricted to providing loans, organic fertilizers, electricity for irrigation, improved crop seeds etc. However, all such measures are inapplicable if soil infertility, plant diseases, poor crop yield etc. continue to persist, and proper knowledge and conditions are not created for agriculture activities (Santos et al. 2012). Practiced of monoculture or single crop plantation has also been considered as a major reason behind soil infertility and poor agricultural production (Araujo et al. 2014). With increasing land scarcity and rising land values, the escalating demand for agricultural land to meet out the food for growing population directly accelerates deforestation and reduces significantly the microbial diversity and ecosystem functions (Singh 2014; Araujo et al. 2014). They have also an adverse effects on soils *i.e.* depletes water holding capacity, soil fertility and disparity in soil nutrients. The fact that terrestrial land masses are the sole provider of the basic necessities of life makes every inch of landmass an important component for food source. Land degradation causes the decline of microbial diversity and influences the environmental, social and economic sustainability (Araujo et al. 2014). The conventional agricultural practices are deteriorating the soil productivity and environmental quality at alarming rate (Seneviratne and Kulasooriya 2013). Hence, use of modern tools and techniques, science behind successful plantation, good knowledge of soil, development of diseases resistance crops and production of high nutrient containing crops or grains are some of the areas to be considered with immediate priority. In this context, application of microbial technology has been highlighted as the most appropriate measure for fulfilling the needs without hampering the nature of the soil or altering the climatic variables (Parr et al. 1994; Singh et al. 2010b; Bagyaraj 2014; Sengupta and Gunri 2015; Schäfer and Adams 2015; Singh et al. 2017).

Microbes perform numerous metabolic functions essential for their own maintenance and in the process can also benefit the biosphere directly or indirectly through nutrient recycling, environmental detoxification, soil health improvement, waste water treatment, etc. (Sengupta and Gunri 2015). Microorganisms has been used in several forms of development; be it in space or soil technology, medicines or waste, animals or plants, food or electricity and are still being explored for different research and development programs (Higa and Parr 1994, Singh et al. 2011a, b; Sarkar et al., 2016). In agriculture, microbial intervention has been be helpful in attaining higher productivity with sustainability in agriculture through increase in soil fertility by different mechanisms such as nitrogen fixation, availability of essential plant nutrients (P, K), decomposition and recycling of wastes (industrial, agricultural), bioaccumulation or microbial leaching of inorganic heavy metals, suppression of soil-borne pathogens, bio-degradation of toxicants (herbicides, pesticides), production of bioactive secondary metabolites, solubilization of nutrient sources and many more (Santos et al. 2012; Sengupta and Gunri 2015; Singh et al. 2017).

In addition to providing benefits to the crop, microbial applications are also the technological imperatives for a sustainable agro-economic regime which can ensure long term food security for a green eco-friendly world irrespective of regional or national boundaries. The chapter is a representation of different forms of microbial techniques that are effective for agricultural sustainability and also discuss the future possibilities of improvement through innovation in the field of agricultural microbiology.

8.2 Role of Bacteria, Cyanobacteria and Fungi in Agriculture

Microorganisms namely bacteria, cyanobacteria and fungi are an integral part of the soil ecosystem. They are an essential component of soil and are responsible for maintenance of soil nutritional balance, bioavailability, water holding capacity, nitrogen cycle etc. However with extensive use of chemical fertilizers, pesticides, herbicides and large scale dumping of industrial waste materials, there has been tremendous loss of beneficial soil microbial communities that ultimately led to a significant decline in soil productivity and ecosystem functioning (Alqarawi et al. 2014). Therefore beneficial microbes along with other management practice are needed to be followed for enrichment of soil and increase crop production. Some beneficial groups of microbe are discussed below;

Bioinoculants are microbial preparations of a single or consortia group of living microorganisms that are actively used for enrichment of soil characteristics. There are numerous bacterial groups that can be helpful for maintaining and regaining the soil fertility and increase crop productivity. One such bacterial group is "Methylotrophs" a sub-population of bacteria that capable of using single carbon compounds like methanol, methylamine etc. and use as their sole sources of energy production. They are a key bacterial group that helps in reduction of green house gases, in addition of their role in cycling of carbon in the soil (Iguchi et al. 2015). Methylotrophs are also known to enhance seed germination of several plant species such as improved yields from red pepper plants, higher yields of rice and tomatoes (Meena et al. 2012). Methylotrophs when present on plant surfaces absorbs harmful chemical substances and gases from the atmosphers and thus acts as a natural air purifier (Kumar et al. 2016). They also regulate the carbon, nitrogen, and phosphorous cycling in the soil ecosystem.

Cyanobacteria are an important group of microbes best known for their active role in purification of water bodies. They carry out vital atmospheric and hydrological services in forms of fixtation of atmospheric N_2, bioleaching of heavy metals and xenobiotics, inhibition of bacterial growth. They have also been reported to produce several bioactive compounds like as Vitamins, phenols, flavonoids etc. of potential medicinal values (Higa 1991; Al-Hasan et al. 2001; Dahms et al. 2006; Singh et al. 2011b). Cyanobacteria's unique abilities to survive in extreme environmental conditions and provide minerals like C, N, P, in forms of dead biomass (manure/humus) for plants in salinated soils is a boost to use degraded soil for cultivation purposes (Singh et al. 2016). The high growth rate, metal sorption capacity and the fact that cyanobacteria does not requires land masses for their growth could play an important role in bioabsorption of heavy metals from several industrial effluents or waste such as oil refinery, paper mill, sugar mill, dye and pharmaceuticals waste etc. and hence mitigate eutrophication and pollution of water bodies or aquatic ecosystems (Vílchez et al. 1997).

Cyanobacteria are amiable biofertilizers and can be used for rice based cropping systems and other legumeous crops as they support mycorrizal association between different fungal, bacterial and plants roots system and provides an ideal rhizosphere. Cyanobacteria are easily available on any water sources as waste and thus serve as the cheapest sources of natural biofertilizers (Ladha and Reddy 2003). The organic matter obtained from the decomposed cyanobacterial biomass can acts as binding agent for soil texture, increase the water holding capacity of the soil, making it more habitable for other plants within few years (Sharma et al. 2012).

Fungi are another widespread and beneficial group of rhizosphere microbes or arbuscular mycorrhizal fungi (AMF), has yet to be fully utilized in agriculture, despite its substantial influence on plant productivity. They contribute to changes in vegetation dynamics, both on the landscape and at the microsite scale (Hart et al. 2003). Reintroducing AMF into agrosystems may improve nutrient use efficiency, water use efficiency, and tolerance to pathogens and herbivores. Arbuscular mycorrhizal fungi (AMF) along with other biological N fixing bacteria fulfill about 5–20% to the total N demand in grassland and savannah all over the world. The contribution of AM fungi to temperate and boreal forests is over 80%. Fungi are also known to improve soil structure by promoting the formation of soil aggregates and pores within through their extra-radical hyphae (mycelium) (Van der Heijden et al. 2008; Peng et al. 2013). N_2 fixatation, Phosphorus mobilization, Potassium solubilization by fungi through release of organic acid anions mainly includes citrate, malate and oxalate (Meena et al. 2012). The mycorrhizal group of fungi also enhances the absorption ability of crop plants for phosphorus and other nutrients by solubilization the relatively immobile and low concentration minerals in the soil. Zinc is the most commonly available minerals being influenced by the association, although uptake of copper (Cu), iron, N, K, Ca and Mg are also reported (Rashid et al. 2016; Singh et al. 2010a; Aggarwal et al. 2011). Some group of microbial culture (bacteria, fungi and cyanobacteria) that can be used as sources of regeneration of degraded soil land and impovement of soil fertility are listed in Table 8.1.

8.3 Biofertilizers

The raising population of the world has resulted in the indiscriminate uses of chemical fertilizer that now shows great threat to nature by polluting air, water, and soil. These chemicals adversely affect soil in terms of depletion of water holding capacity, soil fertility, loss of microbial diversity and disproportion in soil nutrients content (Savci 2012). Application of microbes as biofertilizers is considered as possible remediation to demages caused by chemical fertilizers in agriculture. They possess several beneficial charcteristics for enhancing crop production and food safety in addition to their low cost (Mahanty et al. 2016).

Countering the adverse impact of chemical fertilizers and also maintaining a high productivity of crops has led to the development of biofertilizers. Biofertilizers are a substance containing living micro-organisms which when applied to seeds, plants, or soil, colonizes the rhizosphere or the interior of the plants and promotes plant growth by increasing the supply of nutrients to the host plant (Mahanty et al. 2016; Malusa et al. 2012). Biofertilizers are widely used to support growth of microbes that accelerate the augmention and availability of certain nutrients that plants are incapable to extract from the soil ecosysytem. They improve soil properties through nitrogen fixing, solubilization of insoluble phosphates and other metal ions (Mazid and Khan 2015). Biofertilizers have also been known to promote harvest of naturally available rare and low concentration biological substances through solubilization and mobilization and ultimately, crop yield (Mohapatra et al. 2013). The benefits of using biofertilizers also includes cheap source of nutrients, easily available, ecofriendly in nature, and counter acting the harmful affect caused bt use of chemical fertilizers (Gaur 2010; Bhardwaj et al. 2014).

Biofertilizers can be broadly categorized as N_2 fixing Bio-fertilizers (Free-living, Symbiotic, Associative Symbiotic), Phosphate solubilizing Bio-fertilizers (bacteria, fungi), Phosphate mobilizing Biofertilizers (*Arbuscular mycorrhiza, Ectomycorrhiza, Ericoid mycorrhizae, Orchid mycorrhiza*) and Biofertilizers for micro-nutrients, Silicate and Zinc solubilizers and Plant Growth Promoting Rhizobacteria (PGPR) (Mahanty et al. 2016). Most biofertilizers used till date belongs to one of the following categories *i.e.* nitrogen fixing, phosphate solubilizing and mobilizing, and plant growth promoting rhizobacteria. Some of the microbes that are commercially used as sources of biofertilizers and their role in agriculture sustainability are listed in Table 8.1.

8.3.1 Plant Growth Promoting Rhizobacteria (PGPR)

Plants have always been in a symbiotic relationship with soil microbes (bacteria and fungus) during their growth and development. The symbiotic microorganisms inhabiting the rhizosphere of many plant species have diverse beneficial effects on the host plant (Raza et al. 2016) through different mechanisms such as nitrogen

Table 8.1 Sources of biofertilizers and their role in agriculture sustainability

Microbes	Role	Reference
Microbe (Bacteria)		
Pseudomonas Fluorescens	Siderophores	Mahanty et al. (2016)
Rhizobium leguminosarum	Solubilization of minerals such as phosphorus and cytokinin	Mahanty et al. (2016)
Bradyrhizobium japonicum	Phosphate solubilization, IAA, siderophores	Mahanty et al. (2016)
Sinorhizobium	Chitinase and glucanases Production	Kumar et al. (2010)
Enterobacter asburiae	Phosphate solubilization, IAA, Siderophores etc.	Ahemad and Khan (2010)
Acidothiobacillus Ferrooxidans	Solubilization of Potassium	Liu et al. (2012)
Klebsiella pneumoniae Bacillus polymyxa	Nitrogen fixation	Sahu et al. (2012)
Microbacterium Pseudomonas	Phosphate solubilization	Bhattacharyya and Jha (2012)
Burkholderia	phosphate solubilization	Bhattacharyya and Jha (2012)
Azospirillum. Chroococcum	Improves growth by enhancing seed germination and advancing the root architecture	Bhardwaj et al. (2014)
Microbe (Fungi)		
Aspergillus niger	PSM secrete organic acids, which dissolve unavailable phosphate into soluble form and make it available to the plants	Pal et al. (2015)
Penicillium bilaii	Extracts phosphate from the soil and supply to plants	Mohapatra et al. (2013)
Rhizobium meliloti	Increases N,P,K content	Mohapatra et al. (2013)
Aspergillus fumigatus	Phosphate solubilization	Mosttafiz (2012a, b)
Piriformospora indica	Minerals intake	Pal et al. (2015)
Trichoderma species	Enhance phosphorus absorption	Pal et al. (2015)
Saccharomyces spp.	Zinc solubilization	Martino et al. (2003)
Oidiodendron maius	Zinc solubilization	Martino et al. (2003)
Aspergillus spp	Potash solubilization	Lian et al. (2008)
Microbe (Cyanobacteria/algae)		
Spirulina platensis, Phormidium foveolarum	Provide resistance to plants against pathogens, pests, and diseases	Kumar et al. (2016)
Phaeodactylum Tricornutum, P. lutheri,	Provides chemical defense as it is toxic to grazers.	Singh et al. (2017)
Microcystis panniformis	Show cytotoxic, anti-feedant, insecticidal, and ichthyotoxic responses	Silva-Stenico et al. (2011)

(continued)

Table 8.1 (continued)

Microbes	Role	Reference
Nostoc and Anabaena	Contribute nitrogen and growth promoting substances to plants	Kaushik (2014)
Aulosira fertilisima, Calothrix sp	Water holding and Nitrogen fixatation	Sahu et al. (2012)
Hapalosiphon intricatus K2 and Nostoc sp	Inhibits microbial growth	Kaushik (2014)
Tolypothrix tenuis	Improvement in soil physical properties and help in phosphate solubilization	Kaushik (2014)

fixation. These beneficial free-living soil bacteria are usually referred to as PGPR that defend the health of plants in an eco-friendly manner (Akhtar et al. 2012). PGPR and their interactions with plants are exploited commercially and have scientific applications for sustainable agriculture (Gonzalez et al. 2015). Applications of these associations have been investigated in oat, canola, soy, potato, maize, peas, tomato, lentil, barley, wheat, radicchio, and cucumber (Gray and Smith 2005).

PGPR are involved in a number of activities that are helpful in improvement of soil ecosystem, its dynamic and sustainability in crop production (Gupta et al. 2015). They competitively colonize plant roots system and enhance plant growth by different mechanisms, including phosphate solubilization (Ahemad and Khan 2012) nitrogen fixation, production of indole-3-acetic acid (IAA), siderophores (Jahanian et al. 2012), 1-amino-cyclopropane-1-carboxylate (ACC) deaminase, and hydrogen cyanate (Liu et al. 2016); degradation of environmental pollutants, and production of hormones and antibiotics or lytic enzymes (Xie et al. 2016). In addition, some PGPR may also infer more specific plant growth-promoting traits, such detoxifying of heavy metal, salt tolerance, and control of plant pathogen (Egamberdieva and Lugtenberg 2014).

8.3.2 Types of PGPR

PGPR can be broadly catogerized as extracellular plant growth promoting rhizobacteria (ePGPR) and intracellular plant growth promoting rhizobacteria (iPGPR) (Viveros et al. 2010). ePGPR are group of rhizobium that inhabit the rhizosphere (on the rhizoplane) or in the spaces between the cells of the root cortex, whereas iPGPR mainly inhabit within the specialized nodules in root cells. ePGPR comprises on several bacterial genera such as *Azotobacter*, *Serratia, Azospirillum, Bacillus*, *Caulobacter*, *Chromobacterium*, *Agrobacterium*, *Erwinia*, *Flavobacterium*, *Arthrobacter*, *Micrococcous*, *Pseudomonas,* and *Burkholderia*. The endophytic microbes belonging to iPGPR include *Allorhizobium*, *Bradyrhizobium*, *Mesorhizobium*, *Rhizobium*, and *Frankia*, that can fix atmospheric nitrogen specifically for higher plants (Bhattacharyya and Jha 2012).

8.3.3 Role of PGPR as a Plant Growth Enhancer

PGPR enhance plant growth through direct and indirect mechanisms, which involve enhancing plant physiology and resistance to different phyto-pathogens through various modes and actions (Zakry et al. 2012). These include nutrient fixation, neutralizing biotic and abiotic stress, and producing volatile organic compounds (VOCs) and enzymes to prevent disease. However, the mode of action of different types of PGPR varies according to the type of host plant (Garcia et al. 2015). They are also influenced by a number of biotic factors (plant genotypes, plant developmental stages, plant defense mechanisms, other members of the microbial community) and abiotic factors (soil composition, soil management and climatic conditions) (Vacheron et al. 2013).

8.4 Bio-pesticides

In India, about 30% of the crop yield potential is lost every year as a result of insects, disease and weeds, corresponding to 30 million tons of food grain (Koul 2011). The damage and destruction inflicted on crops by pests have had a serious impact on farming and agricultural practices since long time. Different measures have been adapted to control such pests and the primary strategy employed to eliminate the pests is by using chemical pesticides such as chlorinated hydrocarbons, organophosphates and carbonates (Nawaz et al. 2016). However, considering the harmful effects generated by uses of chemicals pesticides, bio-pesticides are developing to be an appropriate means for eliminating different forms of pests in agriculture sector.

Bio-pesticides are chemicals extracted from natural materials such as plants, animals, bacteria or certain minerals that can be used for controlling pests of different forms and origin. Commercially bio-pesticides consists of microbial pesticides, naturally-occurring substances (biochemical substances), and plant-incorporated-protectants (plants containing added genetic material). Bio-pesticides are employed in agricultural use for the purposes of insect control, disease control, weed control, nematode control and plant physiology and productivity (Gupta and Dikshit 2010; Dutta 2015).

Bio-pesticides have been broadly categorized into three types;

8.4.1 Microbial Pesticides

It consists of microbes such as bacterium, fungus, virus or protozoan as pesticidal component. Microbial pesticides offer protection against diverse group of pests although each type of active component is relatively specific for a single target pest.

Bacterial pesticides offer protection against pest like flies and beetle larvae, caterpillars, fungal and bacterial diseases, soil borne pathogens etc. Fungi are generally helpful in controlling sucking insect pests. Some of the most widely used species include *Beauveria bassiana, Metarhizium anisopilae, Nomuraea rileyi, Paecilomyces farinosus* and *Verticillium lecanii* against pests such as nematodes, aphids, grasshoppers, thrips, mealy bugs, whiteflies, scale insects, mosquitoes, soil borne pathogens, weeds and all types of mites (Pineda et al. 2007).

Viruses used as microbial pesticides are DNA-containing baculoviruses (BVs), Nucleopolyhedrosis viruses (NPVs), granuloviruses (GVs), and the RNA-containing reoviruses, cytoplasmic polyhedrosis viruses, nodaviruses, picrona- like viruses and tetraviruses. However, the main categories used in pest management have been NPVs and GVs. These viruses are widely used for controlling vegetable pests, and shows activity against plant-chewing insects. Viruse microbial pesticides had a substantial impact in different forest habitats against pests such as gypsy moths, pine sawflies, Douglas fir tussock moths and pine caterpillars. Codling moth is controlled by *Cydia pomonella* on fruit trees (Lacey et al. 2008) and potato tuber worm by *Phthorimaea operculella* in stored tubers (Arthurs et al. 2008). Virus-based products are also actively used to control cabbage moths, corn earworms, cotton leaf worms and bollworms, celery loppers and tobacco budworms.

Protozoa, viruses and yeast pesticides are actively used against pests such as grass hoppers, locusts, crickets, caterpillars, leaf spot, fruit drop, greasy spot etc. Nematods mostly the entomopathogenic are used for inhibiting growth of weevils, gnats, white grubs and various species of the Sesiidae family include *Steinernema carpocapsae, S. riobrave, S. glaseri, Heterorhabditis bacteriophora* and *H. megidis* (Shapiro-Ilan et al. 2006).

8.4.2 Plant-Incorporated-Protectants (PIPs)

These are pesticidal substances developed in plants through incoperation of genetic material. The characteristics of protectants against pest or insects are achieved through the manipulation of genetic composition of the organism by adding specific genes (Sudakin 2003). Although target insect or plant resistance properties of PIPs are induced features, the effectiveness of any PIPs is depended on certain parameters like frequency of doses, environmental factors, process of developing resistance, size or amount of the incorporated material, and the rapidity of the organism's reproductive cycle. In insect pest management, a number of plant products derived from neem, custard apple, tobacco, pyrethrum, etc. have been used as safer insecticides. Compounds such as limonene, Pyrethrum /Pyrethrins, Rotenone, Sabadilla, and Ryania are widely used across the globe to control fleas, aphids and mites, ants,, roaches, ticks, beetles, caterpillars and thrips, Squash bugs, harlequin bugs, thrips etc. (Nawaz et al. 2016; Koul 2012). While numbers of plant-incorporated-protectants are in use, concern related to safety of human health and environment are needed to be address before its large scale application in agriculture.

8.4.3 Biochemical Pesticides

Biochemical pesticides are extracts of plants or animals or even a whole plant that are effectively used in controlling pests. Biochemical pesticides consist of secondary metabolites obtained either from plants or animals that may include substances, such as pheromones, leave extracts, juices, latex etc. Extracts obtained from *Maducha longifolia, Derris scandals,* and *Eupholsia antignomum*, pineapple have repulsive smell to insect and are an effective means for controlling crop pests. Blending up Kohomba, Cinnomon and *Croton lexifenio* when mixed with water and sprinkled helps in keeping away warms (Kumari 2016). Biochemical pesticides interfere with mating or breeding cycle of different insects and thereby help in controlling their population. Growing of Napier plants has also proven to be been a successful way of preventing warms and insects from other cultivated crops (Gupta and Dikshit 2010; Dutta 2015).

Some of the bio-pesticides like *Bacillus thuringiensis*, NPV and neem based pesticides, are widely used as commercial sources of bio-pesticides. In India, however, it has been restricted to few parts as proper knowledge on their usage is not made available to farmers. Owing to the specificity of the action, microbes may control only a portion of the pests present in a field and may not control other type of pests present in treated areas, which can cause continuous damage. Heat, UV light and desiccation also reduces the efficacy of bio-pesticides. The delivery systems, special formulations, storage procedures and shelf life have also become an important factor in constraint their pest specificity and limited markets exposure and uses.

8.5 Compost

The practice of producing organic fertilizer through the biological decomposition of organic waste has been carried on for centuries as an art known as composting. The concept of composting though a traditional practice, it has gain momentum in recent times in India through active participation of governmental, non-governmental agencies, academic institutions and commercial places by conversion of waste food to organic fertilizers (Sarkar et al. 2016). Efficient Microorganism (EM) compost is an organic fertilizer developed through assimilation of different microbial culture and which when applied to soil improves soil fertility and stimulates plant growth and crop yield. Composting is seen as a low cost method of diverting low cost materials from landfills or, kitchen wastes while creating a product for agricultural purposes (Saha et al. 2010). It is an aerobic biological process of converting degradable organic waste materials into manure or fertilizers using native microbial cultures for agricultural purposes.

Composting offers several benefits such as enhanced soil fertility and soil health–thereby increased agricultural productivity, improved soil biodiversity, reduced ecological risks and a better environment. It also destroys pathogens and reduces the volume of waste (Tiquia and Tam 2000; Zhu 2006). Furthermore, composting transforms unstable ammonia to stabilize organic forms of nitrogen. When applied to soil, compost provides nutrients to soil and improves its fertility (Lee et al. 2004; Zhu 2006). The use of organic EM compost has been an excellent and cost efficient method of balancing nutrients quality in the soil and enhanced crop productivity among horticultural crops (Sharma et al. 2017). Compost can also be used successfully for biological control of diseases in horticultural crops as it supports a rich diversity of microorganisms, hence extending a potentially antagonistic community to phytopathogens (Coventry et al. 2002; Sharma et al. 2017). Soil enzymes are mainly used to assess soil health as it refers to the sustenance of agricultural productivity of soil. Dehydrogenase activity of different microbes are also been related to EM compost dose and it has been displayed that the oxidative activity of viable soil microflora increases with the increase of EM compost dose (Masto et al. 2006; Gil-Sotres et al. 2005).

EM compost manufacturing plant has been set across the country and different plant adapts different technology or techniques in developing the products. Although there is great variation among agencies on formulation of compost, the general quality of compost depends upon the source and nature of organic waste, design and size of the compositing unit, composting procedure and length of maturation (Hargreaves et al. 2008). In addition, fungal and bacterial cultures are also added to the waste to accelerate the degradation process. Though use of organisms and their effectiveness has been disputed, many commercially available preparations can be used to expedite the process (Sarkar et al. 2016). The commonly used microbial cultures used in development of compost are PGPR's, actinomycetes, fungi, anaerobic bacteria (*Bacillus subtilis, E. coli*), aerobic bacteria (*Pseudomonas aeruginosa*), cyanobacteria, endophytes etc. (Singh et al. 2011a; Mendes et al. 2013; Bhardwaj et al. 2014). Annually of all the fertilizers applied to a crop field about 60–90% of the total applied fertilizers are washed away and only 10–40% are made available for plants. Considering this fact, use of microbial containing compost in forms of fertilizers will be of significance values in nutrient management systems to ensure sustain agricultural productivity and safe environment (Adesemoye and Kloepper 2009). Compost has been a cheap and reliable source of organic manure since long and it has been successfully used in agriculture. In addition to their agriculture values, compost design and usage will also be helpful in cleaning the environmental waste and preventation of several forms of diseases.

8.6 Biochar in Soil Fertility Management and Plant Disease Management

Biochar is the residue obtained from heating agricultural waste in the absence of oxygen or in the presence of little amount of oxygen by pyrolysis or gasification and when buried in soil can act as a long term recalcitrant source of soil organic carbon

(C) (Qian et al. 2015; Kamara et al. 2015; Singh et al. 2010c). The usage of biochar as sources of biofertilizers in agriculture is mainly dependent on its properties such as high electrical conductivity, porosity and stability at lower temperatures. These benefits include the promotion of plant growth, diminishing disease incidence in crops, improvement of soil water-holding capacity, limiting the bioavailability of heavy metals and reducing of nutrient leaching loss, which in turn can reduce fertilizer needs (Elmer and Pignatello 2011).

Researches had proven that upon addition of suitable amount of bioachar to soil it tends to increase soil fertility, pH in acidic soils, soil cation exchange capacity (CEC) and improved soil microbial activity and nutrient retention especially in acidic and coarse textured soils (Jeffery et al. 2011; Kookana et al. 2011). Since biochar contains nutrients, such as N, P, K, it can supply nutrients to the soil directly or in association with different mycorrhizal fungal groups. AMF in combination with biochar plays an important role in acquiring essential nutrients (P, K, Mg, Ca etc.) from soil that has a low mobility in soil and are often poorly accessible by plant. Biochars commonly have negative surface charges and thus, high exchange capacity for base cations such as Ca^{2+}, Mg^{2+} and K^{+} (Liang et al. 2010; Biederman and Harpole 2013). It has been evident from researches that biological nitrogen fixation and beneficial mycorrhizal relationships in association with biochar improves crop yield. It also reduces eutrophication in water bodies by minimizing nutrient losses from soil (Sohi et al. 2010). Biochar is being known to influence microbial activity and community dynamic through alteration of physical and chemical properties like pH, soil texture, release of soluble and availability nutrients (Anderson et al. 2011). The most commonly proposed reasons for biochar having a positive effect on soil microbial activity, is attributed to the numerous pores within the biochar that provide additional habitat for microbes and refuge from their grazers (Quilliam et al. 2013).

Apart from its applications in improvement of soil properties, bioachar are also equally helpful for plant disease management. Applied biochar has been reported to induce resistance against foliar fungal pathogens *Botrytis cinerea* (gray mold) and *Leveillula taurica* (powdery mildew) on pepper and tomato and to the broad mite pest (*Polyphagotarsonemus latus*) on pepper and *Daucus carota*. Biochar concentration between 1% and 5% when supplied on coconut fiber-tuff medium is found to be significantly effective in suppressing diseases in plants of different ages (Elad et al. 2010). Biochar on rice straw had also been reported to improved growth of rice through increased in crop height, tiller number and dry biomass weight (Kamara et al. 2015).

Biochar can also help mitigate environmental issues by removing pollutants from soil and water due to the presence of oxygenated groups *i.e.* carboxyl, hydroxyl, and phenolic function groups on its surface (Qian et al. 2015). Biochar are known to be a potential media for sorption of heavy metal (Pb, Cu, Ni and Cd), phosphorus and antibiotic (Hammer et al. 2014; Kamara et al. 2015). The fact that biochar can remain in the soil for several decades adds an additional advantage in using biochar as possible soil amendment agents. In recent times biochar has been able to attract the interest of researchers, policy makers and farmers across the world owing to their several beneficial properties in improving soil fertility and crop productivity.

8.7 Genetically Modified Organisms (GMOs) and Transgenic Crops

The use of organic crops has been encouraged as an eco-friendly approach in most of the developed countries in the world. Use of organic crops has been successful in such countries mainly due to its low population and highly developed technology. However, in developing and under developed countries where the demand for food is reasonably very high, practice of organic crops might not be sufficient to fulfill the demands of its population. Thus, the introduction of genetically modified organisms (GMOs) is necessary in such parts of the world. Genetically modified organisms is the term mostly used to refer to crop plants or animals modified through *in vivo* techniques for desired traits or improved nutritional content. The transfer process involves shifting the desired gene from the chromosome of a particular plant or animal or any other organism into the cell of desired species (Gasson and Burke 2001; Arya 2015). With increase in demands GMOs have paved its way into fields such as applied biological and medical research programes, pharmaceutical drugs design, experimental drug, and agriculture (Kuruganti and Ramanjaneyulu 2007; Mishra and Singh 2013). GMO's have led to development of insect resistance plants (*Bacillus thuringiensis* (Bt), herbicide resistance, disease resistance, nutritional and other enhancements (golden rice), and other benefits including phytoremediation (use of plants to detoxify soil or groundwater), conserve natural resources, decrease nutrient runoff to rivers, and to help meet the increasing world food demands using a limited amount of land (Dona and Arvanitoyannis 2009; Arya 2015).

Since 1990s several GM products has been developed such as soybean, corn, and cotton seed oil, tomato puree (called Flavr Savr), brinjal, groundnut, sorghum, Hawaiian papaya, potatoes, rapeseed (canola), sugarcane, sugar beet, tobacco, cranberries, raspberries, walnuts, field corn as well as sweet corn and rice have been genetically modified to enhance either their yield, or size, or durability, etc. (Kuruganti and Ramanjaneyulu 2007; Mishra and Singh 2013; Arya 2015). Transgenic plants has also been able to increase the nutrient contents in stable food by transferring it from other natural sources through gene transfer and their expression (more unsaturated fatty acids and proteins from legumes into wheat, augmented content of essential amino acids, and proteins form sunflowers into maize, etc.). Some foods have also been modified to make them resistant to insects and viruses and more able to tolerate herbicides (Royal Society 1998). Genetically modified organisms include micro-organisms (bacteria and yeast), insects, plants, fish, and even mammals (Cabot et al. 2001; Mishra and Singh 2013). Major producers of transgenic crops include USA, Argentina, Brazil, India, Canada, China, Paraguay, South Africa (Kuruganti and Ramanjaneyulu 2007; Mishra and Singh 2013; Arya 2015).

Genetically modified foods are in use since two decades and are deemed generally to be safe, yet they continue to generate controversy from time to time. Such controversies are also in raise due to rapid alterations in diet and lifestyle in urban society that results in heritability variant of phenotypes that are reliant on nutraceuticals or functional food supplementation for their expression. Diseases or disorders like early age diabetes, obesity, coronary artery disease (CAD), and cancer are some of the common diseases that are susceptible to environmental factors and daily lifestyle (Gasson and Burke 2001; Food Safety Department 2005; Mishra and Singh 2013). GMO's have also been reported to have impact on normal plants as they transfer transgenes through cross-pollination or out-crossing, alteration in nucleic composition of soil microorganisms, and long term effect on fauna diversity.

As plants are the most important component of soil ecosystem, concern has also been raised on the long term impact of GMO's on them (Lilley et al. 2006). Impact on soil may include effects on bacterial diversity, number and activity; fungal counts; effects on numbers of protozoa, nematodes and collembola; diversity of nematodes; and woodlice mortality. However, most of these effects are context-dependent and not systematic in character through the season. Most transgenic plants have detectable effects on the soil system, which are relatively minor compared with differences between cultivars or those associated with weather and season (Cartwright and Lilley 2004).

8.7.1 Seedless Crops

The demand for seedless fruits has been on the raise since past few years. Seedless fruits such as banana, watermelon have revolutionized the field of parthenocarpy and have initiated research works for further development of seedless fruits. Seedless fruits are generally achieved by parthenocarpy *i.e.* fruits developed without fertilization and by stenospermocarpy (seeds abort after fertilization) through treatment with chemical substances (Voraquaux et al. 2000). Phytohormones such as auxin, gibberellin, ethylene and cytokinins or mixtures of them in different ratios are normally used for development of seedless in several crop species (Pandolfini 2009; Gillaspy et al. 1993). Absence of seeds in fruits has proven to effectively increase the shelf life of fruits, resulting in improved conservation. Seedless fruits are known to increase other physiochemical values such as fruits taste, texture, flesh/ mesocarp content etc. Advances in microbial technology have allowed seedless fruit development under variable environmental conditions with minimum inputs. They have also made availability of phytohormones easy and cost-efficient (Pandolfini 2009). Some of the commercially available seedless fruits and crops are *Vitis inifera*, *Pistacia vera*, seedless grape, *Brassica rapa*, eggplant, citrus, cucumber and watermelon.

8.7.2 Golden Rice

Vitamin A plays an important role in numerous physiological functions in human beings and its defeciency is one of the major reasons of early deaths across the world. Deficiency of Vitamin A is known to affects over 19 million pregnant women and 190 million preschool-age children across Africa and South-East Asia (Dubock 2014). Although a number of fruits and vegetable contain are rich sources of vitamin, its feasibility to poor section of the society has been a major constrains and root causes of its deficiency. Vitamin A is naturally synthesized in plant components such as plastids, chloroplasts of photosynthetic tissues and chromoplasts of fruits and flowers. Golden rice is one of the best products of microbial technology in recent times. It is a bio-fortified crop that was developed by the fusion of different technology keeping in view the huge requirement of vitamins by the lower section of the society. Bio-fortified staple crops have been a convinent source of micronutrient, vitamins and mineral induced or enhanced through genetic engineering. GMO crops are significantly cheaper, cost effective and more sustainable as they reaches to the needy populations more easily than other supplementation (Vitamin pills) or fortification (minerals or vitamins added to processed food) to address micronutrient deficiencies in the population (Tang et al. 2009; Datta et al. 2007).

Rice is a simple and easily digestible food matrix. Golden rice that contains about 16–35 lg b-carotene gram-^{1} helps in adminstarting the nutrient through easy bioconversion of ß-carotene to vitamin A. Golden rice offers better vision, growth, reproduction, cellular differentiation and proliferation, and integrity of the immune system (Rai et al. 2003). In the near future, safety assessments might use more advance profiling techniques like micro-array, 2D gel electrophoresis, and mass spectrometry for detail studies on expression of nucleic matters (mRNA), protein analysis, and metabolic analysis to determine changes in all metabolites and metabolic intermediates between non GM plants and GM plants and help in establishment of safety parameters. Long-term research is required before these techniques can be applied to safety.

8.8 Future Prospective

Sustainable agriculture is a subject of great interest and lively debate among policy makers, conservationist, ecologist, scientists, farmers and biologist. Agricultural practices have changed dramatically, especially with introduction of "Green Revolution" leading to productivity raised with availability of better technologies, machinery, fertilizers etc. (Abubakar and Attanda 2013). It is essential that innovative technologies are used to ensure sustainable agriculture and productivity using, modern irrigation systems, improved varieties, improved soil quality and conserving the environment using resource conservation technologies (Crosson 1992). However, with every great innovation, there come bigger responsibilities that are needed to be address before adaptation of a new technology. Introduction of chemical fertilizers, herbicides, pesticides have resulted in pollution of surface and ground water resources, depletion of top soil, and reduction of biological diversity, low farm product prices, disintegration of social and economic conditions in rural communities or other reasons that are not self-staining for the small farmers (Brklacich et al. 1991).

Development in agriculture biotechnology and microbial technology had also contributed a lot for increase in crop yield through biofertilizers, bio-pesticides, PGPR's etc. The use of biological approaches is becoming more popular as an additive to chemical fertilizers for improving crop yield in an integrated plant nutrient management system. In this regard, the use of plant growth promoting rhizobium bacteria has found a potential role in developing sustainable systems of crop production (Sturz et al. 2000; Shoebitz et al. 2009). Microorganisms though are very useful for agriculture prosperity, they have certain limitations. They are effective only when presented with favourable and optimized environmental conditions like appropriate moisture, pH, temperature and oxygen (depending on the types of microorganisms) (Parr et al. 1994). Clearly, there remains much to be done before AMF and other microbes can be widely adopted in agrosystems (Hart and Trevors 2005; Rashid et al. 2016).

Biotechnology and microbiology together offers a broad field of research for improvement of crop quality, crop productivity and sustainability of existing system to produce more and better quality of agricultural products through GMO's and transgenic crops. Plants such as BT cotton, tomatoes, potatoes, mustard, brinjal, pumpkin, Golden rice and seedless fruits are some of the best outcome of the collabrative approach between technologies. Further development in agriculture biotechnology can results in development of crop that are resistance to multiple pests, thrive in harsh environmental condition and increased crop yield and reducing usages of harmful chemical fertilizers (Tang et al. 2009; Datta et al. 2007; Mosttafiz et al. 2012a, b). Although its several advantages, the global interest in identifying, stimulating, and transferring practical innovation needs to manifest in visible incentives and investment to encourage systemic innovation and reward breakthroughs across the entire food system and especially at the local level in view of safety assessment of GMO's and other biotechnological products.

8.9 Conclusion

Conventional agriculture has played a significant role in satisfying the demands for food by the large human population for several centuries. The prevalence of diversified cropping systems with petite support from nature has been a key for small farmers in sustaining along with the synergistic interactions between different factors of ecosystem such as soils, plants and animals. However, crucial problem arises with the need to feed the growing popolation. Other constrains for traditional agriculture include drought, nutrient deficits, soil fertility, poor crop yield contamination, plant invasions, etc. Although use of modern agriculture practices and green revolution has been able to eradicate poverty and hunger to some extent, it has also resulted in degradation of soil, presence of large amount of pesticides in crops and other environmental and economic sustainability issues. Organic agriculture has been an excellent initiative for combating chemical fertilizers but issues related to time and low crop yield has been its few limitations. Microbial technology along with advances in biotechnology has helped to sustain environmentally friendly agro-technological practices in recent times. Microbial biotechnology is an important area that promotes for soil fertility, crop protection, food security, value-added products etc. Biofertilizers, PGPR, biopesticides, compost, GMO plants has emerged to be potential microbial technologies for higher crop production besides being good for soil management. Discerning the microbial ecology, environmental science and agricultural biotechnology, "together", they offer promise future for food safety, higher productivity and sustainable agriculture.

References

Abubakar MS, Attanda ML (2013) The concept sustainable agriculture: Challenges and prospects. 5th Int Conf Mechatronics, Materials Sci Engg 53: 012001

Adesemoye AO, Kloepper JW (2009) Plant-microbes interactions in enhanced fertilizer-use efficiency. Appl Microbiol Biotechnol 85:1–12

Aggarwal A, Kadian N, Tanwar A, Yadav A, Gupta KK (2011) Role of arbuscular mycorrhizal fungi (AMF) in global sustainable development. J Appl Natural Sci 3(2):340–351

Ahemad M, Khan MS (2010) Plant growth promoting activities of phosphate-solubilizing *Enterobacter asburiae* as influenced by fungicides. Eurasia. J Biosci 4:88–95

Ahemad M, Khan MS (2012) Evaluation of plant-growth promoting activities of rhizobacterium *Pseudomonas putida* under herbicide stress. Annals Microbio 62:1531–1540

Akhtar N, Qureshi MA, Iqbal A, Ahmad MJ, Khan KH (2012) Influence of Azotobacter and IAA on symbiotic performance of *Rhizobium* and yield parameters of lentil. J Agric Res 50:361–372

Al-Hasan RH, Khanafer M, Eliyas M, Radwan SS (2001) Hydrocarbon accumulation by picocyanobacteria from the Arabian Gulf. J Appl Microbiol 91:533–540

Alqarawi AA, Allah AA, Hashem EFA (2014) Alleviation of salt-induced adverse impact *via* mycorrhizal fungi in *Ephedra aphylla* Forssk. J Plant Interact 9(1):802–810

Anderson CR, Condron LM, Clough TJ, Fiers M, Steward A, Hill RA et al (2011) Biochar induced soil microbial community change: implications for biogeochemical cycling of carbon, nitrogen and phosphorus. Pedobiologia 54:309–320

Araujo ASF, Borges CD, Tsai SM, Cesarz S, Eisenhauer N (2014) Soil bacterial diversity in degraded and restored lands of Northeast Brazil. Antonie Van Leeuwenhoek 106:891–899

Arthurs SP, Lacey LA, Rosa FDL (2008) Evaluation of granulovirus (PoGV) and *Bacillus thuringiensis* subsp. Kurstaki for control of the potato tuberworm (Lepidoptera: Gelechiidae) in stored tubers. J Eco Entomolo 101:1540–1546

Arya D (2015) Genetically modified foods: benefits and risks. Massachusetts Medical Society Committee on nutrition and physical activity

Bagyaraj DJ (2014) Microorganisms in sustainable agriculture. Proc Indian Nat Sci Acad 80(2):357

Bhardwaj D, Ansari MW, Sahoo RK, Tuteja N (2014) Biofertilizers function as key player in sustainable agriculture by improving soil fertility, plant tolerance and crop productivity. Microb Cell Fact 13(66):1–10

Bhattacharyya PN, Jha DK (2012) Plant growth-promoting Rhizobacteria (PGPR): emergence in agriculture. World J Microbiol Biotechnol 28:1327–1350

Biederman LA, Harpole WS (2013) Biochar and its effects on plant productivity and nutrient cycling: a meta-analysis. GCB Bioenergy 5:202–214

Brklacich M, Bryant C, Smit B (1991) Review and appraisal of concept of sustainable food production systems. Environ Manag 15:1–14

Cabot RA, Kühholzer B, Chan A et al (2001) Transgenic pigs produced using in vitro matured oocytes infected with a retroviral vector. Anim Biotechnol 12(2):205–214

Cartwright CD, Lilley AK (2004) Mechanisms for investigating changes in soil ecology due to GMO releases (Defra report EPG 1/5/214) Department for Environment, Food and Rural Affairs

Coventry E, Noble R, Mead A, Whipps JM (2002) Control of Allium white rot (*Sclerotium cepivorum*) with composted onion waste. Soil Biol Biochem 34:1037–1045

Crosson P (1992) Sustainable agriculture. Resources 106:14–17

Dahms HU, Xu Y, Pfeiffer C (2006) Antifouling potential of cyanobacteria: a mini-review. Biofouling 22:317–327

Datta SK, Datta K, Parkhi V, Rai M, Baisakh N, Sahoo G et al (2007) Golden rice: introgression, breeding, and field evaluation. Euphytica 154:271–278

Dona A, Arvanitoyannis IS (2009) Health risks of genetically modified foods. Crit Rev Food Sci Nutr 49(2):164–175

Dubock A (2014) The present status of Golden Rice. J Huazhong Agri Univ 33(6):69–84

Dutta S (2015) Biopesticides: an eco-friendly approach for pest control. World J Pharm Pharm Sci 4(6):250–265

Egamberdieva D, Lugtenberg B (2014) Use of plant growth promoting rhizobacteria to alleviate salinity stress in plants. PGPR to alleviate salinity stress on plant growth. In: Miransari M (ed) Use of microbes for the alleviation of soil stresses. Spinger, New York, pp 73–96

Elad Y, David RD, Harel MY, Borenshtein M, Silber BKA, Graber ER (2010) Induction of systemic resistance in plants by biochar, a soil-applied carbon sequestering agent. Phytopathology 100:913–921

Elmer WH, Pignatello J (2011) Effect of biochar amendments on mycorrhizal associations and *Fusarium* crown and root rot of *Asparagus* in replant soils. Plant Dis 95:960–966

Food Safety Department, World Health Organization (WHO) (2005) Modern food biotechnology, human health and development: an evidence based study. Geneva, Switzerland

Garcia FP, Menendez E, Rivas R (2015) Role of bacterial bio fertilizers in agriculture and forestry. AIMS Bioeng 2:183–205

Gasson M, Burke D (2001) Scientific perspectives on regulating the safety of genetically modified foods. Nat Rev Genet 2:217–222

Gaur V (2010) Biofertilizer–necessity for sustainability. J Adv Dev 1:7–8

Gillaspy G, Ben-David H, Gruissem W (1993) Fruits: a developmental perspective. Plant Cell 5:1439–1451

Gil-Sotres F, Trasar-Cepeda C, Leirós MC, Seoane S (2005) Different approaches to evaluating soil quality using biochemical properties. Soil Biol Biochem 37:877–887

Gonzalez AJ, Larraburu EE, Llorente BE (2015) *Azospirillum brasilense* increased salt tolerance of jojoba during *in vitro* rooting. Ind Crop Prod 76:41–48

Gray EJ, Smith DL (2005) Intracellular and extracellular PGPR: commonalities and distinctions in the plant-bacterium signalling processes. Soil Biol Biochem 37:395–412

Guihéneuf F, Khan A, Tran LSP (2016) Genetic engineering: a promising tool to engender physiological, biochemical, and molecular stress resilience in green microalgae. Front Plant Sci 7:400. https://doi.org/10.3389/fpls.2016.00400

Gupta G, Parihar SS, Ahirwar NK, Snehi SK, Singh V (2015) Plant growth promoting rhizobacteria (PGPR): current and future prospects for development of sustainable agriculture. J Microbio Biochemist 7:96–102

Gupta S, Dikshit AK (2010) Biopesticides: an ecofriendly approach for pest control. J Biopest 3(1):186–188

Hammer EC, Balogh-Brunstad Z, Jakobsen I, Olsson PA, Stipp SLS, Rillig MC (2014) A mycorrhizal fungus grows on biochar and captures phosphorus from its surfaces. Soil Biol Biochem 77:252–260

Hargreaves JC, Adl MS, Warman PR (2008) A review of the use of composted municipal solid waste in agriculture. Agric Ecosyst Environ 123:1–14

Hart MM, Reader RJ, Klironomos JN (2003) Plant coexistence mediated by arbuscular mycorrhizal fungi. Trends Ecol Evol 18:418–423

Hart MM, Trevors JT (2005) Microbe management: Application of mycorrhyzal fungi in sustainable agriculture. Front Ecol Environ 3(10):533–539

Higa T (1991) Effective microorganisms: a biotechnology for mankind. In: Parr JF, Hornick SB, Simpson ME (eds) Proceedings of the first international conference on Kyusei Nature Farming. U.S. Department of Agriculture, Washington, DC, pp 8–14

Higa T, Parr JF (1994) Beneficial and effective microorganisms for a sustainable agriculture and environment. Int Nature Farming Res Centre, Atami, Japan, 16p

Iguchi H, Yurimoto H, Sakai Y (2015) Interactions of methylotrophs with plants and other heterotrophic bacteria. Microorganisms 3:137–151

Jahanian A, Chaichi MR, Rezaei K, Rezayazdi K, Khavazi K (2012) The effect of plant growth promoting rhizobacteria (PGPR) on germination and primary growth of artichoke (*Cynaras colymus*). Int J Agri Crop Sci 4:923–929

Jeffery S, Verheijen FGA, Van der Velde M, Bastos AC (2011) A quantitative review of the effects of biochar application to soils on crop productivity using meta-analysis. Agric Ecosyst Environ 144:175–187

Kamara A, Kamara HS, Kamara MS (2015) Effect of rice straw biochar on soil quality and the early growth and biomass yield of two rice varieties. Agri Sci 6:798–806

Kaushik BD (2014) Developments in cyanobacterial biofertilizer. Proc Indian Natn Sci Acad 80(2):379–388

Kookana RS, Sarmah AK, Van Zwieten L, Krull E, Singh B (2011) Biochar application to soil: agronomic and environmental benefits and unintended consequences. In: Donald LS (ed) Advances in agronomy. Academic Press, San Diego, pp 103–143

Koul O (2011) Microbial biopesticides: opportunities and challenges. CAB Rev: Perspectives in Agri Vet Sci Nutri Nat Res (056):1–26

Koul O (2012) Plant biodiversity as a resource for natural products for insect pest management. In: Gurr GM, Wratten SD, Snyder WE, Read DMY (eds) Biodiversity and insect pests: key issues for sustainable management. Wiley, Sussex, p 85105

Kumar H, Bajpai VK, Dubey RC, Maheswari DK, Kang SC (2010) Wilt disease management and enhancement of growth and yield of *Cajanus cajan* (L) var. manak by bacterial combinations amended with chemical fertilizer. Crop Prot 29(6):591–598

Kumar M, Tomar RS, Lade H, Paul D (2016) Methylotrophic bacteria in sustainable agriculture. World J Microbiol Biotechnol 32:120. https://doi.org/10.1007/s11274-016-2074-8

Kumari P (2016) A study of traditional pest and diseases control methods for sustainable rice cultivation in Sri Lanka Business. IOSR J Manage 18(10):34–36

Kuruganti K, Ramanjaneyulu GV (2007) Genetic engineering in Indian Agriculture – an introductory handbook. Centre for Sustainable Agriculture, Secunderabad

Lacey LA, Headrick HL, Arthurs SP (2008) Effect of temperature on long-term storage of codling moth granulovirus formulations. J Eco Entomolo 101:288–294

Ladha JK, Reddy PM (2003) Nitrogen fixation in rice systems: state of knowledge and future prospects. Plant and Soil 252:151–167

Lee JJ, Park RD, Kim YW, Shim JH, Chae DH, Rim YS, Sohn BK, Kim TH, Kim KY (2004) Effect of food waste compost on microbial population, soil enzyme activity and lettuce growth. Bioresour Technol 93:21–28

Lian B, Wang B, Pan M, Liu C, Teng HH (2008) Microbial release of potassium from K bearing minerals by thermophlic fungus *Aspergillus fumigatus*. Geochim Cosmochim Acta 72:87–98
Liang B, Lehmann J, Solomon D, Kinyangi J, Grossman J, O'Neill B et al (2010) Black carbon increases cation exchange capacity in soils. Soil Sci Soc Am J 70:1719–1730
Lilley AK, Bailey MJ, Cartwright C, Turner SL, Hirsch PR (2006) Life in Earth: the impact of GM plants on soil ecology? Trends Biotechnol 24(1):9–14
Liu D, Lian B, Dong H (2012) Isolation of *Paenibacillus* sp. and assessment of its potential for enhancing mineral weathering. Geomicrobiol J 29(5):413–421
Liu W, Wang Q, Hou J, Tu C, Luo Y, Christie P (2016) Whole genome analysis of halotolerant and alkalotolerant plant growth-promoting rhizobacterium *Klebsiella sp.* D5A. Sci Rep 6:26710
Mahanty T, Bhattacharjee S, Goswami M, Bhattacharyya P, Das B, Ghosh A, Tribedi P (2016) Biofertilizers: a potential approach for sustainable agriculture development. Environ Sci Pollut Res. https://doi.org/10.1007/s11356-016-8104-0
Malusa E, Sas-Paszt L, Ciesielska J (2012) Technologies for beneficial microorganisms inocula used as biofertilizers. Sci World J 12
Martino E, Perotto S, Parsons R (2003) Solubilization of insoluble inorganic zinc compounds by ericoid mycorrhizal fungi derived from heavy metal polluted sites. Soil Biol Biochem 35:133–141
Masto RE, Chhonkar PK, Singh D, Patra AK (2006) Changes in soil biological and biochemical characteristics in a long-term field trial on a sub-tropical inceptisol. Soil Biol Biochem 38:1577–1582
Mazid M, Khan TA (2015) Future of bio-fertilizers in Indian agriculture: an overview. Int J Agric Food Res 3(3):10–23
Meena KK, Kumar M, Kalyuzhnaya MG, Yandigeri MS, Singh DP, Saxena AK, Arora DK (2012) Epiphytic pink-pigmentedmethylotrophic bacteria enhance germination and seedling growth of wheat (*Triticum aestivum*) by producing phytohormone. Antonie Van Leeuwenhoek 101:777–786
Mendes R, Garbeva P, Raaijmakers JM (2013) The rhizosphere microbiome: significance of plant beneficial plant pathogenic and human pathogenic microorganisms. FEMS Microbiol Rev 37:634–663
Mishra S, Singh RB (2013) Physiological and biochemical significance of genetically modified foods: an overview. The Open Nutraceuticals J 6:18–26
Mohapatra B, Verma DK, Sen A, Panda BB, Asthir B (2013) Bio-fertilizers – a gateway to sustainable agriculture. Popular Kheti 1(4):97–106
Mosttafiz S, Rahman M, Rahman M (2012a) Biotechnology: role of microbes in sustainable agriculture and environmental health. The. Internet J Microbiol 10(1):1–6
Mosttafiz S, Rahman M, Rahman M (2012b) Biotechnology: role of microbes in sustainable agriculture and environmental health. The Internet J Microbio 10(1):1–7
Nawaz M, Mabubu JI, Hua H (2016) Current status and advancement of biopesticides: microbial and botanical pesticides. J Entomolo Zoo Stud 4(2):241246
Pal S, Singh HB, Farooqui A, Rakshit A (2015) Fungal biofertilizers in Indian agriculture: perception, demand and promotion. J Eco-friendly Agri 10(2):101–113
Pandolfini T (2009) Seedless fruit production by hormonal regulation of fruit set. Forum Nutr 1:168–177
Parr JF, Hornick SB, Kaufman DD (1994) Use of microbial Inoculants and organic fertilizers in agricultural production. Proc Int Semi Use of Microbial and Organic Fertilizers in Agri Production Taipei, Taiwan
Peng S, Guo T, Liu G (2013) The effects of arbuscular mycorrhizal hyphalnetworks on soil aggregations of purple soil in southwest China. Soil Biol Biochem 57:411–417
Pineda S, Alatorre R, Schneider M, Martinez A (2007) Pathogenicity of two entomopathogenic fungi on *Trialeurodes vaporariorum* and field evaluation of a *Paecilomyces fumosoroseus* isolate. Southwestern Entomolo 32:43–52

Pingali PL (2012) Green revolution: impacts, limits, and the path ahead. Proc Nat Acad USA 109(31):12302–12308

Qian K, Kumar A, Zhang H, Bellmer D, Huhnke R (2015) Recent advances in utilization of biochar. Renew Sustain Energy Rev 42:1055–1064

Quilliam RS, Glanville HC, Wade SC, Jones DL (2013) Life in the 'charosphere' does biochar in agricultural soil provide a significant habitat for microorganisms? Soil Biol Biochem 65:287–293

Rai M, Datta K, Baisakh N, Abrigo E, Oliva N, Datta SK (2003) Agronomic performance of Golden indica rice (cv. IR64). Rice Genet Newsl 20:30–33

Rashid MI, Mujawar LM, Shahzad T, Almeelbi T, Ismail IMI, Oves M (2016) Bacteria and fungi can contribute to nutrients bioavailability and aggregate formation in degraded soils. Microbiol Res 183:26–41

Raza W, Yousaf S, Rajer FU (2016) Plant growth promoting activity of volatile organic compounds produced by bio-control strains. Sci Letters 4(1):40–43

Royal Society (1998) Genetically modified plants for food use. Royal Society, London

Saha JK, Panwar N, Singh MV (2010) An assessment of municipal solid waste compost quality produced in different cities of India in the perspective of developing quality control indices. Waste Manag 30:192–201

Sahu D, Priyadarshani I, Rath B (2012) Cyanobacteria – as potential biofertilizers. CIBTech J Microbiology ISSN:2319–3867

Santos VB, Araujo SF, Leite LF, Nunes LA, Melo JW (2012) Soil microbial biomass and organic matter fractions during transition from conventional to organic farming systems. Geoderma 170:227–231

Sarkar S, Pal S, Chanda S (2016) Optimization of a vegetable waste composting process with a significant thermophilic phase. Int Conf Solid Waste Manag, Proc Env Sci 35:435–440

Savci S (2012) An agricultural pollutant: chemical fertilizer. Int J Env Sci Dev 3(1):73

Schäfer T, Adams T (2015) The importance of microbiology in sustainable agriculture. In: Springer International Publication (ed) Principles of plant-microbe interactions. Lugtenberg, Switzerland

Seneviratne G, Kulasooriya SA (2013) Reinstating soil microbial diversity in agroecosystems: the need of the hour for sustainability and health. Agric Ecosyst Environ 164:181–182

Sengupta A, Gunri SK (2015) Microbial intervention in agriculture: an overview. Afr J Microbiol Res 9(18):1215–1226

Shapiro-Ilan DI, Gouge DH, Piggott SJ, Patterson Fife J (2006) Application technology and environmental considerations for use of entomopathogenic nematodes in biological control. Biol Control 38:124–133

Sharma A, Saha TN, Arora A, Shah R, Nain L (2017) Efficient microorganism compost benefits plant growth and improves soil health in Calendula and Marigold. Horticul Plant J 3(2):67–72

Sharma R, Khokhar MK, Jat RL, Khandelwal SK (2012) Role of algae and cyanobacteria in sustainable agriculture system. Wudpecker J Agri Res 1(9):381–388

Shoebitz M, Ribaudo CM, Pardo MA, Cantore ML, Ciampi L, Cura JA (2009) Plant growth promoting properties of a strain of *Enterobacter ludwigii* isolated from *Lolium perenne* rhizosphere. Soil Biol Biochem 41:1768–1774

Silva-Stenico ME, Silva CSP, Lorenzi AS, Shishido TK, Etchegaray A, Lira SP et al (2011) Nonribosomal peptides produced by Brazilian cyanobacterial isolates with antimicrobial activity. Microbiol Res 166:161–175

Singh B, Singh BP, Cowie AL (2010c) Characterization and evaluation of biochars for their application as a soil amendment. Australian J Soil Res 48:516–525

Singh JS (2014) Cyanobacteria: a vital bio-agent in eco-restoration of degraded lands and sustainable agriculture. Climate Change Env Sustain 2(2):133–137

Singh JS, Kumar A, Rai AN, Singh DP (2016) Cyanobacteria: a precious bio-resource in agriculture, ecosystem, and environmental sustainability. Front Microbiol 7:529. https://doi.org/10.3389/fmicb.2016.00529

Singh JS, Pandey VC, Singh DP (2011a) Efficient soil microorganisms: a new dimension for sustainable agriculture and environmental development. Agric Ecosyst Environ 140:339–353

Singh JS, Pandey VC, Singh DP, Singh RP (2010a) Influence of pyrite and farmyard manure on population dynamics of soil methanotroph and rice yield in saline rain-fed paddy field. Agric Ecosyst Environ 139:74–79

Singh R, Parihar P, Singh M, Bajguz A, Kumar J, Singh S, Singh VP, Prasad SM (2017) Uncovering potential applications of Cyanobacteria and algal metabolites in biology, agriculture and medicine: current status and future prospects. Front Microbiol 8:515

Singh SR, Singh U, Chaubey AK, Bhat M (2010b) Mycorrhizal fungi for sustainable agriculture – a review. Agric Rev 31(2):93–104

Singh JS, Singh DP, Dixit S (2011b) Cyanobacteria: an agent of heavy metal removal. In: Maheshwari DK, Dubey RC (eds) Bioremediation of pollutants. IK International Publisher Co, New Delhi, pp 223–243

Sohi SP, Krull E, Lopez-Capel E, Bol R (2010) A review of biochar and its use and function in soil. In: Donald LS (ed) Advances in agronomy. Academic Press, San Diego, pp 47–82

Sturz AV, Christie BR, Nowak J (2000) Bacterial endophytes: potential role in developing sustainable systems of crop production. Crit Rev Plant Sci 19:1–30

Sudakin DL (2003) Biopesticides. Toxicol Rev 22(2):83–90

Tang G, Qin J, Dolnikowski GG, Russell RM, Grusak MA (2009) Golden Rice is an effective source of Vitamin A1–4. Am J Clin Nutr 89:1776–1783

Tiquia SM, Tam NYF (2000) Co-composting of spent pig litter and sludge with forced-aeration. Bioresour Technol 72:1–7

Vacheron J, Desbrosses G, Bouffaud ML, Touraine B, Moënne-Loccoz Y, Muller D, Legendre L, Wisniewski-Dyé F, Combaret CP (2013) Plant growth promoting rhizobacteria and root system functioning. Front Plant Sci 4(356):1–19

Van der Heijden MG, Bardgett RD, Van Straalen NM (2008) The unseenmajority: soil microbes as drivers of plant diversity and productivity interrestrial ecosystems. Ecol Lett 11(3):296–310

Vílchez C, Garbayo I, Lobato MV, Vega JM (1997) Microalgae- mediated chemicals production and wastes removal. Enzyme Microb Technol 20:562–572

Viveros OM, Jorquera MA, Crowley DE, Gajardo G, Mora ML (2010) Mechanisms and practical considerations involved in plant growth promotion by rhizobacteria. J Soil Sci Plant Nutr 10:293–319

Voraquaux F, Blanvillain R, Delseny M, Gallois P (2000) Less is better: new approaches for seedless fruit production. Trends Biotechnol 18:233–242

Xie J, Shi H, Du Z, Wang T, Liu X, Chen S (2016) Comparative genomic and functional analysis reveals conservation of plant growth promoting traits in *Paenibacillus polymyxa* and its closely related species. Sci Rep 6:21329

Zakry FAA, Shamsuddin ZH, Rahim KA, Zakaria ZZ, Rahim AA (2012) Inoculation of *Bacillus sphaerichus* UPMB-10 to young oil palm and measurement of its uptake of fixed nitrogen using the 15N isotope dilution technique. Microbes Env 27(3):257–262

Zhu N (2006) Composting of high moisture content swine manure with corncob in pilot scale aerated static bin system. Bioresour Technol 97:1870–1875

Chapter 9
Soil Microbial Diversity and Its Utilization in Agriculture in Sri Lanka

S.A. Kulasooriya, Gamini Seneviratne, and E.M.H.G.S. Ekanayake

9.1 Introduction

The soil ecosystem constitutes the loose outermost layer of the Earth's crust including its living and non-living components. Soil formation is a continuous process which depends upon parent material, time, topography, environmental conditions, living organisms and management practices. Being one of the most extensive habitats on Earth, soil sustains different assemblages of plants, animals and microorganisms whose diversity depends largely upon their interactions and environmental conditions that ultimately determine the type of flora and fauna which occupy this habitat. There are close relationships between the non-living soil and the living biosphere and their interactions contribute not only to soil formation but also to its function and sustainment of the entire ecosystem.

9.1.1 Soil Constituents and Their Functions

9.1.1.1 Inorganic Components

The major inorganic component of the soil ecosystem is mineral matter largely formed by the weathering of the underlying bedrock to which process the living component makes a significant contribution, together with the deposition and transfer of particulate matter by the movement of wind and water. The constituents of this mineral fraction are rocks, stones, gravel, sand and clay in descending order of their particle size. The other major inorganic component of the soil is the pore space

S.A. Kulasooriya (✉) • G. Seneviratne • E.M.H.G.S. Ekanayake
National Institute of Fundamental Studies, Kandy, Sri Lanka
e-mail: ananda@kulasooriya.com; gaminis@ifs.ac.lk; ekanayakeifs@gmail.com

J.K. Patra et al. (eds.), *Microbial Biotechnology*,
https://doi.org/10.1007/978-981-10-6847-8_9

which is occupied by water and air whose occupancy volume and distribution depend largely upon the location, climate and weather conditions to which a soil is exposed.

9.1.1.2 Organic Fraction

The organic fraction of a soil is represented by dead animal and plant matter (including leaf litter) together with its living biomass and its excretory products. This fraction generally occupies the top layers of a soil and seldom goes below a depth of one meter, except the deep feeding roots of trees and their associated organisms. Generally it ranges from 1% to 6% in most upland soils, less than 1% in desert soils and 12% to 18% in organic soils, and in extreme cases like peat bogs, marshy wetlands and swamps it could go up to 90%. The organic fraction is generally categorized into three pools: the living biomass, fresh and partially decomposed dead material and well decomposed humus which also includes recalcitrant sequestered carbon under long term storage.

9.1.1.3 The Living Component and Its Diversity

The living component constitutes a tiny fraction of the organic matter ranging between 0.5% and 1%, often less than 10%.This is represented by plant roots and soil fauna and flora. The soil fauna includes micro-fauna such as protozoa, ciliates and nematodes, meso-fauna such as micro-arthropods, ants and insects and macro-fauna such as earthworms, millipedes, centipedes, scorpions, snails and spiders extending up to certain reptiles and burrowing mammals. The micro-flora is represented by microscopic cyanobacteria and certain micro-algae (restricted to the photic surface layers), eubacteria (the most numerous), archaeobacteria, actinobacteria and fungi (having the highest biomass).The bacteria exhibit a very narrow range of morphological diversity represented primarily by three basic cell shapes: spherical (cocci), rod (bacilli) and spiral (spirillum). But they possess a wide range of nutritional diversity ranging from photoautotrophy, chemoautotrophy and heterotrophy which can be saprophytic, parasitic and symbiotic, each one of which can be either facultative or obligate. On the contrary, the cyanobacteria depict only one type of nutrition i.e. photosynthesis. But they exhibit a wide range of morphological diversity ranging from unicells, loose colonies, compact colonies, colonies of definite shape, simple unbranched undifferentiated filaments, unbranched differentiated filaments, tapering filaments, filaments with false branching, filaments with true branching to heterotrichous thalli in which both prostrate and erect systems are uni-seriate, prostrate system multi-seriate and erect system uni-seriate, culminating in both systems being multi-seriate.

9.1.1.4 Functions of the Living Component

Although the living component represents only a small fraction of the soil it plays vital roles in nutrient cycling, decomposition of organic matter, C-sequestration, biological nitrogen fixation, nitrification and de-nitrification, solubilization of phosphorus and production of growth promoting substances. It also contributes to soil mixing and aggregation and maintenance of the community balance under aerobic, micro-aerobic and anaerobic conditions. Biological nitrogen fixation is the conversion of inert dinitrogen (N_2) into a combined form by a process catalyzed by the enzyme nitrogenase. Among the vast assemblage of plants, animals and microorganisms, this process is confined to certain prokaryotic microorganisms which are frequently found in soil.

The combined effects of this soil biota convert soil into a moisture retaining, nutrient rich, cation exchanging, aerobic, healthy mass of spongy substrate through which plant roots can grow and breathe well, absorb nutrients and function actively. This is exemplified in the soil of a natural, undisturbed rain forest that exists in dynamic equilibrium supporting the long term sustenance of its vegetation. The soil microflora has been reported to provide ecosystem services that have been estimated globally to be equivalent to a value of 33 billion US$ annually (Alonso et al. 2001).

9.1.1.5 Interactions Between Soil Microflora and Plants

The soil microflora though largely capable of independent existence, interacts continuously with the below ground vegetation in association with plant roots. Such associations could range from casual relationships to close and intimate interactions as rhizosphere, rhizoplane, endorhizosphere and symbiotic associations such as the nitrogen fixing legume-rhizobial root nodules, non-legume-frankia root nodules, associative nitrogen fixation, symbiotic nitrogen fixing associations between cyanobacteria and certain plants and phosphorus solubilizing endo and ectomycorrhizal associations. It has been observed that certain soil biota produce plant growth promoting substances and also keep off potential pathogens. While there are many such beneficial associations, there are also instances of certain soil organisms becoming pests and pathogens harmful to plants and animals.

9.1.2 Soil Amendments for Better Crop Production

9.1.2.1 Use of Chemical (Synthetic) Fertilizer in High Input Agriculture

The urgent need to accelerate crop production and increase crop yields in order to avoid widespread global starvation immediately after the World War II, made the global community to introduce revolutionary changes to traditional subsistence agriculture prevalent at that time. Crop breeders were tasked to bring about dramatic increases in yield of major food crops. At the same time industrial facilities including factories used for the manufacture of weapons and nitrogen containing explosives became available for other purposes. During this period the Nobel Laureate Norman Borlaug and his team produced high yielding high input responsive wheat and other cereals which resulted in the 'Green Revolution' that ensured the prevention of global starvation. The ready availability of synthetic chemical fertilizers at affordable prices at that time no doubt contributed to the success of these practices and they proliferated even to developing countries like Sri Lanka.

Most soils under traditional farming at that time had a buildup of organic matter. Inorganic fertilizers added to such soils could be retained and supplied to the targeted crops as required. The results were therefore dramatic and significant increases in crop yields were obtained.

However, the continuous application of such chemicals together with other agrochemicals to minimize, weeds, pests and pathogens has resulted in the reduction of soil microbial biodiversity and its pivotal function of building up soil organic matter (Seneviratne et al. 2011). Thus in most agro-ecosystems crop yields have not increased over the years but have either remained stagnant or show trends of decline. Loss of soil organic matter reducing the ability of a soil to retain added soluble nutrients results in massive losses and it has been reported in several studies that not more than 30% of added chemical fertilizers (particularly nitrogen) are taken up by the targeted crop (Seneviratne and Kulasooriya 1994).

Loss of such added nutrients particularly soluble nitrogen and phosphorus results in environmental pollution and nutrient loading into water bodies. Environmental pollution by the excessive use of chemical fertilizers leads to massive nutrient loading into streams and rivers which eventually empty into oceans through large bays like the Gulf of Mexico and the Bay of Bengal leading to the enormous proliferation of algae and cyanobacteria. The rapid death and decomposition of such planktonic blooms cause decreased levels of dissolved oxygen (DO) in bottom waters due to microbial respiration. These depleted DO levels result in massive mortality of aquatic organisms and create so-called "*dead zones*" where little or no aquatic life can be found. Since the 1960's dead zones have increased exponentially worldwide and have now been documented from over 400 systems, affecting more than 245,000 square kilometers of coastal regions (Diaz and Rosenberg 2008). Such information led to a resolution agreed upon at the UN Conference on Environment and Development in 2012 (Rio + 20) to reduce the application of chemical fertilizer by 20% by the year 2020. There is therefore a global interest in looking for alternative technologies to minimize the use of chemical fertilizers and other agro-chemicals without compromising on crop yields.

9.1.3 The Sri Lankan Scenario

9.1.3.1 Diversity of Soil Types and Agriculture in Sri Lanka

Though Sri Lanka is a small island of 6.5 million hectares, it has a wide diversity in topography, climate, soils and vegetation which is represented by seven climatic regions and 46 agro-ecological zones (Wickremasinghe 2013). Of the 12 global soil orders, seven have been identified in Sri Lanka and these have been classified into 14 Great Soil Groups by De Alwis and Panabokke (1972). Ten of these groups located in the dry and the semi-dry low country areas are: Reddish Brown Earths, Low Humic Gley soils, Non-Calcic Brown soils, Red-Yellow Latosols, Alluvial soils, Solodized Solonetz, Regosols, Soils on Old Alluvium, Grumusols, and Immature Brown Loams. Five groups viz.: Red-Yellow Podzolic soils, Reddish Brown Latozolic soils, Immature Brown Loams, Bog and Half Bog soils and Latosols and Regosols on Old Red and Yellow sands, have been identified in the wet and the semi-wet areas of the low-country, mid-country and the central highlands of Sri Lanka.

9.1.3.2 Agriculture in Sri Lanka

The major crop in the lowland flood plains of the wet, intermediate, semi-dry and the dry zones of the country is wetland rice grown under both irrigated and rain-fed conditions, while the major upland plantation crop in these areas is coconut. Soils of the undulating landscape of the dry and semi-dry areas have a good drainage and they are cultivated with upland field crops such as maize and other cereals, short-term vegetables, legume pulses, seasonal fruits, sugar cane and planted forests. Rolling hills and mountainous lands are generally planted with plantation crops such as tea and rubber, fruit trees and exotic vegetables. Some of them are also used for large scale livestock farming. The extent of these different crop cultivations and their productions are given in Table 9.1. It is evident from this table that the principal crop cultivated in Sri Lanka is rice and a large segment of the rural population is engaged in rice production. Research efforts of the Government Department of Agriculture (DoA) are mainly focused on this crop and primarily through breeding and introduction of new improved varieties. Through high input agronomic practices including the extensive use of chemical fertilizer and other agro-chemicals, the country has become self-sufficient in this staple food. Besides this, the DoA has separate divisions dealing with varietal improvement and development of novel agronomic practices in vegetable and horticultural crops. Research and development efforts of the major plantation crops: tea, rubber and coconut are conducted by their own Research Institutes. These traditional plantation crops bring in significant export earnings and sustain a large labor population settled down in these areas. Besides these, cinnamon, spice crops and sugarcane are also cultivated as plantation crops to a lesser extent, under the guidance of their respective Research Institutes.

Table 9.1 Major agricultural and plantation crops of Sri Lanka

Crop	Extent (ha)	Production (m.t.)
Cereals		
Rice (*Oryza sativa*)	1,141,323	4,420,085
Kurakkan *(Eleusine coracana)*	6151	8565
Maize *(Zea mays)*	67,629	243,960
Sorghum (*Sorghum sp)*	79	161
Meneri(Finger millet) (*Panicum sp*).	32	25
Legume pulses		
Green gram (*Vigna radiata*)	11,301	14,546
Cowpea (*Vigna unguiculata*)	8220	13,740
Soya bean (*Glycine max*)	6301	7646
Black gram (*Vigna mungo*)	11,158	11,197
Ground nut (*Arachys hypogea*)	19,975	24,200
Other field crops		
Gingelly *(Sesamum indicum)*	14,044	12,414
Manioc (*Manihotesculenta*)	22,753	324,080
Sweet potato (*Ipomoea batatas*)	4487	44,715
Potato (*Solanum tuberosum)*	5753	95,805
Red onion (*Allium sp)*	4994	63,675
Big onion (*Allium sp*)	3983	65,223
Chilies (*Capsicum sp)*	15,267	72,311
Beans (*Phaseolus vulgaris*)	8279	66,163
Major Plantation crops[a]		
Tea *(Camellia sinensis)*	221,969	290,000
Rubber (*Hevea braziliensis*)	124,000	136,000
Coconut (*Cocos nucifera*)	394,836	2762 million nuts

Source: Agriculture and Environment Statistics Division, Department of Census and Statistics (2016).[a]Census and Statistics (2009)

"*The above Table has been prepared by us obtaining data from the website of the Department of Census and Statistics of the Government of Sri Lanka. This website is in the public domain and it is not copyright protected*"

Following the advent of the 'global green revolution' of the 1950s high input agriculture was initiated in Sri Lanka in the 1960s with the introduction of fertilizer responsive high yielding varieties of crops. Such breeding programs have been quite successful particularly with rice and are continued at present. Over the years these systems have been extended and popularized and today most farmers are accustomed to identifying soil fertility with chemical fertilizers. Such high input systems of agriculture increased crop yields and improved agricultural incomes especially due to the availability and low cost of chemical fertilizers and other agro-chemicals at the time of their introduction. However this scenario changed with the dramatic increase of fossil fuels in the 1970s and its impact on the cost of fossil fuel based chemical fertilizers. These impacts were felt more severely by countries like Sri Lanka which has to import all of its chemical fertilizer requirements and provide

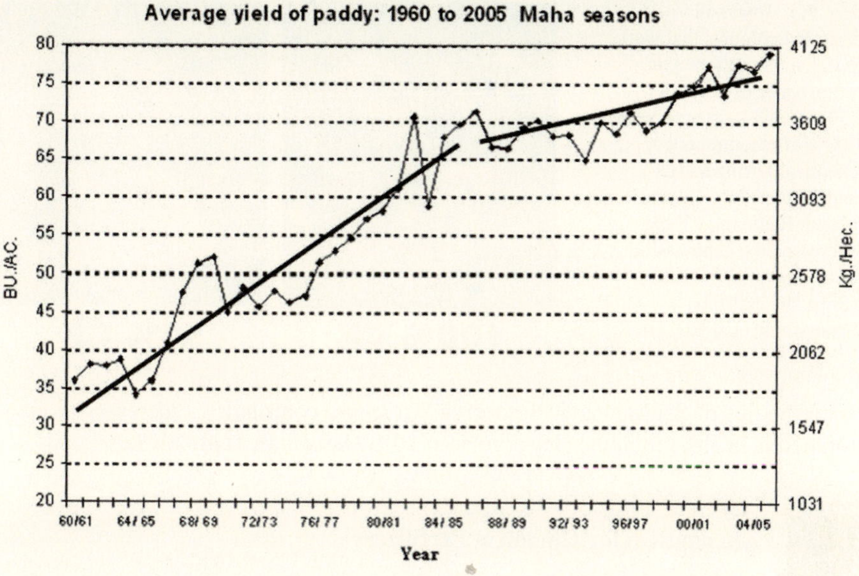

Fig. 9.1 Average yield of paddy during wet or major (Maha) season in Sri Lanka from 1960 to 2005, compiled by Agriculture and Environmental Statistics Division of the Department of Census and Statistics, Colombo, Sri Lanka. Rate of increase in the average yield of paddy over time has slowed down from mid 1980s (http://www.statistics.gov.lk/agriculture/Paddy%20Statistics/PaddyStats.htm) "*This figure has been obtained from the website of the Department of Census and Statistics, Government of Sri Lanka. This website is in public domain and not copyright protected*"

them to resource poor farmers under a heavy government subsidy to ensure an uninterrupted food supply.

9.1.3.3 Consequences of High Input Agriculture

The significant increase in crop yields immediately after the introduction of high input agriculture is also a reflection of the natural fertility of our soils at that time with adequate organic matter that could retain the added inorganic nutrients for efficient absorption by the targeted crop plants. However the continuous use of such inputs has not increased crop yields proportionately (Fig. 9.1) but production costs have gone up decreasing benefits to the farmer.

On the other hand soil microbial populations and diversity have been adversely affected by the indiscriminate use of agro-chemicals (Fig. 9.2).

Furthermore the reduction of organic matter has made these soils less effective in retaining inorganic nutrients added to them resulting in significant pollution of water (Amarasiri 2015). Such pollution has been reported to be responsible for the widespread occurrence and bloom formation of toxigenic cyanobacteria in lentic

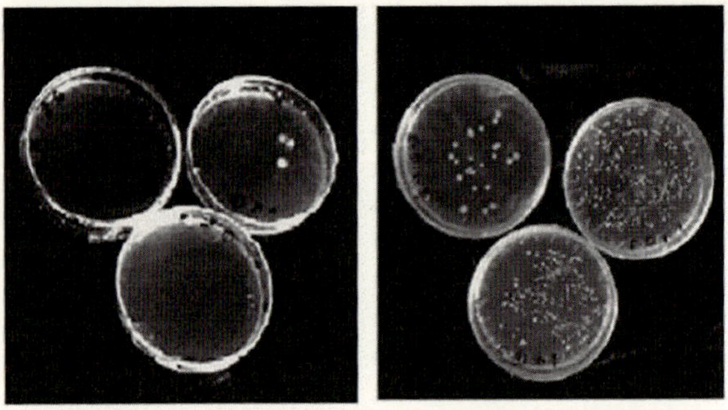

Fig. 9.2 Total soil bacterial colonies that were isolated on nutrient agar, when water extracts of tea soils treated long term with 100% of recommended chemical fertilizers (CF) and 50% of the CF, were plated. Reducing CF has decreased the suppression of the bacterial growth (Reproduced from Seneviratne et al. 2011)

water bodies of Sri Lanka (Kulasooriya 2017) and contribute to increases in environmental health problems (Jeyakumaran 2013; Jayatilake et al. 2013).

9.1.3.4 Alternatives for Chemical Fertilizers

It has become evident that high input agriculture with the continuous application of chemical fertilizer and other agro-chemicals cannot be continued as a cost effective, environmentally benign, sustainable system of crop production. There is therefore a renewed interest to look for alternatives for the reduction of chemical fertilizer applications. The Government of Sri Lanka has taken a policy decision to minimize the use of chemical fertilizer and other agro-chemicals in agriculture. However it is necessary to adopt such radical changes cautiously to ensure that crop yields, farmer incomes and food production levels in the country are not compromised.

The traditional alternatives for chemical fertilizers are the non-synthetic, organic fertilizers such as green manure, animal dung and compost manure prepared either on a domestic or industrial scale. Such alternatives are bulky and not very attractive to farmers used to granulated and/or liquefied chemical fertilizers. Their transport, storage and field application would require additional labor, time and cost. Organic fertilizers also need time to decompose and provide nutrients to the targeted crop plants and this is not compatible with short term crops. A very attractive alternative is the development and application of microbial bio-fertilizers which are living microbial inoculants that continue to live in close association with crop plants and supply nutrients in synchrony and provide other benefits to them. This chapter will deal primarily with studies on the utilization of soil microbial diversity in Sri Lanka with special emphasis on bio-fertilizers and their application in food, forage and plantation crops.

9.2 Utilization of Soil Microbial Diversity

The major role of the vast assemblage of diverse soil microorganisms is the decomposition of organic matter added to soil, thereby participating in the recycling of nutrients that ensures the continuation of life on Earth. In this manner they play crucial roles in the natural Carbon, Nitrogen, Sulfur and other cycles. Certain countries have conducted research studies on the microbial decomposition of organic matter and developed products known as 'efficient microbial inoculants' (EMIs). Such inoculants are applied to increase the rate of decomposition of organic matter, lack of which is a severe constraint for the use of organic matter in the production of short term food crops including all the cereals. A few preliminary trials conducted in Sri Lanka using imported EMIs did not produce the expected outcomes and very little studies are currently in progress in this important area of research.

On the other hand research studies on rhizobial inoculants, biofilm biofertilizers (a novel group introduced by the NIFS), cyanobacterial biofertilizers, root associated nitrogen fixing and plant growth promoting microorganisms, nitrifying and denitrifying microorganisms are in progress and an attempt is made in this chapter to review the status of these studies.

9.2.1 *Rhizobium Bio-fertilizers for Legume Crops*

Most members of the Family Fabaceae (Leguminosae) form root nodules with the soil bacterium of the Genus *Rhizobium* and these nodules are capable of converting atmospheric nitrogen gas (N_2) to a combined form that could be utilized by the host plant. Isolation, purification and identification of naturally occurring rhizobia, authenticating, screening and selection of efficient strains, embedding such strains in suitable carrier material and applying them as bio-fertilizer inoculants to targeted legume crop plants has been an age old practice in both developed and developing countries (Herridge and Gemell 2002; Deaker et al. 2004; Khonje 2014). It has been reported that Australia saves an equivalent of 3.4 billion dollars worth of fertilizer nitrogen annually through the application of rhizobial inoculants (Howeison and Herridge 2005). Surprisingly this technology was not applied in Sri Lanka on an agronomic scale until the turn of the twentieth century. An attempt has been made to popularize *Nitragin-S* (an imported commercial rhizobial inoculant for soybean) in the mid 1980s but this was not successful due to the lack of a suitable locally available, low cost carrier material. Eventually research studies conducted at the National Institute of Fundamental Studies, (NIFS), Kandy, Sri Lanka reported the suitability of modified coir dust (a waste product of the coconut fiber industry) as a carrier material (Seneviratne et al. 1999). This development enabled greenhouse pot and small and large scale field testing of rhizobial strains for soybean and the transfer of this technology to the farmers. A comparison of a nitrogen fertilizer yield response curve with a selected rhizobial inoculant showed that the soybean yield

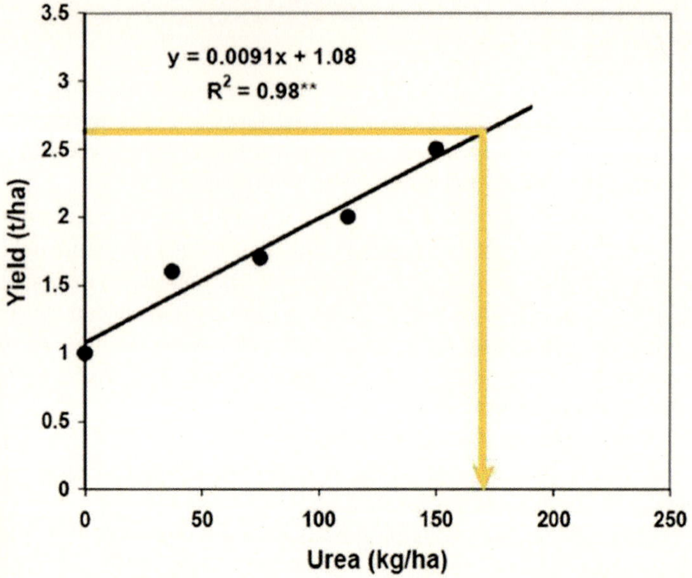

Fig. 9.3 Soybean yield response curve to urea (arrow depicts the yield with inoculant) (Reproduced from Kulasooriya et al. 2014)

obtained with inoculation could even be slightly higher than that obtained with the highest level of recommended urea fertilizer application (Fig. 9.3). Such field trials conducted by farmers in their own fields convinced them that rhizobial bio-fertilizers can replace N-fertilizers completely without any reduction in their yield.

Mean seed yields obtained by four outsourced farmers 5 years after the initial release of inoculants, show that yields have been consistent (Fig. 9.4).

Similarly field testing of rhizobial inoculants in farmers' fields with mung bean (*Vigna radiata*) have given very encouraging results (Fig. 9.5). Field trials with common vegetable bean (*Phaseolus vulgaris*) conducted in several locations have also given promising results (Fig. 9.6).

In all these cases the yields obtained by the complete replacement of chemical N-fertilizer with inoculation have been either equal or better.

Farmer acceptance of rhizobium bio-fertilizers at present is satisfactory and currently around 5000 ha of soybean (*Glycine max*), 1500 ha of mung bean (*Vigna radiata*) and 1000 ha of vegetable bean (*Phaseolus vulgaris*) use rhizobial inoculants annually for their cultivations.

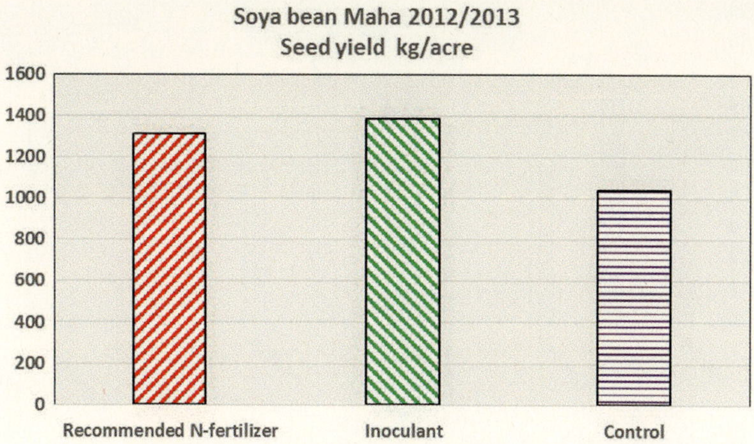

Fig. 9.4 Average soybean yields 5 years after initial release of inoculants (mean values from 4 farmers)

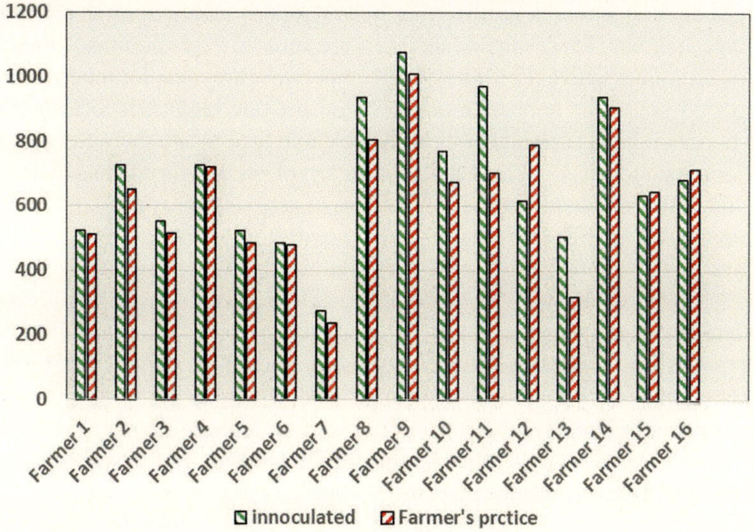

Fig. 9.5 Mung bean yield results from 16 farmers' fields (kg/ha)

9.2.2 Pioneering Studies in Sri Lanka on the Rhizobiology of Clover

Forage legumes such as clover and alfa-alfa commonly used in the livestock industry have been introduced to Sri Lanka as far back as 1942 when livestock farms were established by the National Livestock Development Board of Sri Lanka. However there are no records whether rhizobial inoculants were applied during

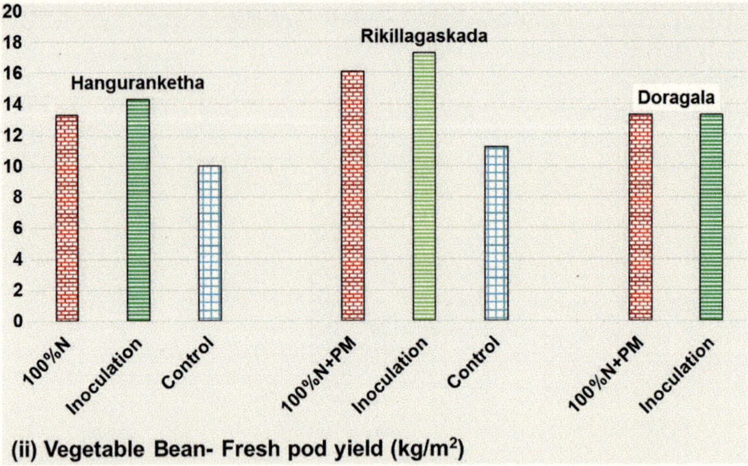

Fig. 9.6 Yield of common bean in three locations tested (PM: poultry manure)

such introduction and urea fertilizer has been regularly added to these nitrogen fixing forage legumes. These large scale farms are situated in the highland cloud forest areas of Sri Lanka where the climate and temperature are ideal for the maintenance of imported breeds of temperate cows. These pristine highlands located 2000 m above mean sea level sustaining our montane cloud forests, are the water catchment areas for the major rivers of Sri Lanka. Pollution of such environments by the regular loading of chemical nitrogen fertilizer was of serious concern and research studies commenced in 2014 to examine the possibility of minimizing N-fertilizer additions by substituting them with rhizobial inoculants. These studies were conducted in collaboration with Ambewela Farms (Pvt) Limited which cultivates red and white clover as its principal forage legume species both in monoculture and in mixed swards with grasses. Preliminary lab studies were carried out to isolate, purify and characterize rhizobia from root nodules collected from clover plants grown at Ambewela Farm. These were followed by greenhouse pot experiments to screen and select efficient rhizobial strains for field testing. The most efficient strain was tested in field trials conducted at the Ambewela Farm. Clover seeds were inoculated with coir dust based rhizobial inoculants at seeding followed by spraying with liquid inoculants after taking crop cuts. Root nodulation and biomass production of these plants were compared with those that received a basal dressing of 40 kg/ha of urea fertilizer at seeding followed by top dresses of 25 kg/ha of urea after obtaining each crop cut. Results on initial root nodulation and the growth of the plants showed that inoculation increased nodulation (Fig. 9.7) and such increases produced better plant growth.

Average results obtained from nine crop cuts obtained during a period of 12 months, showed that the highest biomass was obtained in the treatment with coir dust based seed inoculation as a basal dressing followed by top dressings with liquid

Fig. 9.7 Root nodulation of white clover with inoculation (right), with urea fertilizer (left)

Table 9.2 Dry matter production (total of all the crop cuts)

Treatment	Mean Dry weight (g/m^2)	% increase over control
T1: N-fertilizer (basal + t.d.*)	502^{abc}	9
T2: N-fertilizer (basal + t.d*. liquid inoculation)	520^{abc}	13
T3: N-fertilizer (basal + no t.d*.)	534^{abc}	16
T4: Seed inoculation (basal coir based + t.d*. liquid inoculation)	556^{a}	21
T5: Seed inoculation (basal coir based + no t.d*.)	544^{ab}	19
T6: Seed inoculation (basal liquid + t.d*. liquid inoculation)	534^{abc}	16
T7: Seed inoculation (basal liquid + no t.d*.)	484^{abc}	6
T8: No seed inoculation (t.d*. liquid inoculation)	464^{bc}	1
T9: Control (no N-fertilizer, no inoculation)	459^{c}	–

Three random crop cuts per plot were taken using a 0.5 m^2 wooden frame
Nine such crop cuts were obtained during a period of 12 months
*t.d. top dress with urea or liquid inoculants. CV 5.3%, MSD 79.1

inoculants after each crop cut and this was significantly higher than those under urea treatment (Table 9.2).

It was concluded that urea application to white clover at Ambewela Farm can be replaced by rhizobial inoculation without any reduction in biomass production and this should reduce the cost of clover cultivation and minimize environmental pollution. Currently this farm has adopted large scale application of rhizobial inoculants at seeding and spraying liquid inoculants after obtaining crop cuts.

Research studies are in progress at the NIFS on the development of rhizobial inoculants for food legumes such as cowpea (*Vigna unguiculata*), groundnut (*Arachys hypogea*) and black gram (*Vigna mungo*) and the forage legume alf-alfa (*Medicago sativa*). Adoption of rhizobial bio-fertilizer use by the farmers is encouraging and if government incentives are forthcoming this green technology could make a significant contribution for environmentally benign crop and livestock pasture production in Sri Lanka.

9.2.3 Biofilm Biofertilizer

Certain soil microbiota naturally exists as surface-attached microbial communities in a biofilm mode of growth. They have been shown to be more effective at functioning than monocultures or mixed cultures of microbes. Therefore, such beneficial biofilms have been formulated *in vitro* to be used as biofertilizers called biofilm-biofertilizers (BFBFs) in agriculture and plantations, particularly for non-legumes (Seneviratne et al. 2009). They can address many issues that affect the sustainability of agro-ecosystems. In conventional biofertilizers, it is seen that the importance of surface attachment of microbes and biofilm formation has not been identified, though there are several other reports on the effectiveness of naturally occurring biofilms on soil particles and plant surfaces (West et al. 2007; Nadell et al. 2009; Beauregard et al. 2013). However, the density of such biofilms on plant surfaces, particularly on the root system, is too low to have a significant effect on plant growth, as revealed by improved plant growth with BFBF applications to several crops (Seneviratne et al. 2009). The BFBFs render numerous biochemical and physiological benefits to plant growth, and improve soil quality, thus leading to a reduction of NPK chemical fertilizer (CF) use up to 50% in various crops (Table 9.3). At present, BFBFs are applied in Sri Lanka to thousands of hectares replacing considerable amounts of chemical fertilizers and some agrochemicals, with improved yields of rice, tea and vegetables in a sustainable manner.

The role of BFBFs is to reinstate sustainability of degraded agro-ecosystems through the breaking of dormancy of the soil microbial seed bank, and restoring biodiversity and ecosystem functioning (Seneviratne and Kulasooriya 2013), mainly through biochemical signaling among plants, microbes and fauna. Thus, the concept of BFBFs is not only biofertilization, but also a holistic ecosystem approach (Fig. 9.8).

In undisturbed ecosystems like forests, there is a delicate balance among the interacting counterparts, the center of which is represented by microbes (Fig. 9.8). Their cascading effect and chemical signaling networks, as indicated by arrows support the balance and stability of the ecosystems. This concept is introduced as edaphic ecosystem signal transduction (EST, Seneviratne 2015). However, in disturbed ecosystems like agro-ecosystems, particularly stresses of chemical inputs weaken the interactions through collapsing the signaling networks of the EST, consequently breaking the delicate balance of the ecosystem. Biofilm biofertilizers

Table 9.3 Mean crop yields following application of biofilm biofertilizer (BFBF) combined with 50% of the recommended rate of chemical fertilizer (50% CF) compared with application of the recommended rate of chemical fertilizer (100% CF) in field experiments conducted in different agroecological regions of Sri Lanka[a]

Crop	Mean ± SE crop yield (kg/ha)		Number of sites
	50% CF + BFBF	100% CF	
Tea	4300 ± 606	4100 ± 678	4
Rice	4420 ± 715	3580 ± 1295	5
Maize	2681 ± 322	2502 ± 338	3
Radish	1192 ± 251	992 ± 188	4
Cabbage	1302 ± 342	980 ± 249	4
Biter gourd	1547 ± 445	1563 ± 440	4
Aubergine	748 ± 175	678 ± 260	4
Okra	3107 ± 1719	1739 ± 710	3
Chilli	3478 ± 1754	2350 ± 919	3
Hungarian wax pepper	238 ± 50	152 ± 39	3
Tomato	335 ± 86	397 ± 131	3
Pole bean	2762 ± 886	2396 ± 753	3

Reproduced from Buddhika et al. (2016)

[a]Rice and maize field experiments were conducted during one or two seasons. Field experiments for vegetables were carried out during two consecutive dry and wet seasons. In the case of tea, the yields are annual averages over 4 years. In the same crop, mean yields of the two treatments were not significantly different at 5% probability level, according to Student's t-test. Low yields in some vegetable crops are due to the low plant densities under mixed cropping in subsistence farming

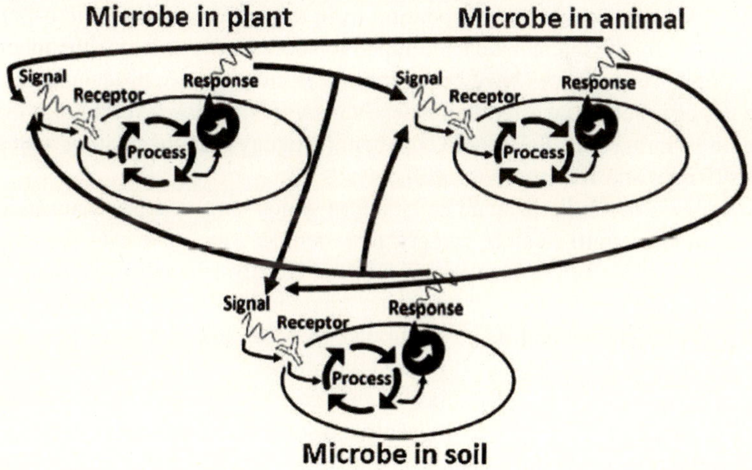

Fig. 9.8 Interactions among biotic counterparts in edaphic ecosystems (Reproduced from Seneviratne 2015)

(BFBFs) reinstate sustainability of the degraded agro-ecosystems through breaking dormancy of the soil microbial seed bank developed under the stress conditions and restoring biodiversity and ecosystem functioning.

9.2.4 Cyanobacterial Biofertilizer

Cyanobacteria, earlier known as blue-green algae include a morphologically diverse group of prokaryotic microorganisms exhibiting oxygenic photosynthesis. Their ancestry can be traced back to 3.8 billion years and it is believed that their photosynthesis contributed to the gradual oxygenation of the archaic anaerobic atmosphere of the Earth (Kulasooriya 2011). Several members of cyanobacteria are capable of nitrogen fixation and they grow luxuriantly in the wetland rice field ecosystems. Extensive research studies have shown their potential as bio-fertilizers for rice and these have been reviewed by Venkataraman (1972), Roger and Kulasooriya (1980) and Kulasooriya and Magana-Arachchi (2016).

Some studies have been conducted in Sri Lanka from the 1970s until the end of the twentieth century to examine the potential of using N_2-fixing cyanobacteria as biofertilizers for rice (Kulasooriya and de Silva 1981; Kulasooriya 1998). These studies showed that inoculants prepared from selected efficient N_2-fixing cyanobacteria grow and proliferate well in pot culture and give encouraging results with respect to growth and yield of rice, but they failed to give similar results under field testing (Kulasooriya and Hirimburegama 1989; Kulasooriya 1991). The inoculants were unable to overcome the competition from the indigenous soil micro-flora and consumption by the micro-fauna and failed to successfully colonize the inoculated fields. Similar experiences have been reported from other rice producing countries and it has also been suggested that amending soil conditions favorable for the proliferation of indigenous flora could be a better strategy than inoculating with exotic species (Roger and Kulasooriya 1980).

Application of *Azolla* (a small aquatic fern which harbors a N_2-fixing endosymbiotic cyanobacteium) as an *in situ* green manure for rice, gave even better results (Kulasooriya et al. 1987). However it was not readily adopted by farmers due to constraints such as the technology being labor intensive, additional demand by *Azolla* for phosphorus and its susceptibility to pests and pathogens (Kulasooriya 1991). Certain studies conducted in Sri Lanka to examine whether some of these constraints could be overcome (Kulasooriya et al. 1994) showed that poultry manure is a better substitute for synthetic concentrated super phosphate and having an unincorporated *Azolla* cover improves the uptake of ^{15}N-labeled chemical N-fertilizer by rice plants. Nevertheless such technological modifications were never adopted by rice farmers. *Azolla* has limited potential as a multi-purpose bio-resource for integrated rice-fish culture systems in irrigated rice fields (Liu 1987) or in combination as a nutrient rich supplementary food for poultry, piggery and ornamental fish culture (Kulasooriya and Magana-Arachchi 2016).

9.2.5 Root Associated Microorganisms

The distribution of the micro-flora in the soil is highly heterogenous and generally heterotrophic organisms are concentrated in close proximity to utilizable substrates. Plant roots continuously slough off their tissues as they grow and they also secrete proteins and carbohydrates. This nutrient rich micro-habitat is teeming with bacteria and fungi which occupy the rhizosphere (zone of root influence), rhizoplane (the root surface) and the endorhizosphere (internal root tissue). Such intimate associations between plant roots and soil organisms have been extensively studied with the objective of making use of such relationships for improving crop production. The discovery by Day and Dobereiner (1976) and Lakshmi Kumari et al. (1976) that bacteria associated with the roots of plants belonging to the family Graminae fix nitrogen was a turning point in research on soil microbiology. Since Graminae includes all the cereal crops such as wheat, corn and rice this finding triggered off a series of studies aimed at utilizing such activity to reduce the application of N-fertilizers in the cultivation of these major cereals. The commonest root associated N_2-fixing bacterium originally named as *Spirillum lipoferum* was subsequently re-named as *Azospirillum lipoferum*. Several other species of *Azospirillum* and other root associated N_2-fixing microorganisms were later isolated and identified. After extensive field studies commercial inoculants of such bacteria were field applied and Okon (1985) concluded that *Azospirillum sp.* contributes to increased yields of cereals and forage grasses by improving root development and increasing their rate of mineral and water uptake from soil and by biological nitrogen fixation. Thus the effects of *Azospirillum* are not confined to nitrogen fixation. Reviewing worldwide data accumulated over 20 years Okon and Labadera-Gonzalez (1994) have reported 60–70% successes with the field application of *Azospirillum* inoculants recording statistically significant increases in crop yields ranging from 5% to 30%. In a recent review chapter Okon et al. (2015) summarizes that root associated bacteria such as *Azospirillum, Herbaspirillum, Gluconacetobacter, Burkholderia, Pseudomonas* and *Paenibacillus* can all be categorized as Plant Growth Promoting Rhizobacteria (de Bruijn 2013).The positive effects of their inoculants are more due to growth promoting effects on the roots of the host plants than due to nitrogen fixation. Among these the genus *Azospirillum* has been the most widely used organism and inoculation effects observed in both pot and field experiments is the result of enhanced mineral and water uptake by the inoculated host plants (Dobbelaere and Okon 2007). They claim that the successes depended upon inoculant quality and it was recommended to prepare high quality inoculants having *Azospirillum* cell densities in the order of 1×10^9 to 1×10^{10} colony forming units per g or per ml.

A preliminary study conducted in Sri Lanka on the inoculation of rice plants with *Azospirillum irakense* showed that under non-sterile conditions, the inoculant could not overcome the competition by indigenous microorganisms and colonize the roots. Root colonization by *A.irakense* was successful only when malate was added as a carbon substrate in pot experiments (Rizvi and Kulasooriya 2002). This could be one explanation why rhizosphere bacterial isolates often do not perform by themselves but perform well when combined with a fungal partner to form a biofilm-biofertilizer. Fungi are well known to decompose recalcitrant carbon substrates and convert them to simple sugars which can then be utilized by most N_2 fixing bacteria.

More recent studies have been done to isolate, characterize and indentify rhizosphere, rhizoplane and endo-rhizosphere bacteria from several traditional, exotic and new improved varieties of rice from Sri Lanka (Dandeniya and Rajapaksha 2013). Out of 21 strains of bacteria belonging to *Pseudomonas* (dominant), *Azospirillum, Enterobacter, Exigubacterium, Kluyvera, Aeromonas, Acinetobacter, Stenotrophomonas, Bacillus* and *Staphylococcus* isolated, 3 were positive for auxin production and P-solubilization, 5 were positive for auxin production and 11 were positive for P-solubilization while 8 were negative for both these traits. Only *Pseudomonas* and *Bacillus* were isolated as endo-rhizosphere organisms and mycorhizal fungi could not be isolated from any of the rice varieties examined. Inoculants produced from some of these isolates are currently being field tested under wetland rice cultivation.

Studies done in Sri Lanka with *Azorhizobium caulinodans* (the stem nodulating, N_2-fixing endo-symboint from the stem nodulated legume *Sesbania rostrata*) has demonstrated colonization of rice roots both under pot and field conditions (Van Holm et al. 1994; Van Nieuwenhove et al. 2000). Recent studies have used molecular techniques such as tagging with Green Fluorescent Protein (GFP) to demonstrate the colonization of rice roots by this microorganism by the formation of biofilms (Perera et al. 2014). Further studies by this group have provided evidence that an *Azorhizobium caulinodans* ORS 571: *Aspergillus sp.* biofilm inoculant together with the flavonoid Naringenin has the ability to replace 50% of recommended urea fertilizer application and support the full vegetative growth of rice plants (Perera et al. 2015a). Furthermore using ^{15}N isotopic techniques a biofilm bio-fertilizer formulated with these components was demonstrated to not only replace 50% of the recommended level of urea fertilizer but also give better rice yields (Perera et al. 2015b, 2016). However, commercial production of this biofilm bio-fertilizer and the transfer of this technology to rice farmers are yet to be achieved.

Another important aspect is the role of soil microorganisms on the utilization efficiency of added nitrogen by rice plants. It has already been observed that not more than 30% of the added fertilizer nitrogen is absorbed by the targeted crop plant (Seneviratne and Kulasooriya 1994). The rest is lost; either volatalized, leached or de-nitrified which is a process mediated by certain soil bacteria. Hardly any studies have been done on this aspect in Sri Lanka although a substantial quantity of applied N-fertilizer is lost in this manner particularly in rice fields which undergo periodic drying and wetting which makes the soil aerobic and anaerobic promoting nitrification and de-nitrification respectively. It is therefore encouraging to note that some preliminary studies have commenced in Sri Lanka on this aspect (Dandeniya 2014). In these studies the response of 10 selected varieties of wetland rice to different mixtures of NH_4^+:NO_3^- ratios as well as the ability of certain varieties to suppress nitrification were examined. Moreover the effect of root derived compounds from rice seedlings on soil nitrifiers was assessed. Such studies could evaluate the inherent ability of certain rice varieties to minimize nitrification:denitrification losses. Based upon such findings it should be possible to develop techniques to inhibit these processes and minimize N-losses under field conditions.

9.3 Conclusions

From the foregoing it is evident that soil supports a vast assemblage of diverse microorganisms. The ancestors of these organisms pioneered the habitation of the original abiotic Earth and their present descendents having evolved through billions of years present an extremely wide range of biodiversity which possess the abilities to utilize almost all the resources available in nature. Besides the ocean ecosystems, it is the soil that contains such a vast assemblage of microorganisms. Over several years of research studies it has been possible to understand the beneficial activities of many of these microorganisms, elucidate and enhance their genetic potential and develop products and techniques which can be applied to improve crop production in an eco-friendly, sustainable manner. Such knowledge and information have given hope and confidence that the negative effects of purely chemical based agriculture on the environment could be curtailed to a considerable extent. Soil biologists have a tremendous responsibility to face the challenges for the future and develop techniques and systems to utilize the microbial world to minimize the use of agrochemicals without compromising on crop yields and ensure increasing farmer incomes and strengthen national, regional and global food security.

Having realized the negative impacts of the indiscriminate use of chemical fertilizers and other agro-chemicals in agriculture, it has been resolved at the 2012 United Nations Conference on Environment and Development held in Rio de Janeiro, Brazil, that all member states should strive to reduce the use of chemicals in crop production at least by 20% by the year 2020.

The Government of Sri Lanka has taken a policy decision to minimize the use of chemicals in agriculture including the plantation sector and this has given an impetus to embark upon research activities to develop farmer friendly non-chemical products to sustain, if not improve crop production. This chapter has reviewed the current status of research and development in Sri Lanka to explore soil microbial diversity to improve agriculture as this is the way forward for sustainable food production in an environmentally benign manner.

References

Alonso A, Dallmeier F, Granek E, Raven P (2001) Biodiversity: connecting with the tapestry of life. Smithsonian Institution/Monitoring and Assessment of Biodiversity Program and President's Committee of Advisors on Science and Technology, Washington, DC

Amarasiri SL (2015) Caring for water, 2nd edn. Greater Kandy Water Supply Project, Worldwide Enterprises (Pvt) Limited Press, Kandy, Sri Lanka, 166 p

Beauregard PB, Chai Y, Vlamakis H, Losick R, Kolter R (2013) *Bacillus subtilis* biofilm induction by plant polysaccharides. Proc Natl Acad Sci U S A 110:1621–1630

Buddhika UVA, Seneviratne G, Ekanayake EMHGS, Senanayake DMN, Igalavithane AD, Weeraratne N, Jayasekara APDA, Weerakoon WL, Indrajith A, Gunaratne HMAC, Kumara RKGK, De Silva MSDL, Kennedy IR (2016) Biofilmed biofertilizers application in agro-ecosystems. In: Gupta VK, Thangdurai D, Sharma GD (eds) Microbial bioresources. CAB International, Wallingford, pp 96–106

Dandeniya WS (2014) The response of selected rice varieties to partial nitrate nutrition and their ability to suppress nitrification. J Soil Sci Soc Sri Lanka 24:9–14

Dandeniya WS, Rajapaksha RMCP (2013) Bacteria with potential plant growth-promoting abilities in the root environment of selected rice varieties grown in Sri Lanka. In: Proceedings of the National Symposium on Soil Biodiversity, 10 & 11th December, 2013, Colombo, Sri Lanka, Biodiversity Secretariat, Ministry of Environment & Renewable Energy, pp 120–126

Day JM, Dobereiner J (1976) Physiological aspects of nitrogen fixation by a spirillum from *Digitaria* roots. Soil Biol Biochem 8:45–50

De Alwis KA, Panabokke CR (1972) Handbook of the soils of Sri Lanka (Ceylon). J Soil Sci Soc Sri Lanka 2:1–97

De Bruijn FJ (2013) Molecular microbial ecology of the Rhizosphere. Wiley-Blackwell Publishers, Hoboken

Deaker R, Roughley RJ, Kennedy IR (2004) Legume seed inoculation technology – a review. Soil Biol Biochem 36:1275–1288

Department of Census and Statistics (2009) Census of agriculture. Agriculture and Environmental Statistics Division of the Department of Census and Statistics, Colombo, Sri Lanka. http://www.statistics.gov.lk/agriculture/Paddy%20Statistics/PaddyStats.htm. Accessed 25 May 2017

Department of Census and Statistics (2016) Census of agriculture. Agriculture and Environmental Statistics Division of the Department of Census and Statistics, Colombo, Sri Lanka. http://www.statistics.gov.lk/agriculture/Paddy%20Statistics/PaddyStats.htm. Accessed 25 May 2017

Diaz RJ, Rosenberg R (2008) Spreading of dead zones and consequences for marine ecosystems. Science 321:926–929

Dobbelaere S, Okon Y (2007) The plant growth promoting effects and plant responses. In: Elmerich C, Newton WE (eds) Nitrogen fixation: origins, applications and research progress, V Associative and endophytic nitrogen-fixing bacteria and cyanobacterial associations. Springer, Heidelberg, pp 145–170

Herridge D, Gemell GHE (2002) Legume inoculants and quality control. In: Herridge D (ed) Inoculants and nitrogen fixation of legumes in Vietnam. ACIAR Proceedings, pp. 105–115

Howeison J, Herridge D (2005) Foreword to 'application of rhizobial inoculants to Australian agriculture'. Aust J Exp Agric 45:ii

Jayatilake N, Mendis S, Maheepala P, Mehta FR (2013) Chronic kidney disease of uncertain etiology prevalence and causative factors in a developing country. BMC Nephrol 14:180. https://doi.org/10.1186/1471-2369-14-180

Jeyakumaran N (2013) Priorities for cancer prevention in northern region of Sri Lanka. In: Voice for Change, Jaffna Managers' Forum, 17, Agrarian Services Lane, Nallur, Jaffna, pp. 72–82

Khonje DJ (2014) Adoption of the rhizobium inoculants technology for pasture improvement in sub-Saharan Africa. Trypanotolerant livestock in West and Central Africa, vol 2. Country Studies. FAO Corporate Document Repository, 13 p

Kulasooriya SA (1991) Constraints for the widespread use of Azolla in rice production. In: Polsinelli M, Materassi R, Vincenzi M (eds) Nitrogen fixation. Kluwer Academic Publishers, Dordrecht, pp 473–479

Kulasooriya SA (1998) Cyanobacteria and Azolla as biofertilizers for rice. In: Subramaniam G, Kaushik BD, Venkataraman GS (eds) Cyanobacterial Biotechnology. Oxford & IBH Publication, New Delhi, pp 201–209

Kulasooriya SA (2011) Cyanobacteria: pioneers of Planet Earth. Ceylon J Sci (Bio Sci) 41:71–88

Kulasooriya SA (2017) Toxin producing freshwater cyanobacteria of Sri Lanka. Ceylon J Sci (Bio Sci) 46:3–16

Kulasooriya SA, de Silva RSY (1981) Multivariate interpretation of the distribution of nitrogen fixing blue-green algae in rice soils of Central Sri Lanka. Ann Bot 47:31–51

Kulasooriya SA, Hirimburegama WK (1989) Phototrophic nitrogen fixation wetland rice fields. In: Datta SK, Sloger C (eds) Nitrogen fixation associated with rice production. Oxford & IBH Publication, New Delhi, pp 191–201

Kulasooriya SA, Magana-Arachchi DN (2016) Nitrogen fixing cyanobacteria: their diversity, ecology and utilization with special reference to rice cultivation. J Natn Sci Foundation Sri Lanka 44:111–128

Kulasooriya SA, Hirimburegama WK, Abeysekera SW (1987) Use of *Azolla*in Sri Lanka. In: Azolla utilization. The International Rice Research Institute, Los Banos, pp 13–140

Kulasooriya SA, Seneviratne G, Wijesundera C (1994) An evaluation of some factors that limit the widespread use of *Azolla* in rice cultivation. In: Hegazi NA, Nayez M, Monib M (eds) Nitrogen fixation in non-legumes. American University Press in Cairo, Cairo, pp 469–474

Kulasooriya SA, Ekanayake EMHGS, Kumara RKGK, Gunarathne HMAC (2014) Use of Rhizobial inoculants could minimize environmental health problems. Chapter 26. In: Krishna Pramanik K, Patra JK (eds) Industrial and environmental biotechnology. Studium Press, New Delhi, pp 433–442

Lakshmi Kumari M, Kavindan SK, SubbaRao NS (1976) Occurrence of nitrogen fixing Spirillum in roots of rice, sorghum, maize and other plants. Indian J Exp Biol 14:638–639

Liu CC (1987) Re-evaluation of *Azolla* utilization in agricultural production. In: Azolla utilization. The International Rice Research Institute, Los Banos, pp 67–76

Nadell CD, Xavier JB, Foster KR (2009) The sociobiology of biofilms. FEMS Microbiol Rev 33:206–224

Okon Y (1985) *Azospirillum* as a potential for agriculture. Trends Biotechnol 3:223–228

Okon Y, Labadera-Gonzalez CA (1994) Agronomic applications of *Azospirillum*: an evaluation of 20 years worldwide field inoculation. Soil Biol Biochem 26:1591–1601

Okon Y, Labandera-Gonzales CA, Lage M, Lage P (2015) Agronomic applications of Azospirillum and other PGPR. In: de Bruijn FJ (ed) Biological nitrogen fixation, vol 2. Wiley-Blackwell Publishers, Hoboken, pp 921–932

Perera TA, Tirimanne TLS, Seneviratne G, Kulasooriya SA (2014) Water stress-induced biofilm formation by inoculated *Azorhizobium caulinodans* ORS 571 on rice roots. In: The 1st Global Soil Biodiversity Conference, Palais de Congres, Dijon, France, Dec 2–5, 2014. P1.124

Perera TA, Tirimanne TLS, Seneviratne G, Kulasooriya SA (2015a) *Azorhizobium caulinodans* ORS 571 – *Aspergillus sp*. biofilm and narigenin: a perfect combination for achieving the full potential of vegetative growth of rice with 50% reduction of urea fertilizer. In: 2nd International Conference on Agriculture and Forestry (ICOAF), 10–12, June 2015, Colombo, Sri Lanka, p. 43

Perera TA, Tirimanne TLS, Seneviratne G, Kulasooriya SA (2015b) Eco-friendly, sustainable and low cost bio-fertilizer to reduce 50% of urea recommendation for rice, with even better yields. In: Proceedings of the Annual Research Symposium, University of Colombo, Sri Lanka, p. 125

Perera TA, Tirimanne TLS, Seneviratne G, Kulasooriya SA (2016) Fifty percent replacement of nitrogen fertilizer recommendation for rice by a biofilm of a diazotroph and rhizosphere fungus. In: 1st International Conference on Bioscience and Biotechnology, 12–13, January 2016, Colombo, Sri Lanka, p. 16

Rizvi EMJM, Kulasooriya SA (2002) Growth of rice variety BW 267-3 as affected by diazotrophs inoculated under different conditions. Ceylon J Sci (Bio Sci) 30:7–19

Roger PA, Kulasooriya SA (1980) Blue-Green Algae and Rice. The International Rice Research Institute, Los Banos, 113 p

Seneviratne G (2015) Signal transduction in edaphic ecosystems governs sustainability. Agric Ecosyst Environ 210:47–49

Seneviratne G, Kulasooriya SA (1994) Fate of applied N in the traditional, modern and conservation farming systems of lowland rice in Sri Lanka. IRRN 19:18

Seneviratne G, Kulasooriya SA (2013) Reinstating sol microbial diversity in agroecosystems: the need of the hour for sustainability and health. Agric Ecosyst Environ 164:181–182

Seneviratne G, Van Holm LHJ, Ekanayake EMHGS (1999) Effect of peat and coir-dust based rhizobial inoculants on the nodulation, plant growth and yield of soybean (*Glycine max* (L) Mettil) Cv PB 1. Trop Agric Res Ext 2:136–139

Seneviratne G, Thilakaratne RMMS, Jayasekara APDA, Seneviratne KACN, Padmathilake KRE, De Silva MSDL (2009) Developing beneficial microbial biofilms on roots of non-legumes: a novel biofertilizing technique. In: Khan MS, Zaidi A, Musarrat J (eds) Microbial strategies for crop improvement. Springer, Heidelberg, pp 51–62

Seneviratne G, Jayasekara APDA, De Silva MSDL, Abeysekera UP (2011) Developed microbial biofilms can restore deteriorated conventional agricultural soils. Soil Biol Biochem 43:1059–1062

Van Holm L, Van Nieuwenhove C, Tamara Kumari IADC, Kulasooriya SA, Vlassak K (1994) Survival and effect of *Azorhizobiumcaulinodans* on rice roots. In: Hegazi NA, Fayez M, Monib M (eds) Nitrogen fixation in Non-legumes. American University Press in Cairo, Cairo, pp 327–332

Van Nieuwenhove C, Van Holm L, Kulasooriya SA, Vlassak K (2000) Establishment of *Azorhizobiumcaulinodans* in the rhizosphere of wetland rice (*Oryzasativa* L.) Biol Fertil Soils 31:143–149

Venkataraman GS (1972) Algal biofertilizers and rice cultivation. Today and Tomorrow's Printers and Publishers, Calcutta, 81 p

West SA, Diggle SP, Buckling A, Gardner A, Griffin AS (2007) The social lives of microbes. Ann Rev Ecol Evol Syst 38:53–77

Wickremasinghe WMDB (2013) Soils of Sri Lanka and Biodiversity. Proceedings of the National Symposium on Soil Biodiversity, 10 & 11 December 2013, Sri Lanka Institute of Development Administration, Colombo, Biodiversity Secretariat, Ministry of Environment and Renewable Energy, pp. 1–2

Chapter 10
Microbial Detoxification of Residual Organophosphate Pesticides in Agricultural Practices

Lata S.B. Upadhyay and Aditya Dutt

10.1 Introduction

What are pesticides? As the name itself suggest, pesticides are the compounds that are used to control pests and their uncontrolled growth. They are used to either kill or deter i.e. discourage the growth of pests. The United Nations Organization for Food and Agriculture (FAO) defines pesticides as a single substance or mixture of various compounds which are intended to be used for destroying or preventing the unwanted species of plants from growing or controlling the population of vectors of both human and animal diseases, and also controlling any pest which messes with any of the agricultural practices or any stage of food production and distribution (i.e. production, storage, processing, transportation and marketing) (Tano 1996; Prieto Garcia et al. 2012).

On one hand, pesticides offer several benefits by eliminating the pests from the undesired places thus protecting the agriculture products (such as crop, fruits and vegetables) and preserving the food for its proper supply. While on the other hand, pesticides have certain disadvantages too. Pesticides pose a serious threat to the environment and ecological balance. Higher concentrations of such entities in the rivers and other water bodies or in soil, adversely affect their ecological balance and thus the whole food chain. The toxic effects of these compounds on humans are due to their mechanism of action in pest. It involves targeting of an enzyme of critical metabolic pathways required for the survival of pest which can be even an enzyme system/metabolic pathway present in human beings.

Pesticides can be both chemical compounds like disinfectants or biological agents like micro-organisms. The natural pesticides or the biological agents working as pesticides are termed as BIOPESTICIDES or BIOCIDES. Broadly, pesticides can be categorized on the basis of "target" pest i.e. insecticides (which target insects),

L.S.B. Upadhyay (✉) • A. Dutt
Department of Biotechnology, National Institute of Technology, Raipur, Chattisgarh, India
e-mail: lupadhyay.bt@nitrr.ac.in

J.K. Patra et al. (eds.), *Microbial Biotechnology*,
https://doi.org/10.1007/978-981-10-6847-8_10

herbicides (which target unwanted plants or weeds), rodenticides (which kill rodents), fungicides (which kill or deter fungi) and others like miticides (which kill moths) and algaecides (which kill or restrict algal growth).

World Health Organization (WHO) has classified pesticides based on their toxicity level and their potential risk for human beings. As per this system of classification, they are grouped as class Ia (extremely hazardous), class Ib (highly hazardous), class II (moderately hazardous), class III (slightly hazardous) and class IV (products unlikely to present acute hazards in normal use) (Moeller 2005). This classification is based on the toxic behavior of pesticides in rats and other laboratory animals. It is done by administering the test compounds orally or dermal and then estimating the median lethal dose (LD50) that produces death in 50% of exposed animals (Prieto Garcia et al. 2012).

Pesticides are also grouped together into majorly four classes based on similarities in their chemical structure and properties:

Organochlorines – Organochlorines were the first chemical pesticides to be practiced in the public health sector and agricultural fields. Since these pesticides possess a stable chemical structure/formulation the usually tend to accumulate and persist in the environment for a prolong time (Moeller 2005). They are mainly used in the eradication of insects or disease vectors of malaria and dengue (Prieto Garcia et al. 2012). In humans these pesticides leads to convulsions as the principle target organ for them is human nervous system on the other hand in case of insect the effect is paralysis which eventually lead to death.

Organophosphates – Organophosphates account for the major share amongst all types of pesticides. All organophosphates contain a phosphate group along with other organic groups. They are highly toxic as a result of their mode of action. Organophosphates act as acetylcholinesterase enzyme potent inhibitors, the role of this enzyme is to hydrolyze acetylcholine in the nervous system of several species, including humans (Moeller 2005). These compounds are very frequently used in agricultureal practices, cosmetics industry, pharma application etc.

Carbamates – These pesticides are derived from carbamic acid and work in the similar way as organophosphates. The only difference being that their action is reversible while organophosphates act irreversibly. They are mainly used as insecticides, fungicides, herbicides and nematacides (Moeller 2005).

Pyrethroids – They are synthetic analogues of naturally occurring pyrethrin which is usually obtained from chrysanthemum flowers. These compounds act on the central nervous system of insects as well as vertebrates (Prieto Garcia et al. 2012).

10.2 Organophosphate Pesticides

Organophosphates are a group of toxic chemical compounds which were discovered in 1938 by few German scientists. They were used as chemical weapons (which could destroy the nervous system) in World War II. Initial entities of this group were found to be very potent and highly toxic. Later on, compounds introduced/used under this class were comparatively less potent with decreased toxicity. Thus they found their application as pesticides in the field of agriculture (Jaga and Dharmani 2003). Their success in preventing crop damage especially from insects led to their commercial production worldwide. This led to a tremendous increase in the scope and reach of organophosphates drastically in several areas of our daily lives.

Organophosphate is the common terminology used for esters of phosphoric acid. These pesticides majorly comprises of a group of insecticides which act on enzyme acetylcholinesterase. This class of pesticides is similar to a nerve agent or neurotoxins as they share the common target enzyme acetylcholinesterase. They target neuro enzyme is essential for the normal functioning and vitality of insect's sensory system (Soltaninejad and Abdollahi 2009).

The issue of concern related to use of organophosphate is that along with the insects, the target enzyme is also equally essential and active in human beings. Thus, long exposures of these pesticides can result in severe adverse effects on the human health like neurotoxicity. Although they have higher rate of biodegradability in comparison to organochlorines and carbamate pesticides still organophosphate pesticide residues is one of the biggest threats to the ecosystem and food industry because their acute toxicities are irreversible (Kanekar et al. 2004). This is one of the major causes of various health hazard conditions associated with consumption of agricultural products and therefore emphasis is being given on the methods of biodegradation or remediation of organophosphates from the environment. In order to proceed in this direction efficiently we first need to understand the classification, application and mechanism of action of Organophosphates along with their harmful effects. All these aspects are discussed in the following sub-sections.

10.2.1 General Applications Organophosphate Compounds

Organophosphates can be generally further divided depending on their applications in various fields in following sub divisions:

Organophosphate Pesticides

Organophosphate compounds are majorly used in the field of agriculture for controlling or /and as a killing agents for pests especially insects. Some of the organophosphate pesticides example include malathion, chlorpyrifos, dimethanoate etc. (Yadav et al. 2015a).

Organophosphate as Nerve Gases

Tabun, cyclosarin, soman etc. are compounds which can irreversibly inhibit enzyme acetylcholinesterase and thus effecting proper functioning of the nervous system. These compounds are so toxic that they can result in loss of vital body functions. They are gaseous in nature and chemically organophosphate, hence referred to as organophosphate nerve gasses. Therefore they are also used as chemical weapons in warfare (Gupta 2006).

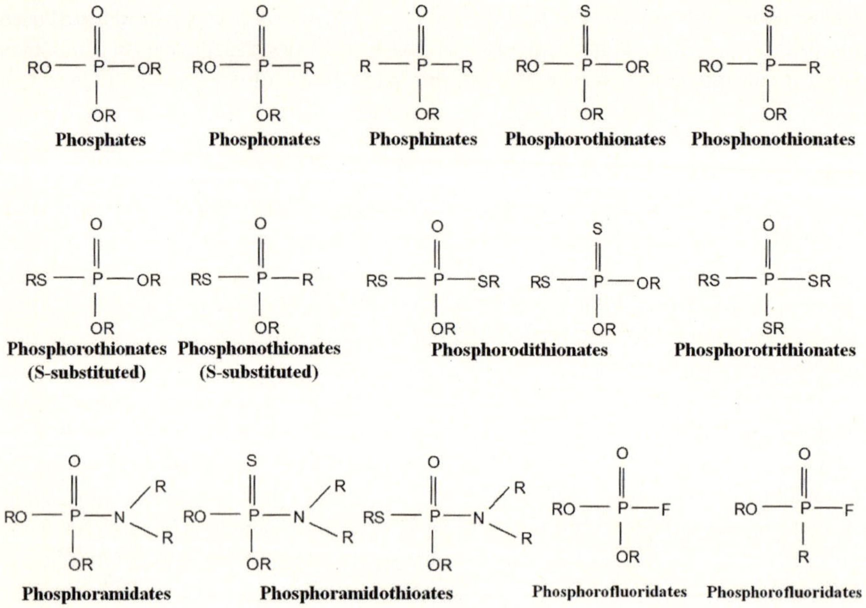

Fig. 10.1 Various structures of organophosphate pesticides

Organophosphates as Flame Retardants

These compounds are haloalkyl phosphates. When used with polymers, they work as flame retardants thus, used in manufacturing of products like textiles, buildings and also used as packaging materials. Some of the examples include tris (2-chloroethyl) phosphate (TCEP) and tris (2-chloropropyl) phosphate (TCPP).

Since, organophosphate compounds have a wider application as pesticides thus; they have been discussed in detail. There are hundreds of organophosphate compounds which fall under this category of pesticides such as trichlorfon, leptophos, mathamidophos etc. These are derivatives of either phosphoric, phosphonic or phosphinic acid (Pehkonen and Zhang 2010). These compounds can be differentiated on the basis of their side chains or any other element attached to phosphorous. Based on their molecular structures, organophosphate pesticides are divided into at least 13 different types or subgroups, structures of 13 groups has been summarized in Fig. 10.1 (Gupta 2006).

10.2.2 Application in Agricultural Practices

Large numbers of chemical compounds belonging to the category of organophosphate pesticides are routinely used in agriculture practices for protection of crops from pests and insects. These pesticides are employed in the field at different stages of growth depending upon the crop type, starting from ploughing of field till harvesting and sometimes even after that. Organophosphate pesticides are also used for crop management during the process of storage, packaging and transportation of fruits, vegetables, pulses and other food supplies.

Some examples of the majorly and widely used organophosphate pesticides in agriculture are discussed below.

Parathion
It is the most widely studied insecticide in respect to its biodegradability. It is used in water flooded rice fields and soil. Apart from rice it is also applied to cotton, and fruit trees like apple and pear. Its biodegradation has also been studied in situ in the fields directly by inoculating the soil with parathion degrading cultures. But due to its high toxicity it has been banned in some countries (Garcia et al. 2003).

Malathion
Malathion is extensively used to control the effect of various insects infecting or damaging vegetable crops and hence considered as a broad-spectrum chemical pesticide. Due to its capability to act on a wide range of insects its application has also been extended to other varieties of food, feed, and ornamental crops. It is used to control a wide range of insects including common garden pests such as scales, thrips and leafhoppers that infect vegetable plants. Being effective against eliminating mosquitoes and small flying insects it is used as an active ingredient in products such as hair oil, spray etc. to control fleas and head lice (Aktar et al. 2009).

Chlorpyrifos
It has been used as pesticide since 1965 and has now become 14th most used pesticide. It is used on corn, soybeans, fruit and nut trees, cranberries, broccoli, and cauliflower, as well as other row crops. It prevents damage from insects like grasshoppers, moths, beetles, scales as well as termites (Gomez et al. 2009).

Dimethoate
Insecticide dimethoate is used to kill mites and insects. It is applied against a wide range of arthrpods, including aphids,, plant hoppers, thrips and whiteflies damaging ornamental plants, alfalfa, apples, corn, tobacco, cotton, grapefruit, grapes, lemons, melons, oranges, pears, pecans, safflower, watermelons, sorghum, soybeans, tangerines, tomatoes, wheat and other vegetables (Van Scoy et al. 2016).

Apart from these, other organophosphate pesticides like diazinon, disulfoton, fonofos, profenofos, phorate are also used for protection of fruits and vegetables crops like grapes, blueberries, potatoes, corn and cotton (Pehkonen and Zhang 2010).

10.2.3 Mechanism of Action as Pesticide

Organophosphate pesticides inhibit the acetylcholinesterase enzyme present in the nervous system of its target pest. This enzyme is crucial for the proper nerve conduction of the target. Affinity of an organophosphate pesticide towards acetylcholinesterase depends on the molecular structure of the pesticide specifically based on

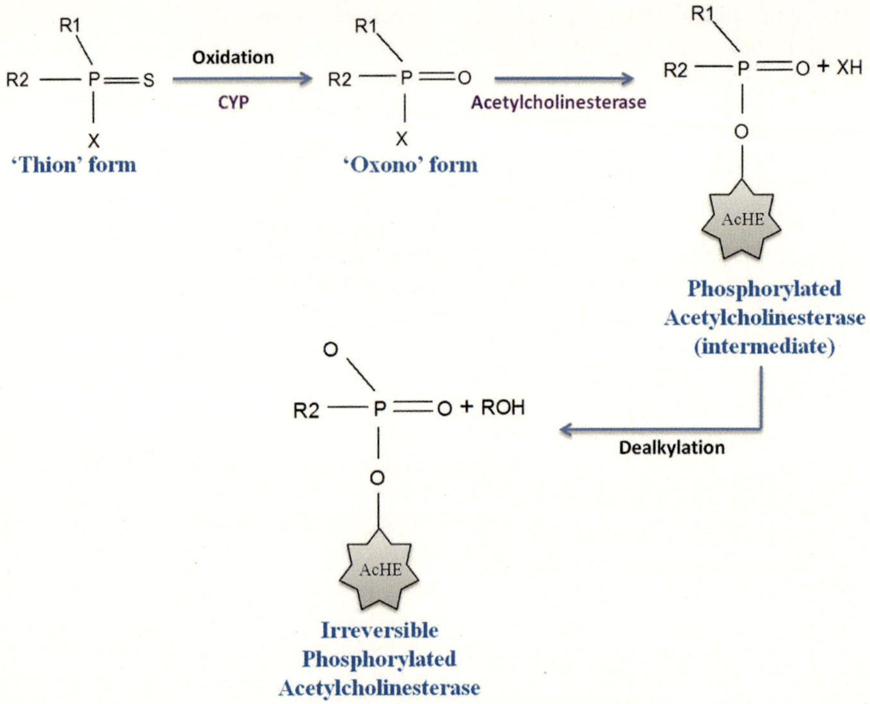

Fig. 10.2 Action of Organophosphate pesticide on Acetylcholinesterase

the side chain functional group. The binding of side chain functional group of pesticides with enzyme catalytic site further determines the time needed by the host resistance mechanism to oxidize, hydrolyze and reduce the level of its toxicity (Moretto 1998; Vale 1998). Higher the binding efficiency of pesticide towards enzyme, less active is the host resistance mechanism.

Organophosphates acquire an active configuration in host body before it can bind to the active site of the enzyme acetylcholinesterase. The central phosphorus atom of the pesticides must be linked to the oxygen atom (known as the oxono group) to acquire active configuration for further metabolism to take place. Which otherwise in majority of cases (organophosphate structure) phosphorous atom is linked to sulphur (possessing the thion group). For organophosphates to act as an acetylcholinesterase inhibitors, the thion group must be oxidized to an oxono group with the help of host cytochrome P450 enzyme (Kwong 2002; Gupta 2006).

The toxicity of organophosphates is primarily associated with the inhibition of the enzyme acetylcholinesterase. The enzyme is found in synaptic membranes. Due to its hydrolytic activity it degrades the neurotransmitter acetylcholine into choline and acetate. This reaction is very crucial for the regulation and proper functioning of synaptic activity in the central and peripheral neural system (Eleršek and Filipič 2011). Organophosphate pesticides which are the inhibitors of acetylcholinesterase block the normal functioning of the enzyme by forming covalent bond with its active site (Fig. 10.2). This leads to accumulation of excessive acetylcholine in

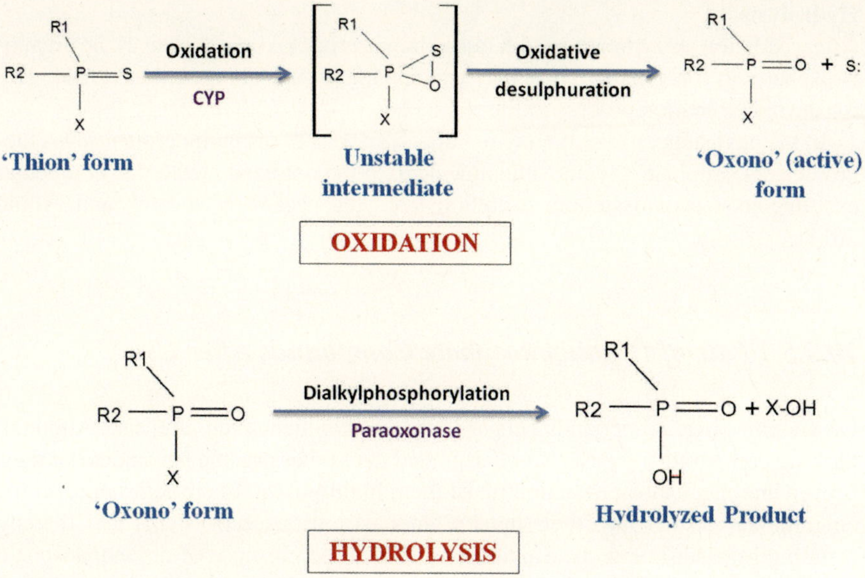

Fig. 10.3 Host defense mechanism against organophosphate compounds

synaptic membranes thereby resulting in neurotoxic effects such as paralysis through the entire body (Gupta 2006). The binding of organophosphate with the active site is slow and irreversible resulting in long term toxic effects.

10.2.4 Host Resistance Mechanism Against Organophosphate Compounds

Xenobiotic compounds, are metabolized in two stages: phase I- metabolic enzymes activate the chemical compound by introduction or exposure of functional groups and then hydrolysis of the activated entity i.e. the oxono form of the organophosphate compound, as represented in Fig. 10.3 (Josephy and Mannervik 2006). While in phase II the hydrolysed product is conjugated and eliminated out of the host system.

Phase I of Organophosphate metabolism involves the following two steps:

Oxidation

Oxidation refers to conversion of non-active form of inhibitors of acetylcholinesterase enzyme (i.e. thion form) to active inhibitors of the enzyme (i.e. oxono form). This is done with the help of cytochrome P450 (CYP) enzymes. The CYP enzymes aids in oxidation of sulphur atom of the thion group to the oxygen atom. This forms an unstable intermediate which further disintegrates to form oxono metabolite of the pesticide along with an active sulphur atom (oxidative desulphuration). Oxono form of the metabolites of organophosphate pesticides are strong inhibitors of acetylcholinesterase. Thus, this reaction is the key reaction for various neurotoxic effects of this class of pesticides.

Hydrolysis

After oxidation, enzyme esterase A (also known as paraoxonase) cleaves the organophosphates to form dialkylphosphate and leaving group. This reaction is necessary for the detoxification process of the pesticide.

In the next stage, phase II enzymes attach various hydrophilic groups, e.g. glucuronic acid, sulphate, glycine, glutamic acid, to the oxidized metabolite compound enabling its fast and smooth excretion from the organism (Eleršek and Filipič 2011).

10.2.5 Fate of Organophosphate Compounds After Use

Organophosphate compounds persist in the environment even after use. Although they are comparatively easier to be degraded than other organic pesticides but their irreversible mechanism of action make them highly toxic. Their persistence in the environment results into their uptake amongst various ecosystems and thereby resulting in harmful effects. Studies have shown the presence of organophosphate residues in agricultural lands, soil, groundwater and surface water bodies like rivers, ponds, lakes etc. More recently organophosphates are increasingly being used in towns and cities for controlling the population of insects causing fatal diseases like malaria and dengue. Thus due to so much widespread use of organophosphate pesticides, they have also been detected in snow, fog and rain water (Chandra and Kumar 2015).

Humans are exposed to these compounds through ingested food, drinking water and through air we breathe. Fate of organophosphates in the environment is affected by its distribution, its chemical and biological processing in the environment.

To determine fate and distribution of these pesticides in environment various mathematical models have been designed and reported that are used to predict the biodegradation of these pesticides taking into account all the three above mentioned processes. The processes that affect fate and distribution of these pesticides are found to be first order reactions as predicted by the mathematical models (Ragnarsdottir 2000).

The change in concentration (C) as a function of time (t) is the product of the concentration of the pollutant ([C]) and rate constant (k):

$$-dC / dt = k[C]$$

Half-life is the time required for removing 50% of the initial concentration of the compound persistent in the environment. The half-life ($t_{1/2}$) of a compound can be calculated by the following expression (Ragnarsdottir 2000):

$$t_{\frac{1}{2}} = \frac{\ln 2}{k}$$

10.2.6 *Harmful Effects*

Organophosphate pesticides attack the nervous system of the target organism by blocking the activity of an essential enzyme acetylcholinesterase. This enzyme is also present in human beings so organophosphate pesticides cause adverse effects in humans too. Human beings are exposed to these pesticides both directly and indirectly (Minton and Murray 1988). Direct exposure is the occupational hazard which is faced by farmers in the agricultural fields, workers in chemical industries and those who spray insecticides for controlling the insects to control diseases like malaria, chikunguniya and dengue. Whereas, indirect exposure is through the contaminated food and water as well as the air (Eddleston et al. 2008).

The harmful effects range from headache, vomiting, spasm to paralysis, lower heart beat and even coma depending upon the duration of exposure (short to prolonged), concentration (low to high) of chemical ingested and level of toxicity of the pesticide (less, mild to highly) (Karami-Mohajeri and Abdollahi 2011).

Immunotoxicity results from inhibition of serine hydrolases of immune complement system or esterases present in membranes of lymphocyte and monocyte. Action of organophosphae compounds effect the immune system and even cause oxidative damage of immune system organs. The effect could be also changes in signal transduction pathways controlling immune cell differentiation and proliferation. Organophosphate pesticides can even lead to genotoxicity and carcinogenicity on chronic and prolonged exposure (Eleršek and Filipič 2011).

10.3 Biodegradation of Organophosphate Pesticides

In earlier days chemical methods of degradation like incineration and chemical hydrolysis were widely used to degrade organophosphate pesticides and are still in practice to some level. But off date due to increasing awareness about harmful effects of these chemical processes, emphasis is being given on finding natural techniques for degradation of pesticides. These natural processes are collectively known as the process of bioremediation. It is a reliable, cost-effective method with advantage like production of less secondary pollutants after the detoxification or removal of organophosphate pesticides.

Bacterial systems are not affected by organophosphate pesticides, which are otherwise very toxic for insects as well as humans. Bacterial system lacks the enzyme acetylcholinesterase on which the organophosphates act. Hence, microorganisms especially bacteria can use organophosphate pesticides as an energy source. *Flavobacterium* sp. ATCC 27551 was the first reported bacteria capable of metabolizing organophosphate pesticides. It was isolated from a soil sample in Philippines in 1973 (Sethunathan and Yoshida 1973).

Since then bacteria have been explored to have novel degradation enzymes and pathways to thrive on these pesticides. It has been found that the bacteria can efficiently utilize these pesticides as additional carbon and nitrogen supplement for their own metabolism before ultimately degrading the pesticide (Singh and Walker 2006). The chemical structure of the organophosphate and nature of soil also influence the biodegradation of the pesticide. For example, studies have shown that alkaline pH of soil is more conducive for higher rate of biodegradation of organophosphate insecticides like fenamiphos (Singh et al. 2003a) and chlorpyrifos (Singh et al. 2003b). Based on the above findings it can be deduced that organophosphate degrading genes in bacteria might have been evolved from the genes of soil basophiles (Singh et al. 2003b). Researchers have also found that if a bacterium was able to degrade any particular organophosphate pesticide then the chances of it rapidly degrading other pesticides with similar chemical structure is high at appreciable rate. This phenomenon is known as cross-enhanced degradation (Singh et al. 2005).

10.3.1 Role of Microorganisms in Degradation of Organophosphate Pesticides

Biodegradation is the term used where organic compounds are broken down by microorganisms. All these organic substances are not necessarily toxic in nature. However study of different microorganisms which could possibly reduce the toxins present in the environment makes them a suitable option for exploitation. A wide variety of bacteria and fungi possess the ability to physically interact with the toxic substances and either completely assimilate them or break them down into smaller nontoxic compounds. In certain cases microbial mediated chemical modification of the pesticide is also responsible for detoxification of these compounds. Microorganisms usually break down the compounds for their growth i.e. to use them as carbon source or to provide themselves with additional nutrient elements required for normal growth and functioning. There are quite a few cases as well where these toxins are not directly used for any purpose (Nutritional benefit) but they are degraded in presence of a growth promoting substrate. This process is known as **co-metabolism**. Co-metabolism however does not completely assimilate the toxins but changes them into a different form which is either non/less toxic or can be utilized as a nutritional source by another organism in the biological niche.

The major limitation with the process of microbial biodegradation of pesticide arises when a microbial culture utilizes the organophosphates as a source of specific element either carbon or sulfur or phosphorus. This is a common phenomenon and often leads to incomplete assimilation of the organophosphate compound. The utilization of a particular pesticide as an elementary source of carbon, phosphorous or sulphur is independent of microorganisms overall nutritional requirement. If a micro-organism use the pesticide as source of carbon and is growing in media deprived of phosphorous then the pesticide cannot be utilized as a source of phosphorous due to lack of genomic response element required i.e. pho regulation (Kertesz et al. 1994). Daughton and Hsieh in 1977 (Daughton and Hsieh 1977) demonstrated metabolism of parathion as sole growth source by strain of

Pseudomonas stutzeri. The cell could not utilize the residual products of parathion under nutrient deprived condition with respect to phosphorous and sulphur. The use of parathion resulted in the formation of diethylphosphorothioanate (DEPT) and p-nitrophenol which remained unutilized. Use of *Pseudomonas aeruginosa* as a mix to the *Pseudomonas stutzeri* culture resulted in using p-nitrophenol as its carbon source. Use of mixed or single culture strain remains a popular choice for the bioremediation process.

10.3.2 Use of Microorganisms for Organophosphate Compound Degradation

There is a wide variety of microorganisms both bacteria and fungi that are currently being exploited for degradation of organophosphate compounds used in agriculture. The selection of microorganisms depends upon the type of organophosphate pesticides to be bioremidiated. The two most widely used organophosphates are parathion and chlorpyrifos. Generally the micro-organism used for degradation of parathion and chlorpyrifos are isolated from area of their extensive application. Table 10.1 summarises different microorganisms (bacteria and fungi) used for microbial degradation of organophosphates and their corresponding degraded product.

Parathion

Parathion is a very toxic insecticides used in the agricultural practices. However the easy detection of its hydrolytic product i.e. p-nitrophenol makes it a popular choice for study. It is easily degraded in the soil under aerobic or anaerobic conditions. Use of a mixed culture was demonstrated by use of a *Bacillus sp. and Pseudomonas sp.* for degradation of parathion into p-nitrophenol. Where the degradation of parathion was done by *Pseudomonas sp.* while *Bacillus sp.* was able to use p-nitrophenol as the carbon source (Siddaramappa et al. 1973). Another bacterial strain capable of metabolizing parathion to p-nitrophenol is *Flavobacterium sp.*

Chlorpyrifos

Chlorpyrifos are the moderately toxic insecticides used in the agricultural practices. They are spayed by mixing them with oily compounds as their solubility in water is low. The degradation of chlorpyrifos includes its conversion into 3,5,6-trichloro-2-pyridinol (TCP). TCP is soluble in water and can be carried away by water bodies which may lead to various health problems associated with humans. While many bacteria and fungi can degrade chlorpyrifos into TCP, only a few can further metabolize TCP e.g. *Cladosporium cladosporioides* Hu-01 (Chen et al. 2012). *Cladosporium cladosporioides* Hu-01 is a fungal strain which uses chlorpyrifos as a carbon source. The intermediate TCP formed is soon completely assimilated by the fungi as well. Some of the examples of chlorpyrifos degrading microorganisms include *Flavobacterium sp.* (bacteria), *Pseudomonas diminuta* (bacteria) *Aspergillus sp.* (fungi) and *Trichoderma harzianum* (fungi).

Table 10.1 List of Organophosphorus degrading micro-organisms

Compound	Degrading micro-organism	Metabolic Product	Reference
Chlorpyrifos	**Bacteria**		
	*Enterobacter*sp.	3,5,6-trichloro-2-pyridinol (TCP)	Singh et al. (2003b)
	*Flavobacterium*sp. ATCC 27551	Diethylthiophosphoric acid	Mallick et al. (1999)
	Micrococcus sp.	NA	Guha et al. (1997)
	Fungi		
	*Aspergillus*sp.	3,5,6-trichloro-2-pyridinol (TCP)	Yadav et al. (2015c)
	Trichodermaharzianum	3,5,6-trichloro-2-pyridinol (TCP)	Dhanya (2014)
	Pencilliumbrevicompactum	3,5,6-trichloro-2-pyridinol (TCP)	Yadav et al. (2015b)
Parathion	**Bacteria**		
	*Flavobacterium*sp. ATCC27551	p-nitrophenol	Sethunathan and Yoshida (1973)
	Pseudomonas diminuta	Diethylthiophosphoric acid	Serdar et al. (1982)
	*Arthrobacter*spp.	p-nitrophenol	Nelson et al. (1982)
	Agrobacterium radiobacter	p-nitrophenol	Horne et al. (2002)
	Bacillus spp.	p-nitrophenol	Nelson et al. (1982)
	*Xanthomonas*sp.	p-nitrophenol	Rosenberg and Alexander (1979)
Methyl parathion	**Bacteria**		
	Pseudomonas sp.	p-nitrophenol	Chaudhry et al. (1988)
	Bacillus sp.	p-nitrphenol and nitrite	Sharmila et al. (1989)
	Plesimonas sp. M6	p-nitrophenol	Cui et al. (2001)
	Pseudomonas putida	p-nitrophenol	Rani and Lalithakumari (1994)
	Pseudomonas sp. A3	p-nitrphenol and nitrite	Cui et al. (2002)
	Flavobacteriumbalustinum	p-nitrophenol	Somara and Siddavattam (1995)
Glyphosate	**Bacteria**		
	Pseudomonas ssp.	Glycine	Kertesz et al. (1994)
	Bacillus megaterium 2BLW	Acetyldehyde and orthophosphate	Quinn et al. (1989)
	Rhizobium sp.	Sarcosine	Liu et al. (1991)
	Agrobacterium sp.	Methane	Wackett et al. (1987)
	Arthrobacter sp. GLP	Glycine	Pipke et al. (1987)
	Flavobacterium sp.	Aminomethylphosphonic acid	Balthazor and Hallas (1986)
	Fungi		
	Penicilliumchrysogenum	Ammonia	Klimek et al. (2001)
	Penicilliumcitrinum	NA	Zboińska et al. (1992)

(continued)

Table 10.1 (continued)

Compound	Degrading micro-organism	Metabolic Product	Reference
Coumaphos	**Bacteria**		
	Nocardiodes simplex NRRL B24074	NA	Mulbry (2000)
	*Agrobacterium radiobacter*P230	NA	Horne et al. (2002)
	Pseudomonas monteilli	Fluorescent Hydrolysis Products	Horne et al. (2002)
	Pseudomonas diminuta	Diethylthiophosphoric acid	Serdar et al. (1982)
Diazinon	**Bacteria**		
	Flavobacterium sp.	2-isopropyl- 6-methyl-4-hydroxy-pyrimidine (IMHP)	Sethunathan and Yoshida (1973)
	Pseudomonas spp.	NA	Rosenberg and Alexander (1979)

10.3.3 Mechanism of Degradation

Organophosphates are degraded by microorganisms by hydrolysis of P-O-alkyl or P-O-aryl bonds. The hydrolysis involves the use of organophosphorus hydrolase enzyme and is the first step in degradation mechanism. While the complete detoxification of organophosphates involves oxidation, alkylation and dealkylation of the compounds carried out by a different set of enzyme system. The degraded product obtained after hydrolysis differs from organophosphate to organophosphate. Organophosphorus hydrolase is a metallo enzyme, the central metal of enzyme binds to the organophosphate moiety and cleaves the P-O or P-S bond. The resultant degraded products are generally less toxic in nature and can be further metabolized with co-culturing technique or by applying microbial consortia.

10.3.4 Level of Remediation Achieved

The remediation process has gained pace in recent times partly because of their protective effect against the toxic effect of pollutants/pesticides on non-target organisms and partly because of the need to save/preserve the biota of the area. Most of the agricultural practices rely on the use of several such pesticides for optimum crop production. But in due course their residue becomes a challenge. The use of chemical hydrolyzing agents and incineration has been a popular method to get rid of the organophosphates. However these methods are facing following limitation (i) require quite an amount of manpower (ii) are expensive in nature and (iii) potential toxic emissions themselves. Use of landfill was another popular measure exploited for the same purpose but that leads to contamination of ground water through leaching. Considering all these points microbial bioremediation looks like an attractive alternative but it has limitations of its own.

Nature provides abundant microorganisms capable of degrading the toxic organophosphate compounds however the process remains slow. The contaminated areas have become the source for isolation of the specific microorganisms which would degrade the organophosphates. The use of whole cell microorganisms is not a feasible option since it requires constant checking of the medium. It also requires fresh inoculum to be incorporated after certain intervals to keep the process fast going. Treatment of agricultural waste water in bioreactor is the only promising application which involves the use of whole cell microorganisms. This has been successfully demonstrated by US department of agriculture where they treat 15,000 liters of cattle dip waste water to remove the coumaphos from the contaminated water using the filter bioreactor and a consortium of microorganisms (Mulbry et al. 1998). Apart from using whole cell for biodegradation of organophosphorus compounds several efforts have also been made to use the cell free organophosphorus hydrolase enzyme (isolated both from natural and recombinant strains). The activity and utility of the cell free organophosphorus hydrolase enzyme has been improved with help of iimmobilization techniques. Organophosphorus hydrolase enzyme has been successfully immobilized on matrix like nylon membrane, silica beads and glass surface for their application in biodegradation of pesticides (Singh and Walker 2006). However, the process could not be popularized due to cost involve in development of such immobilization system thus restricting its wider application but still need attention.

The use of organophosphorus hydrolase is not limited just to the treatment of the polluted area but also to develop effective biosensors which can be used for the detection of the organophosphorus compound contamination in affected areas. This offers a cost effective on site examination of the area and can lead to better application of the remedial strategies involved.

10.3.5 GMO Developed

Researchers have also developed genetically modified organisms showing better activity in terms of pesticide remediation. The genetic modification of the organisms for the detoxification of organophosphate compounds is mainly limited in the production of more active and stable organophosphorus hydrolase. Various studies have been done involving cloning of organophosphorus hydrolase genes from a specific microorganism like *Sphingobium fuliginison, Pseudomonas plecoglocissida, Moraxella Sp. etc.* (Farivar et al. 2017; Horne et al. 2017; Nakayama et al. 2016; Shimazu et al. 2001) on a suitable expression vector and then transforming it back to the original strain. This led to a greater production of the organophosphorus hydrolase. The use of the enzyme so produced can lead to its use for the development of biosensor or can be immobilized onto various surfaces for efficient bioremediation.

10.4 Conclusion

With the current progress in the field of agriculture, the challenge to develop a better yield is the primary concern for everyone. The need for the better yield is to cater the needs of mankind. In order to make the yield better the most obvious first step is to prevent the damage of crop in field by any other sources apart from natural calamities. The use of pesticides offers prevention of agricultural products from being damaged.

While these pesticides are helpful in the beginning, they can cause serious damage to other non-target organisms if not used judiciously and if their residues are not disposed off properly. The conventional methods to degrade pesticides have proven to be not only less cost effective but also pollute the environment by emission of toxic gases as well as these processes could not completely detoxified/removed the compound hence secondary residues are left behind. The conventional methods can be replaced by using the best degrading mechanisms provided by nature i.e. the use of microorganisms. The study of the organophosphorus hydrolase producing microorganisms is a step in the right direction to achieve the desired result. Simple degradation of the organophosphate compounds into less toxic form is not the desired end result of the process but is to develop a process which can completely mineralize the organophosphorus compounds. The process can involve the use of a novel technique to use whole cell organisms directly in the contaminated area or use a consortium of different enzymes in free or immobilized state to achieve the desired end product.

References

Aktar MW, Sengupta D, Chowdhury A (2009) Impact of pesticides use in agriculture: their benefits and hazards. Interdiscip Toxicol 2:1–12. https://doi.org/10.2478/v10102-009-0001-7

Balthazor TM, Hallas LE (1986) Glyphosate-degrading microorganisms from industrial activated sludge. Appl Environ Microbiol 51:432–434

Chandra R, Kumar V (2015) Biotransformation and biodegradation of organophosphates and organohalides. Environ Waste Manag:475–524. https://doi.org/10.1201/b19243-17

Chaudhry GR, Ali AN, Wheeler WB (1988) Isolation of a methyl parathion-degrading Pseudomonas sp. that possesses DNA homologous to the opd gene from a Flavobacterium sp. Appl Environ Microbiol 54:288–293

Chen S, Liu C, Peng C et al (2012) Biodegradation of chlorpyrifos and its hydrolysis product 3,5,6-trichloro-2-pyridinol by a new fungal strain cladosporium cladosporioides Hu-01. PLoS One 7:e47205. https://doi.org/10.1371/journal.pone.0047205

Cui Z, Li S, Fu G (2001) Isolation of methyl parathion-degrading strain M6 and cloning of the methyl parathion hydrolase gene. Appl Environ Microbiol 67:4922–4925

Cui Z, Zhang R, He J, Li S (2002) Isolation and characterization of a p-nitrophenol degradation Pseudomonas sp. strain P3 and construction of a genetically engineered bacterium. Wei Sheng Wu Xue Bao 42:19–26

Daughton CG, Hsieh DPH (1977) Parathion utilization by bacterial symbionts in a chemostat. Appl Environ Microbiol 34:175–184

Dhanya M (2014) Advances in microbial biodegradation of Chlorpyrifos. J Environ Res Develop 9:232–240

Eddleston M, Buckley NA, Eyer P, Dawson AH (2008) Management of acute organophosphorus pesticide poisoning. Lancet 371:597–607. https://doi.org/10.1016/S0140-6736(07)61202-1

Eleršek T, Filipič M (2011) Organophosphorus pesticides – mechanisms of their toxicity. Pesticides- the impacts of pesticide exposure. InTech open access publisher

Farivar TN, Peymani A, Najafipour R (2017) Biodegradation of Paraoxan as an Organophosphate Pesticide with Pseudomonas plecoglocissida Transfected by opd Gene. Biotech Health Sci 4:1–5. 10.17795/bhs-45055.Research

Garcia SJ, Abu-qare AW, Borton AJ et al (2003) Methyl parathion: a review of health effects. North 6:185–210. https://doi.org/10.1080/10937400390155544

Gomez LE, Lassiter R, Mueller J, Olson B (2009) Use and Benefits of Chlorpyrifos in U. S. Agriculture, pp. 317–337.

Guha A, Kumari B, Bora TC, Roy MK (1997) Possible involvement of plasmids in degradation of malathion and chlorpyriphos by Micrococcus sp. Folia Microbiol (Praha) 42(6):574

GuptaRC(2006)Classificationandusesoforganophosphatesandcarbamates.ToxicolOrganophosphate Carbamate Compd:5–24. https://doi.org/10.1016/B978-012088523-7/50003-X

Horne I, Sutherland TD, Harcourt RL et al (2002) Identification of an opd (organophosphate degradation) gene in an Agrobacterium isolate. Appl Environ Microbiol 68:3371–3376. https://doi.org/10.1128/AEM.68.7.3371-3376.2002

Horne I, Sutherland TD, Oakeshott JG, Russell RJ (2017) Cloning and expression of the phosphotriesterase gene hocA from Pseudomonas monteilii C11. Sun Microbiol 2230:14–2687

Jaga K, Dharmani C (2003) Sources of exposure to and public health implications of organophosphate pesticides. Rev Panam Salud Publica 14:171–185. https://doi.org/10.1590/S1020-49892003000800004

Josephy PD, Mannervik B (2006) Molecular toxicology. Oxford University Press, New York

Karami-Mohajeri S, Abdollahi M (2011) Toxic influence of organophosphate, carbamate, and organochlorine pesticides on cellular metabolism of lipids, proteins, and carbohydrates: a systematic review. Hum Exp Toxicol 30:1119–1140. https://doi.org/10.1177/0960327110388959

Kertesz MA, Cook AM, Leisinger T (1994) Microbial metabolism of sulfur and phosphorus-containing xenobiotics. FEMS Microbiol Rev 15:195–215. https://doi.org/10.1111/j.1574-6976.1994.tb00135.x

Klimek M, Lejczak B, Kafarski P, Forlani G (2001) Metabolism of the phosphonate herbicide glyphosate by a non-nitrate-utilizing strain of Penicillium chrysogenum. Pest Manag Sci 57:815–821. https://doi.org/10.1002/ps.366

Kwong TC (2002) Organophosphate pesticides: biochemistry and clinical toxicology. Ther Drug Monit 24:144–149. https://doi.org/10.1097/00007691-200202000-00022

Liu C-M, McLean PA, Sookdeo CC, Cannon FC (1991) Degradation of the herbicide glyphosate by members of the family rhizobiaceae. Appl Environ Microbiol 57:1799–1804

Mallick K, Bharati K, Banerji A et al (1999) Bacterial degradation of chlorpyrifos in pure cultures and in soil. Bull Environ Contam Toxicol 62:48–54. https://doi.org/10.1007/s001289900840

Minton NA, Murray VSG (1988) A review of organophosphate poisoning. Med Toxicol Adverse Drug Exp 3:350–375. https://doi.org/10.1007/BF03259890

Moeller DW (2005) Environmental health. Harvard University Press, Cambridge, MA

Moretto A (1998) Experimental and clinical toxicology of anticholinesterase agents. Toxicol Lett 102–103:509–513

Mulbry W (2000) Characterization of a novel organophosphorus hydrolase from Nocardiodes simplex NRRL B-24074. Microbiol Res 154:285–288. https://doi.org/10.1016/S0944-5013(00)80001-4

Mulbry W, Ahrens E, Karns J (1998) Use of a field-scale biofilter for the degradation of the organophosphate insecticide coumaphos in cattle dip wastes. Pestic Sci 52:268–274

Nakayama K, Ohmori T, Ishikawa S et al (2016) Expression of recombinant organophosphorus hydrolase in the original producer of the enzyme, *Sphingobium fuliginis* ATCC 27551. Biosci Biotechnol Biochem 80:1024–1026. https://doi.org/10.1080/09168451.2015.1123606

Nelson LM, Yaron B, Nye PH (1982) Biologically-induced hydrolysis of parathion in soil: kinetics and modelling. Soil Biol Biochem 14:223–227. https://doi.org/10.1016/0038-0717(82)90029-3

Kanekar P, Bhadbhade BJ, Deshpande NM, Sarnaik SS (2004) Biodegradation of organophosphorous pesticides. Proc Indian Natn Sci Acad B70:57–70

Pehkonen SO, Zhang Q (2010) The degradation of organophosphorus pesticides in natural waters: a critical review. Crit Rev Environ Sci Technol:37–41

Pipke R, Amrhein N, Jacob GS et al (1987) Metabolism of glyphosate in an Arthrobacter sp. GLP-1. Eur J Biochem 165:267–273

Prieto Garcia F, Cortés Ascencio SY, Oyarzun JCG et al (2012) Pesticides: classification, uses and toxicity. Measures of exposure and genotoxic risks. J Res Environ Sci Toxicol 1:2315–5698

Quinn J, Peden JM, Dick RE (1989) Carbon-phosphorus bond cleavage by Gram-positive and Gram-negative soil bacteria. Appl Microbiol Biotechnol 31:283–287. https://doi.org/10.1007/BF00258410

Ragnarsdottir KV (2000) Environmental fate and toxicology of organophosphate pesticides. J Geol Soc London 157:859–876. https://doi.org/10.1144/jgs.157.4.859

Rani NL, Lalithakumari D (1994) Degradation of methyl parathion by *Pseudomonas putida*. Can J Microbiol 40:1000–1006. https://doi.org/10.1139/m94-160

Rosenberg A, Alexander M (1979) Microbial cleavage of various organophosphorus insecticides. Appl Environ Microbiol 37:886–891

Serdar CM, Gibson DT, Munnecke DM (1982) Plasmid involvement in parathion hydrolysis by Pseudomonas diminuta. Appl Environ Microbiol 44:246–249

Sethunathan N, Yoshida T (1973) A Flavobacterium sp. that degrades diazinon and parathion. Can J Microbiol 19:873–875

Sharmila M, Ramanand K, Sethunathan N (1989) Effect of yeast extract on the degradation of organophosphorus insecticides by soil enrichment and bacterial cultures. Can J Microbiol 35:1105–1110. https://doi.org/10.1139/m89-185

Shimazu M, Mulchandani A, Chen W (2001) Simultaneous degradation of organophosphorus pesticides and p-nitrophenol by a genetically engineered Moraxella sp. with surface-expressed organophosphorus hydrolase. Biotechnol Bioeng 76:318–324. https://doi.org/10.1002/bit.10095

Siddaramappa R, Rajaram KP, Sethunathan N (1973) Degradation of parathion by bacteria isolated from flooded soil. Appl Microbiol 26:846–849

Singh BK, Walker A (2006) Microbial degradation of organophosphorus compounds. FEMS Microbiol Rev 30:428–471. https://doi.org/10.1111/j.1574-6976.2006.00018.x

Singh BK, Walker A, Morgan JAW, Wright DJ (2003a) Role of soil pH in the development of enhanced biodegradation of fenamiphos. Appl Environ Microbiol 69:7035–7043

Singh BK, Walker A, Morgan JAW, Wright DJ (2003b) Effects of soil pH on the biodegradation of chlorpyrifos and isolation of a chlorpyrifos-degrading bacterium. Appl Environ Microbiol 69:5198–5206. https://doi.org/10.1128/AEM.69.9.5198-5206.2003

Singh BK, Walker A, Wright DJ (2005) Cross-enhancement of accelerated biodegradation of organophosphorus compounds in soils: dependence on structural similarity of compounds. Soil Biol Biochem 37:1675–1682

Soltaninejad K, Abdollahi M (2009) Current opinion on the science of organophosphate pesticides and toxic stress: a systematic review. Med Sci Monit 15:RA75–RA90

Somara S, Siddavattam D (1995) Plasmid mediated organophosphate pesticide degradation by Flavobacterium balustinum. Biochem Mol Biol Int 36:627–631

Tano J (1996) Identity, physical and chemical properties of pesticides. Pestic Mod World – Trends Pestic Anal 1873:1–18. https://doi.org/10.5772/701

Vale JA (1998) Toxicokinetic and toxicodynamic aspects of organophosphorus (OP) insecticide poisoning. Toxicol Lett 102–103:649–652

Van Scoy A, Pennell A, Zhang X (2016) Environmental fate and toxicology of dimethoate. Rev Environ Contam Toxicol 237:53–70. https://doi.org/10.1007/978-3-319-23573-8_3

Wackett LP, Shames SL, Venditti CP, Walsh CT (1987) Bacterial carbon-phosphorus lyase: products, rates, and regulation of phosphonic and phosphinic acid metabolism. J Bacteriol 169:710–717

Yadav IC, Devi NL, Syed JH et al (2015a) Current status of persistent organic pesticides residues in air, water, and soil, and their possible effect on neighboring countries: a comprehensive review of India. Sci Total Environ 511:123–137. https://doi.org/10.1016/j.scitotenv.2014.12.041

Yadav M, Shukla AK, Srivastva N et al (2015b) Utilization of microbial community potential for removal of chlorpyrifos: a review. Crit Rev Biotechnol:1–16. https://doi.org/10.3109/07388551.2015.1015958

Yadav M, Srivastva N, Shukla AK et al (2015c) Efficacy of Aspergillus sp. for degradation of chlorpyrifos in batch and continuous aerated packed bed bioreactors. Appl Biochem Biotechnol 175:16–24. https://doi.org/10.1007/s12010-014-1244-0

Zboińska E, Maliszewska I, Lejczak B, Kafarski P (1992) Degradation of organophosphonates by Penicillium citrinum. Lett Appl Microbiol 15:269–272. https://doi.org/10.1111/j.1472-765X.1992.tb00781.x

Part II
Application of Microbes for Benefit of Environment

Chapter 11
Bacterial Degradation of Phenol to Control Environmental Pollution

Anna S. Kynadi and T.V. Suchithra

11.1 Introduction

Environmental pollution by a variety of reasons including industrialization, urbanization etc. is a major cause for deterioration of the quality of our lives. This affects even our most indispensable daily requirements including availability of clean water. There are a wide variety of toxic compounds causing such pollution, phenol and its derivatives being a major group amongst them. This group of compounds has been included in the International Register for Potentially Toxic Chemicals (IRPTC) and use is strictly regulated by various government agencies.

11.2 Chemical Properties of Phenol

Phenol or hydroxybenzene is the basic aromatic unit found in a number of synthetic organic products. Some of the other names by which this compound is known are carbolic acid, phenic acid, phenylic acid, phenyl hydroxide or oxybenzene. The chemical formula of the compound is C_6H_5OH. It has a molecular weight of 94.14 g/mol, solubility of 87 g/L in water at 25 °C, melting point of 43 °C, boiling point of 181.8 °C, auto ignition temperature of 715 °C, flash point (open cup) of 87 °C and a pKa value of 9.89×10^{-10} (Busca et al. 2008). The compound is normally crystalline, has a strong odor, is abstemiously volatile under conditions of room temperature, has low evaporation rate when compared to water and is moderately flammable.

Phenol is a parent compound that has a number of derivatives (Michałowicz and Duda 2007). The common derivatives of phenol that are found to contaminate the environment are

A.S. Kynadi • T.V. Suchithra (✉)
School of Biotechnology, National Institute of Technology Calicut, Kozhikode, Kerala, India
e-mail: drsuchithratv@nitc.ac.in

J.K. Patra et al. (eds.), *Microbial Biotechnology*,
https://doi.org/10.1007/978-981-10-6847-8_11

- alkylphenol (short, medium and long chain),
- methylphenol (cresol: ortho, meta and para),
- monochlorophenols: 2-chlorophenol, 3-chlorophenol, 4-chlorophenol
- dichlorophenols: 2,3-dichlorophenol, 2,4-dichlorophenol, etc.
- trichlorophenol
- pentachlorophenol
- mononitrophenols: 4-nitrophenol; 2-nitrophenol; 3-nitrophenol
- 2,4-dinitrophenol
- trinitrophenol (picric acid)
- 2-chloro,4-nitrophenol
- 4-chloro,2-methyl phenol
- 3-methyl,4-nitrophenol

The biggest group of phenol derivatives found in the environment is chlorophenols while the most dangerous group found is nitrophenols. Nitro aromatic compounds are extremely difficult to degrade too.

11.3 Uses of Phenol

Phenol has germicidal properties that are made use of in a number of products. One of the most common uses of phenol is that it is used as in ingredient in cleaning products like disinfectants and slimicides. Phenol has been used as a decontaminator for a long time; the first commercialized product being "carbolic acid" to clean surgical products by Lister. Phenol is effective against many bacteria, fungi and some viruses too. They are very efficient against Gram positive strains.

Phenol is used in the medical and pharmaceutical field in a number of ways. The compound is found in small quantities in pharmaceutical products like aspirin, antiseptics, nasal drops and eye drops. Phenolic compounds are now used as safe disinfectants for medical instruments that do not generally lead to infections; i.e. instruments that are used on skin without any rupture. Phenol also finds its place in veterinary medicine as an internal antiseptic and gastric anesthetic. The compound is also used in various animal husbandry units for equipment sanitation (Joan S. Jeffrey 1997). Some of the cosmetics that we find phenol in are creams and shaving lotions.

Phenol is a component in glue. Bakelite, an industrially important adhesive is a resin containing phenol and formaldehyde. It is low cost and has thermosetting properties which make it a widely accepted industrial grade adhesive. Phenol reacts with acetone to form bisphenol-A, which polymerizes to produce epoxy resins. Phenol is also used in a number of chemical processes for polyester and other polymer manufacture. Refineries and lubricant production units use the compound as a solvent. Phenol-chloroform mixture is routinely used in molecular biology for cell lysis as well as purification of DNA/RNA by removal of proteins (Sambrook and Russell 2001). Rideal walker method for testing potency of disinfectants also uses phenol as a standard. Phenol coefficient is considered one and the relative bactericidal activity of the chemical in question is measured against the strains *Salmonella typhi* and *Staphylococcus aureus*.

11.4 Occurrence of Phenol

Phenols exist in the environment by natural as well as anthropogenic causes (Sikkema et al. 1995). Phenols are naturally produced in the environment by metabolism of a wide range of compounds in plants and these concentrations do not pose a threat to the environment. The concentrations become high enough to induce toxicity by anthropogenic causes. The most common sources of phenol contamination include effluents from industries involved in the production of petroleum related products, glass, fiber, rubber, pharmaceuticals, cleaning products, fertilizers, leather, explosives, steel food and beverages etc. (Field and Lettinga 1991; Aggelis et al. 2002). Some of the industries that deposit phenol in the environment through its effluent streams are coking operation units (28–3900 mg/L), coal processing units (9–6800 mg/L), gas production units (4000 mg/L), petrochemical industry (2.8–1220 mg/L), refineries (6–500 mg/L), pulp and paper industry (0.1–1600 mg/L), pharmaceutical industries (1000 mg/L) and benzene manufacturing (50 mg/L) (Busca et al. 2008).

11.5 Phenol Toxicity

Phenol remains in the water sources it is dumped in to for long periods owing to its high density and low dilution rate (Krastanov et al. 2013), thus leading to dermal exposure as well as ingestion. The harmful effects of phenol on health as well as the environment have been well documented (Calabrese and Kenyon 1991; Lewis 2012). This led to the imposition of a maximum permissible level for phenol in the environment as 0.1 mg/L by the regulations of World Health Organization (Kumaran and Paruchuri 1997; Nuhoglu and Yalcin 2005; Saravanan et al. 2008). There are even reports of phenol concentration being as high as 10 g/L. According to various guidelines, this quantity is higher than the permissible levels in potable water by over thousand fold (Krastanov et al. 2013). The compound also quickly vaporizes, thereby posing a threat of inhalation. Such high concentration of phenol can lead to a variety of physical ailments since it is a protoplasmic poison that denatures proteins. The lethal dose of phenol can be as small as 1 gm. But derivatives of the compound are not as toxic as the parent compound. Some of the ill effects of dermal exposure to phenol are skin lesions, mucosal burns, corneal damage etc. Ingestion leads to local corrosions in the alimentary canal causing nausea, vomiting, diarrhea etc. associated with pain. Phenol vapors can result in respiratory tract irritation and even lead to pneumonia. On higher doses, any of these modes of contact with phenol can lead to severe repercussions including organ damage, coma and even death (Todorović 2003). Phenol has also been studied as a susceptible carcinogen.

11.6 Phenol Degradation

The wide prevalence of phenols and its derivatives in the environment and its toxicity on living beings necessitates its removal from the environment, especially water bodies. The most important step towards this is the treatment of effluent streams from industries before discharge into water bodies. Many different methods have been studied for phenol degradation and its removal from the environment. This includes physical chemical and biological methods. Some of the physicochemical technologies that have been developed include ion exchange, solvent extraction, pervaporation, adsorption, membrane extraction, electrocoagulation, photodecomposition etc. (Lakshmi et al. 2016). But all these methods have various disadvantages including high cost and secondary pollution in the form of toxic intermediate products. In comparison, biological methods facilitate complete degradation of phenol into nontoxic compounds at low cost. Biological methods include microbial degradation as well as enzymatic degradation of which microbial method is more sustainable in the long run. Active research is being carried out in the field to develop systems of effluent treatment using microorganisms.

11.7 Microbial Degradation of Phenols

The focus on the microbial degradation of phenols in recent years has resulted in the isolation, culture, adaptation and enrichment of a number of microorganisms that can grow on the compound as a sole carbon and energy source. Phenol is an antimicrobial agent; many of the microbes are susceptible to this compound. However, there are some microbes, which are resistant to phenol and have the ability to degrade phenol. But this is not a very easy process to standardize and practice since the degradation process is complicated by the presence of phenol derivatives. Some substitutions on phenol like nitro group, methyl group and amino group make degradation difficult while some other substitutions like carboxyl group and hydroxyl group make it easier (Alexander and Lustigman 1966). Microbial degradation of phenol does not only depend upon the substrate to be degraded, but also the microorganism in action as well as the environmental factors involved (Watanabe et al. 1998b; Topalova et al. 2007). Phenol can be degraded by microbes through aerobic as well as anaerobic pathways. Here we can study in detail the mechanism of phenol degradation. Pathways involved in the microbial degradation of phenol have been depicted in the Meta-cyc database (Caspi et al. 2014).[1]

[1]All pathways in this chapter have been adapted from Meta-cyc, the curated database of experimentally elucidated metabolic pathways (Caspi et al. 2014).

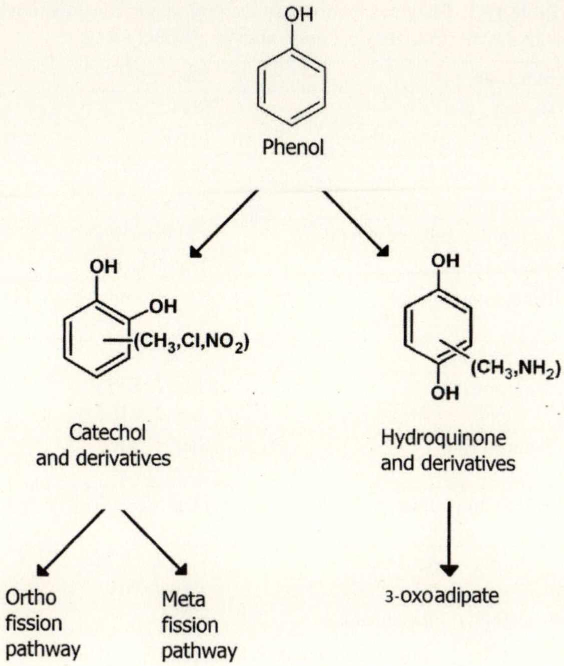

Fig. 11.1 Overview of phenol degradation in bacteria

11.8 Aerobic Degradation of Phenol in Bacteria

A number of bacterial strains have been found to have the ability to metabolize phenol. As shown in Fig. 11.1, the first step in aerobic phenol metabolism is most commonly its hydroxylation to catechol by phenol hydroxylase, and then catechol is metabolized by different strains via either the ortho or meta fission pathways using the enzyme catechol dioxygenase (Schie and Young 2007). The end products enter citric acid cycle thereby facilitating complete degradation. In some cases, instead of catechol, hydroquinone is the intermediate product which is later on degraded to 3-oxoadipate and finally enters the TCA cycle (Kolvenbach and Corvini 2012; Zhang et al. 2013).

11.8.1 Preliminary Degradation of Phenol and its Derivatives

The first step of phenol degradation is its conversion to catechol or hydroquinone. Phenol hydroxylase (PH) is the primary enzyme that catalyzes the addition of a hydroxyl group at the ortho-position of the phenol ring, to produce catechol (Schie and Young 2007). The substrates that are acted upon by this group of enzyme include cresols, chloro-, fluoro-, and aminophenols and are converted into respective catechols. Alkyl phenols with long chains, heavily chlorinated phenols and nitrophenols

Table 11.1 Enzymes involved in the preliminary degradation of phenol and its derivatives along with the substrate they act upon and the product formed

Substrate	Enzyme	Product
Phenol	PH	Catechol
Alkyl phenol (small/medium chain)	PH	Alkyl catechol
Methyl phenols	Methyl PH	Methylcatechol
Chlorophenol (mono/di)	Chloro PH	Chlorocatechol
4-chloro 2- methyl phenol	Chloromethyl PH	Chloromethylcatechol
Trichlorophenol	Trichloro PH	2,6-dichlorohydroxyquinone
Pentachlorophenol	Step 1: Pentachloro PH Step 2: DH	2,6-dichlorohydroxyquinone
4-nitrophenol	NPM	Nitrocatechol
2-nitrophenol	Step 1: NPM Step 2: BR	Catechol
3-nitrophenol	Step 1: NR Step 2: PH	Aminohydroquinone
2,4-dinitrophenol	DN	4-nitrophenol (subsequently converted to nitrocatechol)
2-chloro,4-nitrophenol	Step 1: NPM Step 2: DH	Hydroquinone
3-methyl, 4-nitrophenol	NPM	Methyl hydroquinone

The enzymes are abbreviated as *PH* Phenol hydroxylase, *NPM* Nitrophenol monooxygenase, *DH* Dehalogenase, *BR* Benzoquinone reductase, *DN* Denitratase

are not directly converted into their respective catechols. Instead they are converted into hydroquinones by single step or multistep reactions. The other main enzymes involved in these multistep reactions are Nitrophenol monooxygenase (NPM), Dehalogenase (DH), Benzoquinone reductase (BR) and Denitratase (DN). These reactions have been comprised in Table 11.1.

There are different types of phenol hydroxylases (EC 1.14.13.7) depending upon the number of protein subunits forming the enzyme. This is a group of non-heme diiron monooxygense enzymes (Leahy et al. 2003). Single component PH is a flavin dependent monooxygenase that catalyses the hydroxylation of chlorinated phenols and alkylphenol. It is seen in microbial strains like *Cupriavidus necator* and *Sphingobium chlorophenolicum* (Ledger et al. 2006; Porter et al. 2012). Two-component PH has a monooxygenase subunit and flavin reductase subunit. This enzyme that depends upon NADH and FAD for action is mainly seen in Gram positive bacteria (Saa et al. 2010). Multicomponent PH is an enzyme complex with six subunits. This hexameric monooxygenase component has three subunits in dimeric form; a Fe_2S_2 reductase, a regulator protein and an auxiliary protein. Multicomponent PH is seen mostly in Gram negative bacteria (Murray and Lippard 2007).

Nitrophenol monooxygenase (EC 1.14.13.167) is an FAD containing, NADPH utilizing enzyme that catalyzes the conversion of nitrophenol to nitrocatechol in case of nitro group being at the fourth position (4-nitrophenol) or benzoquinone in the case of nitro group at the second position (2-nitrophenol) (Perry and Zylstra 2007). For the latter reaction, simultaneous removal of nitro group is observed.

Nitroreductases (EC 1.7.1.-) also is an NADPH utilizing enzyme that catalyzes the reduction of nitro group at the third position (3-nitrophenol) or fifth position (2-chloro,5-nitrophenol) of phenol to amino group prior to the action of phenol hydroxylase (Belchik and Xun 2008). This happens either directly or via a nitroso group intermediate.

Benzoquinone reductase (EC 1.6.5.6) is the enzyme that helps in producing catechol or hydroquinone from benzoquinone which is formed as an intermediate of heavily chlorinated or nitrated phenol derivatives (Xiao et al. 2007). Dehalogenases and denitratases work in unison with the enzyme to remove the chlorine and nitrate groups respectively in most microbial strains. In some cases, hydroquinones are further degraded to hydroquinols by the enzyme quinone reductase. Trinitrophenol (picric acid) follows a complex pathway different from all these substrates to reach the TCA cycle (Ebert et al. 1999).

11.8.2 Catechol Degradation

Catechol, the product of aerobic phenol degradation is also a toxic compound like its precursor. The aromatic ring of catechol and its derivatives is cleaved by dioxygenase enzyme thus it being a crucial step in the aerobic degradation of phenol. Two classes of dioxygenases are identified on the basis of aromatic ring cleavage mechanisms giving rise to two different pathways. Catechol 1,2-dioxygenase (EC 1.13.11.1) is an intradiol-dioxygenase using non-heme Fe (III). It cleaves catechol in the ortho position giving rise to the ortho fission pathway (Harwood and Parales 1996). Catechol 2,3-dioxygenase (EC 1.13.11.2) is a extradiol-dioxygenase using non-heme Fe(II) or other two-valent metal ions that cleaves catechol in the meta position giving rise to the meta fission pathway (Perez-Pantoja et al. 2012).

11.8.2.1 Ortho Fission Pathway of Catechol

Ortho fission pathway (Fig. 11.2) is also known as β-ketoadipate pathway (Broderick 1999). According to the pathway, catechol ring is cleaved between the two hydroxyls by the enzyme catechol 1,2-dioxygenase forming cis,cis-muconate which is then further metabolized in three steps to form 3-oxoadipate (MacLean et al. 2006). Further degradation produces acetyl-CoA and succinyl-CoA, both intermediates of TCA cycle. The enzymes involved in this pathway are muconate cycloisomerase, muconolactone isomerase, β-ketoadipate enol-lactone hydrolase, β-ketoadipate: succinyl-CoA transferase and β-ketoadipyl-CoA thiolase.

Muconate cycloisomerase (EC 5.5.1.1) that is also known as muconate lactonizing enzyme is a homo-octamer intramolecular lyase that required Mn^{2+} for its action. Muconolactone isomerase (EC 5.3.3.4), the next enzyme in the pathway is an intramolecular oxidoreductase. β-ketoadipate enol-lactone hydrolase (EC 3.1.1.24) is a carboxylic-ester hydrolase while β-ketoadipate succinyl-CoA

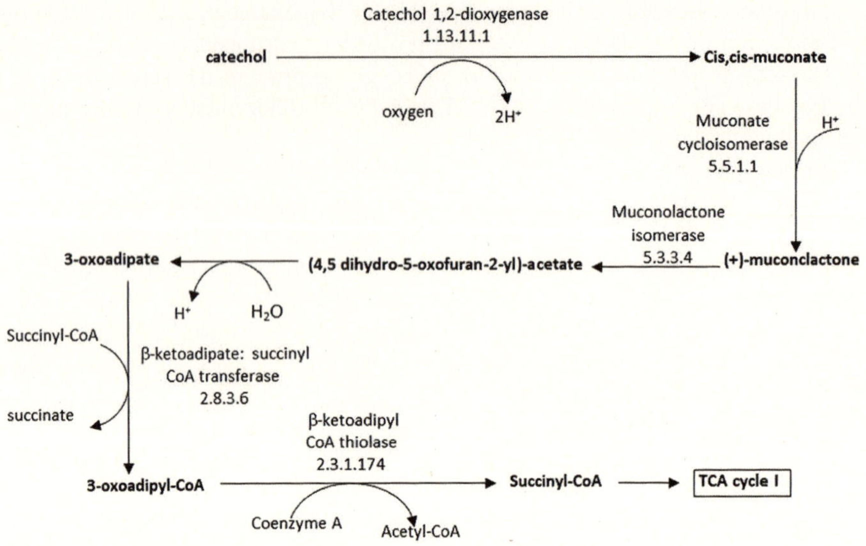

Fig. 11.2 Catechol degradation in bacteria by ortho-fission pathway leading to the formation of end products that enter TCA cycle

transferase (EC 2.8.3.6) is a two component transferase enzyme with an α subunit as well as a β subunit. β-ketoadipyl-CoA thiolase (EC 2.3.1.174) is an acyl transferase that transfers groups other than aminoacyl groups.

A number of bacterial species have been studied well for the catechol ortho cleavage pathway. Some of the species studied are *Cupriavidus necator, Acinetobacter calcoaceticus, Agrobacterium tumefaciens, Pseudomonas aeruginosa, Pseudomonas knackmussii, Pseudomonas putida, Rhizobium leguminosarum, Sinorhizobium meliloti* etc. (Parke and Ornston 1986; Aldrich and Chakrabarty 1988; Kukor et al. 1988; MacLean et al. 2006; Liu et al. 2016).

11.8.2.2 Meta Fission Pathway of Catechol

In the meta-fission pathway, the catechol 2,3-dioxygenase enzyme (EC 1.13.1.2.) cleaves the bond at the meta- position in the aromatic ring as shown in pathways 2 and 3. The product of this cleavage, (2Z,4E)-2-hydroxy-6-oxohexa-2,4-dienoate, is metabolized in one of two routes.

In route 1 (Fig. 11.3) it is hydrolyzed by 2-hydroxymuconic semialdehyde hydrolase (E.C.3.7.1.9) (Harayama and Rekik 1990), generating the important intermediate 2-oxopent-4-enoate in a single step. The other enzymes involved in this route are 2-oxopent-4-enoate hydratase (EC 4.2.1.80), 4-hydroxy-2-oxovalerate aldolase (EC 4.1.3.39) and acetaldehyde dehydrogenase (EC 1.2.1.20). Some of the bacterial species that exhibit this pathway are *Alcaligenes sp., Comamonas sp., Cupriavidus necator, Ralstonia pickettii,* and *Pseudomonas sp.* (Kukor and Olsen 1991; Junker et al. 1994; He and Spain 1999; Tian et al. 2017).

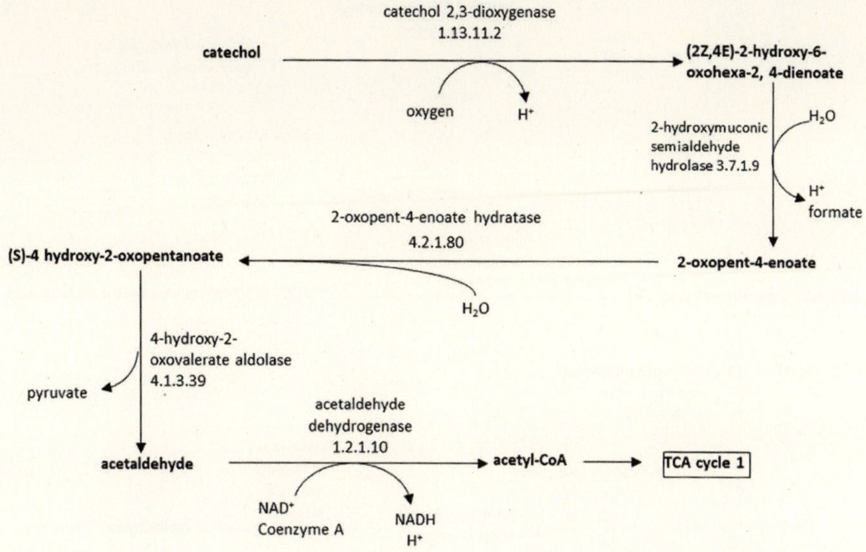

Fig. 11.3 Catechol degradation in bacteria by meta-fission pathway through 2-oxopent- 4-enoate leading to end products that enter TCA cycle

In the second route as seen in Fig. 11.4, 2-oxopent-4-enoate is formed in a multistep reaction (Harayama et al. 1989). The enzymes involved in this route are 2-hydroxymuconic semialdehyde dehydrogenase (EC 1.2.1.85), 2-hydroxymuconate tautomerase (EC 5.3.2.6) and 2-oxo-3-hexenedioate decarboxylase (EC 4.1.1.77); leading to the formation of 2-oxopent-4-enoate. Metabolism of this intermediate is same as that of route 1. A number of bacterial strains have been studied to exhibit this pathway, some of the main genera being *Ralstonia, Azotobacter* and *Pseudomonas* (Sala-Trepat and Evans 1971; Powlowski et al. 1993). Pyruvate and acetaldehyde (and eventually, acetyl-CoA), the end products of these pathways get incorporated to the TCA cycle.

11.8.3 Hydroquinone Degradation

Degradation of hydroquinone formed from nitrophenol and similar derivatives start with hydroquinone dioxygenase enzyme (EC 1.13.11.66) and leads to the production of 3-oxoadipate which is also an intermediate seen in the ortho-fission pathway of catechol as shown in Fig. 11.5.

The other enzymes involved in this pathway leading to the formation of 3-oxoadipate are 4-hydroxymuconic semialdehyde dehydrogenase (EC 1.2.1.61) and maleylacetate reductase (EC 1.3.1.32). 3-oxoadipate is degraded as seen earlier in the ortho-fission pathway of catechol. The genes encoding the enzymes in this pathway most commonly observed in *Pseudomonas sp.* (Zhang et al. 2012). In the degradation of pentachlorophenol, the intermediate formed

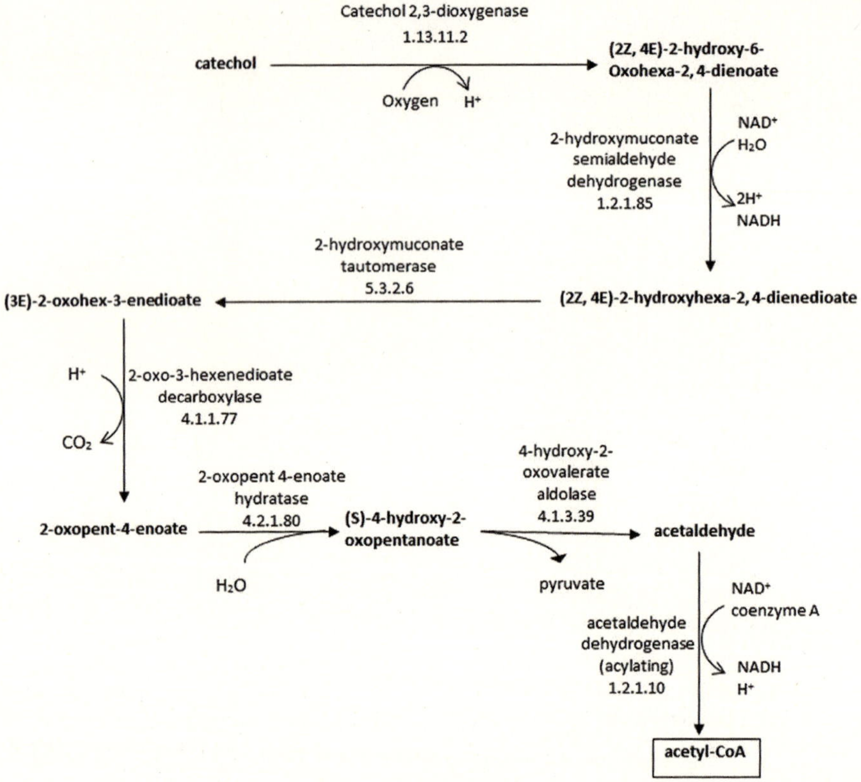

Fig. 11.4 Catechol degradation in bacteria by meta-fission pathway through (2E)-2- hydroxypenta-2,4-dienoate leading to end products that enter TCA cycle

is 2,6-dichlorohydroquinone. 2,6-dichlorohydroquinone dioxygenase cleaves the hydroquinone ring, resulting in the non-aromatic product 2-chloromaleylacetate (Orser and Lange 1994). This compound is dehalogenated resulting in 3-oxoadipate which is then degraded to enter the TCA cycle. A microbial strain that is found to completely degrade pentachlorophenol is *Flavobacterium* sp. (Saber and Crawford 1985).

11.9 Anaerobic Degradation of Phenol in Bacteria

Phenol degradation under anaerobic conditions has been studied though not extensively observed in the microbial community. Some of the bacterial genera that exhibit this pathway are *Thauera, Magnetospirillum* and *Geobactor* (Heider et al. 1998; Nesvera et al. 2015). Under anaerobic conditions, phenol is phosphorylated to phenyl phosphate and then converted to benzoyl-CoA through a three step reaction as shown in Fig. 11.6. In the case of cresols, the methyl group is first oxidized and then converted to benzoyl-CoA. Benzoyl-CoA is then reduced to cyclic

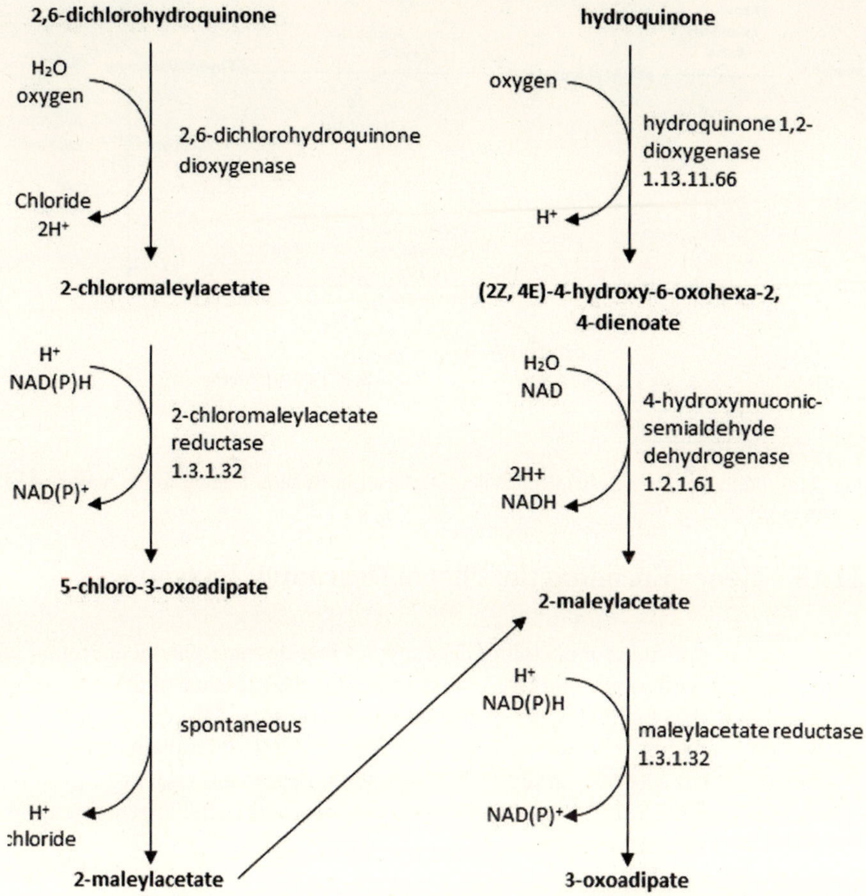

Fig. 11.5 Degradation of hydroquinone and its derivatives in bacteria leading to the formation of 3-oxoadipate

1,5-dienoyl-CoA and then metabolized by a variety of reactions including beta oxidation cycles to produce acetyl-CoA as the end product (Breese et al. 1998).

Phenyl phosphate synthase (EC 2.7.9.-), the primary enzyme in the anaerobic degradation of phenol is a three protein unit out of which two proteins help in transfer of phosphate group from ATP to phenol while the third protein stimulates the reaction (Narmandakh et al. 2006). Phenyl phosphate carboxylase (*EC* 4.1.1.-) uses CO_2 as substrate for the carboxylation in the next reaction (Schuhle and Fuchs 2004). This enzyme with four sub units utilizes divalent cations like Mg^{2+} for its activity. Hydroxybenzoate-CoA ligase (EC 6.2.1.27), catalyzing the conversion of 4-hydroxybenzoate to 4-hydroxybenzoyl-CoA also uses ATP and Mg^{2+} for its action. Hydroxybenzoyl-CoA reductase (EC 1.3.7.9), the final enzyme in the reaction series is a hexamer that helps to remove the hydroxyl group from the substrate by two electron reduction from a reduced ferredoxin, thus forming benzoyl-CoA (Park et al. 2006).

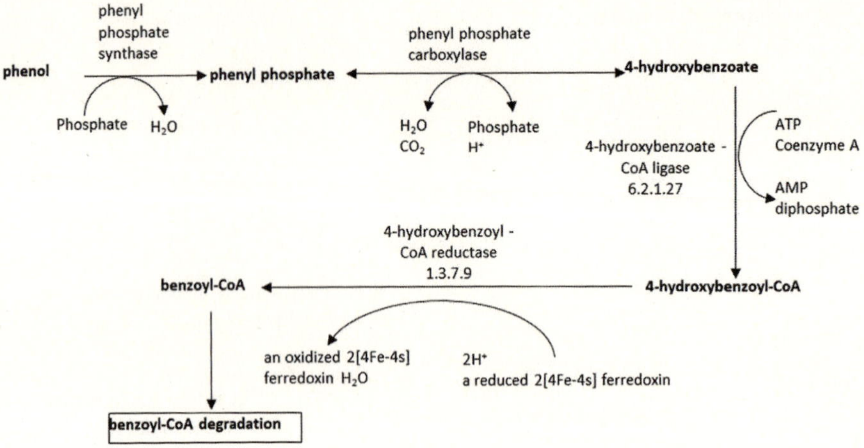

Fig. 11.6 Preliminary anaerobic degradation of phenol in bacteria leading to the formation of benzoyl-CoA

11.10 Genes Encoding the Phenol Degrading Enzymes

Most genes responsible for encoding the enzymes that degrade phenol are found in the genomic as well as plasmid DNA of bacteria. It is the presence of these genes in plasmid DNA that facilitates the exchange of genetic material between different strains; thus making it a very common property of the microbial community. *Pseudomonas* sp. was overserved to harbor phenol degradation genes in plasmid (Kivisaar et al. 1990) while that of *Cupriavidus necator* was found in genomic DNA (Perez-Pantoja et al. 2008).

In case of microbial strains that degrade heavily chlorinated or nitrated phenol derivatives, the genes encoding enzymes for preliminary degradation of the compound are found to be clustered with the genes encoding degradative enzymes of the hydroquinone and 3-oxoadipate. One such example is *Cupriavidus necator*. Genes encoding multiple phenol degrading pathways can be found in the same microbial strain. Depending upon the clustering of genes, a microbial strain might be more adapted to a particular degradation pathway. In *Rhodococcus* sp., the genes encoding the degradation of phenol to catechol *(pheA2A1)* was seen clustered with those encoding the enzymes of the ortho-fission pathway *(catABC)*, thus showing that the strain predominantly utilizes ortho-fission pathway (Szőköl et al. 2014). But these two sets of enzymes were found to have two different regulation systems. At the same time, genes encoding enzymes catalyzing the degradation of phenol to catechol was found to be clustered with those encoding the enzymes of the meta fission pathway *(phlBCDEHI)* in the case of *Bacillus thermoglucosidasius* (Duffner et al. 2000). In both cases two-component phenol hydroxylase was observed while in the case of *Burkholderia* sp., genes encoding multicomponent phenol hydroxylase was found to be clustered with that encoding the genes of the meta fission pathway (Perez-Pantoja et al. 2012).

11.11 Detection of Phenol Degrading Strains

11.11.1 Basic Laboratory Protocol to Select Efficient Phenol Degrading Strains

This section explains a simple laboratory protocol for selection and adaptation of phenol degrading strains. Isolation of phenol degraders starts with a preliminary study for phenol tolerance. To select the strains that are capable of tolerating phenol, the strains can be incubated in nutrient broth fortified with 1 mM phenol. Microbes that exhibit growth in phenol-nutrient broth are those that can tolerate the presence of phenol when all other nutrients for growth are provided. Such strains can be grown in a mineral salts medium (MSM) with 1 mM phenol as the sole carbon source for 24 h. Phenol degrading strains will grow in the phenol MSM media while phenol tolerant non-degraders will not exhibit any growth. Strains that can grow in 1 mM phenol can be transferred to a higher concentration and so on till strains that can thrive in 25 mM phenol are obtained. This step by step procedure also helps not only in isolation of phenol degraders, but also in the gradual adaptation of strains to metabolize the compound.

11.11.2 PCR Based Screening for Phenol Degrading Strains

PCR technique has also been used for rapid screening of phenol degrading strains. The presence of genes encoding phenol hydroxylase is most commonly investigated. Strains harboring multicomponent PH can be selected by amplifying the gene LmPH which encodes the largest unit of the enzyme (Watanabe et al. 1998a). In the case of two-component PH, the gene to be amplified is pheA1, the gene encoding the larger subunit consisting of 542 amino acids (Saa et al. 2010). One-component phenol hydroxylase is rarely observed since the other two are more prominent in the environment (Silva et al. 2013).

11.12 Microbes Commonly Studied to Degrade Phenol and its Derivatives

A wide variety of microorganisms have been studied to degrade phenol. The most explicitly studied genus for phenol degradation is *Pseudomonas* (Agarry et al. 2008). This genus has been studied to harbor a number of non-specific genes that allow it to utilize a wide variety of substrates with high efficiency. *Pseudomonas putida* and *Pseudomonas sp. JS150* have been found to degrade phenol while *Pseudomonas sp. WBC-3* has been studied to degrade its derivative, 4-nitrophenol (Harayama et al. 1989; Mrozik et al. 2011; Chi et al. 2013). *Pseudomonas fluorescens PU1, Pseudomonas aeruginosa, Pseudomonas resinovorans* etc. have been proven to utilize phenol for growth (Ahmed et al. 1995; Yang and Lee 2007; Mahiudddin et al. 2012).

Cupriavidus necator JMP134 has been found to degrade 3-nitrophenol as well as trichlorophenol (Sánchez and González 2007; Chi et al. 2013). This microbial strain is widely studied as a phenol degrader and is also said to be a model organism for degradation of chloro aromatic compounds (Perez-Pantoja et al. 2008). *C. necator* N-1 strain has been observed to carry the genes required for degradation of benzoate, phenol and other chlorinated aromatic compounds along with the genes for polyhydroxyalkanoates metabolism (Mahiudddin et al. 2012). *C. necator* is also explicitly studied for polyhydroxyalkanoates production (Kynadi and Suchithra 2017). This strain was earlier known as *Alcaligenes eutrophus* as well as *Ralstonia eutropha* (Kynadi and Suchithra 2014) and has properties similar to both genera. Therefore it is not surprising that strains of these genera like *Alcaligenes faecalis, Alcaligenes sp.* NyZ215, *Ralstonia taiwanensis* and *Ralstonia sp.* KN1 have been found to have the enzymes to degrade phenol and its derivatives (Nakamura et al. 2000; Thomas et al. 2002; Jiang et al. 2007; Chi et al. 2013).

Strains of genus *Bacillus* also have been found to degrade phenol. Some of the strains that have been observed to degrade this aromatic compound and its derivatives are *B. brevis, B. cereus, B. stearothermophilus* and *B. thermoglucosidasius* (Gurujeyalakshmi and Oriel 1989; Duffner et al. 2000; Arutchelvan et al. 2006; Banerjee and Ghoshal 2010). *Rodococcus* strains also show phenol degrading property. *Rodococcus sp.*CS1 degrades phenol while *R. erythyropolis* M1 is found to degrade a mixture of phenol and 2-chlorophenol (Goswami et al. 2005; Paisio et al. 2012). A consortium of *R. erythyropolis* M1 and *P. fluorescens* P1 has been found to degrade phenol and cresol mixture more efficiently in comparison to individual strains (Goswami et al. 2005). Some of the other strains that have been reported to degrade phenol and its derivatives are *Brevibacillus* sp. strain P6, *Enterobactersp. Acinetobacter calcoaceticus, Arthrobacter citreus etc.* (Thomas et al. 2002; Karigar et al. 2006; Yang and Lee 2007; Liu et al. 2016). Anaerobic phenol degradation has been observed in strains like *Desulfobacterium phenolicum, Desulfobacterium anilini* etc. (Bak and Widdel 1986; Ahn et al. 2009).

Phenol degradation is found in yeast, fungi as well as algae too. Some of the strains that have been found to degrade phenol are *Trichosporon cutaneum* R57, *Trichosporon* sp. LE3, *Candida tropicalis* Z-04 and *Fusarium flocciferium, Fusarium solani, Aspergillus* sp., *Penicillium* sp. *and Graphium* sp. (Santos et al. 2001; Aleksieva et al. 2002; Mendonça et al. 2004; Santos and Linardi 2004; Zhou et al. 2011). Phenol and its methylated homologues have been proven to be degraded by the algal strain *Ochromonas danica* (Semple and Cain 2006). While *Chlorella sp.* was reported to degrade 2,4-dimethylphenol, *Scenedesmus obliquus* acted upon 2,4-dinitrophenol (Klekner and Kosaric 1992).

There has been an increasing awareness about the hazards of phenol contamination in regard to human health in the recent years. This has opened up a vibrant research area in the field of bioremediation. Though a number of microbial strains have been reported to degrade phenol and its derivatives, more studies are required to custom design large scale, cost efficient treatment systems for industrial effluent streams; thus making this an area of research will a great potential.

References

Agarry SE, Durojaiye AO, Solomon BO (2008) Microbial degradation of phenols: a review. Int J Environ Pollut 32:12. https://doi.org/10.1504/IJEP.2008.016895

Aggelis G, Ehaliotis C, Lyberatos F et al (2002) Evaluation of white-rot fungi for detoxification and decolorization of effluents from the green olive debittering process. Appl Microbiol Biotechnol 59:353–360. https://doi.org/10.1007/s00253-002-1005-9

Ahmed AM, Nakhla GF, Farooq S (1995) Phenol degradation by *Pseudomonas aeruginosa*. J Environ Sci Heal Part A Environ Sci Eng Toxicol 30:99–107. https://doi.org/10.1080/10934529509376188

Ahn Y-B, Chae J-C, Zylstra GJ, Häggblom MM (2009) Degradation of phenol via phenylphosphate and carboxylation to 4-hydroxybenzoate by a newly isolated strain of the sulfate-reducing bacterium *Desulfobacterium anilini*. Appl Environ Microbiol 75:4248–4253. https://doi.org/10.1128/AEM.00203-09

Aldrich TL, Chakrabarty AM (1988) Transcriptional regulation, nucleotide sequence, and localization of the promoter of the catBC operon in *Pseudomonas putida*. J Bacteriol 170:1297–1304. https://doi.org/10.1128/jb.170.3.1297-1304.1988

Aleksieva Z, Ivanova D, Godjevargova T, Atanasov B (2002) Degradation of some phenol derivatives by *Trichosporon cutaneum* R57. Process Biochem 37:1215–1219. https://doi.org/10.1016/S0032-9592(01)00336-3

Alexander M, Lustigman BK (1966) Effect of chemical structure on microbial degradation of substituted benzenes. J Agric Food Chem 14:410–413. https://doi.org/10.1021/jf60146a022

Arutchelvan V, Kanakasabai V, Elangovan R et al (2006) Kinetics of high strength phenol degradation using *Bacillus brevis*. J Hazard Mater 129:216–222. https://doi.org/10.1016/j.jhazmat.2005.08.040

Bak F, Widdel F (1986) Anaerobic degradation of phenol and phenol derivatives by *Desulfobacterium phenolicum* sp. nov. Arch Microbiol 146:177–180. https://doi.org/10.1007/BF00402347

Banerjee A, Ghoshal AK (2010) Phenol degradation by *Bacillus cereus*: pathway and kinetic modeling. Bioresour Technol 101:5501–5507. https://doi.org/10.1016/j.biortech.2010.02.018

Belchik SM, Xun L (2008) Functions of flavin reductase and quinone reductase in 2,4,6-trichlorophenol degradation by *Cupriavidus necator* JMP134. J Bacteriol 190:1615–1619. https://doi.org/10.1128/JB.01697-07

Breese K, Boll M, Alt-Mörbe J et al (1998) Genes coding for the benzoyl-CoA pathway of anaerobic aromatic metabolism in the bacterium *Thauera aromatica*. Eur J Biochem 256:148–154. https://doi.org/10.1046/j.1432-1327.1998.2560148.x

Broderick JB (1999) Catechol dioxygenases. Essays Biochem 34:173–189

Busca G, Berardinelli S, Resini C, Arrighi L (2008) Technologies for the removal of phenol from fluid streams: a short review of recent developments. J Hazard Mater 160:265–288. https://doi.org/10.1016/j.jhazmat.2008.03.045

Calabrese EJ, Kenyon EM (1991) Air toxics and risk assessment. Lewis Publishers, Chelsea, MI

Caspi R, Altman T, Billington R et al (2014) The MetaCyc database of metabolic pathways and enzymes and the BioCyc collection of Pathway/Genome Databases. Nucleic Acids Res 42:D459–D471. https://doi.org/10.1093/nar/gkt1103

Chi X-Q, Zhang J-J, Zhao S, Zhou N-Y (2013) Bioaugmentation with a consortium of bacterial nitrophenol-degraders for remediation of soil contaminated with three nitrophenol isomers. Environ Pollut 172:33–41. https://doi.org/10.1016/j.envpol.2012.08.002

Duffner FM, Kirchner U, Bauer MP, Müller R (2000) Phenol/cresol degradation by the thermophilic *Bacillus thermoglucosidasius* A7: cloning and sequence analysis of five genes involved in the pathway. Gene 256:215–221. https://doi.org/10.1016/S0378-1119(00)00352-8

Ebert S, Rieger PG, Knackmuss HJ (1999) Function of coenzyme F420 in aerobic catabolism of 2,4, 6-trinitrophenol and 2,4-dinitrophenol by *Nocardioides simplex* FJ2-1A. J Bacteriol 181:2669–2674

Field JA, Lettinga G (1991) Treatment and detoxification of aqueous spruce bark extracts by *Aspergillus niger*. Water Sci Technol 24:127–137

Goswami M, Shivaraman N, Singh RP (2005) Microbial metabolism of 2-chlorophenol, phenol and para-cresol by *Rhodococcus erythropolis* M1 in co-culture with *Pseudomonas fluorescens* P1. Microbiol Res 160:101–109. https://doi.org/10.1016/j.micres.2004.10.004

Gurujeyalakshmi G, Oriel P (1989) Isolation of phenol-degrading *Bacillus stearothermophilus* and partial characterization of the phenol hydroxylase. Appl Environ Microbiol 55:500–502

Harayama S, Rekik M (1990) The meta cleavage operon of TOL degradative plasmid pWW0 comprises 13 genes. Mol Gen Genet 221:113–120. https://doi.org/10.1007/BF00280375

Harayama S, Rekik M, Ngai KL, Ornston LN (1989) Physically associated enzymes produce and metabolize 2-hydroxy-2,4-dienoate, a chemically unstable intermediate formed in catechol metabolism via meta cleavage in *Pseudomonas putida*. J Bacteriol 171:6251–6258. https://doi.org/10.1128/jb.171.11.6251-6258.1989

Harwood CS, Parales RE (1996) The beta-ketoadipate pathway and the biology of self-identity. Annu Rev Microbiol 50:553–590. https://doi.org/10.1146/annurev.micro.50.1.553

He Z, Spain JC (1999) Comparison of the downstream pathways for degradation of nitrobenzene by *Pseudomonas pseudoalcaligenes* JS45 (2-aminophenol pathway) and by *Comamonas* sp. JS765 (catechol pathway). Arch Microbiol 171:309–316

Heider J, Boll M, Fuchs KB et al (1998) Differential induction of enzymes involved in anaerobic metabolism of aromatic compounds in the denitrifying bacterium *Thauera aromatica*. Arch Microbiol 170:120–131. https://doi.org/10.1007/s002030050623

Jeffrey JS (1997) Sanitation-Disinfection Basics; Poultry fact sheet No. 27 Cooperative Extension; University of California.

Jiang Y, Wen J, Bai J et al (2007) Biodegradation of phenol at high initial concentration by *Alcaligenes faecalis*. J Hazard Mater 147:672–676. https://doi.org/10.1016/j.jhazmat.2007.05.031

Junker F, Leisinger T, Cook AM (1994) 3-Sulphocatechol 2,3-dioxygenase and other dioxygenases (EC 1.13.11.2 and EC 1.14.12.-) in the degradative pathways of 2-aminobenzenesulphonic, benzenesulphonic and 4-toluenesulphonic acids in *Alcaligenes* sp. strain 0-1. Microbiology 140:1713–1722. https://doi.org/10.1099/13500872-140-7-1713

Karigar C, Mahesh A, Nagenahalli M, Yun DJ (2006) Phenol degradation by immobilized cells of *Arthrobacter citreus*. Biodegradation 17:47–55. https://doi.org/10.1007/s10532-005-3048-y

Kivisaar M, Hõrak R, Kasak L et al (1990) Selection of independent plasmids determining phenol degradation in *Pseudomonas putida* and the cloning and expression of genes encoding phenol monooxygenase and catechol 1,2-dioxygenase. Plasmid 24:25–36. https://doi.org/10.1016/0147-619X(90)90022-5

Klekner V, Kosaric N (1992) Degradation of phenols by algae. Environ Technol 13:493–501. https://doi.org/10.1080/09593339209385176

Kolvenbach BA, Corvini PF-X (2012) The degradation of alkylphenols by *Sphingomonas* sp. strain TTNP3 – a review on seven years of research. N Biotechnol 30:88–95. https://doi.org/10.1016/j.nbt.2012.07.008

Krastanov A, Alexieva Z, Yemendzhiev H (2013) Microbial degradation of phenol and phenolic derivatives. Eng Life Sci 13:76–87. https://doi.org/10.1002/elsc.201100227

Kukor JJ, Olsen RH (1991) Genetic organization and regulation of a meta cleavage pathway for catechols produced from catabolism of toluene, benzene, phenol, and cresols by *Pseudomonas pickettii* PKO1. J Bacteriol 173:4587–4594. https://doi.org/10.1128/jb.173.15.4587-4594.1991

Kukor JJ, Olsen RH, Ballou DP (1988) Cloning and expression of the catA and catBC gene clusters from *Pseudomonas aeruginosa* PAO. J Bacteriol 170:4458–4465. https://doi.org/10.1128/jb.170.10.4458-4465.1988

Kumaran P, Paruchuri YL (1997) Kinetics of phenol biotransformation. Water Res 31:11–22. https://doi.org/10.1016/S0043-1354(99)80001-3

Kynadi AS, Suchithra TV (2014) Polyhydroxyalkanoates: biodegradable plastics for environmental conservation. In: Pramanik K, Patra JK (eds) Industrial & environmental biotechnology. Studium Press (India) Pvt. Ltd., New Delhi, pp 1–15

Kynadi AS, Suchithra TV (2017) Formulation and optimization of a novel media comprising rubber seed oil for PHA production. Ind Crops Prod 105:156–163. https://doi.org/10.1016/j.indcrop.2017.04.062

Lakshmi S, Harshitha M, Vaishali G et al (2016) Studies on different methods for removal of phenol in waste water- Review. International Journal of Science Engineering and Technology Research 5:2488–2496

Leahy JG, Batchelor PJ, Morcomb SM (2003) Evolution of the soluble diiron monooxygenases. FEMS Microbiol Rev 27:449–479. https://doi.org/10.1016/S0168-6445(03)00023-8

Ledger T, Pieper DH, González B (2006) Chlorophenol hydroxylases encoded by plasmid pJP4 differentially contribute to chlorophenoxyacetic acid degradation. Appl Environ Microbiol 72:2783–2792. https://doi.org/10.1128/AEM.72.4.2783-2792.2006

Lewis RJ (2012) Sax's dangerous properties of industrial materials. John Wiley, Hoboken, NJ

Liu Z, Xie W, Li D et al (2016) Biodegradation of phenol by bacteria strain *Acinetobacter calcoaceticus* PA isolated from phenolic wastewater. Int J Environ Res Public Health 13(3):300. https://doi.org/10.3390/ijerph13030300

MacLean AM, MacPherson G, Aneja P, Finan TM (2006) Characterization of the β-ketoadipate pathway in *Sinorhizobium meliloti*. Appl Environ Microbiol 72:5403–5413. https://doi.org/10.1128/AEM.00580-06

Mahiudddin M, Fakhruddin ANM, Abdullah-Al-Mahin (2012) Degradation of phenol via meta cleavage pathway by *Pseudomonas fluorescens* PU1. ISRN Microbiol 2012:1–6. https://doi.org/10.5402/2012/741820

Mendonça E, Martins A, Anselmo AM (2004) Biodegradation of natural phenolic compounds as single and mixed substrates by *Fusarium flocciferum*. Electron J Biotechnol 7:38–46. https://doi.org/10.4067/s0717-34582004000100005

Michałowicz J, Duda W (2007) Phenols – Sources and toxicity. Polish J Environ Stud 16:347–362

Mrozik A, Miga S, Piotrowska-Seget Z (2011) Enhancement of phenol degradation by soil bioaugmentation with *Pseudomonas* sp. JS150. J Appl Microbiol 111:1357–1370. https://doi.org/10.1111/j.1365-2672.2011.05140.x

Murray LJ, Lippard SJ (2007) Substrate trafficking and dioxygen activation in bacterial multicomponent monooxygenases. Acc Chem Res 40:466–474. https://doi.org/10.1021/ar600040e

Nakamura K, Ishida H, Iizumi T (2000) Constitutive trichloroethylene degradation led by tac promoter chromosomally integrated upstream of phenol hydroxylase genes of *Ralstonia* sp. KN1 and its nucleotide sequence analysis. J Biosci Bioeng 89:47–54. https://doi.org/10.1016/S1389-1723(00)88049-4

Narmandakh A, Gad'on N, Drepper F et al (2006) Phosphorylation of phenol by phenylphosphate synthase: role of histidine phosphate in catalysis. J Bacteriol 188:7815–7822. https://doi.org/10.1128/JB.00785-06

Nesvera J, Rucka L, Patek M (2015) Catabolism of phenol and its derivatives in bacteria: genes , their regulation , and use in the biodegradation of toxic pollutants. Adv Appl Microbiol 93:107–160. https://doi.org/10.1016/bs.aambs.2015.06.002

Nuhoglu A, Yalcin B (2005) Modelling of phenol removal in a batch reactor. Process Biochem 40:1233–1239. https://doi.org/10.1016/j.procbio.2004.04.003

Orser CS, Lange CC (1994) Molecular analysis of pentachlorophenol degradation. Biodegradation 5:277–288. https://doi.org/10.1007/BF00696465

Paisio CE, Talano MA, Gonzelez PS et al (2012) Isolation and characterization of a Rhodococcus strain with phenol-degrading ability and its potential use for tannery effluent biotreatment. Environ Sci Pollut Res 19:3430–3439. https://doi.org/10.1007/s11356-012-0870-8

Park YJ, Yoo CB, Choi SY, Lee HB (2006) Purifications and characterizations of a ferredoxin and its related 2-oxoacid: Ferredoxin oxidoreductase from the hyperthermophilic archaeon, Sulfolobus solfataricus P1. Journa Biochem Mol Biol 39:46–54

Parke D, Ornston LN (1986) Enzymes of the beta-ketoadipate pathway are inducible in Rhizobium and Agrobacterium spp. and constitutive in *Bradyrhizobium* spp. J Bacteriol 165:288–292. https://doi.org/10.1128/jb.165.1.288-292.1986

Perez-Pantoja D, De la Iglesia R, Pieper DH et al (2008) Metabolic reconstruction of aromatic compounds degradation from the genome of the amazing pollutant-degrading bacterium *Cupriavidus necator* JMP134. FEMS Microbiol Rev 32:736–794. https://doi.org/10.1111/j.1574-6976.2008.00122.x

Perez-Pantoja D, Donoso R, Agulle L et al (2012) Genomic analysis of the potential for aromatic compounds biodegradation in *Burkholderiales*. Environ Microbiol 14:1091–1117. https://doi.org/10.1111/j.1462-2920.2011.02613.x

Perry LL, Zylstra GJ (2007) Cloning of a gene cluster involved in the catabolism of p-nitrophenol by *Arthrobacter* sp. strain JS443 and characterization of the p-nitrophenol monooxygenase. J Bacteriol 189:7563–7572. https://doi.org/10.1128/JB.01849-06

Porter AW, Campbell BR, Kolvenbach BA et al (2012) Identification of the flavin monooxygenase responsible for ipso substitution of alkyl and alkoxyphenols in *Sphingomonas* sp. TTNP3 and *Sphingobium xenophagum* Bayram. Appl Microbiol Biotechnol 94:261–272. https://doi.org/10.1007/s00253-011-3621-8

Powlowski J, Sahlman L, Shingler V (1993) Purification and properties of the physically associated meta-cleavage pathway enzymes 4-hydroxy-2-ketovalerate aldolase and aldehyde dehydrogenase alcylating from *Pseudomonas* Sp. J Bacteriol 175:377–385

Saa L, Jaureguibeitia A, Largo E et al (2010) Cloning, purification and characterization of two components of phenol hydroxylase from *Rhodococcus erythropolis* UPV-1. Appl Microbiol Biotechnol 86:201–211. https://doi.org/10.1007/s00253-009-2251-x

Saber DL, Crawford RL (1985) Isolation and characterization of *Flavobacterium* strains that degrade pentachlorophenol. Appl Environ Microbiol 50:1512–1518

Sala-Trepat JM, Evans WC (1971) The meta cleavage of catechol by *Azotobacter* species. 4-Oxalocrotonate pathway. Eur J Biochem 20:400–413

Sambrook J, Russell DW (2001) Molecular cloning: a laboratory manual. Cold Spring Harbor Laboratory Press, New York

Sánchez MA, González B (2007) Genetic characterization of 2,4,6-trichlorophenol degradation in *Cupriavidus necator* JMP134. Appl Environ Microbiol 73:2769–2776. https://doi.org/10.1128/AEM.02584-06

Santos VL, Linardi VR (2004) Biodegradation of phenol by a filamentous fungi isolated from industrial effluents—identification and degradation potential. Process Biochem 39:1001–1006. https://doi.org/10.1016/S0032-9592(03)00201-2

Santos VL, Heilbuth NM, Linardi VR (2001) Degradation of phenol by *Trichosporon* sp. LE3 cells immobilized in alginate. J Basic Microbiol 41:171–178

Saravanan P, Pakshirajan K, Saha P (2008) Growth kinetics of an indigenous mixed microbial consortium during phenol degradation in a batch reactor. Bioresour Technol 99:205–209. https://doi.org/10.1016/j.biortech.2006.11.045

van Schie PM, Young LY (2007) Biodegradation of phenol: mechanisms and applications. Bioremediat J 4:1–18. https://doi.org/10.1080/10588330008951128

Schuhle K, Fuchs G (2004) Phenylphosphate carboxylase: A new C-C lyase involved in anaerobic phenol metabolism in *Thauera aromatica*. J Bacteriol 186:4556–4567. https://doi.org/10.1128/JB.186.14.4556-4567.2004

Semple KT, Cain RB (2006) Degradation of phenol and its methylated homologues by *Ochromonas danica*. FEMS Microbiol Lett 152:133–139. https://doi.org/10.1111/j.1574-6968.1997.tb10419.x

Sikkema JAN, Jan AM, Poolman B (1995) Mechanisms of membrane toxicity of hydrocarbons. Microbiol Rev 59:201–222

Silva CC, Hayden H, Sawbridge T et al (2013) Identification of genes and pathways related to phenol degradation in metagenomic libraries from petroleum refinery wastewater. PLoS One 8:e61811. https://doi.org/10.1371/journal.pone.0061811

Szőköl J, Rucká L, Šimčíková M et al (2014) Induction and carbon catabolite repression of phenol degradation genes in *Rhodococcus erythropolis* and *Rhodococcus jostii*. Appl Microbiol Biotechnol 98:8267–8279. https://doi.org/10.1007/s00253-014-5881-6

Thomas S, Sarfaraz S, Mishra LC, Iyengar L (2002) Degradation of phenol and phenolic compounds by a defined denitrifying bacterial culture. World J Microbiol Biotechnol 18:57–63. https://doi.org/10.1023/A:1013947722911

Tian M, Du D, Zhou W et al (2017) Phenol degradation and genotypic analysis of dioxygenase genes in bacteria isolated from sediments. Brazilian J Microbiol 48:305–313. https://doi.org/10.1016/j.bjm.2016.12.002

Todorović V (2003) Acute phenol poisoning. Med Pregl 56:37–41

Topalova Y, Ribarova I, Kozuharov D et al (2007) Structural/functional changes in activated sludge in PCP- and ONP-biodegradation technologies. Biotechnol Biotechnol Equip 21:28–33. https://doi.org/10.1080/13102818.2007.10817408

Watanabe K, Teramoto M, Futamata H, Harayama S (1998a) Molecular detection, isolation, and physiological characterization of functionally dominant phenol-degrading bacteria in activated sludge. Appl Environ Microbiol 64:4396–4402

Watanabe K, Yamamoto S, Hino S, Harayama S (1998b) Population dynamics of phenol-degrading bacteria in activated sludge determined by gyrB-targeted quantitative PCR. Appl Environ Microbiol 64:1203–1209

Xiao Y, Zhang J-J, Liu H, Zhou N-Y (2007) Molecular characterization of a novel ortho-nitrophenol catabolic gene cluster in *Alcaligenes* sp. strain NyZ215. J Bacteriol 189:6587–6593. https://doi.org/10.1128/JB.00654-07

Yang C-F, Lee C-M (2007) Enrichment, isolation, and characterization of phenol-degrading Pseudomonas resinovorans strain P-1 and *Brevibacillus* sp. strain P-6. Int Biodeterior Biodegradation 59:206–210. https://doi.org/10.1016/j.ibiod.2006.09.010

Zhang S, Sun W, Xu L et al (2012) Identification of the para-nitrophenol catabolic pathway, and characterization of three enzymes involved in the hydroquinone pathway, in *Pseudomonas* sp. 1-7. BMC Microbiol 12:27. https://doi.org/10.1186/1471-2180-12-27

Zhang W, Yin K, Chen L (2013) Bacteria-mediated bisphenol A degradation. Appl Microbiol Biotechnol 97:5681–5689. https://doi.org/10.1007/s00253-013-4949-z

Zhou J, Yu X, Ding C et al (2011) Optimization of phenol degradation by *Candida tropicalis* Z-04 using Plackett-Burman design and response surface methodology. J Environ Sci 23:22–30. https://doi.org/10.1016/S1001-0742(10)60369-5

Chapter 12
Mineral Solubilization by Microorganism: Mitigating Strategy in Mineral Deficient Soil

Gausiya Bashri, Anuradha Patel, Rachana Singh, Parul Parihar, and Sheo Mohan Prasad

12.1 Introduction

Increased human activities have posed negative impact on the environment. Rapid industrialization is coupled with the release of pollutants in biosphere (Araujo et al. 2013). Soil is the most important interface of the environment, and contamination in soil causes the decline of microbial diversity that questioned the sustainability of crop productivity (Nunes et al. 2012; Sérgio et al. 2014). Soil microbes play an essential role in the environment by contributing to the release of key nutrients from primary minerals by the process of bioleaching not only for their own nutrition but also for the plants. Microorganism are the important component of the soil, as they-maintain the fertility of soil and decline in these micro flora due to soil contamination may decrease the fertility (Alqarawi et al. 2014). The availability of mineral nutrients in soil is essential for the normal growth and development of the plants. Farmers used fertilizers to boost the soil fertility and enhance crop productivity. Conversely, excess use of fertilizers and cropping decreases the arability of land. The people are spending huge money to restore soil fertility aiming to improve crop productivity and environmental security (Leifeld 2012). Beside this, low availability of mineral elements may be mitigated by the different mineral solublizing microorganism. For the successful use of microorganism there must be a good understanding of the all components such as soil, plant and microorganism (Antoun 2012). The coordinated interaction of microorganism with the soil, solubilizes the fixed salts in to available/useful form, which in turn influences the plant growth. This coordinated interaction between the soil and microorganisms occurs in rhizospere (the area around the plant roots) which has strongly adhering soil particles and this zone also has maximum interaction of soil and root (Antoun 2012). Around the rhizosphere

G. Bashri (✉) • A. Patel • R. Singh • P. Parihar • S.M. Prasad
Ranjan Plant Physiology and Biochemistry Laboratory, Department of Botany, University of Allahabad, Allahabad, India
e-mail: gausiya.bashri@gmail.com

J.K. Patra et al. (eds.), *Microbial Biotechnology*,
https://doi.org/10.1007/978-981-10-6847-8_12

plant exudates the organic acids therefore this zone has the maximum microbial activities, termed as Rhizospheric effect by Hiltner (Flaishman et al. 1996). Soil microorganisms balance the quantities of nutrients required for optimum plant growth and development for crop production (Zaidi et al. 2010; Ahemad and Kibret 2014). Among the different nutrients, phosphorus (P), sulphur (S) and potassium (K) are the important minerals for the plants growth and development. Thus, in this chapter we have focussed on the solubilization of these minerals and its mechanism, and importance in plant productivity.

12.2 Phosphate Solubilization by Microorganism

Phosphorus is an essential micronutrient, after nitrogen, for plants growth. The sources of phosphorus in soil are phosphate fertilizers because unlike nitrogen, an atmospheric source does not sufficiently contribute to the phosphorus content in soil (Ezawa et al. 2002). In plants, phosphorus is responsible for carrying out basic metabolic events like photosynthesis and respiration, cell signalling, synthesis of macromolecules (DNA, RNA and protein) (Khan et al. 2010) and in legumes it mediates the atmospheric nitrogen-fixation (Saber et al. 2005) and also involved in root development, flower and seed formation and crop maturity and production. In soil, phosphorus exists in organic as well as in inorganic forms. Among both forms, recurrent application of chemical fertilizers is the chief reservoir of inorganic phosphorus which is insoluble precipitate form and therefore unavailable for uptake in plants (Rengel and Marschner 2005). Only 0.1% of the total phosphorus exists in soluble forms and are easily available for uptake process in plants (Zhou et al. 1992) and remaining exists in insoluble forms and unavailable for plants (Vassileva et al. 2000). In a study it was reported that only 1 μmol L^{-1} phosphorus is available in soil for plants whereas plants require minimum of 30 μmol L^{-1} phosphorus for maximum production. Reason behind this unavailability of phosphorus in soils is improper phosphorus fixation which is also considered as limiting factor in various agricultural and horticultural practices (Daniels et al. 2009). The P-fixation involves conversion of available phosphate into soil solid phase which is correlated with adsorption of phosphorus on soil minerals or precipitated in soils by reaction with free Al^{+3} and Fe^{+3} in the soil solutions (Sharma et al. 2013). Therefore, to full-fill the phosphorus requirement of plants, large amount of phosphate fertilizers is applied, despite this only small fraction of phosphorus is still available for plants by quick transfer into immobile pools, hence continuous application of fertilizer is necessary but on the other side it imposes adverse environmental impacts on soil (Tilman et al. 2001). It is reported that the deficiency of phosphorus in soils is mainly due to phosphorus-fixation, degradation of phosphate rock (source of phosphate fertilizers) which is a non-renewable resource and may be depleted in near future (Cordell et al. 2009) which would make phosphorus unavailable to the plants. Intensive cropping pattern during green revolution has also resulted in widespread deficiency of phosphorus. Therefore, alternative strategies are applied in mineral deficient soil that make insoluble phosphorus derived compounds into soluble forms that quickly taken up by plants and among them phosphate solubilizing bacteria (PSB) are of great importance.

Several reports have described that different bacterial species or microorganisms participate in soil-phosphorus cycle by excreting organic acids which intermediates the conversion of insoluble forms, such as tri-calcium phosphate, di-calcium phosphate, hydroxyapatite, and rock phosphate into soluble forms through solubilization and mineralization and release phosphorus in soil in form of PO_4^{-3} (Hilda and Fraga 1999; He et al. 2002). By this process the bacteria also get benefitted because sugars and organic acids which are necessary for bacterial growth in turn are secreted by plants (Khan et al. 2010). Hence, these phosphate solubilizing bacteria are considered as bio fertilizers and also as phyto-remediant in heavy metal contaminated soil (Ahemad 2015). According to He et al. (1997), some bacteria such as *Pseudomonas* and *Bacillus* (Illmer and Schinner 1992) and fungi like *Aspergillus* and *Penicillium* (Wakelin et al. 2004), actinomycetes and even various algae (cyanobcteria) have the capability of phosphate-solubilization. As compared to fungi, bacteria contribute 1–50% of the total population in phosphate solubilizing microorganisms and among various forms of bacteria, cocci, spiral, bacilli shapes are frequently present in soil while spirilli are rare (Baudoin et al. 2002). Besides, *Pseudomonas* and *Bacillus*, several other PSBs are reported in soil to solubilize P like *Rhodococcus*, *Arthrobacter, Serratia, Chryseobacterium, Gordonia, Phyllobacterium, Delftia* sp. (Wani et al. 2005; Chen et al. 2006), *Azotobacter* (Kumar et al. 2001), *Enterobacter, Pantoea*, and *Klebsiella* (Chung et al. 2005), *Vibrio proteolyticus*, *Xanthobacter agilis* (Vazquez et al. 2000). Furthermore, nitrogen fixing bacteria like *Rhizobium* sp. also showed phosphate solubilization activity that might be beneficial for mitigating P-deficiency of soils by mobilizing inorganic and organic phosphorus (Zaidi et al. 2009).

12.2.1 Mechanism of Phosphate Solubilization

The P-cycle in soil encompasses three components; first is dissolution-precipitation, second is sorption-desorption and last one is mineralization-immobilization, all are mediated by phosphate solubilizing bacteria (PSB). According to McGill and Cole (1981) mechanism of P-solubilization involves three steps: release of mineral dissolving compounds, secretion of extracellular enzymes and release of phosphorus during substrate degradation. During the process of solubilization of insoluble phosphorous, various PSB release organic acids and some enzymes like phosphatase and phytase. Organic and inorganic acids secreted by PSB actively participates in inorganic P-solubilization in which, cations (Al, Fe, Ca) that chelate the phosphorus are captured by –OH and –COOH groups of acids and release P in soluble form in soil system, as a result pH of basic soils decreases (Kpomblekou and Tabatabai 1994), and basic microbial metabolism (respiration or fermentation) are the source of organic acids (Trolove et al. 2003). The action of some acid like gluconic acid, citric acid, oxalic acid and lactic acid easily transform the complex phosphorus compounds like calcium phosphate into easily available forms for plants like mono or dibasic phosphate (Panhwar et al. 2013). Production of H_2S is another strategy which mediates the P-solubilization attributed to release of phosphate from ferric phosphate to yield ferrous sulphate (Swaby and Sperber 1959). *Pseudomonas* doesn't secrete organic acids; hence the P-solubilization is mediated by release of protons by respiration or NH_4^+ assimilation. In this mechanism pumping of protons is considered as major factor mediating the phosphate solubilization and direct role of acids have been implemented (Krishnaraj et al. 1998) and are sole mechanism to promote phosphate solubilisation.

Mineralization of organic phosphorus is strictly mediated by enzyme mainly phosphatases secreted by PSB. Phosphatases may be alkaline or acidic and hydrolyze the soil organic P or split P from organic residues (Nannipieri et al. 2011; Hilda and Fraga 2000). Rodriguez et al. (2006) reported another enzyme named phosphonatases and C–P lyases that cleaves the bond between C-P of organophosphonates.

PSB also secrete siderophores (Greek: "iron carrier") which are small, high-affinity iron-chelating compounds. Siderophores are amongst the strongest soluble Fe^{3+} binding agents known. On an average, around 500 known siderophores are reported that are used by plants as well as micro-organisms (Crowley 2007). Although, siderophores are released by PSM but its application in phosphate solubilization has not been widely known (Hamdali et al. 2008) but their role in phytoremediation and heavy metal chelation of metal such as Mg, Mn, Cr, Cd, Zn, Ni, As and Pb including plutonium chelation has been studied (Schalk et al. 2011).

12.3 Sulphur Solubilization by Microorganism

Sulphur is the fourth major plant nutrient after N, P and K, and is also one of the sixteen nutrient elements that are necessary for the growth and development of plants, especially in crop productivity (Vidyalakshmi and Sridar 2007). In recent decades, due to the reductions in sulphur contributions from atmospheric depositions, a negative sulphur balance in arable soils has occur, increasing dependency on the soil for the requirement of sulphur (Kertesz and Mirleau 2004) which is needed for the synthesis of proteins and a number of essential vitamins and cofactors. So in order to ease this deficiency, sulphur fertilizers are regularly added to soils, generally in a reduced form, such as elemental sulfur. S which is present in reduced form in the fertilizers must be oxidized by bacteria to sulphate (Grayston and Germida 1991; Scherer 2001). From the perspective of plants, inorganic sulphate is the most important form of sulphur, as it is the initial point for cysteine biosynthesis. Though, inorganic sulphate present in soils forms a very small part of the sulphur and, as a result, sulphur deficiency symptoms are now commonly encountered in crop plants (Schnug and Haneklaus 1993).

Generally, S in the form of sulphate (SO_4) is taken up by plant roots and SO_4 go through a chain of transformations prior to its amalgamation into the original compounds (Katyal et al. 1997). Behind this scenario, the key driving force is the microbial biomass, present in the soil. In addition to this, these microbes are also contributing in H_2S (toxic) removal from the environment. Most of the sulphur oxidizing bacteria (SOB) belong to the *Achromatium, Desulfuromonas, Thiobacillus, Thiomicrospira* and *Thiothrix* genera (Das et al. 1996). The sulphur oxidizing microorganisms are mainly the gram negative bacteria and classified as species of *Thiosphaera, Thiomicrospira and Thiobacillus* but heterotrophs, such as some species of *Alcaligens, Paracoccus, Pseudomonas and Xanthobacter* could exhibit chemolithotrophic growth on inorganic sulphur compounds (Kuenen and Beudeker 1982). These chemolithotrophs bears two types of metabolism (i) one can grow only when supplied with oxidizable sulphur compound called obligate chemolithotrophs (ii) and second one also can use the chemolithoautotrophic mode of growth called heterotrophs. The majority of heterotrophic bacteria belong to the genera *Pseudomonas* (Sorokin et al. 1999), *Escherichia coli* (Starkey 1935), *Xanthobacter* (Cho et al. 1992) that participate in sulphur oxidation. The obligate chemolithotrophs are *Thiobacillus neapolitanus, Thiobacillus thioparus, Thiobacillus thiooxidans* (extreme acidophile), *Thiobacillus denitrificans* (facultative denitrifier), *Thiobacillus halophilus* (halophile), *Thiobacillus ferrooxidans* (acidophilic ferrous iron-oxidizer) and some species of *Thiomicrospira*. The heterotrophs are *Thiobacillus novellus, Thiobacillus aquaesulis* (moderate thermophile), *Thiobacillus acidophilus* (acidophile), *Paracoccus denitrificans, Thiobacillus intermedius, Paracoccus versutus, Thiosphaera pantotroph, Xanthobacter tagetidis* and *Thiomicrospira thyasirae.*

12.3.1 Mechanism of Sulphur Solubilization

Sulphur may come from the atmosphere, originally bound sulphur and from the geochemical weathering of soil minerals. This elemental S must transform into SO_4 to be easily available for plant roots. This process of transformation occurs in different steps that may complete in 3–4 weeks (Tandon 1989). The sulphur transformations result from microbial activity that includes the mineralization, immobilization, oxidation and reduction processes. The sulphur (reduced) compounds formed as a result of assimilatory sulphate reduction are the cellular constituents and hence are protected within the cells against reaction with oxygen molecule. The reduction of S-O to H-S is catalyzed by anaerobic sulphate-reducing bacteria. In the magmatic earth crust, S often present in reduced form i.e. sulphide of metals. Oxidation of the magmatic metal sulphides is always facilitated by *Thiobacillus ferrooxidans*, from where it reaches to the crust and finally settled in ore deposits of the ocean. In submerged condition, SO_4 is reduced to H_2S mediated by *Desulfovibrio desulfuricans.* A secondarily formed H_2S when enters into the oxidative zone, its oxidation is carried out by *Thiobacillus thioparus*. The oxidation of H_2S into S is facilitated by the activity of *Thiobacillus thioparus* and *Thiobacillus thiooxidans* in the soil.

12.4 Potassium-Solubilizing Microorganism

Potassium is very advantageous mineral nutrient for agronomy, but is not available to plants for implementation in growth processes because solubilization of potassium in soil is very slow process therefore potassium is directly added in soil in form of fertilizer which on one hand increases the yield of crops and on the other hand its excessive use causes a serious environmental threat, thus a group of microorganism known as Potassium-solubilizing microorganism were identified, which includes bacteria, fungi, and rhizobacteria which rapidly solubilize potassium through their metabolic process and can sustain our prevailing reservoirs and shrink environmental threat caused by excessive use of chemical fertilizers. Potassium-solubilizing or potassium-dissolving bacteria include the class of microorganism which solubilize potassium and make available for plants. Potassium-solubilizing microorganisms (KSM) are extensively used as bio fertilizers, in area where potassium is lacking in soil, or poorly available (Xie 1998). Muentz (1890) firstly described several growth promoting rhizobacteria such as *Bacillus extorquens*, *Aspergillus niger* and *Clostridium pasteurianum* which are now a day's used as a bio fertilizer for sustainable cropping and agricultural practices. In addition, they increase the crop production and resistance against plant diseases and reduce farmers dependency on chemical fertilizers (Reitmeir 1951; Meena et al. 2015a). Potassium-solubilizing bacteria (KSB) secrets numerous organic acid which draws out potassium, silicon and aluminium from insoluble aluminosilicate minerals such as orthoclase, micas, illite. These organic acids interact with the mineral surface in marine environment and dissolves silicon from silicate minerals. Barker et al. (1998) found that this microbial interaction with minerals and their dissolution of organic carbon raise carbonic acid concentration at surfaces of minerals in soils thereby enhancing mineral weathering rates by endorsement of proton dissolution mechanism. Other studies on mineral weathering, reported that some strains of KSB like *Bacillus* species efficiently solubilise K in the culture medium containing K minerals (Vandevivere et al. 1994). *Bacillus* species like *B. mucilaginosus* is known to optimize the releases of K and SiO_2 from primary minerals by secreting organic acids. Potassium solubilizing bacteria (KSB) also plays major role in reducing nutrient deficiency in soil. According to Hutchens, outer envelope of bacterial cell wall consist of mucilaginous sheath i.e. exopolysachharide that chelates the silicon as well as silicate minerals associated with release of K from the complex structures. The solubilization of different crystalline biotite, mica, vermiculite and certain rocks to amorphous form is mediated by production of organic synthesis during the process of microbial metabolism (Weed et al. 1969). Earlier it was reported that many microbial species participate in dissolution of alumina silicate via synthesis of extracellular enzymes, various organic ligands like exudates, metabolic by-products and chelates (Grandstaff 1986; Argelis et al. 1993; Welch and Ullman 1993). These studies reported that among various fungal species, two fungal species, viz., *Cladosporium cladosporioides* and *Penicillium frequentans* have the capability to solubilize the complex sand stone, granite and limestone through production of various acids like oxalic, citric and gluconic acids in broth media during metabolic activities. Similar to the fungal species, some bacteria like *Bacillus megaterium* and *Arthrobacter* spp. are also involved in production of organic acids and mono-hydroxamate siderophore (Chen et al. 2007), and play a vital role in solubilization of elements such as K, Si and Fe from acid-leached soil, muscovite and biotite (Hutchens et al. 2003). During this process the microbial species also get benefitted showing the mutualistic association between microbiota and mineral processes. With the help of these interactions various technologies like biomineralization, bioremediation and biohydrometallurgy (Reitmeir 1951; Ullaman et al. 1996; Requena et al. 1997; Srinivasarao and Takkar 1997; Rawlings 2002; Venkateswarlu et al. 2012; Meena 2015b) have been developed to increase the soil fertility. Organic acids produced by microbial species mediate several roles in soil like increasing the soluble form of minerals in soil, biodegradation of metals and root nutrient recovery (Jones et al. 2003).

In K deficient soils, the K requirement of plants are fulfilled by weathering of illite and feldspar minerals by the production of capsular polysaccharides (Sheng et al. 2006). Apart from this, the biofilm synthesized by bacterial strains, on the mineral surface of rhizospheric soil also participates in mobilization of K (Balogh-Brunstad et al. 2008). For higher productivity and maintenance of soil health, the nutrient requirement of soil is fulfilled by use of fertilizers and mycorrhiza (ectomycorrhiza and arbascular) implanted in biofilms. Hence the microbial biofilms stimulates the solubilization of minerals and make them available to plants. The ectomycorrhizal (rock decaying fungi) species produce organic anions (low molecular weight anions) that makes tiny tunnels on hyphal tips that contains solubilize minerals, easily available for plants (Van Schöll et al. 2008).

12.5 Effect of Mineral Solubilizing Microorganism on Plants

There are many reports which showed that microorganisms which live in soil flora have the ability to promote plant growth (Compant et al. 2010; George et al. 2012; Beneduzi et al. 2013). These microorganisms when added as an inoculant reduced the use of chemical fertilizers and thus considered as important in the context of sustainable agriculture and environmental protection (Lucy et al. 2004; Vale et al. 2010). The addition of microorganisms in soil as inoculum solubilized the nutrient minerals for the use of plants. Beside this, they also support the plant growth through production of hormones such as auxins, gibberellins and zeatin (Cassan et al. 2009), biological nitrogen fixation (Ashraf et al. 2011; Verma et al. 2013) or by means of biological control of pathogens (Wang et al. 2009) (Fig. 12.1). Various minerals solubilizing microorganism has been found to be beneficial in many studies for the growth and development of the plants and has been listed in the Table 12.1 and few have been discussed in detail in following section.

Among different minerals utilization of phosphate solubilizing microorganisms are being used as biofertilizers since 1950's (Krasilinikov 1957). Application of plant growth promoting rhizobacteria (PGPR) together with phosphate solubilizing

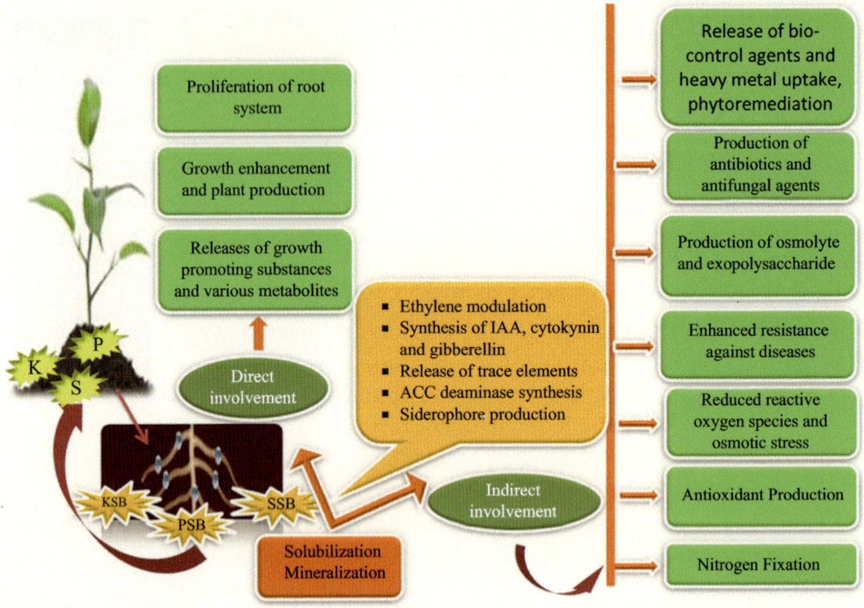

Fig. 12.1 Schematic representation of the direct and indirect effect of mineral solubilizing microorganisms on the growth and development of the plants

microorganisms can reduce application of P fertilizer upto 50% (Jilani et al. 2007). Phosphate solubilizing microorganisms such as *Pseudomonas, Bacillus, Rhizobium, Micrococcus, Flavobacterium, Achromobacter, Erwinia and Agrobacterium* have been reported to increase solubilization of fixed phosphorus (Rodriguez and Fraga 1999). Such as *Pseudomonas* sp. enhanced the P availability by solubilizing the tricalcium phosphate (Dey et al. 2004). *Pseudomonas letiola* increased the mobilization of P and growth of Ligol (young apple trees) by significantly increasing the total shoot length and solubilized insoluble P compounds (Kurek et al. 2013). Verma et al. (2014) showed that *Pseudomonas* and *Trichoderma* enhance the yield and indole-3-acetic acid production, phosphate solubilization and antagonistic activities against *Fusarium oxysporum, Rhizoctonia solani* and *Rhodococcus* sp. Similarly, when EC35, *Pseudomonas* sp. EAV and *Arthrobacter nicotinovorans* EAPAA when inoculated with *Zea mays* and grown in P deficient soils amended with tricalcium phosphate enhanced the plant growth (Pereira and Castro 2014). Therefore, PSM such as *Bacillus megaterium*, *Bacillus circulans*, *Bacillus subtilis* and *Pseudomonas straita* proved to be effective biofertilizers. In addition to this, these microorganism also increasethe growth under various stress conditions, such as *Pseudomonas putida* (Pandey *et al.,* 2006), *Pseudomonas corrugate* (Trivedi and Sa 2008), *Mycobacterium* sp. (Egamberdiyeva and Hoflich (2003) have ability to tolerate the cold conditions, *Pseudomonas* sp. (Shimaila *et al.* 2014) and *Bacillus* and *Hallobacillus* (Dhanushkodi et al. 2013) have capability to enhance the growth of plant under salt stress and *Pseudomonas* sp. (Sandhya et al. 2010), *Arthrobacter* sp., *Bacillus* sp. (Banerjee et al. 2010) showed tolerant nature under drought stress.

Table 12.1 Role of several mineral solubilizing microorganism on plants

Mineral Solubilized	Bacterial inoculant	Plant Species	Role under stress condition	Reference
Phosphate	*Bacillus lentimorbus*	Spinach Carrots, Lettuce	Increased the antioxidant capacity of the edible parts as well as increasing growth and especially improves the nutrient content of crops.	Nautiyal et al. (2008)
	Bacillus megaterium *Pantoea agglomerans*	Banana Maize	Increased the ability of the root to absorb water under saline conditions a plant can survive in hyper saline stress condition	Gond et al. (2015)
	Bacillus cereus	Sword bean	Resistance against plant diseases through various mechanisms, and form endospores, in stressed environmental conditions, thus effective formulation of bio-fertilizer	Garcia et al. (2011), Francis et al. (2010)
	Pseudomonas fluorescens	Beat root, Wheat	High solubilisation of phosphate and synthesis of gibberellic acid which promote growth.	Pandey et al. (2006)
	Pseudomonas Striata and Enterobacter	Chick pea	Powerful phosphate solubilizer	Subbarao (1988)
	Pseudomonas striata	Soybean	Enhanced plant growth under salt stress	Shimaila et al. (2014)
	Bacillus and Hallobacillus	Cashew kernal	Protect plants under salinity stress	Dhanushkodi et al. (2013).
	Rhizobium and Enterobacter	Chick pea	Nitrogen fixation, phosphate and potassiumsolubilization and nitrification, denitrification, production ofplant growth promoters such as indole acetic acid, gibberellic acid, ethylene and cytokinins. Siderophore production	Parul and Dharmendra (2014), Kloepper et al. (2003) Whitelaw (2000), Yahya and Azawi (1998)
	Bacillus megaterium	Maize	Broad spectrum biofertilize, provide resistance against disease, promote growth in plant	Gupta (2004)
	Pseudomonas argeniosa	Mung beans	Improve the growth of plant under drought condition.	Sarma and Saikia (2014)

(continued)

Table 12.1 (continued)

Mineral Solubilized	Bacterial inoculant	Plant Species	Role under stress condition	Reference
	Bacillus subtilis	Cotton, Sunflower, Chick pea	Increased growth and yield of cotton and sunflower	Qureshi et al. (2012)
	Bacillus polymyxa, Bacillus sircalmous, Bacillus circulans and Bacillus species	Cotton	Mobilization of phosphate, better nutrient availability and biosynthesis of growth hormones, promote plant growth	Zaidi et al. (2004)
	Kocuria turfanesis	Lettuce	A phosphate solubilizer, an IAA producer, and a siderophore producer. Improve plant growth under saline condition	Goswami et al. (2014)
	Acetobacter diazotropicus	Rice	Cytokinin synthesis and nitrogen fixation	Kumar et al. (1999), Rodriguez and Fraga (1999)
	Agrobacterium and Alcaligenes	Pea, Turnip, Brassica, Cauliflower	Act as a bio-fertilizer and major role in crop production, releases growth promoting hormone, synthesis of IAA, and major production of citric, succinic and gluconic acid which led to mobilization of phosphate.	Maliha et al. (2004)
	Bacillus species	Barley	Improves the growth and provide resistance against cadmium metal stress, remove cadmium from environment by binding mechanism	Phischik et al. (2002)
	Azospirillium brasilense	Jojoba plant	Improve the salt tolerance; reduce the undesirable effects of saline conditions. The bacteria attenuated salinity's effect on the rooting ability of the jojoba plant and developed tolerance to salt stress.	Gonzalez et al. (2015)
	Gluconacetobacter sp.	Sugarcane	Nitrogen fixation	Chung et al. (2005), Kim et al. (2003)
	Sphingomonas sp.	Tomato	Promote gibberellic acid synthesis and significant increase in growth characteristics	Khan et al. (2014)

	Acinetobacter, Agrobacterium	Cereal crops	Resistance against diseases and antibiotics	Rodriguez and Fraga (1999)
	Mycobacterium	Wheat, Cereal crop	Resistance under cold stress in plant	Egamberdiyeva and Hoflich (2003)
Sulphur	*Thiobacilli*	Sewage sludge	Reduction in pH from 7 to 2. Efficiently reduced the Cr content by 52–58% as well as Cu content by 80%	Chan et al. (2003)
	Thiobacillus	*Raphanus sativus*	Enhancement in the growth of the studied plant as well as enhanced oil production.	Khatibi (2011)
	Thiobacillus	*Brassica napus*	Enhanced the availability of sulphur to the plants, thereby enhancing the growth as well as the oil content.	Salimpour et al. (2010)
	Thiobacillus thiooxidans Thiobacillus thioparu	Sewage sludge	Bioleaching for removal of heavy metals from the sludge.	Jain and Tyagi (1992)
Potassium	*Bacillus pumilus*	Rice (*Oryza sativa*)	Enhancement in glycine betaine-like quaternary compounds and higher shoot biomass	Jha et al. (2010)
	Bacillus megaterium	Maize (*Zea mays* L.)	Enhancement in higher root hydraulic conductance (L) in turn enhancing higher plasma membrane type two (PIP2) aquaporin amount in roots and ZmPIP1;1 protein amount	Marulanda et al. (2010)
	Bacillus subtilis	*Arabidopsis thaliana*	Tissue-specific regulation of high-affinity K^+ transporter (HKT1), thereby lowering Na accumulation	Zhang et al. (2008)
	Bacillus insolitus, Bacillus sp.	Wheat (*Triticum aestivum*)	Enhancement in exo-polysaccharides (EPS) production, dry matter yield and the uptake of K, and Ca, restriction in Na uptake, enhancement in water-insoluble saccharides in the RAS (root adhering soil)	Ashraf et al. (2004)
	Pseudomonas spp.	Maize (*Zea mays* L. cv. Kaveri)	Enhances plant biomass, root growth and mycorrhizal colonization	Sandhya et al. (2010)
	Pseudomonas spp.	Asparagus (*Asparagus officinalis* L.)	Declined the impact of water stress, transplant shock, and disease pressure on the studied plant	Liddycoat et al. (2009)

(continued)

Table 12.1 (continued)

Mineral Solubilized	Bacterial inoculant	Plant Species	Role under stress condition	Reference
	Bacillus	Lettuce (*Lactuca sativa* L.)	Inoculation by bacteria enhanced the shoot cytokinins that enhanced the stomatal conductance and shoot growth	Arkipova et al. (2007)
	Bacillus subtilis	*Arabidopsis*	Enhancement in the transcript level of phosphoethanolamine N-methyltransferase (PEAMT), key enzyme of choline and also increase in glycine betaine synthesis by two and fivefold respectively.	Zhang et al. (2010)
	Bacillus sp.	Pepper	Enhancement in indole-acetic acid production and free proline content as well as up-regulation and down-regulation in stress-inducible genes.	Sziderics et al. (2007)
	Bacillus mucilaginosus	*Zea mays* L.	Increase in mineral availability, uptake and plant growth	Han et al. (2006)
	Bacillus sp.	*Zea mays* L.	Enhanced nitrogen content of the plants, thereby increasing the plant height and yield as well.	Adesemoye et al. (2008)
	Bacillus polymyxa	Maize (*Zea mays* L.(cv. Felix)	Induces stimulatory effect on growth of the plant by enhancing the uptake of nitrogen (N), phosphorus (P) and potassium (K).	Egamberdiyeva (2007)
	Bacillus subtilis	Oat (*Avena sativa*)	Restoration in damage caused by heavy metal as well as enhancement in the productivity.	Pishchik et al. (2009)

In nutrient deficient soils, PGPR mediate the plant growth initiation via exploring the root system throughout the growing season (Defreitas and Germida 1992). A variety of functions in plants is promoted by PGPR under various environmental stresses like promotion in root growth pattern, plant growth, mobilization of non-exchangable K into soluble forms. A number of bacterial strains such as *Bacillus*, *Pseudomonas*, *Azospirillum* and *Enterobacter* are associated with PGPR for the inauspicious responses on plant growth (Kloepper et al. 1991; Hoflich et al. 1994). Some studies have demonstrated the positive role of PGPR in single crop under different environmental conditions (Boelens et al. 1993; Javed and Arshad 1997). Besides K-solubilization, the KSM are also involved in production of phytohormones, siderophores and ammonia production which improves the plant growth as well as soil quality. *Bacillus mucilaginosus*, a potassium solubilizing bacteria promote growth and nutrient uptake in plants (Han et al. 2006). Other study also reported that eroded soil with KSB inoculation results in increase yield of wheat (Mikhailouskaya and Tcherhysh 2005). Potash-solubilizing bacteria (*Frateuria aurantia*) belong to the family Pseudomonadaceae, are found in agricultural land of Tamilnadu region and have high mobility of potassium in soil which considerably promotes crop yield. It has been shown that growth and production of eggplant depend on availability of potassium in soil, as plants grown in soil inoculated with *Frateuria aurantia*, showed high yield (Ramarethinam and Chandra 2005). Han et al. (2006) also observed and reported that compared to non- inoculated treatment the dry weight of maize and wheat, inoculated with (potassium dissolving) bacteria *Bacillus mucilaginosus*, *Azotobacter chroococcum* and *Rhizobium* showed an increment of ~23%, due to higher mobilization of potassium from waste mica (which is a reservoir of potassium for plant growth) (Singh et al. 2010). Basak and Biswas (2009) reported that mica with suitable potassium-solubilizing bacteria, in soil has significantly enhanced the potassium uptake. This was demonstrated by field study which clearly showed that application of mica at 50 mg K kg^{-1} soil showed an increase of 36.0% K content over control in Sudan grass. KSM are great inducer of plant growth and development as they secrete plant growth stimulators which boost and promote seedling initiation and promote root growth and also play substantial role in decomposition of organic materials and compost enrichment (Bahadur et al. 2014).

In conclusion, the deficiencies of nutrients can be overcome by the application of chemical fertilizers to increase the yield of the crop plants. But extensive use of these chemical fertilizers declined the arability of the land and may cause damage to the agricultural soil. Microorganisms are the important component of the soil and plays important role in maintaining the fertility of the soil. Addition of these microorganisms serves as alternative method of fertilizer application instead of chemical fertilizers and development of sustainable agriculture. Hence, there is an urgent need to improve better research in this field for developing this technology and to minimise the use of chemical fertilizers and make use of biofertilizers in large scale in agronomic practices to obtain better results.

References

Adesemoye AO, Torbert HA, Kloepper JW (2008) Enhanced plant nutrient use efficiency with PGPR and AMF in an integrated nutrient management system. Can J Microbiol 54:876–886

Ahemad M (2015) Phosphate solubilising bacteria assisted phytoremediation of metalliferous soil: a review. Biotech 5:111–121

Ahemad M, Kibret M (2014) Mechanisms and applications of plant growth promoting rhizobacteria: current perspective. J King Saud University-Sci 26(1):1–20

Alqarawi AA, Abd Allah EF, Hashem A (2014) Alleviation of salt-induced adverse impact via mycorrhizal fungi in *Ephedra aphylla* Forssk. J Plant Interact 9:802–810

Antoun H (2012) Beneficial microorganisms for the sustainable use of phosphates in agriculture. Process Eng 46:62–67

Araújo AS, Cesarz S, Leite LF, Borges CD, Tsai SM, Eisenhauer N (2013) Soil microbial properties and temporal stability in degraded and restored lands of Northeast Brazil. Soil Biol Biochem 66:175–181

Argelis DT, Gonzala DA, Vizcaino C, Gartia MT (1993) Biochemical mechanism of stone alteration carried out by filamentous fungi living in monuments. Biogeo Chem 19:129–147

Arkipova TN, Prinsen E, Veselov SU, Martinenko EV, Melentiev AI, Kudoyarova GR (2007) Cytokinin producing bacteria enhance plant growth in drying soil. Planta and. Soil 292:305–315

Ashraf M, Hasnain S, Berge O, Mahmood T (2004) Inoculating wheat seeds with exopolysaccharide producing bacteria restricts sodium uptake and stimulates plant growth under salt stress. Biol Fert. Soil 40:157–162

Ashraf MA, Rasool M, Mirza MS (2011) Nitrogen fixation and indole acetic acid production potential of bacteria isolated from rhizosphere of sugarcane (*Saccharum officinarum* L.) Adv Biol Res 5(6):348–355

Bahadur I, Meena VS, Kumar S (2014) Importance and application of potassic biofertilizer in Indian agriculture. Int Res J Biol Sci 12:80–85

Balogh-Brunstad Z, Keller CK, Gill RA, Bormann BT, Li CY (2008) The effect of bacteria and fungi on chemical weathering and chemical denudation fluxes in pine growth experiments. Biogeochem 88:153–167

Banerjee S, Palit R, Sengupta C, Standing D (2010) Stress induces phosphate solubilisation by Arthrobacter sp. and bacillus sp. isolated from tomato rhizosphere. Aust J Crop Sci 4:378–383

Barker WW, Welch SA, Chu S, Banfield F (1998) Experimental observations of the effects of bacteria on aluminosilicates weathering. Am Mineral 83:1551–1563

Basak BB, Biswas DR (2009) Influence of potassium solubilizing microorganism (*Bacillus mucilaginosus*) and waste mica on potassium uptake dynamics by sudan grass (*Sorghum vulgare* Pers.) grown under two Alfisols. Plant Soils 317:235–255

Baudoin E, Benizri E, Guckert A (2002) Impact of growth stages on bacterial community structure along maize roots by metabolic and genetic fingerprinting. Appl Soil Ecol 19:135–145

Beneduzi A, Moreira F, Costa PB, Vargas LK, Lisboa BB, Favreto R, Baldani JI, Passaglia LMP (2013) Diversity and plant growth promoting evaluation abilities of bacteria isolated from sugarcane cultivated in the south of Brazil. Appl Soil Ecol 63:94–104

Boelens J, Zoutmann D, Cambell J, Verstraete W (1993) The use of bioluminescence as a reporter to study the adherence of the plant growth promoting *Rhizopseudomonas* 7NSK2 and ANP15 to canola roots. Can J Microbiol 39:329–334

Cassán F, Maiale S, Masciarelli O, Vidal A, Luna V, Ruiz O (2009) Cadaverine production by *Azospirillum brasilense* and its possible role in plant growth promotion and osmotic stress mitigation. Eur J Soil Biol 45:12–19

Chan LC, XY G, Wong JWC (2003) Comparison of bioleaching of heavy metals from sewage sludge using iron- and sulfur-oxidizing bacteria. Adv Environmen Res 7:603–607

Chen YP, Rekha PD, Arunshen AB, Lai WA, Young CC (2006) Phosphate solubilizing bacteria from subtropical soil and their tricalcium phosphate solubilizing abilities. Appl Soil Ecol 34:3341

Chen Z, Cuin TA, Zhou M, Twomey A, Naidu BP, Shabala S (2007) Compatible solute accumulation and stress-mitigating effects in barley genotypes contrasting in their salt tolerance. J Exp Bot 58:4245–4255

Cho KS, Hirai M, Shoda M (1992) Degradation of hydrogen sulfide by *Xanthomonas* sp. strain DY44 isolated from peat. Appl Environ Microb 58(4):1183–1189

Chung H, Park M, Madhaiyan M, Seshadri S, Song J, Cho H, Sa T (2005) Isolation and characterization of phosphate solubilizing bacteria from the rhizosphere of crop plants of Korea. Soil Biol Biochem 37:1970–1974

Compant S, Clément C, Sessitsch A (2010) Plant growth-promoting bacteria in the rhizo and endosphere of plants: their role, colonization, mechanisms involved and prospects for utilization. Soil Biol Biochem 42:669–678

Cordell D, Drangert JO, White S (2009) The story of phosphorus: global food security and food for thought. Glob Environ Change 19:292–305

Crowley DE (2007) Microbial siderophores in the plant rhizosphere. In: Barton LL, Abadia J (eds) Iron nutrition in plants and rhizospheric microorganisms. Springer, Dordrecht, pp 169–198

Daniels C, Michan C, Ramos JL (2009) New molecular tools for enhancing methane production, explaining thermodynamically limited life styles and other important biotechnological issues. Microb Biotechnol 2:533–536

Das SK, Mishra AK, Tindall BJ, Rainey FA, Stackerbrandt E (1996) Oxidation of thiosulfate by a new bacterium, *Bosea thiooxidans* (strain BI-42) gen. Nov., sp. nov.: analysis of phylogeny based on Chemotaxanomy on 16S ribosomal DNA sequencing. Inter J System Bacteriol 64(4):981–987

Defreitas JR, Germida JJ (1992) Growth promotion of winter wheat by fluorescent *Pseudomonas* under field conditions. Soil Biol Biochem 24:1137–1146

Dey R, Pal KK, Bhatt DM, Chauhan SM (2004) Growth promotion and yield enhancement of peanut (*Arachis hypogaea* L.) by application of plant growth-promoting rhizobacteria. Microbiol Res 159:371–394

Dhanushkodi R, Vithal KL, Pranita B, Sajad A, Kannepalli A (2013) Mitigation of salt stress in wheat seedlings by halo-tolerant bacteria isolated from saline habitats. Springer Plus 2:6

Egamberdiyeva D (2007) The effect of plant growth promoting bacteria on growth and nutrient uptake of maize in two different soils. Appl Soil Ecol 36:184–189

Egamberdiyeva D, Hoflich G (2003) Influence of growth promoting bacteria on the growth ofwheat at different soils and temperatures. Soil Biol Biochem 35:973–978

Ezawa T, Smith SE, Smith FA (2002) Phosphate metabolism and transport in AM fungi. Plant Soil 244:221–230

Flaishman MA, Eyal ZA, Zilberstein A, Voisard C, Hass D (1996) Suppression of *Septoria tritci* blotch and leaf rust of wheat by recombinant cyanide producing strains of *Pseudomonas putida*. Mol Plant-Microbe Interact 9:642–645

Francis I, Holsters M, Vereecke D (2010) The gram-positive side of plant-microbe interaction. Environ Microbial 12:1–12

George P, Gupta A, Gopal M, Thomas L, Thomas GV (2012) Multifarious beneficial traits and plant growth promoting potential of *Serratia marcescens* KiSII and *Enterobacter* sp. RNF 267 isolated from the rhizosphere of coconut palms (*Cocos nucifera* L.) World J Microb Biot 29(1):109–117

Gond SK, Bergen MS, Torres MS, White JF, Kharwar RN (2015) Effect of bacterial endophyte on expression of defense genes in Indian popcorn against *Fusarium moniliforme*. Symbiosis 66:133–140

Gonzalez AJ, Larraburu EE, Llorente BE (2015) *Azospirillum brasilense* increased salt tolerance of jojoba duringin vitro rooting. Ind Crop Prod 76:41–48

Goswami D, Pithwa S, Dhandhukia P, Thakker JN (2014) Delineating *Kocuria turfanensis* 2M4 as a credible PGPR: a novel IAA producing bacteria isolated from saline desert. J Plant Interact 9:566–576

Grandstaff DE (1986) The dissolution rate of Forsteritic olivine from Hawaiian beach sand. In: Colman SM, Dethier DP (eds) Rates of chemical weathering of rocks and minerals. Academic Press, New York, pp 41–59

Grayston SJ, Germida JJ (1991) Sulfur oxidizing bacteria as plant growth promoting rhizobacteria for *Canola*. Can J Microb 37:521–529

Gupta AK (2004) The complete technology book on biofertilizers and organic farming. National Institute of Industrial Research Press, India

Hamdali H, Bouizgarne B, Hafidi M, Lebrihi A, Virolle MJ, Ouhdouch Y (2008) Screening for rock phosphate solubilizing Actinomycetes from Moroccan phosphate mines. Appl Soil Ecol 38:12–19

Han HS, Supanjani P, Lee KD (2006) Effect of co-inoculation with phosphate and potassium solubilizing bacteria an mineral uptake and growth of pepper and cucumber. Plant Soil Environ 52(3):130–136

He ZL, Bian W, Zhu J (2002) Screening and identification of microorganisms capable of utilizing phosphate adsorbed by goethite. Comm Soil Sci Plant Anal 33:647–663

He ZL, Wu J, O'Donnell AG, Syers JK (1997) Seasonal responses in microbial biomass carbon, phosphorus and sulphur in soils under pasture. Biol Fertil Soils 24:421–428

Hilda R, Fraga R (1999) Phosphate solubilizing bacteria and their role in plant growth promotion. Biotechnol Adv 17:319–359

Hilda R, Fraga R (2000) Phosphate solubilizing bacteria and their role in plant growth promotion. Biotech Adv 17:319–359

Hoflich G, Wiehe W, Khn G (1994) Plant growth stimulation with symbiotic and associative rhizosphere microorganisms. Experientia 50:897–905

Hutchens SE, Valsami JE, Eldowney MS (2003) The role of heterotrophic bacteria in feldspar dissolution. Min Mag 67:1151–1170

Illmer PA, Schinner F (1992) Solubilization of inorganic phosphates by microorganisms isolated from forest soil. Soil Biol Biochem 24:389–395

Jain DK, Tyagi RD (1992) Leaching of heavy metals from anaerobic sewage sludge by sulfur-oxidizing bacteria. Enzym Microb Technol 14(5):376–383

Javed NP, Arshad M (1997) Growth promotion of two wheat cultivars by plant growth promoting rhizobacteria. Pak J Bot 29:243–248

Jha Y, Subramanian RB, Patel S (2010) Combination of endophytic and rhizospheric plant growth promoting rhizobacteria in *Oryza sativa* shows higher accumulation of osmoprotectant against saline stress. Acta Physiol Plant. https://doi.org/10.1007/s1173801006049

Jilani G, Akram A, Ali RM, Hafeez FY, Shamsi IH, Chaudhary AN, Chaudhary AG (2007) Enhancing crop growth, nutrients availibility, economics and beneficial rhizosphere microflora through organic and biofertilizers. Ann Microbiol 7:177–183

Jones DL, Shannon DV, Murphy D, Farrar J (2003) Role of dissolved organic nitrogen (DON) in soil cycling in grassland soils. Soil Biol Biochem 36:749–756

Katyal JL, Sharma KL, Srinivas K (1997) ISI/FAI/IFA symposium on sulphur in balanced fertilization. New Delhi, India, Proc, pp 2/1–2/11

Kertesz MA, Mirleau K (2004) The role of soil microbes in plant sulphur nutrition. J Exp Bot 55:1–7

Khan AL, Waqas M, Kang SM (2014) Bacterial endophyte *Sphingomonas* sp. LK11 produces gibberellins and IAA and promotes tomato plant growth. J Microbiol 52:689–695

Khan MS, Zaidi A, Ahemad M, Oves M, Wani PA (2010) Plant growth promotion by phosphate solubilizing fungi – current perspective. Arch Agron Soil Sci 56:73–98

Khatibi R (2011) Using sulfur oxidizing bacteria and P solubilizing for enhancing phosphorous availability to *Raphanus sativus*. African J Plant Sci 5(8):430–435

Kim CH, Han SH, Kim KY, Cho BH, Kim YH, Koo BS, Kim YC (2003) Cloning and expression of pyrroloquinoline quinine (PQQ) genes from a phosphate solubilizing bacterium *Enterobacter intermedium*. Curr Microbiol 47:457–461

Kloepper JW (2003) A review of mechanisms for plant growth promotion by PGPR. In: Reddy MS, Anandaraj M, Eapen SJ, Sarma YR, Kloepper JW (eds) Abstracts and short papers. Mechanisms and applications of plant growth promoting rhizobacteria: current perspective 176th international PGPR workshop, 5–10 October 2003. Indian Institute of Spices Research, Calicut, pp 81–92

Kloepper JW, Zablowicz RM, Tipping B, Lifshitz R (1991) Plant growth mediated by bacterial rhizosphere colonizers. In: Keister DL, Gregan B (eds) The rhizosphere and plant growth, vol 14. BARC symposium, pp 315–326

Kpomblekou K, Tabatabai MA (1994) Effect of organic acids on release of phosphorus from phosphate rocks. Soil Sci 158:442–453

Krasilinikov NA (1957) On the role of soil micro-organism in plant nutrition. Microbiologia 26:659–672

Krishnaraj PU, Khanuja SPS, Sadashivam KV (1998) Mineral phosphate solubilization (MPS) and mps genes -components in eco-friendly P fertilization. Abstracts of Indo US Workshop on Application of Biotechnology for Clean Environment and Energy, National Institute of Advanced Studies, Bangalore, p 27

Kuenen JG, Beudeker RF (1982) Microbiology of *Thiobacilli* and other sulphur oxidising autotrophs mixotrophs and heterotrophs. In: Post Gate JP, Kelly DP (eds) Sulphur bacteria. University Press, Cambridge, pp 473–497

Kumar V, Behl RK, Narula N (2001) Establishment of phosphate- solubilizing strains of *Azotobacter chroococcum* in the rhizosphere and their effect on wheat cultivars under greenhouse conditions. Microbiol Res 156:87–93

Kumar V, Narula N (1999) Solubilization of inorganic phosphates and growth mergence of wheat as affected by *Azotobacter chroococcum* mutants. Biol Fertil Soils 28:301–305

Kurek E, Ozimek E, Sobiczewski P, Słomka A, Jaroszuk-Ściseł J (2013) Effect of Pseudomonas Luteola on mobilization of phosphorus and growth of young apple trees (Ligol)—pot experiment. Sci Hortic 164:270–276

Leifeld J (2012) How sustainable is organic farming? Agric Ecosyst Environ 150:121–122

Liddycoat SM, Greenberg BM, Wolyn DJ (2009) The effect of plant growth promoting rhizobacteria an asparagus seedling and germinating seeds subjected to water stress under greenhouse conditions. Can J Microbiol 55:388–394

Lucy M, Reed E, Glick BR (2004) Application of free living plant growth- promoting rhizobacteria. Antonie van Leewenhoek 86:1–25

Maliha R, Samina K, Najma A, Sadia A, Farooq L (2004) Organic acids production and phosphate solubilization by phosphate solubilizing microorganisms under in vitro conditions. Pak J BiolSci 7:187–196

Marulanda A, Azcon R, Chaumont F, Ruiz-Lozano JM, Aroca R (2010) Regulation of plasma membrane aquaporins by inoculation with *Bacillus megaterium* strain in maize (*Zea mays* L.) plants under unstressed and salt stressed conditions. Planta 232:533–543

McGill WB, Cole CV (1981) Comparative aspects of cycling of organic C, N, S andP through soil organic matter. Geoderma 26:267–268

Meena RK, Singh RK, Singh NP, Meena SK, Meena VS (2015a) Isolation of low temperature surviving plant growth-promoting rhizobacteria (PGPR) from pea (*Pisum sativum* L.) and documentation of their plant growth promoting traits. Biocatal Agric Biotechnol. https://doi.org/10.1016/j.bcab.2015.08.006

Meena VS, Maurya BR, Verma JP, Aeron A, Kumar A, Kim K, Bajpai VK (2015b) Potassium solubilizing rhizobacteria (KSR): isolation, identification, and K-release dynamics from waste mica. Ecol Eng 81:340–347

Mikhailouskaya N, Tcherhysh A (2005) K-mobilizing bacteria and their effect on wheat yield. Latnian. J Agron 8:154–157

Muentz A (1890) Surla decomposition desrochesetla formation de la terre arable. C R Acad Sci 110:1370–1372

Nannipieri P, Giagnoni L, Landi L, Renella G (2011) Role of phosphatase enzymes in soil. In: Bunemann E, Oberson A, Frossard E (eds) Phosphorus in action: biological processes in soil phosphorus cycling. Soil biology, vol 26. Springer, Heidelberg, pp 251–244

Nautiyal CS, Govindarajan R, Lavania M, Pushpangadan P (2008) Novel mechanisms of modulating natural antioxidants in functional foods: involvement of plant growth promoting rhizobacteria. J Agric Food Chem 56:4474–4481

Nunes JS, Araujo ASF, Nunes LAPL, Lima LM, Carneiro RFV, Tsai SM, Salviano AAC (2012) Land degradation on soil microbial biomass and activity in Northeast Brazil. Pedosphere 22:88–95

Pandey A, Trivedi P, Kumar B, Palni LMS (2006) Characterization of a phosphate solubilizing and antagonistic strain of *Pseudomonas putida* isolated from a sub-alpine location in the Indian central Himalaya. Curr Microbiol 53:102–107

Panhwar QA, Jusop S, Naher UA, Othman R, Razi MI (2013) Application of potential phosphate-solubilizing bacteria andorganic acids on phosphate solubilization from phosphate rock in aerobic rice. Sci World J

Parul J, Khichi DS (2014) Phosphate solubilizing microorganism (PSM): an ecofriendly biofertilizer and pollution manager. J Dynamics in Agri Res 1(4):23–28

Pereira SI, Castro PM (2014) Phosphate-solubilizing rhizobacteria enhance *Zea mays* growth in agricultural P-deficient soils. Ecol Eng 73:526–535

Perez-Garcia A, Romero D, de Vicente A (2011) Plant protection and growth stimulation by microorganism: biotechnological applications of bacillus in agriculture. Curr. Open. Biotechnol 22:187–193

Pishchik VN, Provorov NA, Vorobyov NI, Chizevskaya EP, Safronova VI, Tuev AN, Kozhemyakov AP (2009) Interactions between plants and associated bacteria in soils contaminated with heavy metals. Microbiology 78:785–793

Pishchik VN, Vorobyev NJ, Chernyaeva LI, Timofeeva SV, Kazhemyakov AP, Alexeev YV (2002) Experimental and mathematical simulation of plant growth promoting rhizobacteria and plant interaction under cadmium stress. Plant Soil 243:173–186

Qurashi AW, Sabri AN (2012) Bacterial exopolysaccharide and biofilm formation stimulate chickpea growth and soil aggregation under salt stress. Braz J Microbiol 2012:1183–1191

Ramarethinam S, Chandra K (2005) Studies on the effect of potash solubilizing/mobilizing bacteria Frateuria aurantia on brinjal growth and yield. Pestol 11:35–39

Rawlings DE (2002) Heavy metal mining using microbes. Annu Rev Microbiol 56:65–91

Reitmeir RF (1951) Soil potassium. In: Norman AG (ed) Advances in agronomy II. Academic Press, New York, pp 113–164

Rengel Z, Marschner P (2005) Nutrient availability and management in the rhizosphere exploiting genotypic differences. New Phytol 168:305–312

Requena BN, Jimenez I, Toro M, Barea JM (1997) Interactions between plant-growth-promoting rhizobacteria (PGPR), arbuscular mycorrhizal fungi and *Rhizobium* spp. in the rhizosphere of *Anthyllis cytisoides*, a model legume for revegetation in Mediterranean semiarid ecosystems. New Phytol 136:667–677

Rodriguez H, Fraga R (1999) Phosphate solubilizing bacteria and their role in plant growth promotion. Biotechnol Adv 17:319–339

Rodriguez H, Fraga R, Gonzalez T, Bashan Y (2006) Genetics of phosphate solubilization and its potential applications for improving plant growth-promoting bacteria. Plant Soil 287:15–21

Saber K, Nahla LD, Chedly A (2005) Effect of phosphate on nodule formation and N fixation in bean. Agron Sustain Dev 25:389–393

Salimpour S, Khavazi K, Nadian H, Besharati H, Miransari M (2010) Enhancing phosphorous availability to canola (*'Brassica napus'* L.) using P solubilizing and sulfur oxidizing bacteria. Aus. J Crop Sci 4(5):330–334

Sandhya V, Ali SZ, Grover M, Reddy G, Venkateswarlu B (2010) Effect of plant growth promoting *Pseudomonas* spp. on compatible solutes, antioxidant status and plant growth of maize under drought stress. Plant Growth Regul 62:21–30

Sarma RK, Saikia RR (2014) Alleviation of drought stress in mung bean by strain *Pseudomonas aeruginosa* GGRK21. Plant Soils 377:111–126

Schalk IJ, Hannauer M, Braud A (2011) New roles for bacterial siderophores in metal transport and tolerance. Environ Microbiol 13:2844–2854

Scherer HW (2001) Sulphur in crop production. Eur J Agron 14:81–111

Schnug E, Haneklaus S (1993) Physiological backgrounds of different sulfur utilisation in *Brassica napus* varieties. Aspects. Appl Biol 34:235–242

Sérgio A, Araújo F, Borges CD, Tsai SM, Cesarz S, Eisenhauer N (2014) Soil bacterial diversity in degraded and restored lands of Northeast Brazil. Antonie Van Leeuwenhoek 106(5):891–899

Sharma SB, Riyaz ZS, Mrugesh HT, Thivakaran AG (2013) Phosphate solubilizing microbes: a sustainable approach for managing phosphorus deficiency in agricultural soils. SpringerPlus 2:587

Sheng XF, Xia JJ, Cheng J (2006) Mutagenesis of the *Bacillus edaphicus* strain NBT and its effect on growth of chilli and cotton. Agric Sci Chin 37:342–349

Shimaila A, Trevor C, Charles BR (2014) Glick amelioration of high salinity stress damage by plant growth promoting bacterial endophytes that contain ACC deaminase. Plant Physiol Biochem 80:160–167

Singh G, Biswas DR, Marwah TS (2010) Mobilization of potassium from waste mica by plant growth promoting rhizobacteria and its assimilation by maize (*Zea mays*) and wheat (*Triticum aestivum* L.) J Plant Nutri 33:1236–1251

Sorokin DY, Teske A, Robertson LA, Kuenen JG (1999) Anaerobic oxidation of thiosulfate to Tetrathionate by Obligately heterotrophic bacteria, belonging to the *Pseudomonas stutzeri* group. Fed Eur Microb Soci Microbiol Ecol 30:113–123

Srinivasarao CH, Takkar PN (1997) Evaluation of different extractants for measuring the soil potassium and determination of critical levels for plant available K in smectitic soils for sorghum. J Indian Soc Soil Sci 45:113–119

Starkey RL (1935) Isolation of some bacteria which oxidize thiosulfate. Soil Sci 39:197–219

Subbarao NS (1988) Phosphate solubilizing microorganism. In: Biofertilizer in agriculture and forestry regional Biofert. Dev. Centre, Hissar, India, pp 133–142

Swaby R, Sperber JI (1959) Phosphate dissolving microorganisms in the Rhizosphere of legume nutrition of legumes; Proc. Univ. Nottingham 5Th Easter Sch. Agril. Sci. (CSIRO Adelaide). Soils Fert 22(286):289–294

Sziderics AH, Rasche F, Trognitz F, Sessitsch A, Wilhelm E (2007) Bacterial endophytes contribute to abiotic stress adaptation in pepper plants (*Capsicum annum* L.) Can J Microbiol 53:1195–1202

Tandon HLS (1989) Sulphur fertilizers for the tropics. Proc TSI-FAI symp Sulphur in Indian agriculture, New Delhi. pp S IV/2(1–11)

Tilman D, Fargione J, Wolff B, D'Antonio C, Dobson A, Howarth R, Schindler D, Schlesinger WH, Simberloff D, Wackhamer D (2001) Forecasting agriculturally driven global environmental change. Science 292:281–284

Trivedi P, Sa T (2008) *Pseudomonas corrugate* (NRRL B-30409) mutants increased phosphate solubilisation, organic acid production, and plant growth at lower temperatures. Curr Microbiol 56:140–144

Trolove SN, Hedley MJ, Kirk GJD, Bolan NS, Loganathan P (2003) Progress in selected areas of rhizosphere research on P acquisition. Aust J Soil Res 41:471–499

Ullaman WJ, Kirchman DL, Welch WA (1996) Laboratory evidence by microbially mediated silicate mineral dissolution in nature. Chem Geol 132:11–17
Vale M, Seldin L, Araujo FF, Linna R (2010) Plant growth promoting rhizobacteria; fundamentals and applications. In: Maheshwari DK (ed) Plant growth and health promoting bacteria. Springer, Berlin, pp 21–43
Van Schöll, Kuyper TW, Smits MM, Landeweert R, Hoffland E, van Breemen N (2008) Rock-eating mycorrhizas: their role in plant nutrition and biogeochemical cycles. Plant Soil 303:35–40
Vandevivere P, Welch SA, Ullman WJ, Kirchman DJ (1994) Enhanced dissolution of silicate minerals by bacteria at near neutral pH. Microb Ecol 27:241–251
Vassileva M, Azcon R, Barea JM, Vasslev N (2000) Rock phosphate solubilization by free and encapsulated cells of *Yarowiali polytica*. Process Biochem 35:6937
Vazquez P, Holguin G, Puente M, Lopez-cortes A, Bashan Y (2000) Phosphate solubilizing microorganisms associated with the rhizosphere of mangroves in a semi-arid coastal lagoon. Biol Fertil Soils 30:460–468
Venkateswarlu B, Singh AK, Srinivasa R, Rao CG, Kumar KA, Virmani SM (2012) Natural resource management for accelerating agricultural productivity. Studium Press (India) Pvt. Ltd, New Delhi, p 234
Verma JP, Yadav J, Tiwari KN, Jaiswal DK (2014) Evaluation of plant growth promoting activities of microbial strains and their effect on growth and yield of chickpea (*Cicer arietinum* L.) in India. Soil Biol Biochem 70:33–37
Verma JP, Yadav J, Tiwari KN, Kumar A (2013) Effect of indigenous *Mesorhizobium* spp. and plant growth promoting rhizobacteria on yields and nutrients uptake of chickpea (*Cicer aritenium* L.) under sustainable agriculture. Ecol Eng 51:282–286
Vidyalakshmi R, Sridar R (2007) Isolation and characterization of sulphur oxidizing bacteria. J Cult Coll 5:73–77
Wakelin SA, Warren RA, Harvey PR, Ryder MH (2004) Phosphate solubilization by *Penicillium* sp. closely associated with wheat roots. Biol Fertil Soils 40:36–43
Wang S, Huijun W, Junqing Q, Lingli M, Jun L, Yanfei X, Xuewen G (2009) Molecular mechanism of plant growth promotion and induced systemic resistance to tobacco mosaic virus by bacillus spp. J Microbiol Biotechnol 19(10):1250–1258
Wani PA, Zaidi A, Khan AA, Khan MS (2005) Effect of phorate on phosphate solubilization and indole acetic acid (IAA) releasing potentials of rhizospheric microorganisms. Annals Plant Protection Sci 13:139–144
Weed SB, Davey CB, Cook MG (1969) Weathering of mica by fungi. Soil Sci Soc Am 33:702–706
Welch SA, Ullman WJ (1993) The effect of organic acids on plagioclase dissolution rates and stoichiometry. Geochim Cosmochim Acta 57:2725–2736
Whitelaw MA (2000) Growth promotion of plants inoculated with phosphate solubilizing fungi. Adv Agron 69:99–151
Xie JC (1998) Present situation and prospects for the world's fertilizer use. Plant Nutri Fertil Sci 4:321–330
Yahya A, Azawi SKA (1998) Occurrence of phosphate solubilizing bacteria in some Iranian soils. Plant Soil 117:135–141
Zaidi A, Khan MS, Aamil M (2004) Bioassociative effect of rhizospheric microorganisms on growth, yield, and nutrient uptake of green gram. J Plant Nutr 27:601–612
Zaidi A, Khan MS, Ahemad M, Oves M, Wani PA (2009) Recent advances in plant growth promotion by phosphate-solubilizing microbes. In: Microbial strategies for crop improvement. Springer, Berlin/Heidelberg, pp 23–50
Zaïdi I, Ebel C, Touzri M, Herzog E, Evrard JL, Schmit AC et al (2010) TMKP1 is a novel wheat stress responsive MAP kinase phosphatase localized in the nucleus. Plant Mol Biol 73:325–338
Zhang H, Kim MS, Sun Y, Dowd SE, Shi H, Parè W (2008) Soil bacteria confer plant salt tolerance by tissue specific regulation of the sodium transporter HKT1. Mol Plant Microbe In 21:737–744

Zhang H, Murzello C, Sun Y, KimSeongMi XX, Jeter RM, Zak JC, Dowd SE, Paré PW (2010) Choline and osmotic stress tolerance induced in Arabidopsis by the soil microbe *Bacillus subtilis* (GB03). Mol Plant Microbe In 23:1097–1104

Zhou K, Binkley D, Doxtader KG (1992) A new method for estimating gross phosphorus mineralization and immobilization rates in soils. Plant Soil 147:243–250

Chapter 13
Detoxification and Bioremediation of Hexavalent Chromium Using Microbes and Their Genes: An Insight into Genomic, Proteomic and Bioinformatics Studies

H.N. Thatoi and S.K. Pradhan

13.1 Introduction

Chromium exists in several ionic forms out of which, Cr^{6+} is highly mobile and more toxic while on the other hand, the trivalent chromium (Cr^{3+}) is less mobile, thermodynamically more stable and less toxic (Thacker et al. 2006). Cr^{6+} is toxic to living cell systems including microorganisms due to its strong oxidizing power and results in DNA adducts (Kotas and Stasicka 2000; Wise et al. 2004; Ackerley et al. 2006) and exhibits the properties of mutagenic (Puzon et al. 2002), carcinogenic (Codd et al. 2003) as well as, teratogenic (Asmatullah et al. 1998), and considered as an important heavy metal pollutant by United States Environmental Protection Agency (USEPA) (Cheung and Gu 2007). Chromate ion (CrO^{4-}), is a predominant Cr^{6+} oxyanion. Being structurally similar to sulfate (SO^{4-}) CrO^{4-}enters cells through the sulfate-transport system present in the cell membrane, gets reduced to Cr^{3+} by various enzymatic and non- enzymatic processes. Reduction of Cr^{6+} is carried out by glutathione, thiols and other metabolites of the cell, and results in generation of reactive oxygen species (ROS) that exert deleterious effects on protein as well as nucleic acid. The generated reactive species cause DNA adducts and are the cause of the carcinogenic, mutagenic and teratogenic potential of chromate Chardin et al. 2002; Klonowska et al. 2008) (Gibb et al. 2000; Venitt and Levy 1974).

Chemical reduction of Cr^{6+} to Cr^{3+} followed by precipitation is the most common and widely used method of detoxification than those of electrochemical treatment, reverse osmosis, adsorption and ion-exchange. Fe(0), Fe(II), sulphide and organic

H.N. Thatoi (✉)
Department of Biotechnology, North Orissa University, Baripada, Odisha, India
e-mail: hn_thatoi@rediffmail.com

S.K. Pradhan
P.G. Department of Bioinformatics, Orissa University of Agriculture and Technology, Bhubaneswar, Odisha, India

J.K. Patra et al. (eds.), *Microbial Biotechnology*,
https://doi.org/10.1007/978-981-10-6847-8_13

C-based materials are the reductants involved in chromate reduction. Conventional physicochemical treatments are much more expensive for large-scale *in situ* removal of Cr from the contaminated environment. On the other hand, microbial-bioremediation such as sorption, accumulation, and transformation through selective microorganisms are cost effective, ecofriendly as well as sustainable (Shakoori et al. 2000; Eccles 1995). Now a days, researchers across the globe have focussed to decipher the enzymatic role of chromate reductases for detoxification of hexavalent chromium to evaluate its bioremediation potential (Ackerley et al. 2004b). Bacteria have developed different strategies of chromate resistance by employing various detoxification methods most importantly efflux through ion transport, DNA repair and chromate reduction (extracellular/ intracellular) etc. The intracellular chromate reduction is catalysed by cellular reductants like ascorbic acid, glutathione, and flavoenzymes resulting different intermediates such as $Cr^{5+}$and/or Cr^{4+} (Bose et al. 1992; Suzuki et al. 1992; Stearns and Wetterhahn 1997; Lay and Levina 1998; Kalabegishvili et al. 2003), free radicals and less toxic Cr^{3+} (Dillon et al. 1997; Ortega et al. 2005; Kanmani et al. 2012).

But the extracellular chromate reduction is carried out by extracelluar terminal reductases of *c*-type cytochrome (cyt *c*) such as MtrC, and OmcA of *S. oneidensis* MR-1 (Belchik et al. 2011). Chromium resistance is also imparted through efflux of chromate ions from the cell cytoplasm to remove chromium from cells (Nies and Silver 1995). Over the time, numerous Cr^{6+}-resistant/tolerant bacterial strains belonging to *Pseudomonas spp, Bacillus spp, .Bacillus subtilis, Bacillus maroccanus ChrA21 1040, Ochrobactrum intermedium* , *Brevibacterium* sp, *Pseudomonas corrugata (strains 22 and 28), Stenotrophomonas maltophilia, Staphylococcus gallinarum, Pantoea* sp. and *Aeromonas* sp. etc have been reported by different researchers (McLean and Beveridge 2001a, b; Srinath et al. 2001; Camargo et al. 2003a, b; Kamaludeen et al. 2003a, b; Viti et al. 2003; Faisal and Hasnain 2004; Viti and Giovannetti 2007; Alam and Malik 2008; Halpern et al. 2009; He et al. 2009, 2010, 2011; Sturm et al. 2011; Zhang and Li 2011; Verma and Singh 2013). Many Cr^{6+} reducing bacteria have been detected and studied such as *Pseudomonas strain CRB5, Brucella sp.* (Thacker et al. 2007), *Bacillus sp. Strain QC1-2* (Campos et al. 1995), *Burkholderia cepacia MCMB-821* (Wani et al. 2007) and *Thermus scotoductus strain SA-01* (Opperman and Heerden 2007). Members of the genus *Microbacterium* such as *Microbacterium sp.* SUCR140 (Soni et al. 2014), *Microbacterium sp.* chr-3 (Focardi et al. 2013), and *Microbacterium sp.* CR-07 (Liu et al. 2012) have been shown to reduce chromate. Studies on several other bacterial strains like *Shewanella oneidensis MR-1* (Thompson et al. 2007), *Ochrobactrum tritici 5bvl1* (Branco et al. 2008) and *Ralstonia metallidurans strain CH34* (Juhnke et al. 2002) have been shown to contain chromate resistance and reduction genes induced by chromate. In *Rhodobacter sphaeroides*, chromate resistance is carried out by non-specific FADH 2 -dependent metal reductase (Moore and Kaplan 1992, 1994). Study results showed that different members of the genus *Microbacterium* found to be involved in chromate reduction such as *Microbacterium sp.* SUCR140 (Soni et al. 2014), *Microbacterium sp.* chr-3 (Focardi et al. 2013), and *Microbacterium sp.* CR-07 (Liu et al. 2012). Out of the different strategies for heavy metal remediation, microbial transformation is widely practised (Turick et al. 1996; Camargo et al. 2003a, b; Kanmani et al. 2012) which include both pure cultures (Nepple et al. 2000; Park et al. 2000; McLean and Beveridge 2001; Kwak et al. 2003) and microbial consortia (Pattanapipitpaisal et al. 2001).

Bacterial chromate reduction occurs directly through enzymatic activity or indirectly through non-enzymatic means by producing compounds that can reduce Cr (VI).

13.1.1 Enzymatic Chromate Reduction

The enzymatic chromate reduction in bacteria is carried out either through extracellular or intracellular enzymes. The soluble reductases are involved in both extracellular or intracellular reduction of hexavalent chromium (Elangovan et al. 2010; Das et al. 2014) whereas, membrane bound reductases involved in reduction through extracellular means (Wang et al. 1991). These enzymes use chromate as the terminal electron acceptor in the electron transport system while carrying out chromate reduction.

13.1.2 Non-enzymatic Chromate Reduction

Non-enzymatic reduction of hexavalent chromium (Cr^{6+}) is carried out through different chemical substances like ascorbic acid, glutathione (GSH), cysteine, hydrogen peroxide for microbial cells, and ascorbate for higher organisms (Poljsak et al. 2010). Chemical compounds like amino acids, nucleotides, sugars, vitamins, organic acids are also involved in chromate reduction (Dhal et al. 2013). Iron and sulphate-reducing bacteria, can reduce hexavalent chromium through their anaerobic metabolic end products e.g. Fe(II) and HS À .

13.2 Bacterial Bioremediation

Bacteria exhibit various molecular mechanisms responsible for Cr^{6+} reduction (Lovley et al. 1993; Park et al. 2000; Branco et al. 2008). A wide group of bacteria have been found to be involved in reduction of toxic compounds by both aerobic and anaerobic means with desire biochemical pathways.In aerobic bioremediation, bacteria use carbon substrate as the electron donor and oxygen as the electron acceptor and all the biochemical reactions are carried out in the presence of oxygen, whereas anaerobic bioremediation involves reaction occurring in the absence of oxygen where sulphate, nitrate, carbon dioxide, oxidized materials, or organic compounds may replace oxygen as the electron acceptor and the reaction is catalyzed in absence of oxygen.

13.2.1 Aerobic Chromate Reduction

Aerobic Cr^{6+} reduction is mediated through various cytoplasmic enzymes of different group of bacteria. Suzuki et al. (1992) purified NAD(P)H-dependent chromate reductase enzyme from *Pseudomonas ambigua* and charaterised it as the homolog of nitroreductase (Kwak et al. 2003). Park et al. (2000) studied another NAD(P)H and NADH dependent chromate reductase enzyme purified from *P. putida* was found to beinvolved in chromate reduction. Many chromate reduction studies have been carried out on soluble enzymes encoded by chromosomal genes (Mugerfeld et al. 2009). Due to technical difficulties only few of those proteins have been purified and annotated. Most of the aerobic chromium reductases isolated from *Pseudomonas spp*., *Escherichia coli*, and *Bacillus sp. ES29*, found to be soluble in the cytoplasm (Camargo et al. 2003a, b, Ishibashi et al. 1990). Out of those reductases, ChrR in *Pseudomonas putida* (Park et al. 2000) and YieF in *Escherichia coli* (Barak et al. 2006) have been studied in detailed toward the reduction of hexavalent chromium. ChrR reduces Cr^{6+} by transfering one electron and generate Cr^{5+} a reactive intermediate, and then reduced to $Cr^{3+}$by another electron transfer. The Cr^{5+} intermediate can again be oxidized back to Cr^{6+} in the presence of oxygen, ChrR, a reductase enzyme have been reported to prevent cells from the reactive oxygen species (ROS) toxicity (Ackerley et al. 2004a; Cheung and Gu 2007; Gonzalez et al. 2005). YieF, a sequence homolog of ChrR also found to be involved in reduction of Cr^{6+} to Cr^{3+} by utilizing four electron transfer. YieF, an enzyme which gemnerates ROS like that of ChrR but in much lower rate (Ackerley et al. 2004b; Ramirez-Diaz et al. 2008). *P. maltophilia O-2* was the only reported obligatory aerobic chromium reducer which reduces Cr^{6+} by utilising membrane-associated reductase. The microbial reduction of Cr^{6+} to Cr^{3+} also carried out by other groups of facultative aerobes, e.g., *Shewanella alga, Ochrobactrum tritici, Cellulomonas sp.*. The co-occurrence of nitrate in the contaminated environment inhibits metal transformation which is a major drawback in the bioremediation of toxic metals. *Geobacter metallireducens, Desulfovibrio desulfuricans,* and *Sulfurospirillum barnesii* can reduce both chromate Cr^{6+} and nitrate, and found to be beneficial for developing bioremediation strategies. *G. metallireducens* involved in reduction of nitrate to nitrite via the membrane bound nitrate reductase (Nar), whereas a different form of nitrate reductase(nap) is reported in *S. barnesii* and *D. desulfuricans* strain 27,774 (Ball and Nordstrom 1998).

13.2.2 Anerobic Chromate Reduction

Anaerobic chromate reduction in bacteria is carried out with the help of various membrane bound reductases such as flavin reductases, c-type cytochromes and hydrogenases and act as a terminal electron acceptor which is similar to sulfate-reducing bacteria (SRB) and *Enterobacter cloacae* (Wang et al. 1990; Michel et al. 2001; Chardin et al. 2003). Tetraheme c-type cytochrome isolated from *Desulfovibrio* species found to be associated with Cr^{6+} reduction activity (Chardin et al. 2003). Besides the tetra heme c-type cytochrome other heme-c proteins such as octaheme cytochrome c3 (Czjzek et al. 1996), a multiheme (16) cytochrome Hmc (Bruschi et al. 1992), and a non-heme cytochrome (Saraiva et al. 1999) were found to be involved in reduction of hexavalent chromium. The sulfate-reducing bacteria belong to Desulfovibrionaceae family are involved in Cr^{6+} reduction by utilizing hydrogenases enzyme (Lovley and Phillips 1994; Chardin et al. 2003). In some bacteria like *Shewanella, Enterobacter* and SRB, the terminal electron acceptors such as nitrate and sulphate are substituted by chromate and carring out chromate reduction (Chardin et al. 2003; Myers et al. 2000).

13.2.3 Aerobic and Anerobic Chromate Reduction

Bacteria can also reduce chromate under both aerobic and anaerobic conditions. *P. fluorescens* LB300, have been observed in reduction of Cr^{6+} to Cr^{3+} in both aerobically and anaerobically (Bopp and Ehrlich 1988). Under the anaerobic condition *P. fluorescens* LB300 utilized acetate as an electron donor, while different electron donor is used under aerobic condition (Bopp and Ehrlich 1988). Besides that, number of other chromate-resistant bacteria have been reported those have shown Cr^{6+} reduction ability under both aerobic and anaerobic conditions such as *Achromobacter sp.* (Ma et al. 2007), *E. coli, P. ambigua* (McLean and Beveridge 2001), *P. putida* (Barak et al. 2006), *E. cloacae*, *Providencia sp.* (Thacker et al. 2006), *Brucella sp.* (Thacker et al. 2007) and *Bacillus sp.*(Liu et al. 2006). Bacteria such as *P. ambigua* G-1 and *P. putida* PRS2000 can able to reduce Cr^{6+} in faster rate under both aerobic and anaerobic condition while *E. coli ATCC 33456* showed faster reduction rate under aerobic conditions than that of anaerobic conditions (Shen and Wang 1993).

13.2.4 Chromate Reduction and pH

The ionization of chromium is also dependent on the pH of the sample, with chromate (Cr^{6+}) as the dominant ionic form in an aqueous environment at pH 6.5–9 (McLean and Beveridge 2001) and highly mobile in soil–water systems (Losi et al. 1994). Effluents released during various mining and indistrial activities contain toxic metals of alkaline or acidic pH. Most of the bacterial chromate detoxification studies were conducted at neutral/near neutral pH but under alkaline condition very few studies have been reported (Ye et al. 2004; Stewart et al. 2007). When chromate contamination is generally reported at high pH soils condition, high pH condition is considered to be an important factor while designing bioremediation strategies (Kamaludeen et al. 2003a, b; Van Engelen et al. 2008). Thus, those bacteria that can thrive under such high alkaline conditions and involved in metal detoxification need to be characterised and studied (Mangaiyarkarasi et al. 2011). As per the study report of the Ilhan et al. (2004), the optimum temperature and pH for reduction of hexavalent chromium by a strain of *S. saprophyticus* was at 27 °C and 2.0 respectively. At pH 7.0, Mistry et al. (2010) observed the chromate reduction in *S. saprophyticus*. *Oceanobacillus sp.* isolated from the skin of a rainbow trout at pH 9–10 and optimum temperature at 30–36 °C and was identified as a halotolerant obligate alkaliphile (Yumoto et al. 2005). Molokwane et al. (2008) also reported that bacterial culture containing *Oceanobacillus sp.*, could remove Cr^{6+} under anaerobic condition. *Pseudochrobactrum. saccharolyticum* (Kampfer et al. 2006), *Pseudochrobactrum asaccharolyticum* (Kampfer et al. 2006), *Pseudochrobactrum kiredjianiae* (Kampfer et al. 2007), *Pseudochrobactrum lubricantis* (Kampfer et al. 2009), and *Pseudochrobactrum glaciei* (Romanenko et al. 2008) were able to grow at optimum temperature of 25–30 °C and the optimum pH 7.1–7.5 respectively observed in chromium-polluted areas.

13.2.5 Chromate Efflux

Efflux of chromate is mediated by the chromate transporter protein ChrA. The ChrA transporter functions as a chemiosmotic pump that extrudes chromate from the cytoplasm using the proton-motive force and displays a topology of 13 transmembrane segments (TMSs) which has been confirmed in *Pseudomonas aeruginosa* (Alvarez et al. 1999; Pimentel et al. 2002), *Ochrobactrum tritici5bvl1* (Branco et al. 2008), *Cupravidus metallidurans* (Cervantes and Ohtake 1988) and *Shewanella sp.* ANA-3 (Aguilar-Barajas et al. 2008). *Synechococcus sp. IU 625* (S. IU 625), is a Gram-negative bacteria in cell wall structure, cell division and the ability to harbor plasmids and serve as good indicators of environmental pollution, especially heavy metal contamination. ANL48, a plasmid-generated homologue of ChrA, exists in *Synechococcus sp. IU 625* and is encoded by srpC on the pANL plasmid. Both ANL48 and ChrA are members of the CHR chromate ion transport superfamily and are responsible for chromium efflux (Aguilar-Barajas et al. 2011, 2012).

13.3 Molecular Basis of Chromate Reduction

When the bacterial cells are exposed to chromium contaminated environment, these cells will continuously be in oxidative stress to get rid of chromatem toxicity. Genes that confer chromate resistance get activated in those bacterial cells and serve as markers to indicate chromium stress in respective bacteria. These chromate resistant genes vary in different bacteria but involved in same function. Chromate-resistance determinants (CRDs) have been reported in all the kingdom i.e. Archaea, Bacteria and Eukarya (Nies et al. 1998; Flores-Alvarez et al. 2012), confer chromate resistance as it consists of genes belonging to the chromate ion transport (CHR) superfamily (Ramirez-Diaz et al. 2008). The CRDs consists of chrA genes, which encode a putative transporter protein (ChrA) helps in chromate efflux (Pimentel et al. 2002). ChrA protein and its role in chromate detoxification has been studied in many bacteria such as *P. aeruginosa*, *Cupriavidus metallidurans*, and *Shewanella sp.* ANA3 (Cervantes et al. 1990; Aguilar-Barajas et al. 2008), and many of its homologs have been identified in other organisms.

ChrA genes found in bacteria are either chromosomal or plasmid origin (Juhnke et al. 2002), and arranged like that of operons with other chr genes (Fig. 13.1). In *C. metallidurans*, the heavy metal resistance genes are reported in the pMOL28 plasmid and in a chromosomal operon chr2 (chrB2, chrA2, chrF2 genes), but in *Ochrobactrum tritici* 5bv11, chr genes are present in transposable operon (Branco

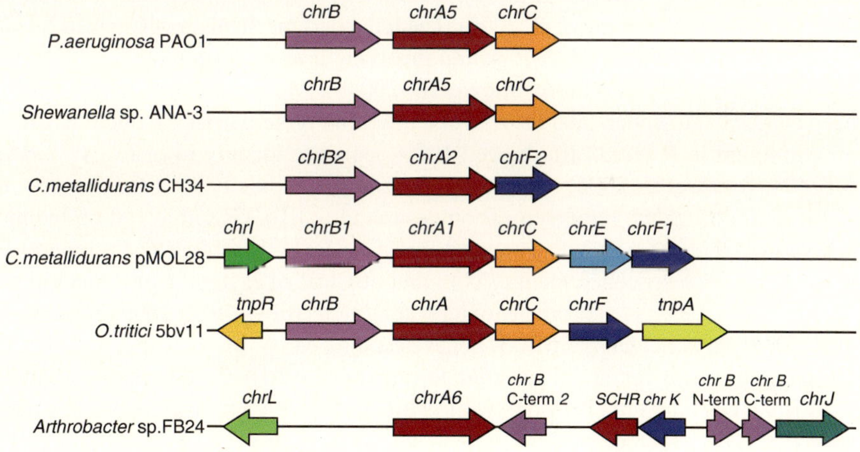

Fig. 13.1 Schematic representation of the main local genomic context of the chr genes analyzed by 'omic' approaches *chrA* chromate ion transporter, *chrB* chromiumsensitiveregulator, *chrC* superoxide dismutase, *chrE* hypothetically involved in cleavage of some chromium-glutathione complexes, *chrF* regulatory protein, *chrI* regulatory protein, *chrJ*, putative malate:quinone reductase, *chrK* YVTN beta-propeller repeat-containing protein, *chrL* probable conserved lipoprotein (LppY/LpqO family), *SCHR* small chromate ion transporter (chrA ortholog), *tnpA* transposase, *tnpR* resolvase

et al. 2008). The genes flanking chrA regions may or may not chromate reduction activity but chrB found to be involved in chromate resistance in *O. tritici* 5bv11 (Branco et al. 2008) and in *C. metallidurans* (Peitzsch et al. 1998; Juhnke et al. 2002).*Cupriavidus metallidurans* strain AE126 lacking chrB1 (Juhnke et al. 2002) showed chromate tolerance activity as compared to wild-type strain (Juhnke et al. 2002). *Shewanella sp.* strain ANA-3 has chrBAC, operon arranged in a similar fashion as that in pMOL28 of *C. metallidurans* CH34 (Aguilar-Barajas et al. 2008). The chromate resistance mechanism is controlled alone by ChrA gene in *E. coli*. ChrBAC operon in high-copy number shown low chromate resistance in comparison to low-copy number vector (Aguilar-Barajas et al. 2008). So there is no direct correlation between chrA flanking genes on chromate resistance and may differ depending on the host omics. CHR-1 protein (homologous to ChrA) is found in several ascomycetes, fungi and yeasts (Flores-Alvarez et al. 2012).Chromate reductase isolated from *Enterobacter cloacaeHO1* membrane bound enzyme which transfers electrons to Cr^{6+} via NADH-dependent cytochromes (Ohtake et al. 1990). NfsA/NfsB der ived from *Vibrio harveyi*, a nitroreductase also showed chromate reductase actiovity (Kwak et al. 2003). Similarly the ferric reductase FerB of *Pseudomonas denitrificans*, found to be involved in chromate reduction (Mazoch et al. 2004). *E. coli* NemA was identified as a highly efficient Cr^{6+} reductase using NADH as a cofactor (Robins et al. 2013). ChrR of *P. putida*, soluble flavin mononucleotide-binding enzyme can catalyze chromate reduction (Park et al. 2000). The ChrR, a NADH-dependent reductase, with broad substrate specificity (Barak et al. 2006). ChrR enzyme of *E. coli* and *P. putida* shares sequence homology and involved in chromate reduction (Barak et al. 2006). The reduction mechanism of ChrA in *E.coli* is different that of the ChrR of *P. putida* involving four-electron reduction of chromate and generating ROS (Ackerley et al. 2004b). More number of studies have been conducted to understand the role of ChrR gene/protein in Cr^{6+} reduction, ChrR protein found in *P. putida* F1 showed 100% sequence identity to that of *P. putida* KT2404 (Barak et al. 2006) but at the acute chromate toxicity level the expression of the genes are down regulated (Thompson et al. 2010). Genomic and proteomic analysis of *S. oneidensis* MR-1 reveled that a NADPH-dependent FMN reductase have shown 28% sequence identity with that of ChrR of *P. putida* and found to be upregulated under acute chromate toxicity condition (Brown et al. 2006; Thompson et al. 2007). In *S. oneidensis* MR-1, genes encoding MtrA, MtrB, MtrC, and OmcA are participated in reduction of solid ferric iron [Fe(III)] (hydr)oxides, uranium [U(VI)] and technetium [Tc(VII)], while MtrC and OmcA are involved in chromate reduction (Belchik et al. 2011).

13.4 Bioinformatics Analysis

13.4.1 Chromate Resistant/Reduction Genes Repository

Genes responsible for chromate resistance/reduction were sequenced and deposited in the gene database of the National Center for Biotechnology Information (NCBI). There are total 3218 numbers of chromate resistant gene sequence entries were downloaded from the NCBI gene database (http://www.ncbi.nlm.nih.gov/gene), belong to different groups of organisms (i.e., bacteria (2921), eukaryotes (274) and archea (23). The total 2921 bacterial gene entries are distributed in 11 different bacterial groups. Of the 11 bacterial groups, majority of the entries are belong to five groups such as Proteobacteria (1694), Firmicutes (586), CFB group bacteria (141), Actinobacteria (107) and Spirochetes (78). In the Proteobacteria group, 1694 gene entries were distributed in four different subgroups such as b-proteobacteria (742), g-proteobacteria (531), a-proteobacteria(369) and others(52). A total of 2921 bacterial gene entries that are involved in chromate resistance or reduction were analyzed. These gene entries belong to more than 500 different bacterial species isolated from cultured bacteria except eight entries are from uncultured bacteria. The major species of cultured bacteria include *E. coli, Cupriavidus metallidurans, Pseudomonas putida, Burkholderia multivorans, Burkholderia pseudomallei, Burkholderia sp., Burkholderia cenocepacia, Ralstonia pickettii, Pseudomonas aeruginosa, Bacillus subtilis, Ralstonia solanacearum, Bacillus cereus, Bacillus thuringiensis, Burkholderia mallei, Acinetobacter baumannii,* and *Burkholderia vietnamiensis.* The highest numbers of gene entries belong to genus Burkholderia, and species *Burkholderia pseudomallei* (Figs. 13.2, 13.3, and 13.4; Table 13.1).

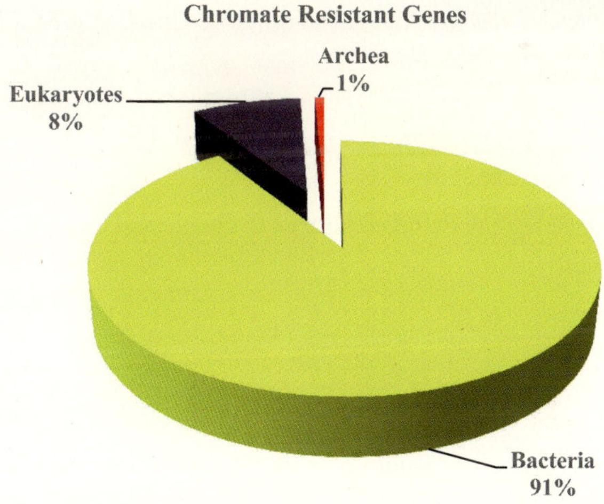

Fig. 13.2 Chromate resistant genes in tree of life

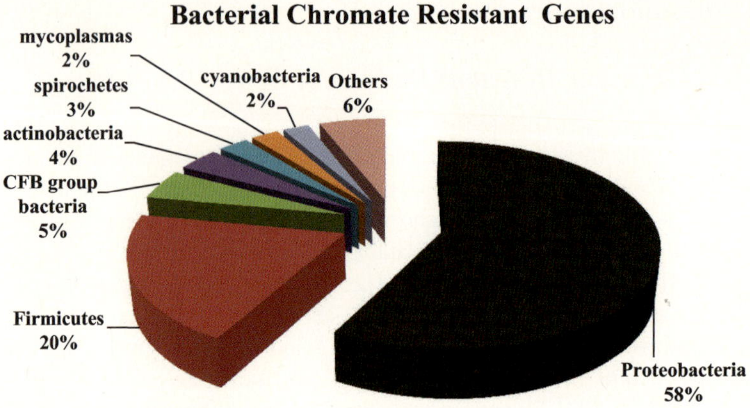

Fig. 13.3 Bacterial chromate resistant genes

13.4.2 Chromate Resistant/Reducing Proteins Repository

The uniprot potein sequence database (http://www.uniprot.org/) was researched to retrieve the list of proteins that confer chromate resistance/reduction. There are 22 numbers of bacterial proteins and one yeast protein involved in chromate detoxification have been reported in the database. Among the bacterial proteins, six proteins are under chromate reductases, nine are belong to transporter group and eight are either FMN/NADPH or others. The major bacterial genera that are involved in chromate detoxification are *Escherichia, Pseudomonas* and *Bacillus*. The *Pseudomonas* group showed all the possible methods of chromate detoxification includes enzymes belong to i.e. reduction, FMN/NADPH and efflux groups (Table 13.2).

13.4.3 Structural and Functional Analysis

Limited studies have been undertaken till now to understand the structure and function of these proteins that are involved in chromate detoxification. The three dimensional structures of these enzymes play an important role in catalytic reduction reaction and the substrate recognition followed by detoxification. The structure of seven reductases such as chromate reductase of *Thermus scotoductus* SA-01, *E. coli* strain K12 and *Gluconacetobacter hansenii* (Gh-ChrR), nitroreductase of *Desulfovibrio alaskensis* (strain G20), *Bacillus subtilis* (strain 168) and *E. coli*, and FMN reductase of *Paracoccus denitrificans* were retrieved from the RCSB Protein Databank (http://www.rcsb.org/pdb). These seven enzymes are categorized as either NADPH-dependent FMN reductase or FMN-dependent nitroreductase. Despite of the structural diversities (e.g., tetramer/dimer, arrangements of helices/sheets/coils)

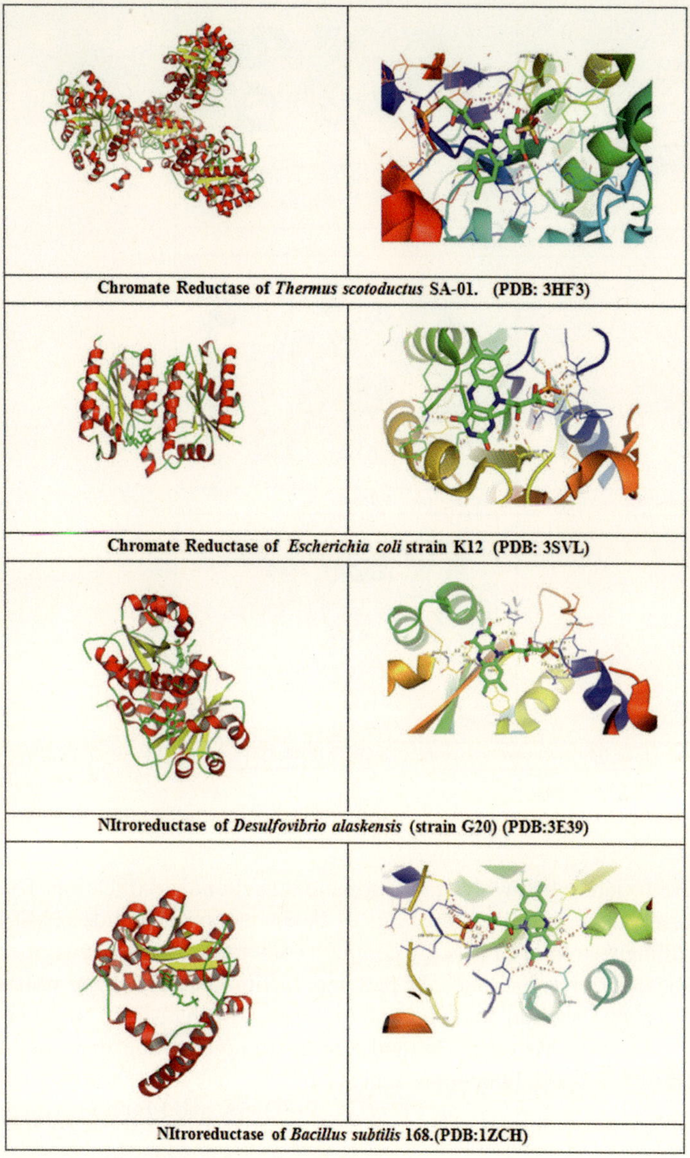

Fig. 13.4 Reductase enzymes from different bacteria species

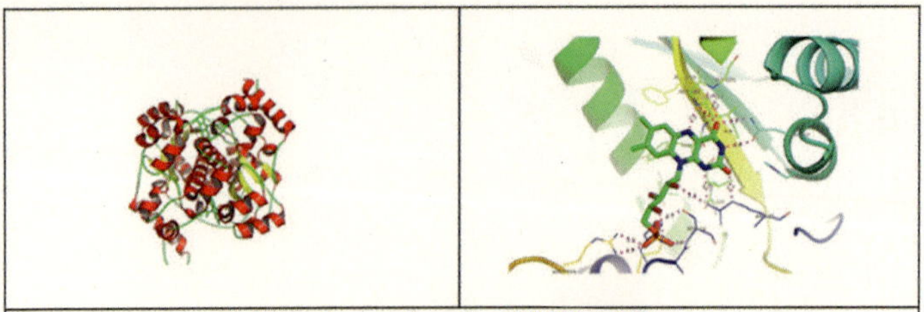

Nitroreductase from *E. coli* (PDB: 1F5V).

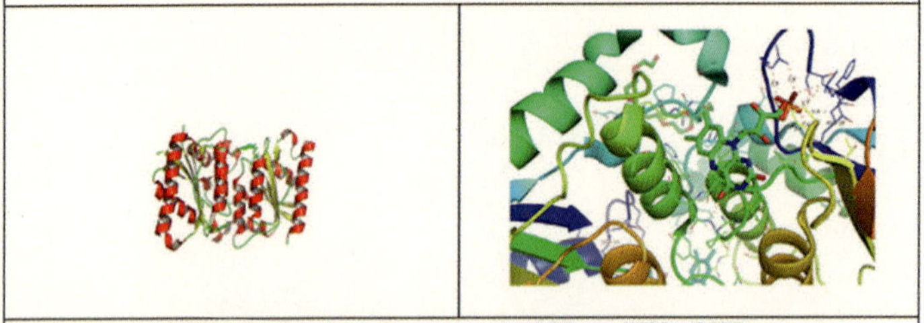

FMN reductase from *Paracoccus denitrificans* (PDB: 3U7R)

Chromate reductase from *Gluconacetobacter hansenii* (Gh-ChrR) (PDB:3S2Y)

Fig. 13.4 (continued)

all the enzymes are found to be involved in chromate reduction/detoxification. The crystal structures revealed the binding pattern of flavin mono nucleotide (FMN) with the enzyme (dimer/tetramer). FMN is a common cofactor for all these enzymes binding to different anchor residues and take part in reduction. The anchoring amino acid residues such as Glu, Tyr, Ser, Asn, Phe, and Arg interact with FMN in NADPH-dependent FMN reductases but Gly, Arg, and Ser are the interacting residues in FMN-dependent nitroreductase. Functional analysis of these enzymes showed presence of enzyme-specific domains such as PF00724, PF03358, and PF00881, which are responsible for chromate reduction.

13.5 Prospect of Chromate Bioremediation

Chromate bioremediation is carried out by both prokaryotic as well as eukaryotic microorganism by employing different types of mechanisms such as reduction, DNA repair, and efflux pumps. In all the cases it has been observed that the

Table 13.1 Chromate resistance gene entries in NCBI gene database

Chromate resistant/reducing organism			Number of chromate genes
Bacteria (2921)	Proteobacteria (1694)	b-proteobacteria	742
		g-proteobacteria	531
		a-proteobacteria	369
		Others	52
		Bacillales	219
	Firmicutes *(586)*	Clostridiales	249
		Others	118
	CFB group bacteria		141
	Actinobacteria		107
	Spirochetes		78
	Mycoplasmas		73
	Cyanobacteria		69
	Thermotogales		28
	Fusobacteria		28
	Brachyspirales		18
	Others		99
Eukaryotes (274)	Fungi		179
	Animal		46
	Others		49
Archea	–		23

resistant/reduction of chromate detoxifications are controlled at molecular level either by genes (plasmid/chromosome) or proteins (extracellular/intracellular/transprter). Many genome wide analysis studies on different microbes have also been carried out in order to understand the underlying mechanisms involved in chromate resistant/reduction. The spread of the 'omic' approaches i.e. genomics, transcriptomic and proteomic methods, have offered the possibilities to the modern day research for carring out the studies on the response of global gene and protein expression of microorganisms. The NextGen sequencing technology now been applied to understand the metabolic processes that until now were unexplained or poorly understood in terms of their correlation with chromate resistance/reduction opening new possibilities to which traditional microbiological methods can contribute. These advances will surely help to identify the key molecules as well as in understanding of the complexity of cell responses to Cr^{6+} toxicity.

Table 13.2 Chromate resistance protein entries in UniProt database

Entry ID	Protein names	Organism	Length
P0AGE6	Chromate reductase	*Escherichia coli (strain K12)*	188
Q88FF8	Chromate reductase	*Pseudomonas putida (strain ATCC 47054/DSM 6125/NCIMB 11950/KT2440)*	186
Q93T20	Chromate reductase	*Pseudomonas putida (Arthrobacter siderocapsulatus)*	186
P0AGE8	Chromate reductase	*Shigella flexneri*	188
P96977	Chromate reductase	*Pseudomonas spp. (strain-G-1)*	243
P0AGE7	Chromate reductase	*Escherichia coli O157:H7*	188
P94424	FMN reductase [NAD(P)H]	*Bacillus subtilis (strain 168)*	249
P74312	NADPH-dependent quinone reductase A..	*Synechocystis sp. (strain PCC 6803/Kazusa)*	206
P17550	Superoxide dismutase [Fe]	*Cupriavidus metallidurans (strain ATCC 43123/ DSM 2839/NBRC 102507/CH34) (Ralstonia metallidurans)*	197
P17117	Oxygen-insensitive NADPH nitroreduc…	*Escherichia coli (strain K12)*	240
P77258	N-ethylmaleimide reductase	*Escherichia coli (strain K12)*	365
P85207	Dihydrolipoyl dehydrogenase	*Thermus scotoductus (strain ATCC 700910/SA-01)*	461
Q51426	Holliday junction ATP-dependent DNA…	*Pseudomonas aeruginosa (strain ATCC 15692/ DSM 22644/CIP 104116/JCM 14847/LMG 12228/1C/PRS 101/PAO1)*	352
P17551	Chromate transport protein	*Cupriavidus metallidurans (strain ATCC 43123/ DSM 2839/NBRC 102507/CH34) (Ralstonia metallidurans)*	/
P14285	Chromate transport protein	*Pseudomonas aeruginosa*	416
Q55027	Probable chromate transport protein	*Synechococcus elongatus (strain PCC 7942) (Anacystis nidulans R2)*	393
O05215	Uncharacterized transporter YwrA	*Bacillus subtilis (strain 168)*	178
O05216	Uncharacterized transporter YwrB	*Bacillus subtilis (strain 168)*	197
Q58128	Putative uncharacterized transporter.	*Methanocaldococcus jannaschii (strain ATCC 43067/DSM 2661/JAL-1/JCM 10045/NBRC 100440) (Methanococcus jannaschii)*	402
O05215	Uncharacterized transporter YwrA	*Bacillus subtilis (strain 168)*	178
O05216	Uncharacterized transporter YwrB	*Bacillus subtilis (strain 168)*	197
Q58128	Putative uncharacterized transporter	*Methanocaldococcus jannaschii (strain ATCC 43067/DSM 2661/JAL-1/JCM 10045/NBRC 100440) (Methanococcus jannaschii)*	402
Q12235	High affinity cysteine transporter	*Saccharomyces cerevisiae (strain ATCC 204508/ S288c) (Baker's yeast)*	531

References

Ackerley DF, Gonzalez CF, Keyhan M, Blake R, Matin A (2004a) Mechanism of chromate reduction by the Escherichia Coli protein, NfsA, and the role of different chromate reductases in minimizing oxidative stress during chromate reduction. Environ Microbiol 6:851–860

Ackerley DF, Gonzalez CF, Park CH, Blake R, Keyhan M, Matin A (2004b) Chromate-reducing properties of soluble flavoproteins from pseudomonas putida and Escherichia Coli. Appl Environ Microbiol 70:873–882

Ackerley DF, Barak Y, Lynch SV, Curtin J, Matin A (2006) Effect of chromate stress on Escherichia Coli K-12. J Bacteriol 188(9):3371e–33381

Aguilar-Barajas E, Paluscio E, Cervantes C, Rensing C, Matin A (2008) Expression of chromate resistance genes from Shewanella sp. strain ANA-3 in Escherichia Coli. FEMS Microbiol Lett 285:97–100

Aguilar-Barajas E, Diaz-Perez C, Ramirez-Diaz MI, Riveros-Rosas H, Cervantes C (2011) Bacterial transport of sulfate, molybdate, and related oxyanions. Biometals 24(4):687–707

Aguilar-Barajas E, Jeronimo-Rodriguez P, Ramirez-Diaz MI, Rensing C, Cervantes C (2012) The ChrA homologue from a sulfur-regulated gene cluster in cyanobacterial plasmid pANL confers chromate resistance. World J Microbiol Biotechnol 28:865–869

Alam MZ, Malik A (2008) Chromate resistance, transport and bioreduction by Exiguobacterium sp. ZM-2 isolated from agricultural soil irrigated with tannery effluent. J Basic Microbiol 48:416–420

Alvarez AH, Moreno-sanchez R, Cervantes C (1999) Chromate efflux by means of the ChrA chromate resistance protein from Pseudomonas Aeruginosa. J Bacteriol 181:7398–7400

Ball JW, Nordstrom DK (1998) Critical evaluation and selection of standard state thermodynamic properties for chromium metal and its aqueous ions, hydrolysis species, oxides, and hydroxides. J Chem Eng Data 43:895–918. https://doi.org/10.1021/je980080a

Barak Y, Ackerley DF, Dodge CJ, Banwari L, Alex C, Francis AJ, Matin A (2006) Analysis of novel soluble chromate and uranyl reductases and generation of an improved enzyme by directed evolution. Appl Environ Microbiol 72:7074–7082. https://doi.org/10.1128/AEM.01334-06

Belchik SM, Kennedy DW, Dohnalkova AC, Wang Y, Sevinc PC, Wu H, Lin Y, Lu HP, Fredrickson JK, Shi L (2011) Extracellular reduction of hexavalent chromium by Cytochromes MtrC and OmcA of Shewanella oneidensis MR-1. Appl Environ Microbiol 77(12):4035–4041. https://doi.org/10.1128/AEM.02463-10

Bopp LH, Ehrlich HL (1988) Chromate resistance and reduction in Pseudomonas Fluorescens strain LB300. Arch Microbiol 150(5):426e431

Bose RN, Moghaddas S, Gelerinter E (1992) Long-lived chromium(IV) and chromium(V) metabolites in the chromium(VI) glutathione reaction – NMR, ESR, HPLC, and kinetic characterization. Inorg Chem 31:1987–1994

Branco R, Chung AP, Johnston T, Gurel V, Morais P, Zhitkovich A (2008) The chromate-inducible chrBACF operon from the transposable element TnOtChr confers resistance to chromium(VI) and superoxide. J Bacteriol 190:6996–7003

Brown SD, Thompson MR, Verberkmoes NC, Chourey K, Shah M, Zhou J, Hettich RL, Thompson DK (2006) Molecular dynamics of the Shewanella oneidensis response to chromate stress. Mol Cell Proteomics 5:1054–1071

Bruschi M, Bertrand P, More C, Leroy G, Bonicel J, Haladjian J et al (1992) Biochemical and spectroscopic characterization of the high molecular weight cytochrome c from *Desulfovibrio vulgaris* Hildenborough expressed in *Desulfovibrio desulfuricans* G200. Biochemistry 31:3281–3288

Camargo FA, Bento FM, Okeke BC, Frankenberger WT (2003a) Chromate reduction by chromium-resistant bacteria isolated from soils contaminated with dichromate. J Environ Qual 32:1228–1233

Camargo FAO, Okeke BC, Bento FM, Frankenberger WT (2003b) In vitro reduction of hexavalent chromium by a cell-free extract of bacillus sp. ES 29 stimulated by Cu2+. Appl Microbiol Biotechnol 62:569–573

Campos J, Martinez-Pacheco M, Cervantes C (1995) Hexavalent-chromium reduction by a chromate-resistant bacillus sp. strain. Antonie Van Leeuwenhoek 68:203–208

Cervantes C, Ohtake H (1988) Plasmid-determined resistance to chromate in *Pseudomonas aeruginosa*. FEMS Microbiol Lett 56:173–176. https://doi.org/10.1111/j.1574-6968.1988.tb03172.x.

Cervantes C, Ohtake H, Chu L, Misra TK, Silver S (1990) Cloning, nucleotide sequence, and expression of the chromate resistance determinant of Pseudomonas Aeruginosa plasmid pUM505. J Bacteriol 172:287–291

Chardin B, Dolla A, Chaspoul F, Fardeau ML, Gallice P, Bruschi M (2002) Bioremediation of chromate: thermodynamic analysis of effects of Cr(VI) on sulfate reducing bacteria. Appl Microbiol Biotechnol 60:352–360

Chardin B, Giudici-Orticoni MT, De Luca G, Guigliarelli B, Bruschi M (2003) Hydrogenases in sulfate-reducing bacteria function as chromium reductase. Appl Microbiol Biotechnol 63:315–321

Cheung KH, Gu J-D (2007) Mechanism of hexavalent chromium detoxification by microorganisms and bioremediation application potential: a review. Int Biodeterior Biodegrad 59:8–15

Codd R, Irwin JA, Lay PA (2003) Sialoglycoprotein and carbohydrate complexes in chromium toxicity. Curr Opin Chem Biol 17(2):213e219

Czjzek M, Guerlesquin F, Bruschi M, Haser R (1996) Crystal structure of a dimeric octaheme cytochrome c3 (M(r) 26,000) from *Desulfovibrio desulfuricans* Norway. Structure 4:395–404

Das S, Mishra J, Das SK, Pandey S, Rao DS, Chakraborty A, Sudarshan M, Das NN, Thatoi HN (2014) Investigation on mechanism of Cr(VI) reduction and removal by bacillus amyloliquefaciens, a novel chromate tolerant bacterium isolated from chromite mine soil. Chemosphere 96:112e121

Dhal B, Thatoi HN, Das N, BD(P (2013) Reduction of hexavalent chromium by bacillus sp. isolated from chromite mine soils and characterization of reduced product. J Chem Technol Biotechnol 85(11):1471e–11479

Dillon CT, Lay PA, Cholewa M, Legge GJF, Bonin AM, Collins TJ, Kostka KL, SheaMcCarthy G (1997) Microprobe X-ray absorption spectroscopic determination of the oxidation state of intracellular chromium following exposure of V79 Chinese hamster lung cells to genotoxic chromium complexes. Chem Res Toxicol 10:533–535

Eccles H (1995) Removal of heavy metals from effluents streams-why select a biological process. Int Biodeterior Biodegrad 160(1e3):5e16

Elangovan R, Philip L, Chandraraj K (2010) Hexavalent chromium reduction by free and immobilized cell-free extract of Arthrobacter rhombi-RE. Appl Biochem Biotechnol 160(1):81e97

Faisal M, Hasnain S (2004) Comparative study of Cr(VI) uptake and reduction in industrial effluent by Ochrobactrum intermedium and Brevibacterium sp. Biotechnol Lett 26:1623–1628

Flores-Alvarez LJ, Corrales-Escobosa AR, Cortes-Penagos C, Martinez-Pacheco M, Wrobel-Zasada K, Wrobel-Kaczmarczyk K, Cervantes C, Gutierrez-Corona F (2012) The Neurospora crassa chr-1 gene is up-regulated by chromate and its encoded CHR-1 protein causes chromate sensitivity and chromium accumulation. Curr Genet 58:281–290

Focardi S, Pepi M, Focardi SE (2013) Microbial reduction of hexavalent chromium as a mechanism of detoxification and possible bioremediation applications. In: Chamy R (ed) Biodegradation—life of science. InTech. Available at http://www.intechopen.com/books/biodegradation-life-of-science/microbial-reduction-of-hexavalent-chromium-as-a-mechanismof-detoxification-and-possible-bioremediat

Gibb HJ, Lees PSJ, Pinsky PF, Rooney BC (2000) Lung cancer among workers in chromium chemical production. Am J Industr Med 38:115–126

Gonzalez CF, Ackerley DF, Lynch SV, Matin A (2005) ChrR, a soluble quinone reductase of pseudomonas putida that defends against H2O2. J Biol Chem 280:22590–22595

Halpern M, Shaked T, Pukall R, Schumann P (2009) Leucobacter chironomi sp nov., a chromate-resistant bacterium isolated from a chironomid egg mass. Int J Syst Evol Microbiol 59:665–670
He ZG, Gao FL, Sha T, YH H, He C (2009) Isolation and characterization of a Cr(VI)-reduction Ochrobactrum sp. strain CSCr-3 from chromium landfill. J Hazard Mater 163:869–873
He MY, Li XY, Guo LA, Miller SJ, Rensing C, Wang GJ (2010) Characterization and genomic analysis of chromate resistant and reducing Bacillus Cereus strain SJ1. BMC Microbiol 10:221
He MY, Li XY, Liu HL, Miller SJ, Wang GJ, Rensing C (2011) Characterization and genomic analysis of a highly chromate resistant and reducing bacterial strain Lysinibacillus fusiformis ZC1. J Hazard Mater 185:682–688
Ilhan S, Nurbas M, Kiliarslan S, Ozdag H (2004) Removal of chromium, lead and copper ions from industrial waste waters by Staphylococcus saprophyticus. Turkish Electronic J Biotechnol 2:50–57
Ishibashi Y, Cervantes C, Silver S (1990) Chromium reduction in Pseudomonas putida. Appl Environ Microbiol 56:2268–2270
Juhnke S, Peitzsch N, H€ubener N, Große C, Nies DH (2002) New genes involved in chromate resistance in Ralstonia metallidurans strain CH34. Arch Microbiol 179:15–25
Kalabegishvili TL, Tsibakhashvili NY, Holman HYN (2003) Electron spin resonance study of chromium(V) formation and decomposition by basalt-inhabiting bacteria. Environ Sci Technol 37:4678–4684
Kamaludeen SP, Megharaj M, Juhasz AL, Sethunathan N, Naidu R (2003a) Chromium-microorganism interactions in soils: remediation implications. Rev Environ Contam Toxicol 178:93–164
Kamaludeen SB, Megharaj M, Naidu R, Singleton I, Juhasz AL, Hawke BG, N(S (2003b) Microbial activity and phospholipid fatty acid pattern in long-term tannery wastecontaminated soil. Ecotoxicol Environ Safety 56:302–310
Kampfer P, Rossello-Mora R, Scholz HC, Welinder-Olsson C, Falsen E, Busse HJ (2006) Description of Pseudochrobactrum gen. nov., with the two species Pseudochrobactrum asaccharolyticum sp. nov. and Pseudochrobactrum saccharolyticum sp. nov. Int J Syst Evol Microbiol 56(8):1823–1829
Kampfer P, Scholz H, Huber B, Thummes K, Busse HJ, Maas EW et al (2007) Desc ription o f Pseudochrobactrum kiredjianiae sp. nov. Int J Syst Evol Microbiol 57(4):755–760
Kampfer P, Huber B, Lodders N, Warfolomeow I, Busse HJ, Scholz HC (2009) Pseudochrobactrum lubricantis sp. nov., isolated from a metal-working fluid. Int J Syst Evol Microbiol 59(10):2464–2467
Kanmani P, Aravind J, Preston D (2012) Remediation of chromium contaminants using bacteria. Int J Environ Sci Technol 9:183–193
Klonowska A, Clark ME, Thieman SB, Giles BJ, Wall JD, Fields MW (2008) Hexavalent chromium reduction in Desulfovibrio vulgaris Hildenborough causes transitory inhibition of sulfate reduction and cell growth. Appl Microbiol Biotechnol 78:1007–1016
Kotas J, Stasicka Z (2000) Chromium occurrence in the environment and methods of its speciation. Environ Pollut 107(3):263e283
Kwak YH, Lee DS, Kim HB (2003) Vibrio harveyi nitroreductase is also a chromate reductase. Appl Environ Microbiol 69:4390–4395
Lay PA, Levina A (1998) Activation of molecular oxygen during the reactions of chromium (VI/V/IV) with biological reductants: implications for chromium-induced genotoxicities. J Am Chem Soc 120:6704–6714
Liu YG, WH X, Zeng GM, Li X, Gao H (2006) Cr (VI) reduction by bacillus sp. isolated from chromium landfill. Process Biochem 41(9):1981e1986
Liu ZM, Wu Y, Lei CF, Liu PM, Gao MY (2012) Chromate reduction by a chromate-resistant bacterium *Microbacterium sp*. World J Microbiol Biotechnol 28:1585–1592. https://doi.org/10.1007/s11274-011-0962-5
Losi ME, Amrhein C, Frankenberger WT (1994) Bioremediation of chromatecontaminated groundwater by reduction and precipitation in surface soils. J Environ Qual 23:1141–1150

Lovley DR, Phillips EJP (1994) Reduction of chromate by *Desulfovibrio vulgaris* and its c3 cytochrome. Appl Environ Microbiol 60:726–728

Lovley DR, Giovannoni SJ, White DC, Champine JE, Phillips EJ, Gorby YA, Goodwin S (1993) *Geobacter metallireducens* gen. Nov. sp. nov., a microorganism capable of coupling the complete oxidation of organic compounds to the reduction of iron and other metals. Arch Microbiol 159:336–344. https://doi.org/10.1007/BF00290916.

Ma Z, Zhu W, Long H, Chai L, Wang Q (2007) Chromate reduction by resting cells of Achromobacter sp. Ch-1 under aerobic conditions. Process Biochem 42(6):1028e1032

Mangaiyarkarasi MM, Vincent S, Janarthanan S, Rao TS, Tata BVR (2011) Bioreduction of Cr(VI) by alkaliphilic *Bacillus subtilis* and interaction of the membrane groups. Saudi J Biol Sci 18:157–167. https://doi.org/10.1016/j.sjbs.2010.12.003

Mazoch J, Tesarik R, Sedlacek V, Kucera I, Turanek J (2004) Isolation and biochemical characterization of two soluble iron(III) reductases from Paracoccus denitrificans. Eur J Biochem 271:553–562

McLean J, Beveridge TJ (2001) Chromate reduction by a pseudomonad isolated from a site contaminated with chromated copper arsenate. Appl Environ Microbiol 67:1076–1084

Michel C, Brugna M, Aubert C, Bernadac A, Bruschi M (2001) Enzymatic reduction of chromate: comparative studies using sulfate-reducing bacteria key role of polyheme cytochromes c and hydrogenases. Appl Microbiol Biotechnol 55:95–100

Mistry K, Desai C, Lal S, Patel K, Patel B (2010) Hexavalent chromium reduction by staphylococcus sp. isolated from Cr(VI) contaminated land fill. Int J Biotechnol Biochem 6(1):117–129

Molokwane PE, Meli KC, Nkhalambayausi-Chirwa EM (2008) Chromium (VI) reduction in activated sludge bacteria exposed to high chromium loading: Brits culture (South Africa). Water Res 42(17):4538–4548

Moore MD, Kaplan S (1992) Identification of intrinsic high-level resistance to rare-earth oxides and oxyanions in members of the class Proteobacteria: characterization of tellurite, selenite, and rhodium sesquioxide reduction in Rhodobacter sphaeroides. J Bacteriol 174:1505–1514

Moore MD, Kaplan S (1994) Members of the family Rhodospirillaceae reduce heavy-metal oxyanions to maintain redox poise during photosynthetic growth. ASM News 60:17–23

Mugerfeld I, Law BA, Wickham GS, Thompson DK (2009) A putative azoreductase gene is involved in the Shewanella oneidensis response to heavy metal stress. Appl Microbiol Biotechnol 82:1131–1141

Myers CR, Carstens BP, Antholine WE, JM(M (2000) Chromium (VI) reductase activity is associated with the cytoplasmic membrane of anaerobically grown Shewanella putrefaciens MR-1. J Appl Microbiol 88(1):98e–106

Nepple BB, Kessi J, Bachofen R (2000) Chromate reduction by Rhodobacter sphaeroides. J Ind Microbiol Biotech 25:198e203

Nies DH, Silver S (1995) Ion efflux systems involved in bacterial metal resistances. J Ind Microbiol 14:186–199

Nies DH, Koch S, Wachi S, Peitzsch N, Saier MH (1998) CHR, a novel family of prokaryotic proton motive force-driven transporters probably containing chromate/ sulfate antiporters. J Bacteriol 180:5799–5802

Ohtake H, Fujii E, Toda K (1990) Reduction of toxic chromate in an industrial effluent by use of a chromate-reducing strain of Enterobacter Cloacae. Environ Technol 11:663–668

Opperman DJ, Heerden EV (2007) Aerobic Cr (VI) reduction by Thermus scotoductus strain SA-01. J Appl Microbiol 103:1364–5072

Ortega R, Fayard B, Salome M, Deves G, Susini J (2005) Chromium oxidation state imaging in mammalian cells exposed in vitro to soluble or particulate chromate compounds. Chem Res Toxicol 18:1512–1519

Park CH, Keyhan M, Wielinga B, Fendorf S, Matin A (2000) Purification to homogeneity and characterization of a novel *Pseudomonas putida* chromate reductase. Appl Environ Microbiol 66:1788–1795. https://doi.org/10.1128/AEM.66.5.1788-1795.2000.

Pattanapipitpaisal P, Brown NL, LF(M (2001) Chromate reduction and 16S rRNA identification of bacteria isolated from a Cr(VI)-contaminated site. Appl Microbiol Biotechnol 57(1e2):257e–2261

Peitzsch N, Eberz G, Nies DH (1998) Alcaligenes Eutrophus as a bacterial chromate sensor. Appl Environ Microbiol 64:453–458

Pimentel BE, Moreno-Sanchez R, Cervantes C (2002) Efflux of chromate by Pseudomonas Aeruginosa cells expressing the ChrA protein. FEMS Microbiol Lett 212:249–254

Poljsak B, Pocs I, Raspor P, Pesti M (2010) Interference of chromium with biological systems in yeast and fungi: a review. J Basic Microbiol 50(1):21e36

Puzon GJ, Petersen JN, Roberts AG, Kramer DM, L(X (2002) A bacterial flavin reductase system reduces chromates(III)eNAD þ complex. Biochem Bio-phys Res 294(1):76e–781

Asmatullah, Qureshi SN, Shakoori AR (1998) Hexavalent chromium induced congenital abnormalities in chick embryos. J Appl Toxicol 18(3):167e171

Ramirez-Diaz MI, Diaz-Perez C, Vargas E, Riveros-Rosas H, Campos-Garcia J, Cervantes C (2008) Mechanisms of bacterial resistance to chromium compounds. Biometals 21:321–332

Robins KJ, Hooks DO, Rehm BHA, Ackerley DF (2013) Escherichia Coli NemA is an efficient chromate reductase that can be biologically immobilized to provide a cell free system for remediation of hexavalent chromium. PLoS One 8:e59200

Romanenko LA, Tanaka N, Frolova GM, Mikhailov VV (2008) Pseudochrobactrum glaciei sp. nov., isolated from sea ice collected from Peter the Great Bay of the Sea of Japan. Int J Syst Evol Microbiol 58(10):2454–2458

Saraiva LM, Da Costa PN, Legall J (1999) Sequencing the gene encoding *Desulfovibrio desulfuricans* ATCC 27774 nine-heme cytochrome c. Biochem Biophys Res Commun 262:629–634

Shakoori AR, Makhdoom M, Haq RU (2000) Hexavalent chromium reduction by a dichromate-resistant gram-positive bacterium isolated from effluents of tanneries. Appl Microbiol Biotechnol 53(3):348e351

Shen H, Wang Y-T (1993) Characterization of enzymatic reduction of hexavalent chromium by Escherichia Coli ATCC 33456. Appl Environ Microbiol 59:3771–3777

Soni SK, Singh R, Awasthi A, Kalra A (2014) A Cr(VI)-reducing *Microbacterium* sp. strain SUCR140 enhances growth and yield of Zea Mays in Cr(VI) amended soil through reduced chromium toxicity and improves colonization of arbuscular mycorrhizal fungi. Environ Sci Pollut Res 21:1971–1979. https://doi.org/10.1007/s11356-013-2098-7

Srinath T, Khare S, Ramteke PW (2001) Isolation of hexavalent chromium-reducing Cr-tolerant facultative anaerobes from tannery effluent. J Gen Appl Microbiol 47:307–312

Stearns DM, Wetterhahn KE (1997) Intermediates produced in the reaction of chromium(VI) with dehydroascorbate cause single-strand breaks in plasmid DNA. Chem Res Toxicol 10:271–278

Stewart DI, Burke IT, Mortimer RJG (2007) Stimulation of microbially mediated chromate reduction in alkaline soil–water systems. Geomicrobiol J 4:655–669

Sturm G, Jacobs J, Sproer C, Schumann P, Gescher J (2011) Leucobacter chromiiresistens sp. nov., a chromate-resistant strain. Int J Syst Evol Microbiol 61:956–960

Suzuki T, Miyata N, Horitsu H, Kawai K, Takamizawa K, Tai Y, Okazaki M (1992) NAD(P) H-dependent chromium(VI) reductase of Pseudomonas Ambigua G-1: Cr(VI) intermediate is formed during the reduction of Cr(VI) to Cr(III). J Bacteriol 174:5340–5345

Thacker U, Parikh R, Shouche Y, Madamwar D (2006) Hexavalent chromium reduction by Providencia sp. Process Biochem 41:1332–1337

Thacker U, Parikh R, Shouche Y, Madamwar D (2007) Reduction of chromate by cell-free extract of Brucella sp. isolated from Cr(VI) contaminated sites. Bioresour Technol 98:1541–1547

Thompson MR, VerBerkmoes NC, Chourey K, Shah M, Thompson DK, Hettich RL (2007) Dosage-dependent proteome response of Shewanella oneidensis MR-1 to acute chromate challenge. J Proteome Res 6:1745–1757

Thompson DK, Chourey K, Wickham GS, Thieman SB, Verberkmoes NC, Zhang B, McCarthy AT, Rudisill MA, Shah M, Hettich RL (2010) Proteomics reveals a core molecular response of pseudomonas putida F1 to acute chromate challenge. BMC Genomics 11:311

Turick CE, Apel WA, Carmiol NS (1996) Isolation of hexavalent chromium contaminated and non-contaminated environments. Appl Microbiol Biotechnol 44(5):683e688

Van Engelen MR, Peyton BM, Mormile MR, Pinkart HC (2008) Fe(III), Cr(VI), and Fe(III) mediated Cr(VI) reduction in alkaline media using a Halomonas isolate from soap Lake, Washington. Biodegradation 19:841–850

Venitt S, Levy LS (1974) Mutagenicity of chromates in bacteria and its relevance to chromate carcinogenesis. Nature (London) 250:493–495

Verma T, Singh N (2013) Isolation and process parameter optimization of Brevibacterium casei for simultaneous bioremediation of hexavalent chromium and pentachlorophenol. J Basic Microbiol 53:277–290

Viti C, Giovannetti L (2007) Bioremediation of soils polluted with hexavalent chromium using bacteria-the challenge. In: Singh SN, Tripathi RD (eds) Environmental bioremediation technologies. Springer, Berlin, pp 57–76

Viti C, Pace A, Giovannetti L (2003) Characterization of Cr (VI)-resistant bacteria isolated from chromium-contaminated soil by tannery activity. Curr Microbiol 46:1–5

Wang P-C, Mori T, Toda K, Ohtake H (1990) Membrane-associated chromate reductase activity from Enterobacter Cloacae. J Bacteriol 172:1670–1672

Wang PC, Toda K, Ohtake H, Kusaka I, Yabe I (1991) Membrane-bound respi-ratory system of Enterobacter cloacae strain HO1 grown anaerobically with chromate. FEMS Microbiol Lett 78(1):11e15

Wani R, Kodam KM, Gawai KR, Dhakephalkar PK (2007) Chromate reduction by Burkholderia cepacia MCMB-821, isolated from the pristine habitat of alkaline crater Lake. Appl Microbiol Biotechnol 75:627–632

Wise SS, Elmore LW, Holt SE, Little JE, Anto Nucci PG, Bryant BH, Pierce WSJ (2004) Telomerase mediated lifespan extension of human bron-chial cells does not affect hexavalent chromium induced cytotoxicity or geno-toxicity. Mol Cell Biochem 255(1e2):103e112

Ye Q, Roh Y, Carroll SL, Blair B, Zhou J, Zhang CL, Fields MW (2004) Alkaline anaerobic respiration: isolation and characterization of a novel alkaliphilic and metal-reducing bacteria. Appl Environ Microbiol 70:5595–5602

Yumoto I, Hirota K, Nodasaka Y, Nakajima K (2005) Oceanobacillus oncorhynchi sp. nov., a halotolerant obligate alkaliphile isolated from the skin of a rainbow trout (Oncorhynchus Mykiss), and emended description of the genus Oceanobacillus. Int J Syst Evol Microbiol 55(4):1521–1524

Zhang KD, Li FL (2011) Isolation and characterization of a chromium-resistant bacterium Serratia sp. Cr-10 from a chromate-contaminated site. Appl Microbiol Biotechnol 90:1163–1169

Chapter 14
Microbial Interaction with Metals and Metalloids: A Prospective Clean Environment

Angana Sarkar, Neha Gupta, Nitya Kumari, and Kriti Gupta

14.1 Introduction

The trace metals and metalloids that are considered priority pollutants are Silver (Ag), Zinc (Zn), Arsenic (As), Titanium (Ti), Beryllium (Be), Selenium (Se), Cadmium (Cd), Lead (Pb), Chromium (Cr), Nickel (Ni), Copper (Cu) and Mercury (Hg) (Dixit et al. 2015). They have essentially emerged from natural sources like rocks, anthropogenic inputs, metallo-ferous minerals, agriculture, mining, waste disposal, metallurgy, etc. Metals like Sodium (Na), Iron (Fe), Potassium (K), Manganese (Mn), Copper (Cu), Magnesium (Mg), Zinc (Zn), Calcium (Ca) and Cobalt (Co) are important for life but can turn toxic if present above the required amount. Whereas metals like Caesium (Cs), Aluminium (Al), Cadmium (Cd), Lead (Pb) and Mercury (Hg) do not hold any known significant metabolic function in the living organisms, but still can cause toxicity if accumulated in the environment. They exert toxicity at a very high level to the environment (Gadd 2010). Comprehensive Environmental Response Compensation and Liability Act (CERCLA), USA, has given maximum permissible limits for heavy metals [Argon (Ar), Silver (Ag), Cd, Hg, Cr and Pb] present in water. The legitimate concentration for Ar, Ag, Cd, Hg, Cr and Pb is 0.01, 0.05, 0.05, 0.002 0.01, and 0.015 mg/L respectively. The Indian standards has established norm for heavy metals in soil for Cd, Zn Cu, Pb and Ni as 3–6, 300–600, 135–270, 250–500 and 75–150 mg/kg, respectively (Nagajyoti et al. 2010).

Microbes alter their physical and chemical state so that they can interact with the heavy metals and metalloids present at the site of contamination, both natural and synthetic environment. The growth, activity and survival of the microorganisms are effected by metals and minerals. The microbes can be considered as the geo-active

A. Sarkar (✉) • N. Gupta • N. Kumari • K. Gupta
Department of Biotechnology & Medical Engineering, NIT Rourkela, Rourkela, Odisha, India
e-mail: sarkar.angana@gmail.com

J.K. Patra et al. (eds.), *Microbial Biotechnology*,
https://doi.org/10.1007/978-981-10-6847-8_14

agents. Minerals having biogenic origin are globally of geological and industrial significance. Along with this, they are also the source of many important structural components for many organisms (Ehrlich 1996; Gadd and Raven 2010).

Bioremediation is a sustainable strategy which utilizes the potential of metabolic activity of microorganisms and plants to clean-up contaminated sites. Bioremediation is a cost effective eco-friendly way of cleaning nature with the help of nature. The most important aspect of bioremediation is the microbiological aspect. The method adopted for bioremediation of contaminants at a polluted site depends upon the type and mechanism of interaction between the metals and the microbes.

Microorganisms being ubiquitous in nature, also dominates the areas containing contamination of heavy metals. They can effortlessly convert heavy metals into their non-lethal forms. In the process of bioremediation, the microorganisms produce metabolic intermediates by mineralizing organic contaminants or certain end products like carbon dioxide and water, which can be utilized for cell growth as primary substrates. Microorganism can act through a two-way defence, which includes enzymes that can degrade the target pollutants and also resistant to the appropriate heavy metal. Bioremediation includes diverse techniques like bioaccumulation, biomineralisation, biosorption, metal-microbe interactions, bioleaching and biotransformation. The forces through which the metal ions can bind to the surface of a cell are covalent bonding, electrostatic interactions, redox potential, extracellular precipitation and van der Waals forces or a combination of all these processes (Blanco 2000). The chemistry behind this binding is that the metal cations are adsorbed by the negatively charged groups (hydroxyl, phosphoryl and carboxyl) present on the cell wall of the microbes. These are then confined by the metal nucleation (Wase and Forster 1997).

14.2 Distribution of Metal and Metalloids in Environments

14.2.1 Global Distribution of Heavy Metals with Respect to Continent

14.2.1.1 Europe

In Europe, environment of different region is contaminated with various heavy metal such as Arsenic, Cadmium, lead, cobalt, nickel, chromium, manganese, cupper, and mercury etc. having concentrations of 2.84, 0.07, 0.27, 0.32, 0.15,0.26, 0.81,... 0.00005, 1.16, 2.12 and 0.8 mg/kg, respectively (Toth et al. 2016). This metal contamination affects region particularly in Western Central Europe, Greece, Central Italy and South-East Ireland. In addition, mining areas also contain toxic metals like As, Cd, Pb and Hg, etc. at elevated concentrations (Toth et al. 2016).

14.2.1.2 Asia

The agricultural soil in the South and Southeast Asian countries such as India, Peninsular Malaysia, Indonesia, Philippines, Pakistan, Vietnam, Bangladesh and Thailand are more heavily polluted by heavy metals and metalloids (Zeng et al. 2015; Sarkar et al. 2016). In Korea, rice paddy soils on the surface area (0-15cms) had moderate concentrations as 0.11 mg kg^{-1} (ranging from 0 to 1.01) for Cd, 0.47 mg kg^{-1} (0–41.6) for Cu, 4.84 mg kg-1 (0–66.4) for Pb, and 4.47 mg kg^{-1} (0–96.7) for Zn. In surface soil of the orchard fields, the average content of Cd, Cu, Pb, Zn, As, and Hg were 0.11 mg kg^{-1} (ranged from 0–0.49), 3.62 mg kg^{-1} (0.03–45.3), 2.30 mg kg^{-1} (0–27.8), 16.60 mg kg^{-1} (0.33–106), 0.44 mg kg^{-1} (0–4.14), and 0.05 mg kg^{-1} (0.01–0.54), respectively (Jo and Koh 2004). In Japan, rice had concentration as 75.9 mg kg^{-1} for Cd, 3.71 mg kg^{-1} Cu, and 22.9 mg kg^{-1} for Zn. Whereas, the rice –fields had an moderate value of 96.4 mg kg^{-1}, 19.5 mg kg^{-1} and 446 mg kg^{-1} for Zn, Cu and Cd, Cu, and Zn, respectively (Herawati et al. 2000). Soil contamination issues are also observed in China, with heavy metal pollution being the most serious issue (Luo and Teng 2006; Brus et al. 2009).

14.2.1.3 Africa

Due to increase of both industrial activities and urbanisation in Africa, heavy metal concentration (mostly heavy metal like Hg, Cd and Pb) also increased in all the parts of continent. Because of the increasing concentration of heavy metal, all environmental matrices of Africa (aquifer, sediment, soil and marine) got contaminated with these heavy metals. Mining activities have become an important source for the entry of heavy metals in the environment in most of the countries of Africa, for example, copper in Zambia, arsenic in Namibia and South Africa, mercury in Algeria and zinc in Nigeria (Fasinu and Orisakwe 2013).

14.2.1.4 North and South America

Heavy metal like chromium is common contaminant found in different countries of America. A study exposed that 75% of drinking water in America contained cancer-causing hexavalent chromium. California scientists labelled it dangerous when it surpasses a mere 0.02 parts per billion (Andrews and Walker 2016). Agricultural soils and mangrove forest of US, America also reported by contaminated with Cd, Pb, Zn, Cu and Ni (Holmgren et al. 1993; Fernández-Cadena et al. 2014).

14.2.1.5 Australia

Heavy metals like Cd, Pb, Zn and Cu has contaminated environmental matrices like soil, water, sediments, marine etc. in several regions of Australia. A study on origin and enormity of contamination, in vegetable and soil samples, of heavy metal at 46 locations across four vegetable growing areas in New South Wales, Australia reported that the area located in the proximity of smelters, such as in Port Kembla and Boolaroo, were highly contaminated with heavy metals. The data and information collected by this study suggested that for metals of interest to workshop, natural sources are surpassed by anthropogenic sources in case of Cd and Pb (Kachenko and Singh 2005).

14.2.1.6 Antarctica

Many regions of Antarctica are contaminated with heavy metals like Pb, Zn, nickel, Cu etc. The study done in soils and coastal sediments near the Brazilian Antarctic Station, King George Island revealed that the heightened concentration of metals in Ferraz surface sediments were as followed: boron, molybdenum, and lead (>90%); vanadium and zinc (70–80%); nickel, copper, magnesium, and manganese (30–40%) (Santos et al. 2005).

14.2.2 Distribution of Heavy Metals in India

Several places of India is highly contaminated with hexavalent Cr (Tamil Nadu, Uttar Pradesh, Gujarat and Orissa), Pb (Andhra Pradesh, Madhya Pradesh, Chattisgarh and Gujarat), Hg (Odisha, Madhya Pradesh and Tamil Nadu), Cu (Jharkhand, Tamil Nadu and Madhya Pradesh) and As (West Bengal, Bihar, Tamil Nadu, Uttar Pradesh and other North Eastern States of India) (Marg 2011). While anthropogenic sources like industrial activities (industrial coolants, E-waste, lead acid batteries, chromium salts manufacturing, leather tanning, bangle industry, paints, smelting operations, ceramics, coal-based thermal power plants, hospital waste, spent catalyst, sulphuric acid plant, pesticides plants etc.,) and mining (Coal, iron, chromium, uranium etc.,) play lead role to disseminate metals in the aquifer, sediments, soil or marine environments, where the geogenic source acts as the origin (Mohankumar et al. 2016). The presence of different heavy metals in India is provided in Table 14.1.

Table 14.1 Distribution of metals and metalloid in aquifer with respect to India

Sl. no	States	Heavy Metals and Metalloids above permissible limits (groundwater)	References	Heavy Metals and Metalloids above permissible limits (Sediment)	References	Heavy Metals and Metalloids above permissible limits (soil)	References
1.	Andhra Pradesh	Pb, Ni	Srinivas et al. (2013)	Cu, Cr, Ni, Zn	Machender et al. (2012)	Cr, As, Cd, Zn and Cu	Govil et al. (2001)
2.	Arunachal Pradesh	–	–	Cu	Sharma et al. 2015	–	–
3.	Assam	As, Mn	Haloi and Sarma (2012)	Cu, Ni, Pb, Co	Das et al. (2015)	Cd, Cr, Ni, Pb	Nath (2013)
4.	Bihar	As	Thakur, Gupta (2015)	Cd	Dey and Choudhary (2015)	Cd	Dey and Choudhary (2015)
5.	Chhattisgarh	Cr, Pb	Sharma et al. (2013)	lead	Patel et al. (2006)	lead	Patel et al. (2006)
6.	Goa	Fe	Singh and Kamal (2017)	–	–		
7.	Gujarat	Ni, Cd	Patel and Manoj (2015)	–	–	Cu, Cr, Ni, Zn	Krishna and Govil (2007)
8.	Haryana	–	–	–	–	Cu, Cd	Urmila et al. (2016)
9.	Himachal Pradesh	–	–	Cd	Kashyap and Vera (2015)	Pb, Cr, Cd	Sharma et al. (2014)
10.	Jammu & Kashmir	As, Ni, Cd, Cu, Pb	Raja et al. (2013)	–	–	Zn	Ali and Bhat (2014)
11.	Jharkhand	Mg	Giri et al. (2012)	Pb, Ni, Cd	Banerjee et al. (2016)	–	–
12.	Karnataka	–	–	Cr, Zn, Ni	Hejabi et al.	Cu	Kumar and Srikantaswamy (2012)
13.	Kerala	Cu, Pb	Varghese and Jaya (2014)	Cu, Pb, Zn	Sheela et al. (2012)	Zn, Pb	Prasanth et al. (2013)

(continued)

Table 14.1 (continued)

Sl. no	States	Heavy Metals and Metalloids above permissible limits (groundwater)	References	Heavy Metals and Metalloids above permissible limits (Sediment)	References	Heavy Metals and Metalloids above permissible limits (soil)	References
14.	Madhya Pradesh	Cr, Ni	Jinwal and Dixit (2008)	Cu, Pb	Shrivastava (2014)	Pb	Ahuja (2016)
15.	Maharashtra	Cu, Cr	Fazil et al. (2012)	As	Singare et al. (2012)	Pb	Sonawane et al. (2013)
16.	Manipur	Cu	Meitei and Rakesh (2013)	–	–	Cu, Cd, Pb	
17	Meghalaya	Cr, Zi, Pb, Co, Ni		–	–	Zn, Pb, Ni	Talukdar et al. (2015)
18.	Nagaland	Pb	Jamir et al. (2011)				
19.	Odisha	Ni	Dash et al. (2016)	Cd, Hg	Banerjee et al. (2017)		
20.	Punjab	Ni	Sekhon and Singh (2013)			Cd	Vanita et al. (2014)
21.	Rajasthan	Cd, Pb	Charan et al. (2015)	Cd, Cr, Cu, Ni, Pb, Zn	Sharma and Kumar (2016)	Pb, Cr, Cu, Zn	Sharma and Kumar (2016)
22.	Tamil Nadu	Pb, Cr, Cd, Hg	Kumar et al. (2012)	Cr, Ni, Zn.	Harikrishnan et al. (2015)	Cr	Dheeba and Sampathkumar (2012)
23.	Tripura	As, Mn	Banerjee et al. (2011)				
24.	Uttarakhand	Ni	Jain et al. (2010)	Cd, Cr, Pb, Cu	Maurya and Malik (2016)	Cr	Kumar and Chopra (2015)
25.	Uttar Pradesh	As, Pb, Cr	Kulshreshtha et al. (2015)			Zn, Cd	Yadav et al. (2013)
26.	West Bengal	As				As	Roychowdhury et al. (2002)

14.2.2.1 Aquifer

An aquifer is a form of saturated rock through which water can easily pass and groundwater can be harvested from there. Aquifer generally gets contaminated by several aspects like uncontrolled hazardous waste, septic system, landfill, chemical and road salts, atmospheric contaminants etc. Among all these, aquifer is mostly contaminated by heavy metals and metalloids because of their broad spectrum of sources, toxicity of many metals and difficulties in remediating metal-contaminated sites. Several reports reveal presence of extreme concentrations of many heavy metals including Hg, As, Ni, Zn, Cd, Pb, Cr, Cu, Fe and Mn, in the groundwater of rural and urban areas of Kakinada, East Godavari District, Andhra Pradesh and Olpad Taluka, Surat Dist, Gujarat (Srinivas et al. 2013; Patel and Manoj (2015). According to WHO, the recommended levels of Mn and As should be 0.4 mg/L. But in a study conducted by Haloi and Sarma (2012), 22.5% of the groundwater samples from Brahmaputra flood plain Barpeta District, Assam, India reported exceeded heavy metal concentration. Therefore, it is necessary to subject water to biological or chemical treatments, to especially keep As and Mn under arbitrary levels. Thakur and Gupta (2015) reported that the values of trace elements like As are above the permissible limit in the groundwater of some part of Bihar, India. The reason for the presence of As in natural waters is weathering of volcanic rocks, sedimentary rocks of marine origin, fossil fuels, mining wastes, mineral deposits, agricultural waste and irrigation practices. Chromium and lead are present in elevated amount in the groundwater of (Agrang Block) Raipur District, Chhattisgarh, India (Sharma et al. 2013). The study reveals famous Dal Lake in Kashmir is heavily contaminated with As, Ni, Cd, Cu, Pb (Raja et al. 2013). Detail distribution of metals/metalloids is shown in Fig. 14.1.

The study of groundwater from various locations of the mining sites (Bagjata and Banduhurang) in Jharkhand, India for heavy metal, showed that the heavy metals were generally present in amounts below the permissible limits except for Mn and Fe, which were above the threshold values for drinking water at both the study area. Using Heavy Metal Pollution Index (HPI), high concentrations of Pb and Cu were found in the groundwater of Valiathura Sewage Farm in Thiruvananthapuram district, Kerala (Varghese and Jaya 2014). Seven heavy metals: Zn, Ni, Mn, Cu, Pb, Fe and Cr were determined in the groundwater of different sites of two districts (Bhopal and Sehore) of Madhya Pradesh and it was concluded that Cr, Ni and Fe had slightly higher concentration than the limits enlisted by WHO (Jinwal and Dixit 2008). Fazil et al. (2012) study evaluated the heavy metals present in the groundwater of Beed City, Maharashtra, India and it was found that copper and chromium were present in very high concentration. From source and various tanks and tap, around 50 samples were collected from each district. Mn, Fe, Co, Li, Mg, Na, K, Cr, Ca, SE, Cd, Pb, Ni, Au, Cu, Ag, Zn, Rb and As were determined. After treatment, Cr, Pb, Fe, Mo, Co, Zn and Ni decreased but Cd, Mn and Cu increased slightly. Thus, concentration higher than expectation was found. This increase might be due to the chemicals that are used to treat water. The concentration of iron falls within the limit which is considered safe for drinking water. Certain toxic elements like Cu

Fig. 14.1 Distribution of heavy metals with respect to India in Aquifer

and Cd have a strong positive correlation with other elements, which indicates that the increase of one element may increase the concentration of other element in treated water. Dash et al. (2016) & Sekhon and Singh (2013) carried out study and revealed that concentration of Ni is present in high concentration in groundwater of Odisha and Punjab respectively. From seven sites in Bikaner city of Rajasthan, groundwater samples were collected to assess seven heavy metals viz.: Fe, Cd, Zn, Pb, Mn, Cu and Ni and it was revealed that Cd and Pb were present in high amount

(Charan et al. 2015). Assessment of groundwater quality of Tuticorin City, Tamil Nadu with respect to trace metal contamination was carried out by Kumar et al. (2012). Underground water samples were taken from Indo Bangla Border Districts of Tripura, India and studied for contamination of metals manganese and iron. The groundwater quality of District Nainital, Uttarakhand, India was assessed and it was found that due to some localized effect, the iron and lead concentrations exceeded the threshold value in one sample only (Jain et al. 2010). During pre- and post-monsoon seasons, the nickel concentration was observed to exceed the threshold value in 60 and 32.5% samples, respectively. In all the samples of the study area, the concentration of other metals, viz., Mn, Cu, Cr, Cd, and Zn were within the permissible limits. The study carried out by Kulshreshtha et al. (2015) reveals that Ground Water of Meerut Region in Uttar Pradesh is contaminated with heavy metals like Pb, Cd, As, etc.

14.2.2.2 Sediment

Sediment is the part of Earth crust that usually has been broken down due to the process of weathering and erosion. They carry elevated concentration of different heavy metals when they are transported via force of wind, water, ice, gravity and ultimately form a layer of suspended solid particles on the bottom or bed of the water bodies. Since, most of the heavy metals have high adsorption/absorption ability; they easily get accumulated on the subsurface sediment. They can also act as a nonpoint source and have potential to release the sediment-bound metals to overlying waters, and in turn adversely affect aquatic organisms (Goher et al. 2014). Detail distribution of metals/metalloids is shown in Fig. 14.2. The study carried out by Machender et al. 2012 on sediment contamination to determine range and dispersion of heavy metals (Cu, As, Cr, Zn, Ni, Pb) in Balanagar industrial area of Andhra Pradesh revealed that Cu, Cr, Ni, Zn were present in high concentrations. Sharma et al. 2015 carried out a study to measure the contamination status due to heavy metals in the sediment of Kameng river of Arunachal Pradesh and reported that the concentration of Cu levels, recommended by WHO, were exceeding. Sediment of Kabar wetland in Begusarai district of Bihar is highly contaminated by toxic Cd (Dey and Choudhary 2015) Patel et al. 2006 carried out a study to calculate the heavy metal (especially Pb) concentration in sediment of Chhattisgarh State, central India and revealed that Pb concentration in sediment is ranged from 6 to 1410 ppb. The sediment of Kuntbhyog lake of Himachal Pradesh, India and observed that Cd is present in high concentration during pre-monsoon and monsoon seasons having a value of (0.021 ± 0.004) mg/l and (0.013 ± 0.002) mg/l subsequently (Kashyap and Vera 2015). The appearance of heavy metals like Pb, Ni, and Cd found in elevated amounts in sediment in the Subarnarekha river, Jamshedpur (Banerjee et al. 2016). The research carried out by Hejabi et al. 2011 on sediment of Kabini River, Karnataka, India and found that heavy metals such that Cr Zn and Ni are present in high concentration. Sheela et al. 2012 study revealed that there was heavy metal contamination of Cu, Pb and Zn in coastal lake sediments of Southern Kerala, India.

Fig. 14.2 Distribution of heavy metals with respect to India in Sediment

Kunda River Sediment at Khargone District, Madhya Pradesh, was having high Cu and Pb contamination (Shrivastava 2014). Singare et al. 2012 conducted a study to evaluate heavy metal contaminants in sediment of Vasai Creek of Mumbai and found As in elevated amount. The experiment carried out by Banerjee et al. 2017 revealed the presence of toxic Pb and Co in Chilika Lake, Odisha India. Soil Sediments of Jaipur and Kota Industrial Areas, Rajasthan, was associated with heavy metal such as Zn,Cd, Pb, Cr, Ni and Cu having concentrations as 508.98 ppm, 58.18 ppm, 940 ppm, 417.81 ppm, 182.87 ppm and 423.98 ppm respectively

(Sharma and Kumar 2016). The study by Harikrishnan et al. (2015) showed heavy metal pollution (Cr, Ni and Zn) and potential ecological risk of sediments of East Coast of Tamilnadu. Maurya and Malik (2016) conducted a study for assessment of heavy metal distribution of sediments in Kali River of western U.P. and observed that concentration of Cd, Cr, Pb, Cu Mn and Zn were 3.38, 20.11, 81.53, 258.48, 258.48, 3.40 μg/g in sediment, respectively.

14.2.2.3 Soil

Soil is defined as the organic and inorganic material on the surface of the Earth that can act as a filtration system for surface water as well as a medium for growth of the plants. Inorganic material includes weathered rocks and minerals and organic materials include nutrients released from decomposed plants and animals. The soil pollution because of the heavy metals results in biomagnification and ultimately cause harm to the human beings. Detail distribution of metals/metalloids is shown in Fig. 14.3. The study carried out by Govil et al. (2001) revealed that load of heavy metal in the soils in the Patancheru industrial development area, Andhra Pradesh were 500 mg/kg, 200 mg/kg and 240 mg/kg, and for Cu, Cd and Cr, respectively. The heavy metal e.g. Cd, Pb, Cr, Ni, etc. contamination in soil in Sivasagar and Dibrugarh district of Assam was reported exceeding WHO recommended levels (Nath 2013). The study conducted by Krishna and Govil 2007 in Industrial Area of Surat, Gujarat, Western India, revealed that the soil in the study area contained heavy metal loads of 139.0 mg/kg, 137.5 mg/kg, 79.0 mg/kg, 305.2 mg/kg and 51.3 mg/kg, for Zn, Cu, Ni, Cr and Co. Urmila et al. 2016 carried a study for assessing heavy metal pollution in soil of Jhajjar, Haryana-India, and revealed that concentration of Pb varied from 17.82 mg/kg to 93.25 mg/kg; Cu from 17.00 mg/kg to 74.13 mg/kg; Zn from 11.31 mg/kg to 71.93 mg/kg; Cd from 0.28 mg/kg to 4.08 mg/kg and Ni from 14.05 mg/kg to 52.87 mg/kg of soil. Soil in the proximity of a paper industry located in Nahan Area of Himachal Pradesh was found to be contaminated with heavy metals like Pb, Cd and Cr (Sharma et al. 2014). Polluted soils of industrial belts of Jammu, was found to be containing Cd concentration between 0.78 to 5.11 and 0.0 to 3.8 μg g^{-1}, respectively (Ali and Bhat 2014). The range of different metals in industrial area of Mysore city, Karnataka were 6.8 mg/kg to 20.3 mg/kg for Cu, 6.6 mg/kg to 22.0 mg/kg for Cr, 66 mg/kg to 121 mg/kg for Zn and Ni from 10 mg/kg to 18.1 mg/kg for Ni(Kumar and Srikantaswamy 2012). Prasanth et al. 2013 revealed elevated amounts of Zn, and Pb in soils of Koratty Region, Kerala. The study carried out by Ahuja (2016) revealed that the surface agricultural soils had an average (Cu, Cd, As and Zn) concentration of given metals that were over two times higher than the background values and Cu, Cd and Zn were added by anthropogenic sources. Presence of lead was observed in agricultural soil in and around Toranmal (Triable Region) of Maharashtra Sonawane et al. (2013). In North Eastern States, Kakching-Wabagai Area, Thoubal District Manipur concentration of Fe and Zn were found at high limit, while in Meghalaya high concentration of Ni, Pb and Zn were detected. The study conducted by Vanita et al. (2014) in the

Fig. 14.3 Distribution of heavy metals with respect to India in Soil

agricultural soil of Amritsar, Punjab, India found that Zn (73–320 $mgkg^{-1}$) was predominant followed by Pb (7.7–118 $mgkg^{-1}$), Ni (9.67–24.32 $mgkg^{-1}$), Cu (8.4–24 $mgkg^{-1}$) and Cd (0.55–1.39 $mgkg^{-1}$). The study for measuring the contamination heavy metal in the surface soil around industrial Area, Tamil Nadu and Haridwar (Uttarakhand) revealed that heavy metals (Zn, Pb and Cr) dominated the industrial area (Dheeba and Sampathkumar 2012, **Kumar and Chopra** 2015). **The agricultural soil** in urban area of Allahabad, Uttar Pradesh, India and revealed that concentration of Zn, Cd are present in high amount (Yadav et al. 2013). The study carried out by Roychowdhury et al. 2002 for assessment of soil in arsenic affected area of West Bengal, India showed elevated amount of arsenic.

14.2.2.4 Marine

Marine ecosystem is the Earth's aquatic system which covers two-third of the surface of the Earth and thus it is important to protect it from any contamination or pollution. Marine ecosystem is polluted by heavy metals through weathering, erosion of rocks and dust particles from volcano. Heavy metals are generally associated with particles which can stay in solution for very long time and hence when introduced in marine ecosystem it can stay for longer period of time, which ultimately effects the life of aquatic ecosystem. Various heavy metals that include Zn, Cu, Cr, Cd, Pb are present at high concentration at different locations of the marine environment (Nanda 2015; Chaitanya et al. 2016).

14.3 Microbial Interaction with Metal and Metalloids

Metals like Na, Fe, K, Mn, Cu, Mg, Zn, Ca and Co are important for life but can turn toxic if present above the required amount. Whereas metals like Pb, Hg, Cs, Cd and Al do not hold any known significant metabolic function in the living organisms, but still can cause toxicity if accumulated in the environment. They exert toxicity at a very high level to the environment (Gadd 2010). Detail microbe metal interaction is shown in Fig. 14.4.

14.3.1 Metal-Microbes Interaction

Microbes use an array of mechanisms with minerals and metals present in synthetic and natural environments. Alteration in physical and chemical state helps to convert the heavy metals into non-toxic forms. The metals and minerals affect the growth of microbes, their activity and survival. Microbes are dominant in heavy metal contaminated soil and work via two-way defense that includes enzyme production that degrade the target pollutants and resistance to appropriate heavy metals (Dixit et al. 2015). There are different forms of bioremediation namely biosorption, biomineralisation, bioleaching, metal-microbe interactions, biotransformations and bioaccumulation (Table 14.2). Depending upon the mechanism of interaction of metals with the microorganisms at the contaminated site, the bioremediation strategy is adopted. Microbes can dissolve the metal and oxidise or reduce the transition metals. The metal-microbes interaction occurs via oxidizing, binding, volatizing, reducing, immobilizing and transformation of heavy metals (Table 14.2).

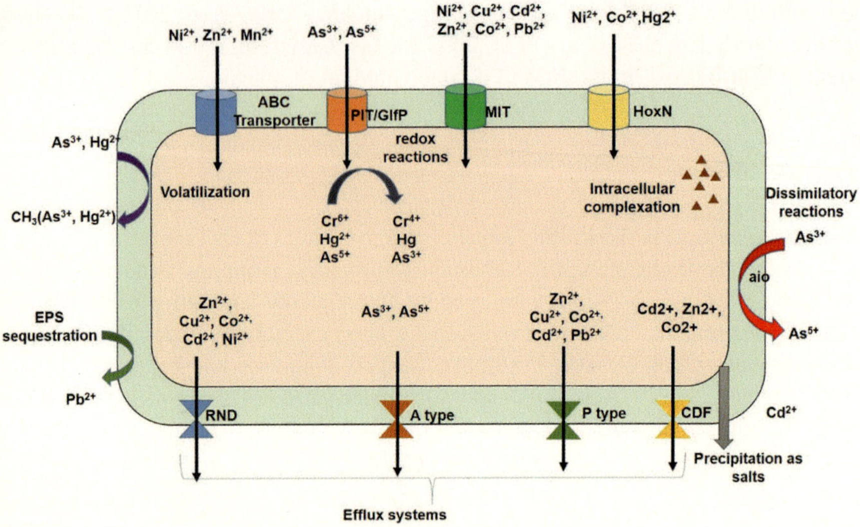

Fig. 14.4 General mechanism of metal tolerance in bacteria

Table 14.2 Microbial role and activities in biogeochemical process

Sl. No.	Metals and metalloids	Role of microorganisms	References
1.	Iron (Fe)	Iron solubilisation by organic acids, metabolites etc. Reduction: Fe^{3+} to Fe^{2+}: [redox mobilisation] Oxidation: Fe^{2+} to Fe^{3+}: [redox immobilization] Metallic sorption to Fe oxides (immobilization)	Gadd et al. (2007) Kim and Gadd (2008)
2.	Chromium(Cr)	Oxidation: Cr^{3+} to Cr^{5+} Reduction: Cr^{5+} to Cr^{3+}: [redox immobilization] Accumulation of Cr oxyanions	Gadd (2010)
3.	Lead(Pb)	Biomethylation (solubilisation) Biosorption (immobilization) Lead oxalate formation (immobilisation)	Gadd et al. (2007) Gadd (2010)
4.	Cobalt(Co), Nickel(Ni), Zinc(Zn), Cadmium(Cd)	Bioprecipitation (immobilization) Uptake and accumulation (immobilization) Biosorption (immobilization) Biomineral formation: Formation of sulphides, phosphates and carbonates (immobilisation) Co^{3+} reduction	Banfield et al. (2005) Gadd (2010)

(continued)

Table 14.2 (continued)

Sl. No.	Metals and metalloids	Role of microorganisms	References
5.	Silver(Ag)	Reduction: Ag(I) to Ag(0): [redox immobilisation] Biosorption (immobilization) Bioaccumulation	Warren and Haack (2001)
6.	Copper(Cu)	Mobilization from Cu containing minerals in rocks. Uptake and accumulation (immobilization) Biosorption (immobilization) Biopreciptation (immobilization)	Gadd et al. (2007) Gadd (2010) Warren and Haack (2001)
7.	Arsenic(As)	Biomethylation of As species (solubilisation) Reduction of As oxyanions Oxidation of As oxyanions	Adriano (2001) Adriano et al. (2004a)
8.	Mercury (Hg)	Hg biomethylation (Solubilisation) Reduction: Hg^{2+} to Hg(0): [redox mobilisation] Oxidation: Hg(0) to Hg^{2+} Volatilisation: Hg as Hg(0) Biosorption (immobilization) Accumulation	Adriano (2001) Gadd (2010)
9.	Manganese (Mn)	Oxidation: Mn^{2+} to Mn^{5+}: [redox immobilisation] Mn^{5+} reduction [redox mobilisation], indirect $Mn^{5+}O_2$ reduction by metabolites Bioaccumulation of Mn to surfaces Biosorption (immobilization) Bioprecipitation, intracellular precipitation (immobilization)	Ehrlich and Newman (2009) Gadd et al. (2007) Kim and Gadd (2008)
10.	Selenium (Se)	Reductive transformation of Se oxyanions, e.g. Se^{6+} to Se^{5+} to Se(0) [redox immobilisation] Se(0) oxidation (redox mobilisation) Biomethylation and demethylation of Se compounds (Solubilisation) assimilation of organic and inorganic Se compounds	Ehrlich (1996) Gadd (2010)
11.	Uranium(U)	Biosorption (immobilization) Intracellular precipitation (immobilisation) Reduction: U^{5+} to U^{6+}: [redox immobilization] Oxidation: U^{5+} to U^{6+} Bio mineralisation of uranium (immobilization)	Banfield et al. (2005) Gadd (2010)

14.3.1.1 Bioremediation by Adsorption

The forces through which the metal ions can bind to the surface of cell are electrostatic interactions, covalent bonding, van der Waals forces, redox interactions and extracellular precipitation or a combination of all these processes (Blanco 2000). The chemistry behind this binding is that the metal cations are adsorbed by the negatively charged groups (carboxyl, hydroxyl and phosphoryl) present on the cell wall of the microbes. These are then confined by the metal nucleation (Wase and Forster 1997). Binding of metals to extracellular surfaces immobilise the metal thus preventing their entry into the cell. Phosphoryl groups and phospholipids in bacterial lipopolysaccharides in the outer membrane also strongly interact with cationic metals.

14.3.1.2 Biosorption

Biosorption is the process that involves a biosorbent that has higher affinity towards the sorbate (metal ions) and it is extended until equilibrium occurs between the two components (Dixit et al. 2015). The extent of biosorption varies with the level of metal and microorganisms. The metals that can be extracted through this technique are U, Mn, Ni, Hg, Au, Zn, Pb, Cd, Cu, Th, Cs, Ag, and Sn.

14.3.1.3 Natural Occurrences of Metal-Microbe Interaction

Microbes are closely correlated with the biogeochemical cycle of metals and metalloids and depending upon the interaction and mechanism involved in an environment, metals are either immobilized or mobilized (Gadd 2010). Metal immobilization, in nature, can occur through extra-cellular precipitation or cellular sequestration and accumulation. Dissolution of insoluble metal-containing phases results in metal mobilization. One of the practical examples is the bioleaching of metals from the ores (Ehrlich and Brierley 1990). These methods act as an essential part for controlling biological availability of metals in soils, sediments, and water.

14.3.1.3.1 Metal Mobilisation

Metals can be mobilised by protonolysis, Fe^{3+}-binding siderophores, redox reactions, methylation and indirect Fe^{3+} attack (Gadd 2010). This can result in volatilisation of metals. Microbes can assemble metals and through redox processes can outbreak the mineral surfaces (Ehrlich 1996; Lloyd and Lovley 2001). The solubility of Fe^{3+} and Mn^{4+} is raised by reduction to Fe^{2+} and Mn^{2+} respectively. As a result of methylation, methylated derivatives of some metals and metalloids are formed that increases the mobility of these elements. Various microbes like methanogens, sulphate-reducing bacteria and clostridia act under anaerobic state and fungi such as Alternaria and Penicillium spp. act under aerobic state. Such microorganism can mediate methylation of Pb, Hg and Sn and the metalloids Te, Se and As (Gadd 2010). The methylated derivatives of these elements differ in their volatility, toxicity and solubility. The methylated compounds having a volatile nature are often lost from the soil. But the methylation of some heavy metals may not remediate the soil. For e.g. bacteria and fungi can methylate Mercuric ion (Hg^{2+}) to more toxic compound, methyl mercury [$(CH_3)Hg^+$] (Barkay and Wagner-Dobler 2005). But methyl mercury can be further methylated to dimethyl mercury, by certain bacteria, which is volatile in nature. Similarly phenyl mercury can also be converted to volatile diphenyl mercury by microbial interactions.

14.3.1.3.2 Metal Immobilization

This technique is used to decrease mobilization of metals by changing their physical and chemical states. It can be done in two ways: Ex situ and in situ immobilization, in order to remediate contaminated soil. In ex situ technique, the contaminated soil is removed from the original place but its storage is highly risky (e.g., in the case of radio nuclides). The in situ technique is applied on unexcavated soil.

The primary role of immobilisation carried out by the method biosorption, is to change the original soil metals to more geochemically stable phases. Each and every microbial material can act an effective biosorbent for the metals, excluding the alkali metal that are cations like K+ and Na+, and which can be a crucial passive method in dead and living organisms (Gadd 2010). Sorption plays an important role as bioremediation technique by influencing bioavailability and thus useful in microbe-metal-mineral interactions. Heavy metals can be reduced to lower redox state by microbes and this can reduce the mobility and toxicity for various elements. The reduction of $U^{4+\ to}$ U^{6+} forms the base for removal of uranium from the contaminated leachates and waters and also the uranium ores are formed such as uraninite (UO_2) (Lovley & Coates 1997; Landa 2005). For reductive precipitation of metals like U^{6+}, Cr^{6+}, Tc^{7+} and Pd^{2+}, sulphur- and sulphate-reducing bacteria are important (Gadd 2010). The microbial reduction of gold species and ionic silver results in formation of gold Au (0) and elemental silver Ag (0) (Kierans et al. 1991; Southam et al. 2009). Various organic and inorganic biominerals like phosphates, oxalates, oxides, sulfides and carbonates are formed by microbes, which results in the metal immobilization (Gadd et al. 2007). Iron-containing minerals in rocks, sediments and soils are weathered partially by chemical activity and partially by fungal and bacterial action. Eradication of ferric iron may cause precipitation of Fe^{3+}. Hydrous iron oxides formed by microbes can accumulate metals by co-precipitation or adsorption, in aqueous environment. Reduction of iron oxides or acidification can end up in remobilization of the adsorbed metals (Ehrlich and Newman 2009).

14.3.1.4 Mechanisms of Metal Tolerance by Microbes

It is hard to clear away heavy metals from the environments as they don't degrade easily and thus persist in the environment. The toxicity of heavy metals is generally because of the interaction of metals with enzymes (proteins) which results in inhibition of the metabolic procedures. If the presence of these metals in the environment exceeds the threshold value, it poses toxicity to all forms of life (Kumar et al. 2014). If a bacterial strain can grow in a contaminated area containing high concentration of heavy metal, it states the tolerance of metal by the present microbe. There are some microbes that have tolerance towards metals. The reason behind this is the early exposure of the microbe to the heavy metals. But in some microbes, it is believed that this property has been developed due to genetic changes after exposure to heavy metals in past years. To remove the toxic metals from contaminated area, new techniques are implemented. Biosorption is one of the important methods which is based on the ability of different materials to bind to a metal. In the process of bioremediation, the microorganisms mineralize organic contaminants to metabolic intermediates or certain end products like carbon dioxide and water, which can be utilized for cell growth as primary substrates. Microorganism can act through a two-way defense, which includes enzymes that can degrade the target pollutants and also resistant to the appropriate heavy metal (Dixit et al. 2015). Due to presence of metal in the environment, the microorganisms have developed ways for metal resistance (eg. use of a specific plasmid for a particular metal) and detoxification. (Gomathy and Sabarinathan 2010). Metal resistance mechanism can be broadly categorised into two parts:

(a) General mechanism of metal resistance
(b) Metal dependent mechanism of metal resistance

14.3.1.4.1 General Mechanism of Metal Resistance by Microbes

Binding of metals to extra cellular materials immobilizes the metals and prevents the entry into the cell. Binding of metals with microbial cells is important both ecologically and practically. Ecologically cell surface binding plays considerable part in the distribution of metals, especially in the aquatic environment. Practically, this capacity of microbes to sorb metals have been exploited for the purpose of bioremediation i.e. removal of metal contaminants from nature. Four phenomena have been observed that contribute to the general mechanism of metal resistance.

(a) **Exopolymer binding:** Extracellular polymeric substances (EPSs) or exopolymers are prevalent in nature and render defense against desiccation, phagocytosis and parasitism. Exopolymers include polysaccharides, carbohydrates, and sometimes fatty acids, nucleic acid, fatty which are responsible for extracellular binding (Schiewer and Volesky 2000). EPSs are produced by many microorganisms rendering a strong binding to the metals. EPSs prevent the toxic metals to enter the cell by mobilizing or immobilizing them and thus play a crucial role in metal cycling (Gomathy and Sabarinathan 2010). These interactions can efficiently bind lead, cadmium and uranium. The presence of negatively charged groups on exopolymer such as hydroxyl, succinyl, phosphate, amine, amide and uronic acids, contribute to binding of metals (as metals are positively charged). It results in immobilisation of metals and thus prevents their entry into the cell.

(b) **Siderophores:** These compounds that belong to the biggest well-known compounds can connect and transfer or shuttle Fe. They are highly specific Fe^{3+} ligands. Their main function is to increase the concentration of iron in the areas having very low concentrations of iron and then transfer it into the cell. Fe^{2+} along with Fe^{3+} is mediated into the cell by the siderophore. Siderophores interact with metals having similar chemical structure like iron (eg. aluminium; gallium and chromium etc.) that is, forms similar size of trivalent ions as irons. The metal bioavailability is reduced when siderophore binds to the metals and thus results in reduction of metal toxicity. For example, siderophores reduces copper toxicity in cyanobacteria (Roane and pepper 2000). The organisms have developed certain methods that can ensure Fe demand is completed either by attachment to solid iron mineral (for example Fe oxides or by production of species-specific siderophores (Gomathy and Sabarinathan 2010).

(c) **Biosurfactant complexation:** Some compounds produced by the microbes are excreted out. This class of compounds is classified as biosurfactant. They can form complex with the metals such as lead and cadmium. It increases the mobility of the resultant complex and thus increases the solubility. These complexes are non-toxic to the cells. Various researches have concluded that metal contaminated sites provide a better isolation site than the uncontaminated sites, for the biosurfactant-producing microorganism (Gomathy and Sabarinathan 2010).

(d) **Precipitation:** It causes immobilisation of metals or heavy metals, which results the soluble metals to become insoluble in nature. It may be dependent or independent on the cellular metabolism. In case of dependent precipitation, the metal removed from the solution is involved with the dynamic defense system of the microbes. While independent precipitation results from the chemical interplay between metal and the cell surface.

14.3.1.4.2 Metal Dependent Mechanism of Metal Resistance

(a) **Metallothioneins:** These are cysteine-rich proteins which have low molecular weight. They are divided into three different types according to their cysteine structure and function i.e., Cys-Cys, Cys-X-Cys and Cys-X-X-Cys. These motifs are characteristic and invariant for metallothioneins. Metallothioneins are classified into class-I metallothioneins (MTs) and they include all which are found in animals and class-II MTs includes those which are present in plants and other microbes (Fowler et al. 1987).The thiol group for the mercaptide bonds is obtained from the invariant alignment of the cysteine group (cys) and an arrangement of metal-thiolate clusters is made. In cadmium (Cd) and zinc (Zn), the alpha domain in carboxy terminal region is a metal-cys cluster.

(b) **Methylation of metals:** In this mechanism of interaction, only certain metals are involved. Hence they are considered as metal dependent mechanism of resistance. Methylation generally increases the toxicity of metals due to increased lipophilicity and thus has increased permeability through the cell membrane (Gomathy and Sabarinathan 2010). But with the aid of metal volatilisation, they are easily diffused away from the cell and metal toxicity is decreased. This phenomenon of metal volatilisation has been noticed in lead (Pb), mercury (Hg), tin (Sn), selenium (Se) and arsenic (As). For example, mercury (Hg^{2+}) is oxidized to methyl mercury and dimethyl mercury, which are volatile and very toxic forms of mercury and can rapidly spread away from the cell (Roane and pepper 2000). Metals from the contaminated surface waters can be significantly removed via methylation of metals. In gram negative and gram positive bacteria, the resistance to mercury might involve reduction of Hg^{2+} to Hg^0 (elemental form of mercury).

(c) **Biosorption:** the process of biosorption involves a biosorbent having high affinity towards sorbate (metal ions) and the interaction is extended until equilibrium is achieved between both the components (Dixit et al. 2015). The extent of biosorption varies with the level of metal and microorganisms. Biosorption mechanism can be divided into two categories on the basis of dependence on the cell's metabolism: Metabolism dependent and non-metabolism dependent. Metabolism dependent biosorption occur due to intracellular accumulation of the metal when it is transported across the cell membrane. Non-metabolism dependent biosorption occurs when the uptake of the metal is due to the physico-chemical interaction between the metal and the functional group present on the surface of the microbe. It can be observed from the biosorption of U, Cu, Ni, Zn, Pb, Cd, Hg, Th, Cs, Au, Ag, Sn and Mn. Successful remediation of Zn^{2+} and Cd^{2+} can be done by this method through the ion exchange mechanism.

(d) **Efflux system:** Plasmid-encoded energy-dependent metal efflux systems are used by certain microbes to remove the metals from the cell. They include chemiosmotic ion/ proton pumps and ATPases system which are correlated to Cadmium (Cd), Chromium (Cr) and Arsenic (Ar) resistance.

14.3.2 Interaction at Molecular Level

Mostly, efflux forms the basis for the resistance system to metals by mircobes (Fig. 14.4). Two groups of efflux are known: P-type ATPases (e.g., the Cu^{2+}, Cd^{2+}) and Zn^{2+}ATPases of gram-negative bacteria. Chemiosmotic pumps, e.g., the three constituent divalent cation efflux order czc, ncc and cnr, of *R.metallidurans* CH34 (Taghavi et al. 1997).

(a) **Lead:** Lead resistant bacterium *Ralstonia metallidurans* CH34, contains an active lead resistance operon *pbr*. The unique property of this operon is that it blend the functions included in uptake, efflux and Pb^{2+} accumulation (Borremans et al. 2001). Metallothioneins (MTs) are encrypted by the *smt* locus which consists of two individually transcribed genes *smtb* and *smtA*. The elementary role of MTs is zinc homeostasis, but Pb^{2+} is also competent of switching on the expression of *smtA*. Efflux of Pb^{2+} is mediated mostly via P-type ATPases from P_{IB} family. Some P_{IB} pumps are: CadA from *S. aureus*, ZntA from *E. coli*, CadA2 from *P. putida* KT2440 and PbrA from *C. metallidurans* (Jarosławiecka and Piotrowska-Seget 2014).

(b) **Arsenic:** Three or Five membered operons involved in arsenate and arsenite resistance contains both *ars1* and *ars2*. These operons were recognized in either the chromosomal DNA or Plasmid of several bacteria including *Corynebacterium glutamicum. Achromobacter xyloxidan, Staphylococcus aureus, Pseudomonas putida, Bacillus* sp. etc. These operons are placed at certain distance from each other, in the bacterial chromosome (Sarkar et al. 2016; Roychowdhury et al. 2002). Both of them contain genes encoding a regulatory protein *arsR*, an arsenite reductase *arsC1'* and an arsenite permease *arsB*. Arsenate reductase genes and arsenite permease and (*arsC4* and *arsB3*) were also identified scattered on the chromosome. Another type of periplasmic dissimilatory reductase, *arr* is involved in reduction of arsenate and use it in respiratory metabolism. Arsenite oxidation is facilitated by a periplasmic enzyme aio.

(c) **Zinc:** Zinc resistance is conferred by czrC operon. This gene was identified in methicillin-resistant *Staphylococcus hyicus* (Slifierz et al. 2014). From several studies, it was found that czrC gene conferred widespread zinc resistance in several microorganisms (Cavacco et al. 2010).

(d) **Copper:** *cop* is the copper-resistance operon. Three protein products of *cop* operon were characterised which provides a better understanding of copper resistance mechanism. The *cop* proteins are *copA*(72 kDa)*, copB*(39 kDa) *and copC*(12 kDa). *copA and copC* are periplasmic proteins and *copB* is an outer membrane protein. The *cop* proteins serve in the copper resistant mechanism by mediating the sequestration of copper out of the cytoplasm (Cha and Cooksey 1991).

(e) **Nickel and Cobalt:** *rcnA*(*yohM)* gene is responsible for nickel and cobalt resistance. Different studies were conducted and it was inferred that membrane bound polypeptide is encoded by the gene *yohM* that shows increased nickel and cobalt resistance in *E. coli* (Rodrigue et al. 2005). This gene was specifically induced by Co and Ni only, not by other metals like Cu, Zn or Cd. *rcnA* is proposed as the new denomination to *yohM.*

(f) **Chromium:** Several chromium resistance species that belong to different genera have been isolated with five or seven member operon. One of such strain, *Ochrobactrum tritici* strain 5bvl1 was found to contain transposon-located (TnOtChr) chromate resistance operon with five numbers of genes *chrB, chrA, chrC and chrF*. The *chrA* and *chrB* contributed to high resistance but this was not found in *chrC* or *chrF* genes (Morais et al. 2011).

(g) **Cadmium:** *cadA* and *cadC* genes conferred cadmium resistance. These genes seem to be organised in an operon and their transcription occurs in vivo and it is cadmium dependent. $cadC_{st}$ and $cadA_{st}$ were the two genes located on the chromosome of *Streptococcus thermophilus* 4134 that constituted a cadmium resistance cassette (Viti et al. 2014)

(h) **Mercury:** The reduction of Hg^{2+} to Hg(0) is mediated by mercuric reductase (*MerA*). The diversity of *MerA* is not much known. From places like sea-ice brine bacteria, freshwater and high arctic snows were isolated and seven *merA* determinants were identified (Møller et al. 2014). The two classes of mercury resistance are: narrow spectrum specifies resistance to the inorganic mercury and broad-spectrum specifies resistance to organomercurials, which is encoded by gene *merB*.

(i) **Iron:** Siderophores have the largest subgroup of known compounds that can adhere and transport, or shuffle iron. *E. coli* has six identified siderophores receptors (*Cir, Fiu, FecA, FepA. FhuE, FhuA*) which provide specificity for several ferri-siderophores. Many bacteria can take up ferrous iron anaerobically via *FeoB*. Along with this, *E. coli* contains three iron storage proteins (*FtnA, FtnB* and *Bfr,*). Out of these *FtnA* plays the major storage role (Sarkar et al. 2016).

14.3.3 Role of Microbes in Dissemination of Metals and Metalloids in Environment

All the metals/ metalloids are present, either in elemental form or in mineral form in the subsurface earth crust, from a long period of time. Metal corrosion, sediment re-suspension, atmospheric deposition, metal evaporation from water resources to soil and ground water and soil erosion of metal ions and leaching of heavy metals leads to contamination (Nriagu and Pacyna 1988). Furthermore, there are reports that reveal that natural phenomena like weathering and volcanic eruptions can also cause heavy metal pollution (Jung 2008; He et al. 2005; Shallari et al. 1998; Nriagu and Pacyna 1988). The role of microbes in metal bioremediation is well known and most desired approach. At the same time, these microbes play lead role in metal/ metalloids dissemination from subsurface earth crust. The detail process is given in Table 14.3. Microbes play a major role in biological weathering, due to which dissemination of metals and metalloids occur in environment. Organisms those care for decomposition of rocks are bacteria, fungi and other soil microbes. Microbial metabolisms directs the dissolution of minerals, including oxidation of different metal bound oxides/hydroxides results in the generation of acid mine drainage which, in turn, triggers the heavy metal contamination during mining activities. Moreover, microbial metabolism helps in the formation of different minerals over geological time. In the presence of moisture some microbes secrete carbonic acid or other different acids which corrodes the rock. For example, the microbial transformation of As where bacteria can trigger reduction and methylation of As that can result in formation of gaseous arsines, which can either cause organic As compound to mineralize into inorganic As or can lead to volatilization of As. Such conversions arouse As cycling and accumulation in the soil. Arsenic accumulation in soil leads

Table 14.3 Processes that leads to mobilization and immobilization of metals and metalloids

Mobilization	Immobilization
1. Leaching: chemolithotrophic and chemo organotrophic	1. Biosorption
2. Redox reactions resulting in mobilisation: $Fe^{3+} \rightarrow Fe^{2+}$ $Hg^{2+} \rightarrow Hg$ $Se \rightarrow Se^{4+}$, se^{6+} $Mn^{4+} \rightarrow Mn^{2+}$	2. Intracellular precipitation, intracellular accumulation, sequestration, biomineralisation
3. Methylation: Metals: Hg, Sn, Pb Metalloids: As, Se, Te	3. Redox reactions resulting in immobilisation: $Cr^{6+} \rightarrow Cr^{3+}$ $Mn^{2+} \rightarrow Mn^{4+}$ Se^{6+}, $Se^{4+} \rightarrow Se$ $Fe^{2+} \rightarrow Fe^{3+}$ $Ag(I) \rightarrow Ag$ $U^{6+} \rightarrow U^{4+}$
4. Bio corrosion of metals	4. Biomineral formation [organic and inorganic precipitation]
	5. Metal sorption to biogenic minerals.

to toxicity and contamination of the groundwater thus becomes an important issue to be looked upon. Also, arsines are the most toxic forms of As, thus their evaluation in contaminated environment is of great concern (Turpeinen et al. 2002).

14.4 Cleaning up Strategies

14.4.1 Physical Remediation

The physico-chemical feature of the pollutants present in the soil is not altered by this method. Soil washing and soil vapour extraction are the two methods used under this technique. In soil washing method, contaminated soil is physically removed and then treatment is done at a mill or off-site. After removal of the contaminants, the clarified soil is brought back to the site. Soil vapour extraction technique extracts the contaminants in vapour form, by the use of wells and pipes.

14.4.1.1 Advantages of Physical Remediation

- Soil washing technique is very cost effective since in this method the volume of the soil for treatment is reduced.
- It is effective in removing wide range of contaminants from the soil, both organic and inorganic at the same time (Khan et al. 2000)
- Broad reasonable application usually preferred in limited or on a regional scale.

14.4.1.2 Disadvantages of Physical Remediation

- It cannot be implemented in-situ.
- This method of remediated requires large area in order to set up the cleaning system. The field set up for physical remediation may vary from project to project.
- May result in production of waste that is needed to be disposed of properly (Khan et al. 2000).
- Operational skills are required which calls for high level of training.
- Soil washing technique is predominantly effective with the soil that is coarse in nature. It must have comparatively high percentage of coarse grains.
- If the wastewater or soil contains some chemical additives, some specialised treatment may be required which are rather expensive and requires skills.
- Air emissions from cleaning equipment may lead an increment in the cost of operation.

14.4.2 Chemical Remediation

In this method, chemicals are used to separate the pollutants from the contaminated media. Generally, solvent extraction and chemical oxidation is used as the remediation methods under this mechanism. Solvent extraction technique cleans the chemicals that are unable to dissolve in water and tend to sorb to sediment and soil. The solvents used desorb such chemicals and efficiently remove them from the polluted areas. In the method of chemical oxidation, oxidants are pumped into the ground and mixed with harmful chemicals. Chemicals are broken down into harmless substances by the help of oxidants, like water and carbon dioxide.

14.4.2.1 Advantages of Chemical Remediation

- They can be implemented over wide range of contaminants.
- Rapid action.
- Multiple contaminants can be treated simultaneously.
- Results in complete contaminant destruction.
- Chemical oxidation can be done in-situ. Only a small area for unit setup is required.

14.4.2.2 Disadvantages of Chemical Remediation

- Higher initial and overall costs are there.
- Involves use of hazardous chemicals and it can also result in high waste volumes.
- Equipment maintenance required.

14.4.3 Biological Remediation

Biological remediation or bioremediation proves out to be more economical over physical and chemical remediation techniques, since the pollutants here can be treated on-site. Moreover the effluent volumes generated by bioremediation are smaller to a great extent and thus it can be disposed of easily. Moreover since these practise is based upon natural processes (makes use of microbes to neutralise contaminants at a site), it is highly acceptable. As a result the hazardous substances are broken down to less toxic or non-toxic substances. Certain microorganisms require heavy metals, as essential micronutrient, in different amount to facilitate their growth and development. Microorganisms are regarded as metal accumulators due to presence of distinctive original property of remediation of toxic metals in the soil. Genetic engineering of such metal accumulators can help in expressing a missing trait and thus resulting in a differentially expressed gene. Some of the microbes used to remediate the heavy metals and metalloids from the soil are given in Table 14.4.

14.4.3.1 Zinc

Saccharomyces cerevisiae, through ion exchange mechanism, acts as a biosorbent and removes Zn^{2+}. Ectomycorrhizal fungi (*Paxillus involutus* and *Suillus granulatus* and) are able to deliver elements from wood and apatite (K, Pb, Ca, Mn and Ti) and gather them in the mycelia (Wallander et al. 2003). ectomycorrhizal fungi and Ericoid mycorrhizal have the potential to dissolve a variety of zinc-bearing minerals (Leyval and Joner 2001). *Synechococcus* sp. (cynobacterial strains) has been reported with the expression of the smtA gene and production of metal-binding protein (Gadd 2010). *Penicillium chrysogenum* and *Aspergillus niger* and provide zinc resistance and are also included in leaching.

14.4.3.2 Copper

Due to the increased uptake of metal, *Saccharomyces cerevisiae* mutants (pmr1D) are highly receptive to heavy metals. By combining biosorption with continuous metabolic uptake after physical adsorption, the mutants are able to remove Cu^{2+} from synthetic effluents. The capability of *Citrobacter spp.* to generate phosphate enzymatically results in copper precipitation (Gomathy and Sabarinathan 2010). Biosorption of copper, by Z. ramigera and C. vulgaris, occurs through both formation of coordination bonds between metals and amino and carboxyl groups of cell wall polysaccharides and adsorption.

Table 14.4 Wild type and genetically engineered bacteria for remediation of metals causing contamination

Sl. No	Metals remediated	Microbes Used		Bioremediation strategy involved	References
		Wild strain	Genetically engineered strain		
1.	Zinc(Zn)	*Saccharomyces cerevisiae, Mucor nigeri* *Aspergillus niger and penicillium chrysogenum* Ectomycorrhizal fungi, e.g.: *Suillus granulates, Paxillus involutus*	*Synechococcus* sp. (smtA gene expression)	Biosorption Leaching Mobilisation Binding	Gadd (2010)
2.	Copper (Cu)	*Aspergillus niger and penicillium chrysogenum Acidithiobacillus* sp. E.g.: *A. ferrooxidans* and *A. thiooxidans.* *Bacillus megaterium* *Citrobacter* spp. *Phanerochaete chrysosporium, Phormidium valderianum, Pseudomonas syringae* *C. vulgaris* and *Z. Ramigera*	*Saccharomyces cerevisiae* mutants(pmr1D) (Gadd G. 2010)	Leaching Siderophore Precipitation Adsorption of copper Immobilisation Biosorption	Gadd (2010) Gomathy and Sabarinathan (2010) Mohsenzadeh and Shahrokhi (2014) Rajendran et al. (2003)
3.	Nickel (Ni)	*Phormidium valderianum* *Zooglea* spp. *Aspergillus* and *Penicillium* spp. *Chromobacterium violaceum* and *Pseudomonas fluorescens*	*E.coli* (nixA gene introduced from *Helicobacter pylori*) Recombinant *Staphlococcus xylosus and S. Carnosus* *P. fluorescens* 4F39	Immobilisation Leaching Binding	Rajendran et al. (2003) Gadd (2010) Dixit et al. (2015)
4.	Cobalt (Co)	*Phormidium valderianum* *Zooglea* spp. *Aspergillus* and *Penicillium* spp.		Immobilisation Leaching	Rajendran et al. (2003) Gadd (2010)
5.	Arsenic (As)	*Escherichia coli* *Staphylococcus aureus, B.subtilis, P. putida* clostridia, methanogens and sulfate-reducing bacteria. *Cunninghamella elegans*	*E. coli* (*Arabidopsis thaliana* PC synthase gene (AtPCS) was expressed) *Arabidopsis thaliana* (*c*-ECS (GSH synthase) were expressed)	Efflux system Methylation Biosorption Accumulation	Rajendran et al. (2003) Gadd (2010) Dixit et al. (2015) Dhankher et al. (2002)

6.	Lead (Pb)	*P. aeruginosa* clostridia, methanogens and sulfate-reducing bacteria *Cunninghamella elegans* *Aspergillus parasitica* and *Cephalosporium aphidicola* Ectomycorrhizal fungi, e.g.: *Suillus granulates, Paxillus involutus*	*Salmonella choleraesuis* strain 4A *Proteus penneri* strain GM10	Methylation Biosorption	Gadd (2010) Dixit et al. (2015) Naik et al. (2012)
7.	Chromium (Cr)	*Staphylococcus aureus, B.subtilis,E. coli,P.putida*	Genetically modified *Deinococcus radiodurans*	Reduction from $Cr^{5+}$to Cr^{3+}	Rajendran et al. (2003) Brim et al. (2006)
8.	Cadmium (Cd)	*P. aeruginosa* *Citrobacter* spp., *Staphylococcus aureus, Escherichia coli, P.putida* *Bacillus subtilis* Ectomycorrhizal fungi *Suillus granulatus* and *Pisolithus tinctorius*	*Ralstonia eutropha* (czc operon), *A.xylosoxidans* (ncc operon) *Salmonella enteritica* (expression of thiosulphate reductase gene) Recombinant *Staphlococcus xylosus and S. Carnosus* *E. coli* (*Arabidopsis thaliana* PC synthase gene (AtPCS) was expressed)	Efflux system Binding	Rajendran et al. (2003) Gadd (2010) Leyval and Joner (2001)

(continued)

Table 14.4 (continued)

Sl. No	Metals remediated	Microbes Used		Bioremediation strategy involved	References
		Wild strain	Genetically engineered strain		
9.	Iron(Fe)	*Acidithiobacillus ferrooxidans* *Leptospirillum ferrooxidans* *Sulfolobus* spp. *Acidianus brierleyi* *Sulfobacillus thermosulfidooxidans* *Aspergillus, Alternaria* and *Cladosporium* *Gallionella* spp. and *Leptothrix* spp.		Oxidation Leaching Precipitation	
10.	Mercury (Hg)	*Rhizopus arrhizus* clostridia, methanogens and sulfate-reducing bacteria *Cunninghamella elegans*	*Deinococcus geothemalis* (expression of mer operon) *Cupriavidus metallidurans* strain MSR33 (genetically modified by introduction of pTP6 plasmid) Genetically modified *Pseudomonas* strain e.g.: *Pseudomonas putida* Genetically modified *Escherichia coli*	Methylation Biosorption Mercury reduction Biodegradation of Hg	
11.	Uranium (U)	*Citrobacter* spp. *Glomus intraradices* *Geobaccter* spp.		Reduction	

14.4.3.3 Cobalt

Saccharomyces cerevisiae mutants (pmr1D) have the ability to remove Co^+ from synthetic effluents by a combination of biosorption and continuous metabolic uptake after physical adsorption. *Zooglea spp.* is involved in cobalt metal uptake.

14.4.3.4 Nickel

Phormidium valderianum helps in immobilisation of nickel. *Penicillium* and *Aspergillus* spp. helps in nickel resistance. Hydrogen cyanide forming bacteria, e.g. *Pseudomonas fluorescens* and *Chromobacterium violaceum*, is able to mobilize nickel as various cyanide compounds and complexes (Gadd 2010). Phytochelatin synthase (PCS) is the expressed gene present in genetically engineered *P. fluorescens* 4F39 provides an efficient way of bioremediating nickel (Dixit et al. 2015).

14.4.3.5 Arsenic

Alcaligenes faecalis helps in bacterial oxidation of AsO_2^- to AsO_4^{3-}. *Escherichia coli, Staphylococcus aureus, Pseudomonas putida, Bacillus subtilis* are associated in arsenic resistance. These are associated with efflux systems that are plasmid-encoded energy dependent and include the chemiosmotic ion/pumps and ATPases (Rajendran et al. 2003).

14.4.3.6 Lead

The soil contaminated by lead Pb^{2+} can be biodegraded by biosorption method by utilizing the fungal species like *Cephalosporium aphidicola* and *Aspergillus parasitica* and (Dixit et al. 2015). Ectomycorrhizal fungi (*Paxillus involutus* and *Suillus granulatus*) are able to deliver elements from wood ash and apatite (Pb, K, Mn, Ca, Ti,) and gather them in the mycelia (Wallander et al. 2003). Ectomycorrhizal fungi and ericoid mycorrhizal have the potential to dissolve a variety of lead-bearing minerals (Leyval and Joner 2001). The immobilization of lead is noticed in many bacterial species, including *Azotobacter spp., Staphylococcus aureus, Citrobacter spp.*, and *Micrococcus luteus*, which can produce phosphate enzymatically and result in the precipitation of lead (Gomathy and Sabarinathan 2010).

14.4.3.7 Chromium

Enterobacter clocae or *Pseudomonas fluorescens* are involved in the reduction of CrO_4^{2-} to Cr $(OH)_3$. *Pseudomonas putida, Staphylococcus aureus, Escherichia coli, Bacillus subtilis* are related to chromium resistance. These are associated with efflux systems that are plasmid-encoded energy dependent and include the chemiosmotic ion/pumps and ATPases (Rajendran et al. 2003).

14.4.3.8 Cadmium

Ion-exchange mechanism is utilized *Saccharomyces cerevisiae* for the removal of Cd^{2+} and thus acts as biosorbent. Genetically modified *Ralstonia eutropha* is used to decrease the toxic effect of Cd^{2+} by expressing mouse metallothionein on the cell surface. Ectomycorrhizal fungi (*Paxillus involutus* and *Suillus granulatus*) are able to release elements from wood ash and apatite (Ti, K, Mn, Pb, Ca) and then gather them in the mycelia (Wallander et al. 2003). Ectomycorrhizal fungi and ericoid mycorrhizal and have the potential to dissolve a variety of cadmium-bearing minerals (Leyval and Joner 2001). *Deinococcus radiodurans* (radiation resistant bacterium) is genetically engineered to naturally reduces Cr^{4+} to Cr^{3+}. It has been done for complete toluene (fuel hydrocarbon) degradation by cloned genes of *xyl* and *tod* operons of *Pseudomonas putida* (Brim et al. 2006). *Bacillus subtilis, Staphylococcus aureus, Pseudomonas putida, Escherichia coli,* are associated with cadmium resistance. These are associated with efflux systems that are plasmid-encoded energy dependent and include the chemiosmotic ion/pumps and ATPases (Rajendran et al. 2003).

14.4.3.9 Iron

Ferrous iron can be oxidized enzymatically by certain bacteria, e.g. acidophiles such *as Acidithiobacillus ferrooxidans, Sulfobacillus thermosulfidooxidans, Acidianus brierleyi, Leptospirillum ferrooxidans* and *Sulfolobus spp.*, (Gadd 2010). *Leptothrix* spp., *Aspergillus, Cladosporium Gallionella* spp. and *Alternaria,* also display iron resistance (Ehrlich and Newman 2009).

14.4.3.10 Mercury

Mercury resistant fungi *Verticillum terrestre, Neocosmospora vasinfecta* and *Hymenoscyphus ericae,* convert Hg^{2+} to a non-toxic state, from a toxic state. Bacterium *Deinococcus geothemalis* was genetically engineered and mer operon from E.coli was added that coded for Hg^{2+} reduction. The reports showed reduction of Hg at elevated temperatures (Brim et al. 2003). Mercury resistant bacteria *Cupriavidus metallidurans* strain MSR33 was genetically modified by inserting a pTP6 plasmid. It provided genes (merB and merG) that help in the regulation of Hg biodegradation along with the synthesis of mercuric reductase (MerA) and organomercurial lyase protein (MerB) (Dixit et al. 2015). Generally, two different methods for Hg degradation by bacteria like *Klebsiella pneumonia* M426 are present: mercury volatilization by reduction of Hg^{2+} to Hg (0) and mercury precipitation as insoluble Hg. Phytochelatin 20 expression on the cell surface of *Escherichia coli* and *Moreaxella* sp. gathers 25 times more Hg than the wild-type strains (Bae et al. 2001).

14.4.3.11 Uranium

Geobacter species can change the state of Uranium from its soluble state U^{6+} to insoluble state U^{4+} and immobilise them. *Citrobacter* spp. and *Glomus intraradices* also confer uranium resistance in contaminated soil (Gadd 2010).

14.5 Conclusion

Heavy metals are essential and important trace elements but, as the concentration of these heavy metals increases due to natural or industrial activities, it becomes toxic to many microbes. Microbes, on the other hand, have adapted to tolerate the metals or can even use them to grow. Hence this interaction between microbes and metals on environmental matrices is an essential part of Earth's biogeochemical cycle. Such type of activities has both negative as well as positive effects on the environment. Both processes, mobilization and immobilization, dissociation and association are governed by microbial metabolisms/catabolisms. These activities are the basis of microbe mediated bioremediation. Bioremediation using microbes shows excellent implication of interaction between metal and microbes which will be more promising when genetic engineering will come in picture.

References

Adriano DC (2001) Trace elements in the terrestrial environment: biogeochemistry, bioavailability and risks of metals, 2nd edn. Springe, New York

Ahuja R (2016) Characterization of heavy metals in contaminated agricultural soil. Int J Theor Appl Sci 8(1):38–40

Ali M, Bhat AK (2014) Soil microbiological indices of polluted soils of industrial belts of Jammu, India. Int J Curr Microbiol App Sci 3(1):559–576

Andrews D, Walker B (2016) U.S. EPA, Occurrence data for the unregulated contaminant monitoring rule. www.epa.gov/dwucmr/occurrence-data-unregulated-contaminant-monitoring-rule#3

Bae W, Mehra RK, Mulchandani A, Chen W (2001) Genetic engineering of Escherichia coli for enhanced uptake and bioaccumulation of mercury. Appl Environ Microbiol 67(11):5335–5338

Banerjee S, Das B, Umlong IM, Devi RR, Kalita H, Saikia LB, Borah K, Raul PK, Singh L (2011) Heavy metal contaminants of underground water in Indo Bangla Border Districts of Tripura, India. Int J ChemTech Res 3(1):516–522

Banerjee S, Kumar A, Maiti SK, Chowdhury A (2016) Seasonal variation in heavy metal contaminations in water and sediments of Jamshedpur stretch of Subarnarekha river, India. Environ Earth Sci 75(3):1–12

Banerjee S, Pramanik A, Sengupta S, Chattopadhyay D, Bhattacharyya M (2017) Distribution and source identification of heavy metal concentration in Chilika Lake, Odisha India: an assessment over salinity gradient. Curr Sci 112(1):87

Banfield JF, Cervini-Silva J, Nealson KH (eds) (2005) Molecular geomicrobiology, Reviews in Mineralogy and Geochemistry, vol 59. Mineralogical Society of America, Washington, DC

Barkay T, Wagner-Dobler I (2005) Microbial transformations of mercury: potentials, challenges, and achievements in controlling mercury toxicity in the environment. Adv Appl Microbiol 57:1–52

Blanco A (2000) Immobilization of nonviable cyanobacteria and their use for heavy metal adsorption from water. In: Oluguin EJ, Sanchez G, Hernandez E (eds) Environmental biotechnology and cleaner bioprocesses. Taylor and Amp Francis, Philadelphia, p 135

Borremans B, Hobman JL, Provoost A, Brown NL, van Der Lelie D (2001) Cloning and functional analysis of thepbr lead resistance determinant of *Ralstonia metallidurans* CH34. J Bacteriol 183(19):5651–5658

Brim H, Venkateswaran A, Kostandarithes HM, Fredrickson JK, Daly MJ (2003) Engineering Deinococcus geothermalis for bioremediation of high-temperature radioactive waste environments. Appl Environ Microbiol 69(8):4575–4582

Brim H, Osborne JP, Kostandarithes HM, Fredrickson JK, Wackett LP, Daly MJ (2006) Deinococcus radiodurans engineered for complete toluene degradation facilities Cr(IV) reduction. Microbiology 152:2469–2477

Brus DJ, Li Z, Song J, Koopmans GF, Temminghoff EJ, Yin X, Yao C, Zhang H, Luo Y, Japenga J (2009) Predictions of spatially averaged cadmium contents in rice grains in the Fuyang Valley, PR China. J Environ Qual 38(3):1126–1136

Cavaco LM, Hasman H, Stegger M, Andersen PS, Skov R, Fluit AC, Aarestrup FM (2010) Cloning and occurrence of czrC, a gene conferring cadmium and zinc resistance in methicillin-resistant Staphylococcus aureus CC398 isolates. Antimicrob Agents Chemother 54(9):3605–3608

Cha JS, Cooksey DA (1991) Copper resistance in pseudomonas syringae mediated by periplasmic and outer membrane proteins. Proc Natl Acad Sci 88(20):8915–8919

Chaitanya I, Satyaprakash M, Reddy TB (2016) Bioaccumulation of heavy metals in marine fish samples at Visakhapatnam and Bheemili region, north east coast of Andhra pradesh, India. Int J Sci Environ Technol 5:1718–1729

Charan PD, Singh M, Rakhecha P, Jakhar AK, Bithoo KS, Meena MK (2015) Study of heavy metals concentration in ground water samples collected from Bikaner city, Rajasthan. Int J Eng Res Manag Technol 2(5):16–17

Das P, Kumar M, Sarma KP (2015) Speciation of heavy metals in surface sediment of the Brahmaputra river, Assam, India. J Environ Res Dev 9.3A:944–952

Dash A, Das HK, Mishra B (2016) Heavy metals contamination of ground water in and around Joda of Keonjhar district, Odisha, India. IOSR J Environ Sci 10(10):44–50

Dey R, Choudhary SK (2015) Heavy metal in sediments of Kabar Lake, a tropical wetland in Begusarai district of Bihar. Ecol Environ Cons 22(3):1509–1515

Dhankher OP, Li YJ, Rosen BP, Shi J, Salt D, Senecoff JF, Sashti NA, Meagher RB (2002) Engineering tolerance and hyperaccumulation of arsenic in plants by combining arsenate reductase and gamma-glutamylcysteine synthetase expression. Nat Biotechnol 20:1140–1145

Dheeba B, Sampathkumar P (2012) Evaluation of heavy metal contamination in surface soil around industrial area, Tamil Nadu, India. Int J ChemTech Res 4(3):1229–1240

Dixit R, Malaviya D, Pandiyan K, Singh UB, Sahu A, Shukla R, Singh BP, Rai JP, Sharma PK, Lade H, Paul D (2015) Bioremediation of heavy metals from soil and aquatic environment: an overview of principles and criteria of fundamental processes. Sustainability 7(2):2189–2212

Ehrlich HL (1996) Microbes and metals. Appl Microbiol Biotechnol 48(6):687–692

Ehrlich HL, Brierley CL (1990) Microbial mineral recovery. McGrawHill, New York

Ehrlich HL, Newman DK (2009) Geomicrobiology, 5th edn. CRC Press/Taylor & Francis, Boca Raton

Fasinu PS, Orisakwe OE (2013) Heavy metal pollution in sub-Saharan Africa and possible implications in cancer epidemiology. Asian Pac J Cancer Prev 14(6):3393–3402

Fazil MI, Iqbalb MA, Abdullah S (2012) A study on heavy metal ion contamination of groundwater reserves in Beed City, Maharashtra, India. Bull Environ Pharmacol Life Sci 1(1):18–21

Fernández-Cadena JC, Andrade S, Silva-Coello CL, De la Iglesia R (2014) Heavy metal concentration in mangrove surface sediments from the north-west coast of South America. Mar Pollut Bull 82(1):221–226

Fowler BA, Hildebrand CE, Kojima Y, Webb M (1987) Nomenclature of metallothionein. In: Kagi JHR, Kojima Y (eds) Metallothionein II, pp 19–22
Gadd GM (2010) Metals, minerals and microbes: geomicrobiology and bioremediation. Microbiology 156(3):609–643
Gadd GM, Raven JA (2010) Geomicrobiology of eukaryotic microorganisms. Geomicrobiol J 27(6-7):491–519
Gadd GM, Burford EP, Fomina M, Melville K (2007) Mineral transformation and biogeochemical cycles: a geomycological perspective. In: Gadd GM, Dyer P, Watkinson S (eds) Fungi in the environment. Cambridge University Press, Cambridge, pp 78–111
Giri S, Mahato MK, Singh G, Jha VN (2012) Risk assessment due to intake of heavy metals through the ingestion of groundwater around two proposed uranium mining areas in Jharkhand, India. Environ Monit Assess 184(3):1351–1358
Goher ME, Farhat HI, Abdo MH, Salem SG (2014) Metal pollution assessment in the surface sediment of Lake Nasser, Egypt. Egyptian J Aquatic Res 40(3):213–224
Gomathy M, Sabarinathan KG (2010) Microbial mechanisms of heavy metal tolerance-a review. Agric Rev 31(2):133–138
Govil P, Reddy G, Krishna A (2001) Contamination of soil due to heavy metals in the Patancheru industrial development area, Andhra Pradesh, India. Environ Geol 41(3):461–469
Gupta S, Bhatnagar M, Jain R (2003) Physico-chemical characteristics and analysis of Fe and Zn in tubewell water and sewage water of Bikaner City. Asian J Chem 15(2):727
Haloi N, Sarma HP (2012) Heavy metal contaminations in the groundwater of Brahmaputra flood plain: an assessment of water quality in Barpeta District, Assam (India). Environ Monit Assess 184(10):6229–6237
Harikrishnan N, Suresh Gandhi M, Chandrasekaran A, Ravisankar R (2015) Assessment of heavy metal pollution and potential ecological risk of sediments of East Coast of Tamilnadu by Energy Dispersive XRay Fluorescence Spectroscopy (EDXRF) and Sediment Quality Guidelines (SQGS). J Heavy Metal Toxicity Dis 3:1–7
He ZL, Yang XE, Stoffella PJ (2005) Trace elements in agroecosystems and impacts on the environment. J Trace Elem Med Biol 19(2):125–140
Hejabi AT, Basavarajappa HT, Karbassi AR, Monavari SM (2011) Heavy metal pollution in water and sediments in the Kabini River, Karnataka, India. Environ Monit Assess 182(1):1–13
Herawati N, Suzuki S, Hayashi K, Rivai IF, Koyama H (2000) Cadmium, copper, and zinc levels in rice and soil of Japan, Indonesia, and China by soil type. Bull Environ Contam Toxicol 64(1):33–39
Hoffland E, Kuyper TW, Wallander H, Plassard C, Gorbushina AA, Haselwandter K, Holmstrom S, Landeweert R, Lundstrom US, authors o (2004) The role of fungi in weathering. Front Ecol Environ 2(5):258–264
Holmgren GGS, Meyer MW, Chaney RL, Daniels RB (1993) Cadmium, lead, zinc, copper, and nickel in agricultural soils of the United States of America. J Environ Qual 22(2):335–348
Jain CK, Bandyopadhyay A, Bhadra A (2010) Assessment of ground water quality for drinking purpose, District Nainital, Uttarakhand, India. Environ Monit Assess 166(1):663–676
Jamir TT, Devi WB, Singh UI, Singh RB (2011) Lead, iron and manganese contamination in spring pond and well water in Nagaland, one of the seven North-Eastern States of India, A future danger. J Chem Pharama Res 3:403–411
Jarosławiecka A, Piotrowska-Seget Z (2014) Lead resistance in microorganisms. Microbiology 160(1):12–25
Jinwal A, Dixit S (2008) Pre and post monsoon variation in physio-chemical characteristic in groundwater quality in Bhopal, India. Asian J Exp Sci 22:311–316
Jo IS, Koh MH (2004) Chemical change in agricultural soils of Korea: date review and suggested countermeasures. Environ Geochem Health 26:105–107
Jung MC (2008) Heavy metal concentrations in soils and factors affecting metal uptake by plants in the vicinity of a Korean Cu-W mine. Sensors 8(4):2413–2423
Kachenko AG, Singh B (2005) Heavy metals contamination in vegetables grown in urban and metal smelter contaminated sites in Australia. Water Air Soil Pollut 169(1):101–123

Kashyap R, Vera KS (2015) Seasonal variation of certain heavy metals in kuntbhyog lake of Himachal Pradesh, India. J Environ Ecol Fam Urb Stud 1(1):15–26
Khan AG, Kuek C, Chaudhry TM, Khoo CS, Hayes WJ (2000) Role of plants, mycorrhizae and phytochelators in heavy metal contaminated land remediation. Chemosphere 41(1):197–207
Kim BH, Gadd GM (2008) Bacterial physiology and metabolism. Cambridge University Press, Cambridge
Kierans M, Staines AM, Bennett H, Gadd GM (1991) Silver tolerance and accumulation in yeasts. Biol Met 4:100–106
Krishna AK, Govil PK (2007) Soil contamination due to heavy metals from an industrial area of Surat, Gujarat, Western India. Environ Monit Assess 124(1):263–275
Kulshreshtha A, Soni RK, Shinde CP (2015) Quantitative estimation of heavy metals in ground water in Meerut Region in Uttar Pradesh. J Appl Pharm Sci 8(8):46–49
Kumar SD, Srikantaswamy S (2012) Heavy metals pollution assessment in industrial area soil of Mysore city, Karnataka, India. Int J Appl Sci Eng Res 1(4):604–611
Kumar V, Chopra AK (2015) Heavy metals accumulation in soil and agricultural crops grown in the province of Asahi India Glass Ltd., Haridwar (Uttarakhand), India. Adv Crop Sci Technol:1–6
Kumar SK, Magesh NS, Chandrasekar N (2012) Trace Element Concentration in Groundwater, Tuticorin City, Tamil Nadu, India. Bull Environ Contam Toxicol 88(6):876–879
Kumar B, Kumari S, Flores LC (2014) Plant mediated detoxification of mercury and lead. Arabian J Chem; In Press
Landa ER (2005) Microbial biogeochemistry of uranium mill tailings. Adv Appl Microbiol 57:113–130
Leyval C, Joner EJ (2001) Bioavailability of heavy metals in the mycorrhizosphere. In: Gobran GR, Wenzel WW, Lombi E (eds) Trace elements in the rhizosphere. CRC Press, Boca Raton, pp 165–185
Lovley DR, Coates JD (1997) Bioremediation of metal contamination. Curr Opin Biotechnol 8:285–289
Lloyd JR, Lovley DR (2001) Microbial detoxification of metals and radionuclides. Curr Opin Biotechnol 12(3):248–253
Luo YM, Teng Y (2006) Status of soil pollution degradation and countermeasures in China. Soil Sci 38(5):505–508
Machender G, Dhakate R, Mallikharjuna ST, Mangaraja RB, Prasanna RL (2012) Heavy metal contamination in sediments of Balanagar industrial area, Hyderabad, Andra Pradesh, India. Arab J Geosci 7(2):513–525
Marg BZ (2011) Hazardous metals and minerals pollution in India: sources, toxicity and management. A position paper, Indian National Science Academy, New Delhi
Maurya PK, Malik DS (2016) Distribution of heavy metals in water, sediments and fish tissue (*Heteropneustis fossilis*) in Kali River of western U.P. India. Int J Fish Aquat Stud 4(2):208–215
Meitei LS, Rakesh K (2013) A comparative study of the ground and surface water quality with reference to heavy metal concentrations in the Imphal valley Manipur, India. Int J Environ Sci 3(6):1857
Mohankumar K, Hariharan V, Rao NP (2016) Heavy metal contamination in groundwater around industrial estate vs residential areas in Coimbatore, India. J Clin Diagn Res JCDR 10(4):BC05
Mohsenzadeh F, Shahrokhi F (2014) Biological removing of Cadmium from contaminated media by fungal biomass of Trichoderma species. J Environ Health Sci Eng 12(1):102
Møller AK, Barkay T, Hansen MA, Norman A, Hansen LH, Sørensen SJ, Boyd ES, Kroer N (2014) Mercuric reductase genes (merA) and mercury resistance plasmids in High Arctic snow, freshwater and sea-ice brine. FEMS Microbiol Ecol 87(1):52–63
Morais PV, Branco R, Francisco R (2011) Chromium resistance strategies and toxicity: what makes Ochrobactrum tritici 5bvl1 a strain highly resistant. Biometals 24(3):401–410
Nagajyoti PC, Lee KD, Sreekanth TVM (2010) Heavy metals, occurrence and toxicity for plants: a review. Environ Chem Lett 8(3):199–216

Naik MM, Pandey A, Dubey SK (2012) Biological characterization of leadenhanced exopolysaccharide produced by a lead resistant Enterobacter cloacae strain P2B. Biodegradation 23:775–783

Nanda P (2015) Bioaccumulation of heavy metals and physiological response in anabas testudineus on exposure to paper mill effluent. J Environ Anal Toxicol 5(1):1

Nath TN (2013) Heavy metals contamination of tea estates soil in Sivasagar and Dibrugarh district of Assam, India. Int J Adv Res Technol 2(4):2278–7763

Pacyna JM, Nriagu JO (1988) Quantitative assessment of worldwide contamination of air, water and soils by trace metals. Nature 333(6169):134–139

Patel M, Manoj K (2015) Assessment of heavy metals in potable ground water of Olpad Taluka, Surat, Gujarat, India. Int J Curr Microbiol App Sci 4:124–128

Patel KS, Shrivas K, Hoffmann P, Jakubowski N (2006) A survey of lead pollution in Chhattisgarh State, central India. Environ Geochem Health 28(1):11–17

Prasanth KM, Sreekala PP, Sandeep S, Kripa PK, Sreejesh KK (2013) Heavy metals and its fractions in soils of Koratty Region, Kerala. Res J Recent Sci 2:171–176

Raja IA, Khan MY, Khan NA, Wani MR, Bhat AA (2013) Assessment of some metals in the drinking water of Dal Lake Kashmir. Nature Sci 11(3):63–64

Rajendran P, Muthukrishnan J, Gunasekaran P (2003) Microbes in heavy metal remediation. Indian J Exp Biol 41(9):935–944

Roane TM, Pepper IL (2000) Microorganisms and metal pollutants. Environ Microbiol:421–441

Rodrigue A, Effantin G, Mandrand-Berthelot MA (2005) Identification of rcnA (yohM), a nickel and cobalt resistance gene in Escherichia coli. J Bacteriol 187(8):2912–2916

Roychowdhury T, Uchino T, Tokunaga H, Ando M (2002) Arsenic and other heavy metals in soils from an arsenicaffected area of West Bengal, India. Chemosphere 49(6):605–618

Santos IR, ilho EVS, Schaefer C, Filho MRA, Campos LS (2005) Heavy metal contamination in coastal sediments and soils near the Brazilian Antarctic Station, King George Island. Mar Pollut Bull 50(2):185–194

Sarkar A, Paul D, Kazy SK, Sar P (2016) Molecular analysis of microbial community in arsenic-rich groundwater of Kolsor, West Bengal. J Environ Sci Health A 51(3):229–239

Schiewer S, Volesky B (2000) Biosorption processes for heavy metal removal. In: Environmental microbe-metal interactions. American Society of Microbiology, pp 329–362

Sekhon GS, Singh B (2013) Estimation of heavy metals in the groundwater of Patiala District of Punjab, India. Earth Resour 1(1):1–4

Shallari S, Schwartz C, Hasko A, Morel JL (1998) Heavy metals in soils and plants of serpentine and industrial sites of Albania. Sci Total Environ 209(2-3):133–142

Sharma A, Kumar A (2016) Assessment of heavy metal contamination in soil sediments of Jaipur and Kota Industrial Areas, Rajasthan, India. International Journal of Engineering, Management & Sciences (IJEMS) 3(10):1–7

Sharma P, Dubey A, Chatterjee SK (2013) Determination of heavy metals in surface and ground water in an around (Agrang Block) Raipur District, Chhattisgarh, India. Int J Scient Eng Res 4:722–724

Sharma MC, Baxi S, Sharma KK, Singh M, Patel S (2014) Heavy metal ions levels and related physicochemical parameters in soils in the vicinity of a paper industry location in Nahan Area of Himachal Pradesh. J Environ Anal Toxicol 4(6):1

Sharma BB, Sarma HP, Borah L (2015) Chemical speciation of copper and cadmium in Kameng river sediments using sequential extraction procedure. Int J Environ Sci 6(1):88–96

Sheela AM, Letha J, Joseph S, Thomas J (2012) Assessment of heavy metal contamination in coastal lake sediments associated with urbanization: Southern Kerala, India. Lakes Reserv Res Manag 17(2):97–112

Shrivastava V (2014) Geochemical assessment of heavy metal pollution and toxicity of Kunda River Sediment at Khargone District, Madhya Pradesh, India. Int J Eng Res Technol (IJERT) 3(2):329–333

Singare PU, Trivedi MP, Ravindra M (2012) Sediment heavy metal contaminants in Vasai Creek of Mumbai: pollution impacts. Am J Chem 2(3):171180

Singh G, Kamal RK (2017) Heavy metal contamination and its indexing approach for groundwater of Goa mining region, India. Appl Water Sci 7(3):1479–1485

Slifierz MJ, Friendship RM, Weese JS (2014) Methicillin-resistant Staphylococcus aureus in commercial swine herds is associated with disinfectant and zinc usage. Appl Environ Microbiol 81(8):2690–2695

Sonawane NS, Sawant CP, Patil RV (2013) Soil quality assessment and heavy metal contamination in agricultural soil in and around Toranmal (Triable Region) of Maharashtra. Arch Appl Sci Res 5(2):294–298

Southam G, Lengke MF, Fairbrother L, Reith F (2009) The biogeochemistry of gold. Elements 5:303–307

Srinivas J, Purushotham AV, Murali Krishna KVSG (2013) A study of heavy metals contamination in surface and groundwater of rural and urban areas of Kakinada, East Godavari district, A. P. Int J Civil, Structur, Environ Infrastructur Engineer Res Develop 3:231–236

Taghavi S, Mergeay M, Van der Lelie D (1997) Genetic and physical maps of the Alcaligenes eutrophusCH34 Megaplasmid pMOL28 and its derivative pMOL50 obtained after temperature-induced mutagenesis and mortality. Plasmid 37(1):22–34

Talukdar B, Basumatary S, Kalita HK, Baishya RA, Dutta A, Srivastava SK, Sarma D (2015) Histopathological alternations in liver and kidney of Tor tor (Ham) inhabited in coal mining affected areas of Simsang River, Garohills; Meghalaya. Natl Acad Sci Lett 38(4):321–324

Thakur BK, Gupta V (2015) Groundwater arsenic contamination in Bihar: causes, issues and challenges. MANTHAN J Commer Manag 2(1)

Toth G, Hermann T, Szatmari G, Pasztor L (2016) Maps of heavy metals in the soils of the European Union and proposed priority areas for detailed assessment. Sci Total Environ 565:1054–1062

Turpeinen R, Pantsar-Kallio M, Kairesalo T (2002) Role of microbes in controlling the speciation of arsenic and production of arsines in contaminated soils. Sci Total Environ 285(1):133–145

Urmila, Garg A, Annu (2016) Assessment of heavy metal pollution in soil of Jhajjar, Haryana-India. 8(5):629–634

Vanita C, Piar C, Avinash N, Kaur KJ, Yogesh BP (2014) Evaluation of heavy metals contamination and its genotoxicity in agricultural soil of Amritsar, Punjab, India. Int J Res Chem Environ 4(4):20–28

Varghese J, Jaya DS (2014) Metal pollution of groundwater in the vicinity of Valiathura sewage farm in Kerala, South India. Bull Environ Contam Toxicol 93(6):694–698

Viti C, Marchi E, Decorosi F, Giovannetti L (2014) Molecular mechanisms of Cr (VI) resistance in bacteria and fungi. FEMS Microbiol Rev 38(4):633–659

Wallander H, Mahmood S, Hagerberg D, Johansson L, Pallon J (2003) Elemental composition of ectomycorrhizal mycelia identified by PCR-RFLP analysis and grown in contact with apatite or wood ash in forest soil. FEMS Microbiol Ecol 44:57–65

Warren LA, Haack EA (2001) Biogeochemical controls on metal behaviour in freshwater environments. Earth Sci Rev 54:261–320

Wase J, Forster CF (1997) Biosorbents for metal ions. Taylor & Francis, London

Yadav A, Yadav PK, Shukla DN (2013) Investigation of heavy metal status in soil and vegetables grown in urban area of Allahabad, Uttar Pradesh, India. J Sci Res Pub 3(9):1–7

Zeng F, Wei W, Li M, Huang R, Yang F, Duan Y (2015) Heavy metal contamination in rice-producing soils of Hunan province, China and potential health risks. Int J Environ Res Public Health 12(12):15584–15593

Chapter 15
Biodetoxification of Toxic Heavy Metals by Marine Metal Resistant Bacteria- A Novel Approach for Bioremediation of the Polluted Saline Environment

Ranjan Kumar Mohapatra, Pankaj Kumar Parhi, Jayanta Kumar Patra, Chitta Ranjan Panda, and H.N. Thatoi

15.1 Introduction

Heavy metal contamination of the environment is considered as a major threat to ecology and health of the living being due to their toxicity, widespread occurrence and accumulation in the food chain (Iyer et al. 2005; Aryal and Liakopoulou-Kyriakides 2015; Alvarez et al. 2017).With the rapidly increasing industrialization and urbanisation activities, these pollutants are directly or indirectly being discharged to the environment causing serious pollution in the ecosystem and threatening to the biological life (Wang 2002; Das et al. 2008; Devika et al. 2013). Elements with atomic masses more than 50 amu are known as heavy metals (Weast 1984; Voica et al. 2016). At concentrations higher than the prescribed limits, these are not only poisonous to the ecosystem but also toxic to the human body in various ways, such as disruption of cell membranes, denaturation of DNA, alteration of enzymatic activity, induction of carcinogenicity etc. (Voica et al. 2016). Some heavy metals (e.g. Cu, Ni, Co, Mn, Zn and Fe) are essential elements for the existence of living organisms as they are associated with several cellular and biochemical reactions. However, many other metals (e.g. Cr, Cd, As, Pb, Hg etc) are harmful due to their

R.K. Mohapatra (✉) • C.R. Panda
Environment and Sustainability Department, CSIR-Institute of Minerals and Materials Technology, Bhubaneswar, Odisha, India
e-mail: mranjankumar.biotech@gmail.com

P.K. Parhi
School of Applied Sciences and School of Chemical Technology, School of Biotechnology, KIIT University, Bhubaneswar, India

J.K. Patra
Research Institute of Biotechnology and Medical Converged Science, Dongguk University, Goyang-si, Gyeonggi-do, South Korea

H.N. Thatoi
Department of Biotechnology, North Orissa University, Baripada, Odisha, India

J.K. Patra et al. (eds.), *Microbial Biotechnology*,
https://doi.org/10.1007/978-981-10-6847-8_15

toxicity, xenobiotic nature and lack of any biological function (Singh et al. 2011; Fan et al. 2014; Aryal and Liakopoulou-Kyriakides 2015; Pepi et al. 2016; Alvarez et al. 2017). The heavy metals (e.g. Cr(VI), Pb(II), Cd(II), As(III/V)) cannot be degraded easily by the natural processes and therefore can exist for a long time in the environment resulting in bioaccumulations and biomagnifications inside organisms (Tchounwou et al. 2012; Chaudhary et al. 2014; Voica et al. 2016).

In the detoxification process, heavy metals could not be degraded by any physical or chemical means but some of them could be transformed from more toxic forms to relatively less toxic form under certain conditions (Aryal and Liakopoulou-Kyriakides 2015). Physicochemical remediation of heavy metal contaminated environments is not convenient due to high energy consumption, high cost and incomplete metal removal (Malik 2004; Voica et al. 2016). Whereas bioremediation using metal resistant bacteria have been considered as potential alternative for clean up and biodetoxification of such areas in an efficient cost effective and environmental friendly way (Naik et al. 2012; Voica et al. 2016). Bacteria possesses varieties of mechanisms (Fig. 15.1) to tolerate and bioremediate high concentration of toxic heavy metals in various ways such as precipitation (as phosphate, sulphides and carbonates), volatisation (via methylation/ ethylation/ reduction), ATP mediated efflux system, intracellular bioaccumulation (mediated by metallothionein proteins), biosorption at cell surface and sequestration in extra cellular polymeric substances

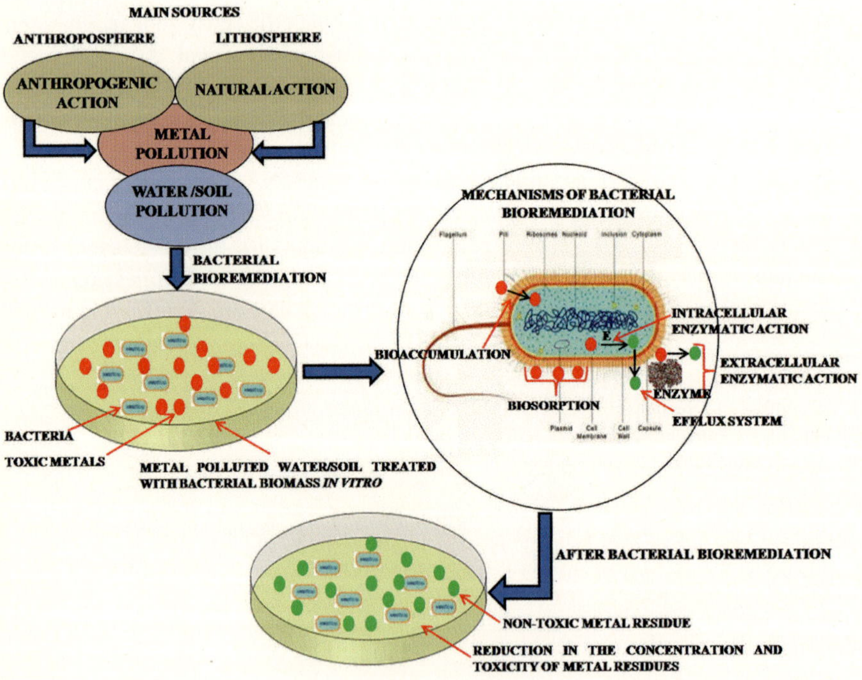

Fig. 15.1 Mechanisms involved in bacterial bioremediation in metal contaminated environments

(EPS) (Ahalya et al. 2003; Das et al. 2008; Naik et al. 2012). Marine bacteria are more suitable candidate for potential heavy metal remediation for maintaining sustainability as they are more capable for quick adjustment to the changing environmental factors such pH, salinity, temperature etc. (Dash et al. 2013).

15.2 Sources of Heavy Metals

Heavy metals present in the environment (water, sediment, soil, air and living organisms) come from both anthropogenic and natural sources. Anthropogenic or manmade activities create a constant and permanent pollution, while naturally occurring pollution due to seasonal influence, geogenic activities and weathering does not affect more (Alvarez et al. 2017; Bradl 2005). Anthropogenic pollution consists of three major sources such as industrialization, urbanisation and agriculture.

Industries associated with mining, metallurgic, surface finishing, electroplating, electrolysis, electro-osmosis, distilleries, tanneries, photography, manufacturing electrical appliance, production of iron, steel, energy, fertilizer, pesticide, paints, varnishes and pharmaceuticals, aerospace and atomic energy installations etc. produce huge amount of metal contaminants during their operation (Bradl 2005; Wuana and Okieimen 2011; Chabukdhara and Nema 2012; Alvarez et al. 2017; Mohapatra et al. 2017a). Mining and metallurgical industries caused direct metal contamination during extraction, processing of raw materials whereas other industries create indirect metal pollution from the burning of fossil fuel for boiler operation (Bradl 2005; Wuana and Okieimen 2011; Li et al. 2015). Textile and Tanneries industries produce highly metal contaminated effluents rich in Cr(VI), As(III/V), Pb(II) and Cd(II) and create water pollution (Bhuiyan et al. 2010). Electronic waste released high concentration of heavy metals to the environment when not treated properly (Wu et al. 2015).

Agricultural practices contribute to heavy metal pollution through the use of pesticides, fertilizers and soil amendments. Natural phosphate fertilisers also contains Ni(II), Cd(II), Zn(II), Pb(II), As(III/V) and Cr(VI) (Wuana and Okieimen 2011). Inorganic pesticides may also contain heavy metals like Hg(II), Pb(II), As(III/V) and Cu(II) as active ingredients (Paranjape et al. 2014).Urbanisation generates huge amounts of solid waste, waste water and sewage sludge which are the ultimate source of heavy metals like Fe(III), Mn(II) and Cr(VI) (Kothe et al. 2010). Inadequate disposal of municipal solid waste, long term irrigation of waste water and leachate of municipal land fill are associated with soil, surface water and ground water pollution (Fernandez et al. 2014; Alvarez et al. 2017). Natural and geogenic activities are also involved in heavy metal contaminations from various sources including different rocks, volcanic eruption, mineral deposits erosion, oceanic evaporation and general pedogenic processes (Bradl 2005; Zeng et al. 2014).

15.2.1 Sources of Chromium(VI) Contamination

Extensive use chromium in various industrial and manufacturing processes such as for metal decoration, dyeing in textile industries, silk printing, tanning in the leather industries, ink, paints, green varnishes as catalysts for halogenations, alkylation and catalytic cracking of hydrocarbons, in the ceramic industry, fuel and propellant additive, electroplating, wood preservatives, glass and plastic colorants, anticorrosive coating agents, anodizing of aluminium, water-cooling, nuclear power production, e-waste etc. leads to Cr(VI) pollution in the environment (Subramanian et al. 2012; Ergul-Ulger et al. 2014; Malaviya and Singh 2016; Swapna et al. 2016; Mohapatra et al. 2017a).

15.2.2 Sources of Cadmium(II) Contamination

Cadmium enters to the environment through various industrial processes such as mining, electroplating, coatings, stabilizing plastics, nickel-cadmium batteries manufacturing, alloy and specific electronics compounds (e.g. Cadmium tellurium, CdTe), pigments, cement, fossil fuel combustion, municipal and sewage sludge incineration and high phosphate fertilizers (Zouboulis et al. 2004; Khan et al. 2016).

15.2.3 Sources of Lead(II) Contamination

Lead can originate from different anthropogenic sources as well as geochemical processes. Major sources of lead which can contaminate the environments are various industrial applications such as ceramics, batteries, printing pigments manufacturing, pesticide production, fuel additive, photographic materials, explosive manufacturing, coating, automotive, aeronautical, metal smelting plants and incinerators (Selatnia et al. 2004a; Bueno et al. 2008; Ren et al. 2015; Ma et al. 2015; Pepi et al. 2016).

15.2.4 Sources of Arsenic(III/V) Contamination

Natural geogenic activities (volcanic, weathering, marine sedimentary rocks, fossil fuels, minerals) and anthropogenic activity (mining, agricultural chemicals such as pesticides, fertilisers, herbicides, wood preservatives, smelting operations, coal combustion, medical products, industrial activity) are the major sources of arsenic contamination (Aksornchu et al. 2008; Taran et al. 2013; Vishnoi and Singh 2014).

15.3 Toxic Effects of Heavy Metals

Heavy metals accumulation through food chain, leads to serious ecological hazards as a result of solubility and mobility nature (Amoozegar et al. 2012). Heavy metal ions when entered the cell became toxic since it could compete with or replace a functional metal ion as well as cause conformational modification, denaturation, inactivation of enzymes and disruption of cellular and organelles integrity immediately it enters the cell (Blackwell et al. 1995; Elsilk et al. 2014).

15.3.1 Chromium(VI) Toxicity

United state environmental protection agency (USEPA) has been reported Cr(VI) to be mutagenic and carcinogenic to living organisms and recommended maximum permissible limit for drinking water is 0.05 g/l (Thatheyus and Ramya 2016; Swapna et al. 2016). Cr(VI) compounds are strong oxidising agents, highly soluble in water and considered to be highly mutagenicity and carcinogenicity due to quick permeability through biological membranes and interact with intramolecular protein and nucleic acids causing DNA damage, alter gene expression (Cervantes et al. 2001; Focardi et al. 2012; Malaviya and Singh 2016; Mohapatra et al. 2017a). In addition to carcinogenic and mutagenic effect, and the toxicity of Cr(VI) is also contributed by generation of huge amount of free radicals and reactive oxygen species (ROS) from the intracellular oxidation-reduction process of chromium compound (Megharaj et al. 2003; Kiliç et al. 2010; Gu et al. 2015; Jin et al. 2014). Cr(VI) compounds can cause serious injury to animals and humans including allergies, eczema, irritations and respiratory track disorders, mental disorders, spasms, genotoxicity etc. (Bagchi et al. 2002; Zhitkovich 2011; Das et al. 2014; Swapna et al. 2016).

15.3.2 Lead(II) Toxicity

According to the European Directive (Mohapatra et al. 2017b) (2008/105/EC), the maximum permissible limit of lead and its compound in surface water is 0.0072 mg/l and less than 10 μg/l is recommended safe permissible level of drinking water (Muñoz et al. 2015). Pb can induce conformational changes of nucleic acids and proteins, inhibits enzyme activity resulting membrane dysfunction, modify the osmotic balance of cells for its high affinity for thiol and oxygen groups and alters oxidative phosphorylation (Vallee and Ulmer 1972; Pepi et al. 2016; Bruins et al. 2000). Lead decreases the efficiency of ATPase pump by affecting concentration of sodium, potassium and calcium as well as the activity of protein kinase, which maintain the concentration gradient of these ions in the cells. Inclusion bodies formation is often stimulated by Pb in cells that may help to translocate metals into the nuclei, cause DNA damage and alter the gene expression. Lead can also replace essential metal ions such as Zn, Ca and Fe from enzymes. (Hu et al. 1998; Watt et al. 2000; Lam et al. 2007; Shahid et al. 2012; Murthy et al. 2012; Wasi et al. 2013; Naik and Dubey 2013). Lead can enters into human body through inhalation, ingestion, dermal contact or transfer via the placenta accumulation in the living tissues may cause serious health problems such as encephalopathy, hepatitis and nephritic syndrome, damage the nervous system, kidney and reproductive system particularly in children due to extremely toxicness (Tunali et al. 2006; Ren et al. 2015). Lead is a mutagenic and teratogenic metal causing severe health hazards by accumulating in the human body mainly in the internal organs such as liver, spleen, pancreas and stomach (Siripongvutikorn et al. 2016). High concentration of lead exposer may cause neurodegenerative impairment, central and peripheral nervous system damage, renal failure, reproductive damage, skeletal disorder, mental retardation, decrease in children's IQ levels and cancer (AAP 2005; Bhakta et al. 2012; Järup 2003). Higher levels of blood Pb (>70 mg dl^{-1}) induces nephropathy, behavioural disturbance, learning disabilities, deficit in fine and gross motor development, reduced fertility both in men and women, neuropathy, Alzheimer disease, increased intracranial pressure, seizure and death (Navas-Acien et al. 2007; Iqbal et al. 2008; Woodruff et al. 2008; Sowmya et al. 2014).

15.3.3 Cadmium(II) Toxicity

Cadmium and its compounds are water soluble and hence easily entered in to human food chain and accumulate in living organisms. Even at very less concentration (0.001–0.1 mg/l), Cd can cause serious toxicity to the living cells (Huang et al. 2014). Cd disrupts protein function through binding to glutathione and the protein sulphydryl groups and displaces zinc and iron from proteins (Banjerdkij et al. 2005; Flora et al. 2008; Sabdono 2010; Mathivanan and Rajaram 2014a, b). Cadmium ions are extremely hazardous to human health causing kidney and liver damage, respiratory tract problems, chronic pulmonary problems, cardiovascular problems, nervous system problems, osteoporosis and fractures, anemia, eosinophilia, anosmia, apoptosis, diabetes mellitus, renal failure, oncogenes activation, Itai-Itai disease and lead to death (Waisberg et al. 2003; Edwards and Prozialeck 2009; Semerjian 2010; Abd-Elnaby et al. 2011; Khan et al. 2016). Toxic nature of Cd also affects the aquatic organisms including severe inhibition of their physiological processes, photosynthesis, and nitrogen fixation (Shamim and Rehman 2012).

15.3.4 *Arsenic(III/V) Toxicity*

Arsenic exists in the environment mainly in two form such as arsenite(III) and arsenate(V). Both are toxic, but arsenite being more toxic than arsenate. Arsenite can cause harmful health effect to humans, animals and other organisms by entering food chain (through soil, food, water, air) as organoarsenic compounds (Mateos et al. 2006). As(III) can inactivate proteins by binding to the sulfhydryl groups of its cysteine residues (Cavalca et al. 2013). Arsenic can penetrate into the cell by the same mechanism as phosphate transported through the cell membrane due to its structural analogy with inorganic phosphate resulting disrupting the metabolic reactions of phosphorylation, inhibit the synthesis of adenosine triphophate (Shrestha et al. 2008). Maximum permissible limit of arsenic concentration in drinking water is 0.01 mg/l (WHO 2011; Aksornchu et al. 2008). Long term expose of small amount of arsenic just above permissible limit, it causes various disease of arsenic toxicity like skin itching, hyperpigmentation, hyperkeratosis, weight loss, loss of appetite, gastrointestinal disorder (e.g. nausea, anorexia, stomach irritation, abdominal pain, enlarged liver and spleen), moderate to severe anaemia, decrease production of red and white blood cells, weakness, lethargy, chronic respiratory disorder, lung irritation, immune-suppression, arsenicosis and cancer due to DNA damage (Tchounwou et al. 2004; Banerjee et al. 2011; Ahsan et al. 2012; Taran et al. 2013; Cavalca et al. 2013; Dey et al. 2016). Expose to high inorganic arsenic concentration can cause infertility, miscarriages in women, type II diabetes, declining resistant to infection, brain damage and cardiovascular effect including hypertension, coronary artery disease, peripheral vascular disease and atherosclerosis. (Mateos et al. 2006; Hopenhayn 2006; Walton et al. 2004; Yoshida et al. 2004).

15.4 Conventional Methods of Heavy Metal Removal

The presence of heavy metals above critical values in the environment is unacceptable and their remediation/removal should be necessary (Aryal and Liakopoulou-Kyriakides 2015). Various commonly used conventional methods such as reverse omosis, electrodialysis, evaporative recovery, ultrafiltration, ion-exchange, phytoremediation are used in their removal/remediation programme. These methods have various disadvantages like inefficient, uneconomical, generation of secondary wastes, creation of environmental problems as described below (Table 15.1) (Ahalya et al. 2003; Aryal and Liakopoulou-Kyriakides 2015). These techniques are also inconvenient for treating of industrial effluents having less than 100 mg/l of dissolved toxic metal ions (Gabr et al. 2008).

Table 15.1 Disadvantages of conventional metal removal methods

Conventional methods	Disadvantages	References
Reverse osmosis	Expensive	Ahalya et al. (2003)
Electrodialysis	Formation of metal hydroxides, which clog the membrane	Ahalya et al. (2003)
Ultrafiltration	Generation of sludge	Ahalya et al. (2003)
Ion-exchange	High cost and partial removal of certain ions	Ahalya et al. (2003); Aryal and Liakopoulou-Kyriakides (2015)
Chemical precipitation	Large amount of sludge containing toxic compounds produced during the process	Ahalya et al. (2003); Aryal and Liakopoulou-Kyriakides (2015)
Phytoremediation	It takes a long time for removal of metals and the regeneration of the plant for further biosorption is difficult	Ahalya et al. (2003); Aryal and Liakopoulou-Kyriakides (2015)
Evaporative recovery	Limitation in applicability, cost effective, less efficient	Ahalya et al. (2003); Aryal and Liakopoulou-Kyriakides (2015)

15.5 Bioremediation of Heavy Metals

To overcome the disadvantages of above conventional techniques, new economic and effective separation technologies have been evolved for the removal of toxic heavy metals. Biosorption/bioaccumulation is considered as one of the potential method for the removal of heavy metals and bioremediation of the metal contaminated sites (Veglio and Beolchini 1997). Metal biosorption is defined as the removal or accumulation of metals ions from waste solution by using biological materials through the variety of metabolically mediated or physico-chemical uptake processes (Fourest and Roux 1992; Ahalya et al. 2003). The retaining of heavy metal ions within microbial cell by means of biosorption is referred to as bioaccumulation (Liang et al. 2014). The use of various microorganisms such as algae, bacteria, fungi and yeasts used for heavy metal removal from contaminated sites has been emerged as a potential technique and proved to be potential biosorbent for metal recovery (Volesky 1986; Rajendran et al. 2003; Wasi et al. 2013; Kordialik-Bogacka and Diowksz 2014; Raja Rao et al. 2014).

The major *advantages* of biosorption/bioaccumulation over conventional treatment methods are low cost of the biosorbent, high metal binding efficiency, rapid process, selectivity for specific metals of interest, less/no generation of toxic sludge, no need of costly growth media, recovery of metal ions in concentrated form, regeneration of biosorbent (Kratochvil and Volesky 1998; Sari and Tuzen 2009; Rangabhashiyam et al. 2013; Javanbakht et al. 2014).

There are some *disadvantages* associated with the metal biosorption process. These include req. continuous supply of nutrients for microbial growth needed, affect of requirement of pH, temperature and heavy metal toxicity to their metabolic process and recovery of metals and regeneration of bacterial biomass may be more complicated. The above drawbacks of biosorption phenomenon somehow may be overcome by applying inactive or dead biomass for removing of heavy metals from aqueous solutions via physical or chemical interaction of metal ions on microbial cell surface (Volesky 2001; Ilamathi et al. 2014; Koduru et al. 2014).

15.6 Bacterial Bio-sorption of Heavy Metals

Bacteria are widespread in nature and have the capacity to adapt and perform their various physiological cellular activities in any extreme environmental conditions. Bacteria are more capable of rapid adjustment towards environmental changes and deterioration and act as a potential candidate by playing major role in protection and develop sustainability of the spoiled ecosystem (Dash et al. 2013). Bacteria from metal polluted habitats possess a high levels of metal tolerant capacity and can be useful in toxic metal bioremediation (biosorption/ bioaccumulation/ biotransformation) by variety of inherent mechanisms including precipitation of metals as phosphate, sulphide, carbonate; volatilization via methylation/ ethylation; physical exclusion in membranes and extracellular polymeric substances(EPS); energy driven metal efflux system and intracellular sequestration mediated by metallothionein like proteins (Naik et al. 2012; Alencar et al. 2017) (shown in Fig. 15.1). Bacterial biomass has a tendency to remove the heavy metals from aqueous solutions in very dilute conditions. Both active (live) and inactive (dead) bacterial cells have the capability to remove heavy metals from aqueous solutions and also potentially used by various researchers (Srinath et al. 2002; Gabr et al. 2008; Wierzba and Latała 2010; Huang et al. 2013; Jaafarzadeh et al. 2014; Bakyayita et al. 2014).

15.7 Diversity and Distinctive Features of Marine Bacteria

Marine environment comprises more than 90% of total biosphere and become the largest habitat on the earth. The microorganisms present in that marine habitat are responsible for half of the total global primary production and nutrient cycling (Lauro et al. 2009). Marine ecosystem is a huge storage of native marine bacteria occurring naturally i.e. 3.6×10^{29} cells (Sogin et al. 2006). Systemic documentation of bacterial diversity in marine environment was started with characterizing 60 marine species by ZoBell and Upham (1944). Benthic bacterial population of Indian Ocean basin was found to be higher level in the range of 0.48–1.21 × 10 5 CFU/g categorised under six major taxonomic groups such as α, β and γ proteobacteria, actinobacteria, bacilli and flavobacteria (Loka Bharathi and Nair 2005). In Pacific Ocean, dominant bacterial genera like *Desulfobacterium*, *Desulforhopalus*, *Desulfococcus*, *Desulfosarcina*, *Pelobacter* and *Syntrophus* are found (Inagaki et al. 2006). γ-proteobacteria was also reported as the most widespread bacterial entity in the cobalt-rich crust deposit region of the Pacific Ocean (Liao et al. 2011). In the Arctic Ocean, both photoautotrophic and photoheterotrophic prokaryotic bacterial population was decreased about three fold by the effects of global warming resulted increasing water column mixing (due to loss of ice cover) and altering current patterns (Lovejoy et al. 2006; Cottrell and Kirchman 2009). In the polar Ocean's environment, a group of psychrophilic bacteria such as *Colwellia sp.*, *Marinobacter sp.*, *Planococcus sp.*, and *Shewanella sp.* are found to inhabitate as the sea-ice microbial community (Bowman et al. 1997; Hollibaugh et al. 2007). Different physiological groups of marine bacteria isolated from deep and inshore seas, mangroves and coral reef ecosystems has been reviewed by Das et al. (2006) is presented in Table 15.2.

Both autotrophic and heterotrophic bacteria are present in abundant numbers in the marine environment (Stanley 2005) and can be actively used in variety of broad prospects like production of antibiotics and enzyme (Okami 1986), biosurfactant production (Maneerat and Phetrong 2007), marine light absorption (Stramski and Kiefer 1998), bioremediation of heavy metals and hydrocarbons (Rainbow 1995; Margesin and Schinner 2001), oil biodegradation (Nweke and Okpokwasili 2003), bioremediation of diesel-contaminated soils (Gallego et al. 2001), degradation of metatoluic acid and agar (Prakash et al. 2008; Vijayaraghavan and Rajendran 2011), polyphosphate accumulation (Ohtake et al. 1985), degradation of plastic debris (Derraik 2002), and antibiofilm activity (Jiang et al. 2011).

Table 15.2 Different physiological groups of marine bacteria isolated from marine environment

Group	Physiology	Example
Archaebacteria	Chemoautotrophic,anaerobic, thermophillic and mesophillic.	*Desulfomonas, Desulfovibrio, Desulfobulbus, Desulfotomaculum and Desulfococcus*
Methanogenic bacteria	Chemoautotrophs, strictly anaerobes, utilise a limited number of simple carbon compounds (hydrogen, carbon dioxide, formate, acetateand methanol) as their carbon and energy sources for methanogenesis.	*Metahnococcus, Methanocercina, Methanomicrobium, Methanogenium, Methanoplanus, Methanococcoides and Methanobolus*
Halophillic bacteria	Requirs about 12–15% NaCl to survive and grow well even at concentration up to saturation.	*Haloarcula, Halobacterium, Haloferax and Halococcus*
Eubacteria Luminous bacteria	Produce light by a simple protein-like substance called luciferin in contact with the oxygen molecule; gram negative and motile heterotrophic rods.	*Photobacterium leiognathi, Photobacterium phosphorium, Vibrio Fischeri, Vibrio Harveyi*
Nitrifying bacteria	Oxidise either ammonia to nitrite (*Nitrosococcus*) or nitrite to nitrate (*Nitrococcus*) and convert nitrogen to a form readily available for other biological processes.	*Nitrosococcus, Nitrococcus*

15.8 Sources of Heavy Metal Resistant Marine Bacteria

Marine metal resistant bacteria can be isolated from the water, soil and sediments of the marine ecosystem, mangroves associated with the marine habitats, normal flora of the marine organisms and deep sea hydrothermal vents (Naik et al. 2012; Dash et al. 2013). Marine sediments are more contaminated and can accumulate heavy metals several orders of magnitude than surface water (Mohapatra and Panda 2017). Therefore, several benthic sediment bacteria attached to sediment particles show more resistant towards toxic heavy metals (Naik et al. 2012; Mohapatra et al. 2016). Deep-sea hydrothermal vent fluids are enriched with toxic metals and a likely place to find heavy metal resistant bacteria adapted to that local vicinity (Naik et al. 2012; Devika et al. 2013). Coral reef and fauna are also exposed to high metal concentration from industrial and mine effluent discharged to marine ecosystem through river. Hence, coral reef, sponges having symbiotic bacteria potentially resistant to toxic heavy metals (Naik et al. 2012).

15.9 Importance of Marine Halophilic/Halotolerant Bacteria Over Non-saline Bacteria

Most of the polluted environments are characterized by elevated or low temperature, alkaline or acidic pH, high pressure and high salt concentration. Marine bacteria are under halophilic/halotolerant group which get exposure to such unfavourable conditions naturally such as varying temperature, pH, salinity, conductance, sea water temperature, water currents, precipitation regimes and wind patterns. Due to suitably adapted to the most adverse environment conditions and possessing complex characteristic features of adaptation, the bacteria isolated from the marine sources are believed to be better utilized in bioremediation of toxic metals and many other recalcitrant xenobiotics compounds. Direct use of native indigenous marine bacteria without any genetic manipulation for *in situ* bioremediation in any adverse conditions is the main advantage (Dash et al. 2013). Various potential halophilic/halotolerant bacteria isolated from different metal contaminated sites reported by various researchers for bioremediation and detoxification of the toxic heavy metals such as Cr(VI), Pb(II), Cd(II) and As(III/V) are presented in Tables 15.3, 15.4, 15.5, and 15.6.

Table 15.3 Cr(VI) remediation by Halophilic/Halotolerant bacteria

Cr(VI)-resistant bacteria	Isolation source	Tolerance	pH/NaCl	References
Halomonas sp. TA-04	Sediment, Taranto gulf, Italy	4 mM	pH: 6.5, NaCl: 8%	Focardi et al. (2012)
Vigribacillus sp.	Mangrove soil, Bhitarkanika, India	1000 mg/l	pH: 8.0, NaCl: 6%	Mishra et al. (2012)
Planococcus maritimus VITP21	Kumta costal, Karnatak, India	1000 mg/l	pH: 7.0; NaCl: 4%	Subramanian et al. (2012)
Bacillus subtilis	Tannery effluent contaminated soil	100–4000 mg/l	pH: 9.0	Mangaiyarkarasi et al. (2011)
Halomonas sp. CB5	Sambhar salt lake, Rajasthan, India	1000 mg/l	pH:8.0, NaCl: 25%	Chandra and Singh (2014)
Bacillus subtilis SHB 13	Soil, sludge and swage samples	1000 mg/l	pH: 7.0, NaCl: 4%	Swapna et al. (2016)
Exiguabacterium indicum MW-1	Marine water, Paradip port, bay of Bengal	1500 mg/l	pH: 8.0, NaCl: 1–5%	Mohapatra et al. (2017a)

15.10 Characteristics of Bacterial Cell Wall and Their Metal Binding/Sorption

The bacterial cell walls play an important role in heavy metal biosorption. The functional groups of bacteria such as of peptidoglycan, teichoic and teichuronic acids, phospholipids, lipopolysaccharides, and various proteins are responsible for metal binding (Vijayaraghavan and Yun 2008; Din et al. 2014). Gram-positive bacteria exhibit lower levels of surface complexation due to the heavily cross-linked peptidoglycan layer and Gram-negative bacteria expose most of their lipopolysaccharides, phospholipids, and proteins on cell walls (Joo et al. 2010). Gram-positive bacteria cell wall contain a thick layer of peptidoglycan layer having polymer of sugars (N- acetylglucosamine and N-acetylmuramic acid), teichioic acid, teichuronic acid, lipoteichoic acid and cross links of short peptides (Yee and Fein 2001). Gram-negative cell wall lack of thick peptidoglycan layer and contain a highly permeable porous outer membrane rich in protein and lipopolysaccharides (Yee and Fein 2001). At pH values greater than dissociation constant (pKa), the surface functional groups are protonated and the negatively charged ligands can participate for metal cations interactions, whereas at pH values lower than pKa, complexation can also occur specifically for carboxylic groups (Esposito et al. 2001). At lower pH values, bacterial cell wall becomes

Table 15.4 Lead(II) remediation by Halophilic/Halotolerant bacteria

Pb(II)-resistant bacteria	Isolation source	Tolerance	pH/NaCl	References
Halomonas sp.	Hypersaline soil and water, Iran	5 mM	pH: 3.0–6.0, NaCl: 5%	Amoozegar et al. (2012)
Bacillus sp. Pb15	Marine sediment, Sarno river mouth, gulf of Naples, Italy	4.82 $mmol^{-1}$	pH: 7.0, NaCl: 2.4%	Pepi et al. (2016)
Alcaligens sp., *Enterobacteriaceae* sp., *Kurthia* sp., *Staphylococcus* sp., *Vibrio* sp.	Sediments of Vembanad Lake, Kerala, India	0.1–12 mM	pH: -, NaCl: 5–15%	Sowmya et al. (2014)
Halomonas elongate, Tetragenococcus halophilus	ATCC 33173 ATCC 33315	1 mg/l	pH: 7, NaCl: 10–20% pH: 7, NaCl: 10%	Siripongvutikorn et al. (2016)
Micrococcus luteus DE2008	Microcoleus consortium	3 mM	pH: 6.5–7.0. NaCl: 8%	Puyen et al. (2012); Maldonado et al. (2010)
Alcanivorax consortia	Sepetiba Bay, Brazil	6 μg/ml	Sea water	Waite et al. (2016)
Acinetobacter sp. THKPS16	Sediment, meiliang bay, Taihu lake, China	100 mg/l	pH: 5, NaCl: -	Ma et al. (2015)
Klebsiella sp. 3S1	Waste water treatment plant	3.4 mM	pH: 5, NaCl: -	Muñoz et al. (2012, 2015)

positive due to the deprotonation of surface functional groups, and thus, formed positively charged active sites are responsible for metal anion binding. Transition metal ions can coordinate three to eight ligands and often exhibit an octahedral coordination because of free d-orbital presence in their electronic structure (Aryal and Liakopoulou-Kyriakides 2013a). Thus, the three-dimensional network structure of bacterial cell surface peptidoglycan layer can be responsible for metal binding (Wang et al. 2006; Fomina and Gadd 2014).

Table 15.5 Cadmium(II) remediation by Halophilic/Halotolerant bacteria

Cd(II)-resistant bacteria	Isolation source	Tolerance	pH/NaCl	References
Halomonas sp.	Hypersaline soil and water, Iran	5 mM	pH: 3.0, NaCl: 1%	Amoozegar et al. (2012)
Alcaligens sp., *Enterobacteriaceae* sp., *Kurthia* sp., *Staphylococcus* sp., *Vibrio* sp.	Sediments of Vembanad Lake, Kerala, India	0.05–4.0 mM	pH: -, NaCl: 5–15%	Sowmya et al. (2014)
Halomonas elongate *Tetragenococcus halophilus*	ATCC 33173 ATCC 33315	3 mg/l	pH: 7, NaCl: 10–20% pH: 7, NaCl: 10%	Siripongvutikorn et al. (2016)
Pseudoalteromonas sp.SCSE709–6	Deep sea sediment, south China	100 mg/l	pH: 6.5–7.5, NaCL: 3%	Zhou et al. (2013)
Vibrio harveyi 5S-2	Sediment, Alexander eastern harbour, Egypt	> 60 mg/l	pH: 7.2	Abd-Elnaby et al. (2011)
Pseudoalteromonas sp. CD15	Coral tissue, Awur Bay, Jepera waters	5 mg/l	–	Sabdono (2010)
Alteromonas macleodii ASC1	Sediment, Hurghada harbour, Red Sea	150 mg/l	pH: 6.0	El-Moselhy et al. (2013)
Bacillus sp. NT-1 *Enterobacter* sp. NT-5 *Aeromonas* sp. NT-10 *Pseudomonas* sp. TT-10	Water and sediment of industrially polluted estuarine	400 mg/l 400 mg/l 400 mg/l 500 mg/l	pH: 8.0, NaCl: 0.5%	Mathivanan and Rajaram (2014a)
Pseudomonas stutzeri N-1 *Pseudomonas mendocina* C-1 *Alcaligens faecalis* C-8 *Acinetobacter baumannii* C-10 *Bacillus licheniformis* C-12 *Lysinibacillus fusiformis* C-14	Surface water of Cuddalore coastal ecosystem, Tamil Nadu, India	350 mg/l 250 mg/l 200 mg/l 200 mg/l 400 mg/l	–	(Mathivanan and Rajaram 2014b)

Table 15.6 Arsenic (III) remediation by Halophilic/Halotolerant bacteria

As(III)-resistant bacteria	Isolation source	Tolerance	pH/NaCl	References
Bacillus sp. KM02 *Aneurinibacillus aneurinilyticus*	Ground water alkaline nature, Burdwan, WB, India	As(III): 500 mg/l, as(V): 4500 mg/l	NaCl: 8–10% pH:7.23	Dey et al. (2016)
Halorcula sp. IRU1	Hypersaline Urmia lake, Iran	As(III): 90 mg/l	NaCl: 25% pH: 8.0	Taran et al. (2013)
Vibrio alginolyticus NCIMB *Halomonas marina* IAM 14107 *Alteromonas macleodii* IAM 12914 *Marinomonas communis* IAM12914	NCIMB, Aberdeen, Scotland IAM, Tokyo, Japan	As: 730 mg/l As: 310 mg/l As: 210 mg/l As: 510 mg/l	NaCl: 25% pH: --	Takeuchi et al. (2007)

15.11 Modifications of Bacterial Biomass (Biosorbent) for Enhanced Biosorption

15.11.1 Chemical Modifications

Chemically modification of bacterial biomass by treating with inorganic and organic substances enhanced the biosorption capacity of heavy metals, which may be due to the rupture of bacterial cell wall and/or formation of additional binding sites for metal ions. Pre-treatment of bacterial biomass with various chemicals such as NaOH, $(Na)_2CO_3$, NH_4OH, KOH, TritonX-100, C_2H_5OH, CH_3OH, HCl, H_2SO_4, Na_2CO_3, $(NH_4)_2SO_4$, polyacrylic-acid, aceton, toluene, chloroform-methanol, etc. may enhance the bisorption of metals (Mameri et al. 1999; Puranik and Paknikar 1999; Sar et al. 1999; Nakajima et al. 2001; Wang et al. 2003; Liu et al. 2004; Selatnia et al. 2004b; Mao et al. 2013).

15.11.2 Biological Modifications

Enhanced biosorption of heavy metals can be achieved by modifying binding sites and increasing affinity for metal ions by applying genetic manipulation techniques (Dash et al. 2013). Due to significant development in the field of molecular biology and genetics, genetic engineering and protein engineering application may lead to the development of new genetically engineered microbes, peptides or biopolymers, over expression of metal binding proteins (MerP) (Goyal et al. 2003; Kao et al. 2008). The above practice can be achieved by insertion of novel genes into genome, insertion new plasmid, alteration of metabolic pathways and chemotaxis (Pieper and Reineke 2000).

15.12 Biosorbtion Process

The biosorption process involves a biosorbent (solid biological material) and a solvent (liquid) containing a dissolved species of metal ions (sorbate) to be sorbed. The biosorbent attracted the metal ions for binding due to its higher affinity. Strong biosorbent behaviour of certain bacteria towards metal ions is a function of the chemical constitute of the cell membrane. The degree of sorbent affinity for the metal ions determines its distribution between the solid and liquid phases. The process continues till equilibrium is established between the amount of biosorbent-bound metal species (sorbate) and its portion remaining in the solution (Das et al. 2008).

15.13 Biosorption Mechanisms

The complex structure of the bacterial cell involves in metal ion up take is associated with various mechanisms of absorption. The bacterial metal absorption may be classified according to various criteria as follows:

15.13.1 On the Basis of Cellular metabolism Dependence

According to the dependence on the cellular metabolism, biosorption mechanisms can be categorised into two types. One is metabolism dependent (Active) and another is Non-metabolism dependent (Passive) (Shown in Fig. 15.2).

15.13.1.1 Metabolism Dependent (Active)

Metabolism dependent biosorption can take place only in the viable cells and metal ions are transported into the cell across the cell membrane during cellular metabolism, resulting intracellular accumulation. Heavy metal transportation across bacterial cell membrane mediated by the same mechanism as it is used for metabolically important potassium, magnesium and sodium ions (Veglio and Beolchini 1997; Ahalya et al. 2003).

15.13.1.2 Non -Metabolism Dependent (Passive)

During non-metabolism dependent biosorption, metal uptake takes place by means of physico-chemical interaction between the metal ions and the cell surface functional groups. Cell wall components of bacterial biomass such as polysaccharides,

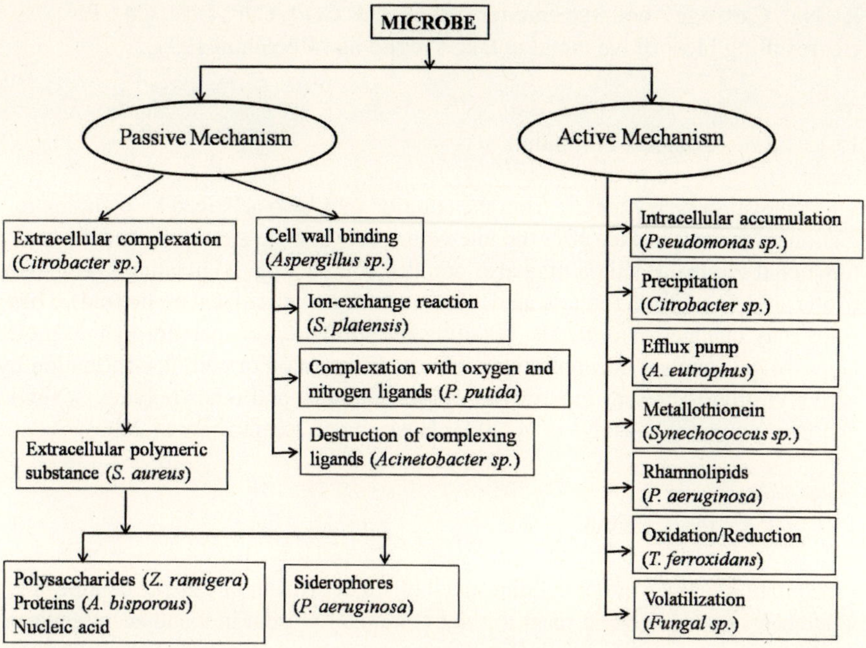

Fig. 15.2 Mechanisms (Active and Passive) associated with microbial heavy metal sorption process

proteins and lipids carrying intense metal binding functional groups like amino, carboxyl, sulphate and phosphate are mainly involved in this process. Physical adsorption, ion exchange, complexation and precipitation etc., are coming under non-metabolism dependent biosorption which are comparatively rapid and reversible processes (Kuyucak and Volesky 1988).

15.13.1.2.1 Physical Adsorption

Physical adsorption metal ions take place by the effect of Vander Waals' forces of attraction. Metal biosorption by dead bacterial biomass takes place through electrostatic force of interactions between the bacterial cell wall components and metal ions present in aqueous solutions (Veglio and Beolchini 1997; Ahalya et al. 2003).

15.13.1.2.2 Ion Exchange

Cell walls of bacteria contain polysaccharides and bivalent metal ions exchange with the counter ions present the aqueous solution (Ahalya et al. 2003). The metallic ion exchange takes place with the naturally occurring cellular ions like

K^+, Na^+, Ca^{2+}, Mg^{2+} with the counter ions such as CO^{2+}, Cu^{2+}, Cd^{2+}, Cr^{6+}, Pb^{2+}, As^{3+} etc. resulting biosorptive metal uptake (Veglio and Beolchini 1997).

15.13.1.2.3 Complex Formation

The removal of heavy metals from the solution may also acheived by complex formation on the cell surface after the interaction between the metal ions and the active functional groups. Bacteria may also produce complexing or chelating agents (e.g. oxalic acid, citric acid, fumaric acid, gluonic acid, lactic acid and malic acid). These acids may chelate toxic metals, resulting the formation of metallo-organic molecules. Metal removal occured may be due to biosorbtion or complex formation by carboxyl groups present in bacterial polysaccharides and other polymers (Veglio and Beolchini 1997; Ahalya et al. 2003; Hossain and Anantharaman 2006),

15.13.1.2.4 Precipitation

Precipitation may be either cellular metabolism dependent or independent of it. As a defence system, bacteria react to the toxic metal present in its adjacent environment and produce some compound, which favour the precipitation process. In case of the precipitation independent of cellular metabolism, there may be a consequence of the chemical interaction between the metal ions and the bacterial cell surface (Veglio and Beolchini 1997; Ahalya et al. 2003).

15.13.2 On the Basis of Cellular Location of Metal Removal

According to the location of the bacterial cell where the metal ion removal takes place from the solution, biosorption can be classified as follows and shown in Fig. 15.3.

(i) Extra cellular accumulation/ precipitation.
(ii) Cell surface sorption/Precipitation.
(iii) Intracellular accumulation.

Heavy metal biosorption in living bacterial cell occur in different cellular regions is a multiple step process. The bacterial cell wall consisting variety of polysaccharides and proteins functional groups provide active binding sites for metal ions. The metal ions are fist adsorbed to the cell surface by the interactions of metals with cell surface functional groups resulting extracellular and cell surface accumulation/precipitation. Then the adsorbed metal ions come across the cell membrane and get entre into the cell cytoplasm resulting intracellular accumulation (White et al. 1995; Das et al. 2008).

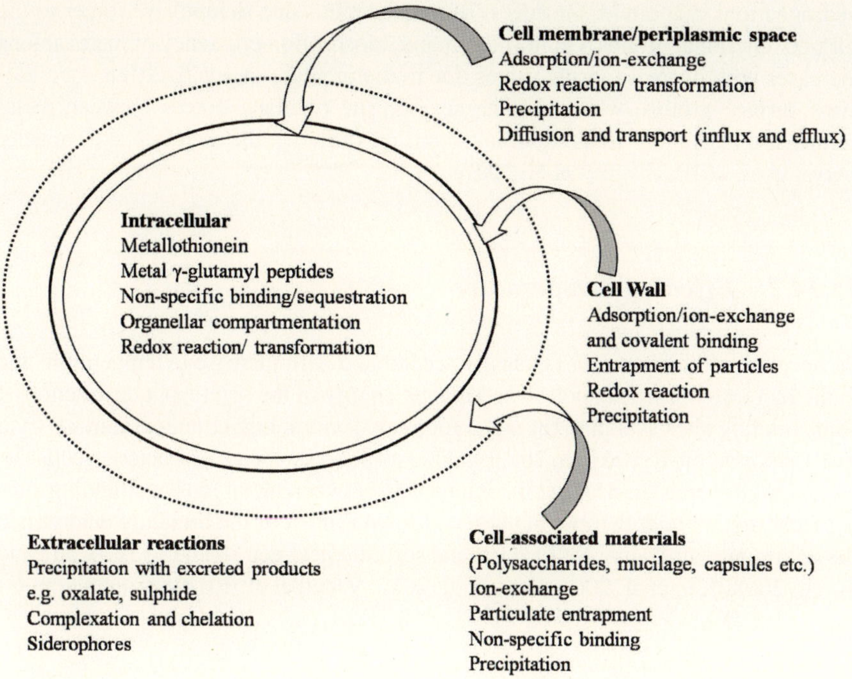

Fig. 15.3 Heavy metal bio-sorption in different cellular components of the microbes during the detoxification process

15.14 Factors Affecting Bacterial biosorption of Heavy Metals

Biosorption of heavy metals by bacterial cell is affected by different operating parameters such as pH, temperature, concentration of biomass, contact time, initial metal ions concentration and interfering co-ions (Liang et al. 2014; Chathuranga et al. 2014; Gupta et al. 2014). The details of the factors affecting bacterial biosorption have been discussed below.

15.14.1 Effect of pH

The pH is an important parameter which significantly affects the solution chemistry of metal ions and the surface functional groups of the bacterial cell wall (Friis and Myers-Keith 1986; Galun et al. 1987; Long et al. 2014). The biosorption capacity of metal cations increases with increase in pH values, and this may be due to the more negative binding sites exposed on biomass surface (Aksu and Gülen 2002). At low pH values, the binding sites of the cell wall are blocked and associated with hydrogen ions that hinder the access of metal cations due to repulsive forces to the surface functional groups. On the other hand, biosorption efficiency of metal anions increases with decrease in pH values due to the increase in positively charged biomass surface groups, whereas at higher pH, the repulsive forces between metal anions and negatively charged biomass surface diminish the metal uptake capacity (Aryal et al. 2010; Ziagova et al. 2007).

15.14.2 Effect of Temperature

Biosorption of heavy metals is usually accelerated with increase in temperature due to the increase in surface motion and kinetic energy of the solute, but destruction of some binding sites available for metal ions can occur at higher temperatures (Aryal and Liakopoulou-Kyriakides 2013b). The sorption capacity of bacterial cells for metal ions depends on whether the interaction between metal ions and binding sites is exothermic or endothermic in nature. However, most of the literature determined the optimum temperature for heavy metal sorption between 20 and 35 °C (Calfa and Torem 2008; Kao et al. 2008; Oves et al. 2013; Veneu et al. 2013).

15.14.3 Effect of Biomass Concentration

Biosorption of heavy metals dependents on biomass concentration used as the sorption medium. An increase in biomass concentration usually results in increase of biosorption efficiency, probably due to the availability of more numbers of binding sites in biosorbent surface. It was observed that the sorption efficiency increased with increase in biomass concentration, but biomass concentrations above 1.0 and 2.0 g/l had lower impact in sorption efficiency (Aryal et al. 2010). At above optimum biomass concentration concentrations, uptake capacity is almost constant, may be due to the saturationof potential binding sites and interference between active sites (Kang et al. 2007). Some studies have pointed out that uptake capacity of heavy metals decreases with increasing the biomass concentration as a result of strong limitations of ionic species mobility in the biosorption medium, leaving less binding sites for metal ions (Aryal et al. 2012; Tangaromsuk et al. 2002).

15.14.4 Effect of Contact Time

The contact time is also one of the most important factors for metal biosorption process. The contact time (equilibrium time) indicates the sorption-desorption processes occurring after saturation of metal ions on biomass surface. After the equilibrium time, equilibrium capacities are almost constant, suggesting an equilibrium balance for sorption process (Aryal and Liakopoulou-Kyriakides 2015).

15.14.5 *Effect of Initial Metal Ion Concentration*

The initial metal ion concentration is an important parameter for biosorption process. The metal ions per unit mass of biomass is increased upon an increase in metal ion concentrations, since the initial metal ion concentration provides the necessary driving force to overcome the resistance to the mass transfer of metal ions between aqueous and solid phases, but decreases the sorption percentage (Aryal et al. 2010). At lower metal ion concentrations, all metal ions present in thesolution may enhance the interaction between metal ions and bacterial binding sites and thus result in higherbiosorption efficiency (Aryal and Liakopoulou-Kyriakides 2015).

15.14.6 *Effect of Interfering Ions*

During the treatment of waste water containing multiple metal, biosorption process may be influenced by the each and every metal ions present. The presence of co-ions including metal cations and anions in wastewater may cause interference and competition phenomena for biosorption sites. In some cases, metal cations as competent ions may increase biosorption of anionic species by enhancing the additional binding sites (Aryal and Liakopoulou-Kyriakides 2011). The decrease in sorption performance in the presence of co-metallic ions may be due to the competition of both metal ions for the same binding sites on the cell surface (Aryal and Liakopoulou-Kyriakides 2015).

15.15 Equilibrium Studies

Biosorption isotherm is the relationship between quantities of metal ions per unit of bacterial biomass and the concentration of these metal ions in the solution. The isotherm parameters give information about surface properties and the affinity of binding sites of bacterial cells as well as biosorption mechanism. Determination of isotherm parameters provides important information about the design of biosorption systems.

Langmuir and Freundlich models are the two widely accepted and linearised equilibrium adsorption isotherm models among the best suitable models for sorption of heavy metal ions (Lu et al. 2006; Gabr et al. 2008; Joo et al. 2010; Huang and Liu 2013). The theoretical basis of the Langmuir model (Das et al. 2008; Ahalya et al. 2003) relies that there are a finite number of binding sites on the adsorbent surface with a single molecular layer adsorption and there is no interaction between the adsorbed molecules. Whereas in case of Freundlich isotherm model (Das et al. 2008; Ahalya et al. 2003), it describes about the adsorption site of an adsorbent which is available or not for the incoming adsorbet. The Langmuir and Freundlich model equations are as given in equation (15.1 and 15.2), respectively.

$$q = \frac{Q_{max} bC_{eq}}{1 + bC_{eq}} \tag{15.1}$$

Where, 'q' is expressed as milligrams of metal accumulated per gram of the biosorbent; 'Q_{max}' is the maximum specific metal uptake corresponding to the site saturation; 'b' is the ratio of adsorption and desorption rates and 'C_{eq}' is the residual metal concentration in solution.

$$Q = K_F C_{eq}^{\ 1/n} \tag{15.2}$$

Where, specific uptake 'Q' is reported as a function of the metal concentration 'C_{eq}'; 'K_F' and 'n' are constants.

These models can be applied at a constant pH. These two models are used for modelling of biosorption equilibrium in the presence of single metal.

15.16 Thermodynamic Studies

The thermodynamic studies can provide valuable information of the overall heat and energies associated with complex multipath biosorption processes and biosorption mechanisms. Standard Gibbs free energy change (ΔG^o) describes the degree of spontaneity during the sorption process. The negative and positive ΔG^o values indicate the spontaneous and non-spontaneous sorption process, whereas ΔG^o values up to −20 kJ /mol and greater than −40 kJ/mol indicate the physical and chemical adsorption. Based on the magnitudes of ΔGo values, spontaneous (Lin and Lai 2006; Yan et al. 2010; Aryal et al. 2012; Chakravarty and Banerjee 2012; Prasad et al. 2013) and non-spontaneous (Uslu and Tanyol 2006; Gialamouidis et al. 2010) biosorption processes were reported. The negative and positive values of ΔH^o indicate the exothermic and the endothermic nature of biosorption process. During the heavy metal sorption process, positive and negative values of entropy change (ΔS^o) indicates the increase and decrease of randomness at the solid-solution interface.

15.17 Characterization of Treated Biomass (Biosorbent)

In order to understand the surface binding mechanism, it is important to identify the functional groups on the biomass surface and the conformational change after treatment with metal ions. Hence, various techniques have been used to characterize the metal treated biomass (biosorbent) and understand the biosorption mechanisms. Some of the commonly used important techniques are described below.

15.17.1 pKa Values

The pKa values of surface functional groups in bacterial species have been determined using potentiometric titration as the addition of acid or alkali solutions (Seki et al. 1998; Pagnanelli et al. 2000; Esposito et al. 2001; Kang et al. 2007; Gabr et al. 2008; Aryal et al. 2012). However, it cannot be solely considered for determination of binding sites responsible for metal interaction, since some pKa values are overlapped due to the complex nature of bacterial biomass.

15.17.2 Fourier Transform Infrared (FTIR)

The shifting, disappearance and reappearance of spectral bands from raw biomass to metal-treated biomass was used to demonstrate the binding of metal ions on biomass surface. FTIR spectroscopic analysis of the metal treated bacterial biomass revealed that carboxylic, amine, amide, phosphate, hydroxyl, carbonyl, phosphoryl, sulphonate, aldehyde, and amide sites of bacterial cells are the key functional groups for metal ion interaction (Gabr et al. 2008; Aryal et al. 2010;; Masood and Malik 2011; Guo et al. 2012; Hasan et al. 2012; Mohapatra et al. 2017a, b).

15.17.3 Zeta Potential

The zeta potential study provides the valuable information of the net effective charges present on the bacterial cell surface (Aryal and Liakopoulou-Kyriakides 2015). Huang et al. (2013) reported that dead *Bacillus cereus* RC-1 cells exhibited a greater amount of negative surface charges than active cells.

15.17.4 SEM & TEM

Scanning electron microscopy (SEM) analysis has been used for the interaction of metal ions on biomass surface. Whereas, transmission electron microscopy (TEM) analysis can be generally used to determine the metal ions present inside the living cell (Aryal and Liakopoulou-Kyriakides 2015). Mohapatra et al. (2017a, b) reported that morphological changes such as cell size, cell surface properties of the *Exiguobacterium indicum* and *Brevibacillus laterosporus* during Cr(VI) reduction process has been established by the SEM analysis of treated biomass.

15.17.5 XRD, XPS and EPR

The X-ray powder diffraction (XRD) is a method using X-ray on microcrystalline samples for structural characterization of materials (Aryal and Liakopoulou-Kyriakides 2015). Das et al. (2014) reported that XRD pattern of Cr(VI) treated *Bacillus amyloliquefaciens* biomass mainly amorphous nature and some peaks were observed due to polysaccharides and fatty acids in the cell wall. X-ray photoelectric spectroscopy (XPS) can be performed to characterize the surface chemistry of metal ions (Aryal and Liakopoulou-Kyriakides 2015). Ye et al. (2013) found that Cu(II) ion adsorbed and reduced to Cu(I) via ion-exchange mechanism by the proteins and polysaccharides present on the *Stenotrophomonas maltophilia* surface. Electron paramagnetic resonance (EPR) spectroscopy has also been used for studying the material with unpaired electrons. Paul et al. (2012) conducted the EPR analysis in order to detect the possible valence states of chromium before and after sorption on *Acinetobacter*.

15.17.6 EDX and EXAFS

Energy-dispersive X-ray (EDX) is a useful analytical method to estimate the elemental or chemical characterization of sample or to examine the binding of metal ions on bacterial cell membrane (Huang et al. 2013; Aryal and Liakopoulou-Kyriakides 2015). During Cr(VI) reduction by *Exiguobacterium indicum* (Mohapatra et al. 2017a) and *Brevibacillus laterosporus* (Mohapatra et al. 2017b), presence of Cr(VI/III) ions in the bacterial surface by the mechanism of complexion, precipitation, adsorption etc. were confirmed by the EDX analysis. Extended X-ray absorption fine structure (EXAFS) methodology can be applied to find out the metal ion sorption mechanism on bacterial cells (Aryal and Liakopoulou-Kyriakides 2015). Cu(II) biosorption on the *Bacillus subtilis* biomass surface as a $(CuO5H_n)^{n-8}$ mono-den-tate, inner-sphere surface complex via carboxyl groups was confirmed by the Moon and Peacock (2011) through EXAFS study.

15.18 Conclusions and Future Prospectives

Marine ecosystem has continuously polluted by heavy metals due to various anthropogenic and geochemical activities. Marine bacteria present in that metal polluted environment gain resistant against various toxic metals and became more potential to tolerate, sequester and remove the toxic metal ions from contaminated environment. Bio-removal of toxic metals using marine bacteria now gained a major attention in the field of bioremediation research. Rapid adaptation of the marine bacteria to the changing unfavourable environment made them more potential and a solution to overcome various disadvantages of other modes of bioremediation. Large number of marine bacteria has been discovered and identified by number of researchers but still there is need to explore better resistant marine bacterial community present in different marine habitats for their application in bioremediation programmes. There is a great scope for application of genetic engineering to make marine metal resistant bacteria more efficient for heavy metal bioremediation. The practical application of the newly discovered native and genetically modified highly metal resistant marine bacteria would be a future bioremediation prospective. The recent development of research in marine metal resistant bacteria reveals that, the sustainable and ecofriendly biotechnological applications of marine metal resistant bacteria for metal bioremediation of various metal contaminated environments could be a suitable approach for future environmental detoxification process.

References

AAP: American Academy of Pediatrics, Committee on Environmental Health (2005) Lead exposure in children: prevention, detection, and management. Pediatrics 116(4):1036–1046

Abd-Elnaby H, Abou-Elela GM, El-Sersy NA (2011) Cadmium resisting bacteria in Alexandria Eastern Harbor (Egypt) and optimization of cadmium bioaccumulation by *Vibrio harveyi*. Afr J Biotechnol 10(17):3412–3423

Ahalya N, Ramachandra TV, Kanamadi RD (2003) Biosorption of heavy metals. Res J Chem Environ 7(4):71–79

Ahsan N, Faruque K, Shamma F, Islam N, Akhand AA (2012) Arsenic adsorption by bacterial extracellular polymeric substances. Bangladesh J Microbiol 28(2):80–83

Aksornchu P, Prasertsan P, Sobhon V (2008) Isolation of arsenic-tolerant bacteria from arsenic-contaminated soil. Sonklanakarin J Sci Technol 30(1):95–102

Aksu Z, Gülen H (2002) Binary biosorption of iron (III) and iron (III)-cyanide complex ions on Rhizopusarrhizus: modelling of synergistic interaction. Process Biochem 38(2):161–173

Alencar FLS, Navoni JA, do Amaral VS (2017) The use of bacterial bioremediation of metals in aquatic environments in the twenty-first century: a systematic review. Environ Sci Pollut Res 24:16545–16559. https://doi.org/10.1007/s11356-017-9129-8

Alvarez A, Saez JM, Costa JSD, Colin VL, Fuentes MS, Cuozzo SA, Amoroso MJ (2017) Actinobacteria: current research and perspectives for bioremediation of pesticides and heavy metals. Chemosphere 166:41–62

Amoozegar MA, Ghazanfari N, Didari M (2012) Lead and cadmium bioremoval by Halomonas sp., an exopolysaccharide-producing halophilic bacterium. Prog Biol Sci 2(1):1–11

Aryal M, Liakopoulou-Kyriakides M (2011) Equilibrium, kinetics and thermodynamic studies on phosphate biosorption from aqueous solutions by Fe(III)-treated Staphylococusxylosus biomass: common ion effect. Colloids Surf A Physicochem Eng Asp 387:43–49

Aryal M, Liakopoulou-Kyriakides M (2013a) Binding mechanism and biosorption characteristics of Fe(III) by pseudomonas sp. cells. J Water Sustain 3:117–131

Aryal M, Liakopoulou-Kyriakides M (2013b) Characterization of mycobacterium sp. strain Spyr1 biomass and its biosorptionbehavior towards Cr(III) and Cr(VI) in single, binary and multi ion aqueous systems. J Chem Technol Biotechnol 89:559–568

Aryal M, Ziagova M, Liakopoulou-Kyriakides M (2010) Study on arsenic biosorption using Fe(III)-treated biomass of staphylococcus xylosus. Chem Eng J 162:178–185

Aryal M, Ziagova MG, Liakopoulou-Kyriakides M (2012) Cu(II) biosorption and competitive studies in multi-ions aqueous systems by *Arthrobacter* sp. Sphe3 and *Bacillus sphaericus* cells: Equillibrium and thermodynamic studies. Water Air Soil Pollut 223(8):5119–5130

Aryal M, Liakopoulou-Kyriakides M (2015) Bioremoval of heavy metals by bacterial biomass. Environ Monit Assess 187(1):4173

Bagchi D, Stohs SJ, Downs BW, Bagchi M, Preuss HG (2002) Cytotoxicity and oxidative mechanisms of different forms of chromium. Toxicology 180(1):5–22

Bakyayita GK, Norrström AC, Nalubega M, Kulabako RN (2014) Kinetic studies of cd(II) and Pb(II) ions biosorption from aqueous media using untreated and chemically treated biosorbents. Water Sci Technol 69:2230–2236

Banerjee S, Datta S, Chattyopadhyay D, Sarkar P (2011) Arsenic accumulating and transforming bacteria isolated from contaminated soil for potential use in bioremediation. J Environ Sci Health A 46(14):1736–1747

Banjerdkij P, Vattanaviboon P, Mongkolsuk S (2005) Exposure to cadmium elevates expression of genes in the OxyR and OhrR regulons and induces cross-resistance to peroxide killing treatment in Xanthomonascampestris. Appl Environ Microbiol 71(4):1843–1849

Bhakta JN, Munekage Y, Ohnishi K, Jana BB (2012) Isolation and identification of cadmium-and lead-resistant lactic acid bacteria for application as metal removing probiotic. Int J Environ Sci Technol 9(3):433–440

Bhuiyan MA, Parvez L, Islam MA, Dampare SB, Suzuki S (2010) Heavy metal pollution of coal mine-affected agricultural soils in the northern part of Bangladesh. J Hazard Mater 173(1):384–392

Blackwell KJ, Singleton I, Tobin JM (1995) Metal cation uptake by yeast: a review. Appl Microbiol Biotechnol 43(4):579–584

Bowman JP, McCammon SA, Brown MV, Nichols DS, McMeekin TA (1997) Diversity and association of psychrophilic bacteria in Antarctic Sea ice. Appl Environ Microbiol 63:3068–3078

Bradl H (2005) Heavy metals in the environment: origin, interaction and remediation: origin, interaction and remediation, vol 6. Academic Press, Amsterdam

Bruins MR, Kapil S, Oehme FW (2000) Microbial resistance to metals in the environment. Ecotoxicol Environ Saf 45(3):198–207

Bueno BYM, Torem ML, Molina F, De Mesquita LMS (2008) Biosorption of lead (II), chromium (III) and copper (II) by R. Opacus: equilibrium and kinetic studies. Miner Eng 21(1):65–75

Calfa BA, Torem ML (2008) On the fundamentals of Cr(III) removal from liquid streams by a bacterial strain. Miner Eng 21:48–54

Cavalca L, Corsini A, Zaccheo P, Andreoni V, Muyzer G (2013) Microbial transformations of arsenic: perspectives for biological removal of arsenic from water. Future Microbiol 8(6):753–768

Cervantes C, Campos-García J, Devars S, Gutiérrez-Corona F, Loza-Tavera H, Torres-Guzmán JC, Moreno-Sánchez R (2001) Interactions of chromium with microorganisms and plants. FEMS Microbiol Rev 25(3):335–347

Chabukdhara M, Nema AK (2012) Assessment of heavy metal contamination in Hindon River sediments: a chemometric and geochemical approach. Chemosphere 87(8):945–953

Chakravarty R, Banerjee PC (2012) Mechanism of cadmium binding on the cell wall of an acidophilic bacterium. Bioresour Technol 108:176–183

Chandra P, Singh DP (2014) Removal of Cr (VI) by a halotolerant bacterium Halomonas sp. CSB 5 isolated from sāmbhar salt lakeRajastha (India). Cell Mol Biol 60(5):64–72

Chathuranga PKD, Dissanayake DMREA, Priyantha N, Iqbal SS, Mohamed Iqbal MC (2014) Biosorption and desorption of lead(II) by Hydrillaverticillata. Biorem J 181:92–203

Chaudhary A, Salgaonkar BB, Braganca JM (2014) Cadmium tolerance by haloarchaeal strains isolated from solar salterns of Goa, India. Int J Biosci, Biochem Bioinforma 4(1):1–6

Cottrell MT, Kirchman DL (2009) Photoheterotrophic microbes in the Arctic Ocean in summer and winter. Appl Environ Microbiol 75:4958–4966

Das N, Vimala R, Karthika P (2008) Biosorption of heavy metals–an overview. Indian J Biotechnol 7:159–169

Das S, Lyla PS, Khan SA (2006) Marine microbial diversity and ecology: present status and future perspectives. Curr Sci 90:1325–1335

Das S, Mishra J, Das SK, Pandey S, Rao DS, Chakraborty A, Sudarshan M, Das N, Thatoi H (2014) Investigation on mechanism of Cr (VI) reduction and removal by *Bacillus amyloliquefaciens*, a novel chromate tolerant bacterium isolated from chromite mine soil. Chemosphere 96:112–121

Dash HR, Mangwani N, Chakraborty J, Kumari S, Das S (2013) Marine bacteria: potential candidates for enhanced bioremediation. Appl Microbiol Biotechnol 97(2):561–571

Derraik JGB (2002) The pollution of the marine environment by plastic debris: a review. Mar Poll Bull 44:842–852

Devika L, Rajaram R, Mathivanan K (2013) Multiple heavy metal and antibiotic tolerance bacteria isolated from equatorial Indian Ocean. Int J Microbiol Res 4:212–218

Dey U, Chatterjee S, Mondal NK (2016) Isolation and characterization of arsenic-resistant bacteria and possible application in bioremediation. Biotechnol Rep 10:1–7

Din MI, Hussain Z, Mirza ML, Shah AT, Athar MM (2014) Adsorption optimization of lead (II) using Saccharumbengalense as a non-conventional low cost biosorbent: isotherm and thermodynamics modeling. Int J Phytoremediation 16:889–908

Edwards JR, Prozialeck WC (2009) Cadmium, diabetes and chronic kidney disease. Toxicol Appl Pharmacol 238(3):289–293

El-Moselhy KM, Shaaban MT, Ibrahim HA, Abdel-Mongy AS (2013) Biosorption of cadmium by the multiple-metal resistant marine bacterium Alteromonasmacleodii ASC1 isolated from Hurghada harbour, Red Sea. Arch Sci 66(2):259–272

Elsilk SE, El-Shanshoury AERR, Ateya PS (2014) Accumulation of some heavy metals by metal resistant avirulent *Bacillus anthracis* PS2010 isolated from Egypt. Afr J Microbiol Res 8(12):1266–1276

Ergul-Ulger Z, Ozkan AD, Tunca E, Atasagun S, Tekinay T (2014) Chromium (VI) biosorption and bioaccumulation by live and acid-modified biomass of a novel Morganellamorganii isolate. Sep Sci Technol 49(6):907–914

Esposito A, Pagnanelli F, Lodi A, Solisio C, Veglio F (2001) Biosorption of heavy metals by Sphaerotilusnatans: an equilibrium study at different pH and biomass concentrations. Hydrometallurgy 60:129–141

Fan J, Okyay TO, Rodrigues DF (2014) The synergism of temperature, pH and growth phases on heavy metal biosorption by two environmental isolates. J Hazard Mater 279:236–243

Fernandez DS, Puchulu ME, Georgieff SM (2014) Identification and assessment of water pollution as a consequence of a leachate plume migration from a municipal landfill site (Tucum an, Argentina). Environ Geochem Health 36:489–503

Flora SJS, Mittal M, Mehta A (2008) Heavy metal induced oxidative stress & its possible reversal by chelation therapy. Indian J Med Res 128(4):501–523

Focardi S, Pepi M, Landi G, Gasperini S, Ruta M, Di Biasio P, Focardi SE (2012) Hexavalent chromium reduction by whole cells and cell free extract of the moderate halophilic bacterial strain *Halomonas* sp. TA-04. Int Biodeterior Biodegrad 66(1):63–70

Fomina M, Gadd GM (2014) Biosorption: current perspectives on concept, definition and application. Bioresour Technol 160:3–14

Fourest E, Roux JC (1992) Heavy metal biosorption by fungal mycelial by-products: mechanisms and influence of pH. Appl Microbiol Biotechnol 37(3):399–403

Friis N, Myers-Keith P (1986) Biosorption of uranium and lead by Streptomyces longwoodensis. Biotechnol Bioeng 28(1):21–28

Gabr RM, Hassan SHA, Shoreit AAM (2008) Biosorption of lead and nickel by living and non-living cells of Pseudomonas Aeruginosa ASU 6a. Int Biodeter Biodegr 62:195–203

Gallego JLR, Loredo J, Lamas JF, Azquez FV, Anchez JS (2001) Bioremediation of diesel-contaminated soils: evaluation of potential in situ techniques by study of bacterial degradation. Biodegradation 12:325–335

Galun M, Galun E, Siegel BZ, Keller P, Lehr H, Siegel SM (1987) Removal of metal ions from aqueous solutions by *Penicillium* biomass: kinetic and uptake parameters. Water Air Soil Pollut 33(3–4):359–371

Gialamouidis D, Mitrakas M, Liakopoulou-Kyriakides M (2010) Equilibrium, thermodynamic and kinetic studies on biosorption of Mn(II) from aqueous solution by *Pseudomonas* sp., *Staphylococcus xylosus* and *Blakeslea trispora* cells. J Hazard Mater 182:672–680

Goyal N, Jain SC, Banerjee UC (2003) Comparative studies on the microbial adsorption of heavy metals. Adv Environ Res 7(2):311–319

Gu Y, Xu W, Liu Y, Zeng G, Huang J, Tan X, Wang D (2015) Mechanism of Cr (VI) reduction by aspergillus niger: enzymatic characteristic, oxidative stress response, and reduction product. Environ Sci Pollut Res 22(8):6271–6279

Guo J, Zheng XD, Chen QB, Zhang L, XP X (2012) Biosorption of cd(II) from aqueous solution by pseudomonas plecoglossicida: kinetics and mechanism. Curr Microbiol 65:350–355

Gupta VK, Bhushan R, Nayak A, Singh P, Bhushan B (2014) Biosorption and reuse potential of a blue green alga for the removal of hazardous reactive dyes from aqueous solutions. Biorem J 18:179–191

Hasan HA, Abdullah SRS, Kofli NT, Kamarudin SK (2012) Isotherm equilibria of Mn^{+2}biosorption in drinking water treatment by locally isolated bacillus species and sewage activated sludge. J Environ Manag 111:34–43

Hollibaugh JT, Lovejoy C, Murray AE (2007) Microbiology in polar oceans. Oceanography 20:140–145

Hopenhayn C (2006) Arsenic in drinking water: impact on human health. Elements 2(2):103–107

Hossain SM, Anantharaman N (2006) Studies on bacterial growth and arsenic (III) biosorption using *Bacillus subtilis*. Chem Biochem Eng Q 20(2):209–216

Hu H, Rabinowitz M, Smith D (1998) Bone lead as a biological marker in epidemiologic studies of chronic toxicity: conceptual paradigms. Environ Health Perspect 106(1):1–7

Huang F, Dang Z, Guo CL, GN L, RR G, Liu HJ, Zhang H (2013) Biosorption of cd (II) by live and dead cells of Bacillus Cereus RC-1 isolated from cadmium-contaminated soil. Colloids Surf B: Biointerfaces 107:11–18

Huang F, Guo CL, GN L, Yi XY, Zhu LD, Dang Z (2014) Bioaccumulation characterization of cadmium by growing Bacillus Cereus RC-1 and its mechanism. Chemosphere 109:134–142

Huang W, Liu ZM (2013) Biosorption of cd(II)/Pb(II) from aqueous solution by biosurfactant-producing bacteria: isotherm kinetic characteristic and mechanism studies. Colloids Surf B: Biointerfaces 105:113–119

Ilamathi R, Nirmala GS, Muruganandam L (2014) Heavy metals biosorption in liquid solid fluidized bed by immobilized consortia in alginate beads. Int J Chem Tech Res 6:652–662

Inagaki F, Nunoura T, Nakagawa S, Teske A, Lever M, Lauer A, Suzuki M, Takai K, Delwiche M, Colwell FS, Nealson KH, Horikoshi K, D'Hondt S, Jorgensen BB (2006) Biogeographical distribution and diversity of microbes in methane hydrate-bearing deep marine sediments on the Pacific Ocean margin. PNAS Microbiol 103:2815–2820

Iqbal S, Muntner P, Batuman V, Rabito FA (2008) Estimated burden of blood lead levels ⩾5μg/dl in 1999-2002 and declines from 1988 to 1994. Environ Res 107(3):305–311

Iyer A, Mody K, Jha B (2005) Biosorption of heavy metals by a marine bacterium. Mar Pollut Bull 50(3):340–343

Jaafarzadeh N, Teymouri P, Babaei AA, Alavi N, Ahmadi M (2014) Biosorption of cadmium(II) from aqeous solution by NaCl-treated *Ceratophylum demersum*. Environ Eng J 13:763–773

Järup L (2003) Hazards of heavy metal contamination. Br Med Bull 68(1):167–182

Javanbakht V, Alavi SA, Zilouei H (2014) Mechanisms of heavy metal removal using microorganisms as biosorbent. Water Sci Technol 69:1775–1787

Jiang P, Li J, Han F, Duan G, Lu X, Gu Y, Yu W (2011) Antibiofilm activity of an exopolysaccharide from marine bacterium *Vibrio* sp. QY101. PLoSOne 6(4):e18514

Jin W, Zhang Z, Wu G, Tolba R, Chen A (2014) Integrated lignin-mediated adsorption-release process and electrochemical reduction for the removal of trace Cr (VI). RSC Adv 4(53):27843–27849

Joo JH, Hassan SHA, SE O (2010) Comparative study of biosorption of Zn^{+2} by *Pseudomonas aeruginosa* and *Bacillus cereus*. Int Biodeterior Biodegrad 64:734–741

Kang S, Lee J, Kima K (2007) Biosorption of Cr(III) and Cr(VI) onto the cell surface of pseudomonas aeruginosa. Biochem Eng J 36:54–58

Kao WC, Huang CC, Chang JS (2008) Biosorption of nickel, chromium and zinc by MerP-expressing recombinant Escherichia Coli. J Hazard Mater 158(1):100–106

Khan Z, Rehman A, Hussain SZ, Nisar MA, Zulfiqar S, Shakoori AR (2016) Cadmium resistance and uptake by bacterium, *Salmonella enterica* 43C, isolated from industrial effluent. AMB Express 6(1):54

Kiliç NK, Stensballe A, Otzen DE, Dönmez G (2010) Proteomic changes in response to chromium (VI) toxicity in Pseudomonas Aeruginosa. Bioresour Technol 101(7):2134–2140

Koduru JR, Chang YY, Kim IS (2014) Low-cost schizandra chinesis fruit peel for co(II) removal from aqueous environment: adsorption properties and mechanism. Asian J Chem 26:289–297

Kordialik-Bogacka E, Diowksz A (2014) Metal uptake capacity of modified *Saccharomyces pastorianus* biomass from different types of solution. Environ Sci Pollut Res 21:2223–2229

Kothe E, Dimkpa C, Haferburg G, Schmidt A, Schmidt A et al (2010) Streptomycete heavy metal resistance: extracellular and intracellular mechanisms. In: Sherameti I, Varma A (eds) Soil biology. Springer, Berlin/Heidelberg, pp 225–235

Kratochvil D, Volesky B (1998) Advances in the biosorption of heavy metals. Trends Biotechnol 16(7):291–300

Kuyucak N, Volesky B (1988) Biosorbents for recovery of metals from industrial solutions. Biotechnol Lett 10(2):137–142

Lam TV, Agovino P, Niu X, Roché L (2007) Linkage study of cancer risk among lead-exposed workers in New Jersey. Sci Total Environ 372(2):455–462

Lauro FM, McDougald D, Thomas T, Williams TJ, Egan S, Rice S, DeMaere MZ, Ting L, Ertan H, Johnson J, Ferriera S, Lapidus A, Anderson I, Kyrpides N, Munk AC, Detter C, Hang CS, Brown MV, Robb FT, Kjelleberga S, Cavicchiol R (2009) The genomic basis of trophic strategy in marine bacteria. PNAS 106:15527–15533

Li JL, Zhang CX, Wang YX, Liao XP, Yao LL, Liu M, Xu L (2015) Pollution characteristics and distribution of polycyclic aromatic hydrocarbons and organochlorine pesticides in groundwater at Xiaodian sewage irrigation area, Taiyuan City. Huan jing ke xue= Huanjing kexue 36(1):172–178

Liang S, Guo X, Lautner S, Saake B (2014) Removal of hexavalent chromium by different modified spruce bark adsorbents. J Wood Chem Technol 34:273–290

Liao L, Xu XW, Jiang XW, Wang CS, Zhang DS, Ni JY, Wu M (2011) Microbial diversity in deep-sea sediment from the cobalt-rich crust deposit region in the Pacific Ocean. FEMS Microbiol Ecol 78(3):565–585

Lin CC, Lai YT (2006) Adsorption and recovery of lead (II) from aqueous solutions by immobilized Pseudomonas Aeruginosa PU21 beads. J Hazard Mater 137:99–105

Liu HL, Chen BY, Lan YW, Cheng YC (2004) Biosorption of Zn (II) and cu (II) by the indigenous Thiobacillusthiooxidans. Chem Eng J 97(2):195–201

Loka Bharathi PA, Nair S (2005) Rise of the dormant: simulated disturbance improves culturable abundance, diversity and functions of deep-sea bacteria of Central Indian Ocean Basin. Mar Georesour Geotechnol 23:419–428

Long J, Luo D, Chen Y (2014) Identification and biosorption characterization of a thallium-resistant strain. Chin J App Environ Biol 20:426–430

Lovejoy C, Massana R, Pedros-Alio C (2006) Diversity and distribution of marine microbial eukaryotes in the Arctic Ocean and adjacent seas. Appl Environ Microbiol 72:3085–3095

Lu WB, Shi JJ, Wang CH, Chang JS (2006) Biosorption of lead, copper and cadmium by an indigenous isolate *Enterobacter* sp. J1 possessing high heavy-metal resistance. J Hazard Mater 134:80–86

Ma Y, Wang P, Wang C, Zhang SH, Cheng S (2015) Isolation and characterization of Pb-resistant strains and the removal of Pb(II). Fresenius Environ Bull 24(3b):1150–1157

Malaviya P, Singh A (2016) Bioremediation of chromium solutions and chromium containing wastewaters. Crit Rev Microbiol 42(4):607–633

Maldonado J, Diestra E, Huang L, Domènech AM, Villagrasa E, Puyen ZM, Duran R, Esteve I, Solé A (2010) Isolation and identification of a bacterium with high tolerance to lead and copper from a marine microbial mat in Spain. Ann Microbiol 60(1):113–120

Malik A (2004) Metal bioremediation through growing cells. Environ Int 30(2):261–278

Mameri N, Boudries N, Addour L, Belhocine D, Lounici H, Grib H, Pauss A (1999) Batch zinc biosorption by a bacterial nonliving Streptomyces Rimosus biomass. Water Res 33(6):1347–1354

Maneerat S, Phetrong K (2007) Isolation of biosurfactant-producing marine bacteria and characteristics of selected biosurfactant. Songklanakarin J Sci Technol 29:781–791

Mangaiyarkarasi MM, Vincent S, Janarthanan S, Rao TS, Tata BVR (2011) Bioreduction of Cr (VI) by alkaliphilic Bacillus Subtilis and interaction of the membrane groups. Saudi J Biol Sci 18(2):157–167

Mao J, Won SW, Yun YS (2013) Development of poly (acrylic acid)-modified bacterial biomass as a high-performance biosorbent for removal of cd (II) from aqueous solution. Ind Eng Chem Res 52(19):6446–6452

Margesin R, Schinner F (2001) Biodegradation and bioremediation of hydrocarbons in extreme environments. Appl Microbiol Biotechnol 56:650–663

Masood F, Malik A (2011) Biosorption of metal ions from aqueous solution and tannery effluent by bacillus sp. FM1. J Environ Sci Health A 46:1667–1674

Mateos LM, Ordóñez E, Letek M, Gil JA (2006) Corynebacterium glutamicum as a model bacterium for the bioremediation of arsenic. Int Microbiol 9(3):207–215

Mathivanan K, Rajaram R (2014a) Tolerance and biosorption of cadmium (II) ions by highly cadmium resistant bacteria isolated from industrially polluted estuarine environment. Indian journal of geo-marine. Sciences 43(4):580–588

Mathivanan K, Rajaram R (2014b) Isolation and characterisation of cadmium-resistant bacteria from an industrially polluted coastal ecosystem on the southeast coast of India. Chem Ecol 30(7):622–635

Megharaj M, Avudainayagam S, Naidu R (2003) Toxicity of hexavalent chromium and its reduction by bacteria isolated from soil contaminated with tannery waste. Curr Microbiol 47(1):0051–0054

Mishra RR, Dhal B, Dutta SK, Dangar TK, Das NN, Thatoi HN (2012) Optimization and characterization of chromium (VI) reduction in saline condition by moderately halophilic *Vigribacillus* sp. isolated from mangrove soil of Bhitarkanika, India. J Hazard Mater 227-228:219–226

Mohapatra RK, Panda CR (2017) Spatiotemporal variation of water quality and assessment of pollution potential in Paradip port due to port activities. Indian J Geo Marine Sci 46(07):1274–1286

Mohapatra RK, Pandey S, Thatoi H, Panda CR (2017b) Reduction of chromium (VI) by marine bacterium *Brevibacillus laterosporus* under varying saline and pH conditions. Environ Eng Sci 34(9):617–626

Mohapatra RK, Pandey S, Thatoi HN, Panda CR (2016) Screening and evaluation of multi-metal tolerance of chromate resistant marine bacteria isolated from water and sediment samples of Paradip port, Odisha Coast. J Adv Microbiol 2(3):135–147

Mohapatra RK, Parhi PK, Thatoi H, Panda CR (2017a) Bioreduction of hexavalent chromium by Exiguobacteriumindicum strain MW1 isolated from marine water of Paradip port, Odisha, India. Chem Ecol 33(2):114–130

Moon EM, Peacock CL (2011) Adsorption of cu (II) to *Bacillus subtilis*: a pH-dependent EXAFS and thermodynamic modelling study. Geochim Cosmochim Acta 75(21):6705–6719

Muñoz AJ, Espínola F, Moya M, Ruiz E (2015) Biosorption of Pb (II) ions by *Klebsiella* sp. 3S1 isolated from a wastewater treatment plant: kinetics and mechanisms studies. Biomed Res Int 2015:1–12

Muñoz AJ, Ruiz E, Abriouel H, Gálvez A, Ezzouhri L, Lairini K, Espínola F (2012) Heavy metal tolerance of microorganisms isolated from wastewaters: identification and evaluation of its potential for biosorption. Chem Eng J 210:325–332

Murthy S, Bali G, Sarangi SK (2012) Biosorption of lead by Bacillus Cereus isolated from industrial effluents. British Biotechnol J 2(2):73–84

Naik MM, Dubey SK (2013) Lead resistant bacteria: lead resistance mechanisms, their applications in lead bioremediation and biomonitoring. Ecotoxicol Environ Saf 98:1–7

Naik MM, Pandey A, Dubey SK (2012) Bioremediation of metals mediated by marine bacteria. In: Microorganisms in Environmental Management. Springer, Netherlands, pp 665–682

Nakajima A, Yasuda M, Yokoyama H, Ohya-Nishiguchi H, Kamada H (2001) Copper biosorption by chemically treated Micrococcus Luteus cells. World J Microbiol Biotechnol 17(4):343–347

Navas-Acien A, Guallar E, Silbergeld EK, Rothenberg SJ (2007) Lead exposure and cardiovascular disease: a systematic review. Environ Health Perspect 115:472–482

Nweke CO, Okpokwasili GC (2003) Drilling fluid base oil biodegradation potential of a soil staphylococcus species. African J Biotechnol 2:293–295

Ohtake H, Takahashi K, Tsuzuki Y, Toda K (1985) Uptake and release of phosphate by a pure culture of *Acinetobacter calcoaceticus*. Water Res 19:1587–1594

Okami Y (1986) Marine microorganisms as a source of bioactive agents. Microbial Ecol 12:65–78

Oves M, Khan MS, Zaidi A (2013) Biosorption of heavy metals by bacillus thuringiensis strain OSM29 originating from industrial effluent contaminated north Indian soil. Saudi J Biol Sci 20:121–129

Pagnanelli F, Petrangeli PM, Trifoni M, Toro L, Veglio F (2000) Biosorption of metal ions on Arthrobacter sp.: biomass characterization and biosorptionmodeling. Environ Sci Technol 34:2773–2778

Paranjape K, Gowariker V, Krishnamurthy V, Gowariker S (2014) The pesticide encyclopedia. Cabi, Oxfordshire

Paul ML, Samuel J, Chandrasekaran N, Mukherjee A (2012) Comparative kinetics, equilibrium, thermodynamic and mechanistic studies on biosorption of hexavalent chromium by live and heat killed biomass of *Acinetobacter junii* VITSUKMW2, an indigenous chromite mine isolate. Chem Eng J 187:104–113

Pepi M, Borra M, Tamburrino S, Saggiomo M, Viola A, Biffali E, Balestra C, Sprovieri M, Casotti R (2016) A bacillus sp. isolated from sediments of the Sarno River mouth, gulf of Naples (Italy) produces a biofilm biosorbingPb (II). Sci Total Environ 562:588–595

Pieper DH, Reineke W (2000) Engineering bacteria for bioremediation. Curr Opinion Biotechnol 11:262–270

Prakash D, Raushan RK, Sangodkar UX (2008) Isolation and characterization of meta-toluic acid degrading marine bacterium. Indian. J Mar Sci 37:322–325

Prasad KS, Ramanathan AL, Paul J, Subramanian V, Prasad R (2013) Biosorption of arsenite (as^{+3}) and arsenate (as^{+5}) from aqueous solution by *Arthrobacter* sp. biomass. Environ Technol 34(19):2701–2708

Puranik PR, Paknikar KM (1999) Biosorption of lead, cadmium, and zinc by Citrobacter strain MCM B-181: characterization studies. Biotechnol Prog 15(2):228–237

Puyen ZM, Villagrasa E, Maldonado J, Diestra E, Esteve I, Solé A (2012) Biosorption of lead and copper by heavy-metal tolerant Micrococcus Luteus DE2008. Bioresour Technol 126:233–237

Rainbow PS (1995) Bio monitoring of heavy metal availability in the marine environment. Mar Poll Bull 31:183–192

Raja Rao P, Pallavi D, Venkateshwarlu T (2014) Removal of heavy metals in fly ash by using Saccharomyces Cerevisiae. Int J Appl Eng Res 9:107–114

Rajendran P, Muthukrishnan J, Gunasekaran P (2003) Microbes in heavy metal remediation. Indian J Environ Biol 41:935–944

Rangabhashiyam S, Anu N, Selvaraju N (2013) Biosorption of heavy metals using low cost agricultural by products. Res J Chem Environ 17:112–123

Ren G, Jin Y, Zhang C, Gu H, Qu J (2015) Characteristics of bacillus sp. PZ-1 and its biosorption to Pb (II). Ecotoxicol Environ Saf 117:141–148

Sabdono A (2010) Cadmium removal by a bioreducpiun coral bacterium Pseudoalteromonas sp. strain CD15 isolated from the tissue of coral Goniastreaaspera, Jepara waters. J Coast Dev 13(2):15–25

Sar P, Kazy SK, Asthana RK, Singh SP (1999) Metal adsorption and desorption by lyophilized *Pseudomonas aeruginosa*. International biodeterioration. Biodegradation 44(2):101–110

Sari A, Tuzen M (2009) Biosorption of as(III) and as(V) from aqueous solution by macrofungus (*Inonotus hipidus*) biomass: equilibrium and kinetics studies. J Hazard Mater 164:1372–1378

Seki H, Suzuki A, Mitsueda SI (1998) Biosorption of heavy metal ions on *Rhodobactersphaeroides* and *Alcaligeneseutrophus* H16. J Colloid Interface Sci 197:185–190

Selatnia A, Bakhti MZ, Madani A, Kertous L, Mansouri Y (2004b) Biosorption of cd^{2+} from aqueous solution by a NaOH-treated bacterial dead Streptomyces Rimosus biomass. Hydrometallurgy 75(1):11–24

Selatnia A, Boukazoula A, Kechid N, Bakhti MZ, Chergui A, Kerchich Y (2004a) Biosorption of lead(II) from aqueous solution by a bacterial dead *Streptomyces rimosus* biomass. Biochem Eng J 19(2):127–135

Semerjian L (2010) Equilibrium and kinetics of cadmium adsorption from aqueous solutions using untreated Pinushalepensis sawdust. J Hazard Mater 173(1):236–242

Shahid M, Pinelli E, Dumat C (2012) Review of Pb availability and toxicity to plants in relation with metal speciation; role of synthetic and natural organic ligands. J Hazard Mater 219:1–12

Shamim S, Rehman A (2012) Cadmium resistance and accumulation potential of Klebsiella Pneumoniae strain CBL-1 isolated from industrial wastewater. Pak J Zool 44:203–208

Shrestha RA, Lama B, Joshi J, Sillanpää M (2008) Effects of Mn (II) and Fe (II) on microbial removal of arsenic (III). Environ Sci Pollut Res 15(4):303–307

Singh R, Gautam N, Mishra A, Gupta R (2011) Heavy metals and living systems: an overview. Indian J Pharmacol 43:246–253

Siripongvutikorn S, Asksonthong R, Usawakesmanee W (2016) Evaluation of harmful heavy metal (hg, Pb and cd) reduction using *Halomonaselongata* and *Tetragenococcushalophilus* for protein hydrolysate product. Funct Foods Health Dis 6(4):195–205

Sogin ML, Morrison HG, Huber JA, Welch DM, Huse SM, Neal PR, Arrieta JM, Hernd GJ (2006) Microbial diversity in the deep sea and the underexplored 'rare biosphere. Proc Nat Acad Sci 103:12115–12120

Sowmya M, Rejula MP, Rejith PG, Mohan M, Karuppiah M, Hatha AM (2014) Heavy metal tolerant halophilic bacteria from Vembanad Lake as possible source for bioremediation of lead and cadmium. J Environ Biol 35(4):655–660

Srinath T, Verma T, Ramteke PW, Garg SK (2002) Chromium (VI) biosorption and bioaccumulation by chromate resistant bacteria. Chemosphere 48:427–435

Stanley ME (2005) Environmental chemistry. CRC, Boca Raton. ISBN 1-56670-633-5

Stramski D, Kiefer DA (1998) Can heterotrophic bacteria be important to marine light absorption? J Plankton Res 20:1489–1500

Subramanian S, Sam S, Jayaraman G (2012) Hexavalent chromium reduction by metal resistant and halotolerant *Planococcusmaritimus* VITP21. Afr J Microbiol Res 6(47):7339–7349

Swapna TH, Papathoti NK, Khan MY, Reddy G, Hameeda B (2016) Bioreduction of Cr (VI) by biosurfactant producing marine bacterium *Bacillus subtilis* SHB 13. J Sci Ind Res 75:432–438
Takeuchi M, Kawahata H, Gupta LP, Kita N, Morishita Y, Ono Y, Komai T (2007) Arsenic resistance and removal by marine and non-marine bacteria. J Biotechnol 127(3):434–442
Tangaromsuk J, Pokethitiyook P, Kruatrachue M, Upatham ES (2002) Cadmium biosorption by *Sphingomonas paucimobilis* biomass. Bioresour Technol 85:103–105
Taran M, Safari M, Monaza A, Reza JZ, Bakhtiyari S (2013) Optimal conditions for the biological removal of arsenic by a novel halophilic archaea in different conditions and its process optimization. Polish journal of. Chem Technol 15(2):7–9
Tchounwou PB, Centeno JA, Patlolla AK (2004) Arsenic toxicity, mutagenesis, and carcinogenesis-a health risk assessment and management approach. Mol Cell Biochem 255(1–2):47–55
Tchounwou PB, Yedjou CG, Patlolla AK, Sutton DJ (2012) Heavy metal toxicity and the environment. In: Molecular, clinical and environmental toxicology. Springer, Basel, pp 133–164
Thatheyus AJ, Ramya D (2016) Biosorption of chromium using bacteria: an overview. Sci Inter 4(7):74–79
Tunali S, Cabuk A, Akar T (2006) Removal of lead and copper ions from aqueous solutions by bacterial strain isolated from soil. Chem Eng J 115(3):203–211
Uslu G, Tanyol M (2006) Equilibrium and thermodynamic parameters of single and binary mixture biosorption of lead(II) and copper(II) ions onto pseudomonas putida: effect of temperature. J Hazard Mater 135:87–93
Vallee BL, Ulmer DD (1972) Biochemical effects of mercury, cadmium, and lead. Annu Rev Biochem 41(1):91–128
Veglio F, Beolchini F (1997) Removal of metals by biosorption: a review. Hydrometallurgy 44(3):301–316
Veneu DM, Torem ML, Pino GAH (2013) Fundamental aspects of copper and zinc removal from aqueous solutions using a *Streptomyces lunalinharesii* strain. Miner Eng 48:44–50
Vijayaraghavan K, Yun YS (2008) Bacterial biosorbents and biosorption. Biotechnol Adv 26:266–291
Vijayaraghavan R, Rajendran S (2011) Studies on agar degrading Salegentibacter sp. and characterization of its agarase. Int J Biosci 1:56–64
Vishnoi N, Singh DP (2014) Biotransformation of arsenic by bacterial strains mediated by oxidoreductase enzyme system. Cell Mol Biol 60(5):7–14
Voica DM, Bartha L, Banciu HL, Oren A (2016) Heavy metal resistance in halophilic bacteria and archaea. FEMS Microbiol Lett 363(14):fnw146
Volesky B (1986) Biosorbent materials. Biotechnol Bioeng Symp 16:121–126
Volesky B (2001) Detoxification of metal-bearing effluents: biosorption for the next century. Hydrometallurgy 59:203–216
Waisberg M, Joseph P, Hale B, Beyersmann D (2003) Molecular and cellular mechanisms of cadmium carcinogenesis. Toxicology 192(2):95–117
Waite CCC, da Silva GOA, Bitencourt JAP, Sabadini-Santos E, Crapez MAC (2016) Copper and lead removal from aqueous solutions by bacterial consortia acting as biosorbents. Mar Pollut Bull 109(1):386–392
Walton FS, Harmon AW, Paul DS, Drobná Z, Patel YM, Styblo M (2004) Inhibition of insulin-dependent glucose uptake by trivalent arsenicals: possible mechanism of arsenic-induced diabetes. Toxicol Appl Pharmacol 198(3):424–433
Wang JL (2002) Immobilization techniques for biocatalysts and water pollution control. Science press, Beijing
Wang L, Chua H, Wong PK, Lo WH, PHF Y (2003) Ni2+ removal and recovery from electroplating effluent by *Pseudomonas putida* 5-x cell biomass. J Environ Sci Health A 38(3):521–531
Wang L, Li FT, Zhou Q (2006) Contribution of cellsurface components to Cu^{2+} adsorption by pseudomonas putida 5-x. Appl Biochem Biotechnol 128:33–46
Wasi S, Tabrez S, Ahmad M (2013) Use of *Pseudomonas* spp. for the bioremediation of environmental pollutants: a review. Environ Monit Assess 185(10):8147–8155

Watt GCM, Britton A, Gilmour HG, Moore MR, Murray GD, Robertson SJ (2000) Public health implications of new guidelines for lead in drinking water: a case study in an area with historically high water lead levels. Food Chem Toxicol 38:S73–S79

Weast RC (1984) CRC handbook of chemistry and physics, 64th edn. CRC Press, Boca Raton

White C, Wilkinson SC, Gadd GM (1995) The role of microorganisms in biosorption of toxic metals and radionuclides. International biodeterioration. Biodegradation 35(1–3):17–40

WHO (2011) Guidelines for drinking-water quality. WHO Chron 38:104–108

Wierzba S, Latała A (2010) Biosorption lead (II) and nikel (II) from an aqueous solution by bacterial biomass. Pol J Chem Technol 12(3):72–78

Woodruff TJ, Carlson A, Schwartz JM, Giudice LC (2008) Proceedings of the summit on environmental challenges to reproductive health and fertility: executive summary. Fertil Steril 89(2):e1–e20

Wu Q, Leung JYS, Geng X, Chen S, Huang X et al (2015) Heavy metal contamination of soil and water in the vicinity of an abandoned e-waste recycling site: implications for dissemination of heavy metals. Sci Total Environ 506:217–225

Wuana RA, Okieimen FE (2011) Heavy metals in contaminated soils: a review of sources, chemistry, risks and best available strategies for remediation. ISRN Ecology 2011(402647). https://doi.org/10.5402/2011/402647

Yan L, Yin H, Zhang S, Leng F, Nan W, Li H (2010) Biosorption of inorganic and organic arsenic from aqueous solution by Acidithiobacillusferrooxidans BY-3. J Hazard Mater 178:209–217

Ye J, Yin H, Xie D, Peng H, Huang J, Liang W (2013) Copper biosorption and ions release by *Stenotrophomonas maltophilia* in the presence of benzo [a] pyrene. Chem Eng J 219:1–9

Yee N, Fein J (2001) Cd adsorption onto bacterial surfaces: a universal adsorption edge? Geochim Cosmochim Acta 65(13):2037–2042

Yoshida T, Yamauchi H, Sun GF (2004) Chronic health effects in people exposed to arsenic via the drinking water: dose–response relationships in review. Toxicol Appl Pharmacol 198(3):243–252

Zeng D, Zhu K, Pei X (2014) Characteristics of heavy metal circulation in biosphere. Agric Sci Technol 15(4):642–647

Zhitkovich A (2011) Chromium in drinking water: sources, metabolism, and cancer risks. Chem Res Toxicol 24(10):1617–1629

Zhou W, Ma Y, Zhou J, Zhang Y (2013) Bio-removal of cadmium by growing deep-sea bacterium Pseudoalteromonas sp. SCSE709-6. Extremophiles 17(5):723–731

Ziagova M, Dimitriadis G, Aslanidou D, Papaioannou X, Tzannetaki EL, Liakopoulou-Kyriakides M (2007) Comparative study of cd(II) and Cr(VI) biosorption on staphylococcus xylosus and *Pseudomonas* sp. in single and binary mixtures. Bioresour Technol 98:2859–2865

Zobell CE, Upham HC (1944) A list of marine bacteria including descriptions of sixty new species. Bull Scripps Inst Oceanog 5:239–292

Zouboulis AI, Loukidou MX, Matis KA (2004) Biosorption of toxic metals from aqueous solutions by bacteria strains isolated from metal-polluted soils. Process Biochem 39(8):909–916

Chapter 16
Role of Biofilms in Environment Pollution and Control

Mukesh Kumar Yadav

16.1 Introduction

Biofilm is an irreversible assemblage of microbes to biotic or abiotic surfaces and enclosed in self-produced matrix of extra polymeric substances (EPS). The microbes could adherent to various material depends on environment in which biofilm develops. In environment, the microbes could attach on the non-cellular material such as corrosion particles, clay or slit particles and mineral crystals (Donlan 2002). In medical setting microbial biofilms has been detected on medical indwelling devices and on human tissues (Mack et al. 2006). Biofilms cause various diseases in humans such as cystic fibrosis (CF), native valve endocarditis, otitis media (OM), periodontitis, wound infections, superficial skin infections and chronic bacterial prostatitis (Agarwal et al. 2010; Costerton et al. 1999; Maki et al. 2006; Otto 2010; Thornton et al. 2013; Donlan 2001). The microbes with-in biofilms are surrounded by self-produced matrix of extracellular polymeric substances made up of proteins, polysaccharides, e-DNA and lipids. The matrix protects the bacteria with-in the biofilms and give rise to persistent strains (Hoiby et al. 2011; Hengzhuang et al. 2011, 2012). The matrix prevents the diffusion of antibiotics and hence increases resistance of bacteria with in biofilms. In-deed at sub-MIC concentration most bacteria adopt biofilm mode of growth. The eradication and treatment of biofilm related infections is difficult using conventional antibiotics and results increased treatment expense and recovery time.

Most of microbes adopt biofilm mode of growth and cycle the nutrients. Many bacteria are know that forms biofilm on plants roots, shoots and on leaves and help

M.K. Yadav (✉)
Department of Otorhinolaryngology-Head and Neck Surgery & Institute for Medical Device Clinical Trials, Korea University College of Medicine, Seoul, South Korea

Korea University Guro Hospital, Seoul, South Korea
e-mail: mukiyadav@gmail.com

J.K. Patra et al. (eds.), *Microbial Biotechnology*,
https://doi.org/10.1007/978-981-10-6847-8_16

in plant growth promotion and pest control. The biofilms of bio-control agents could be utilize for the pest control in agriculture. The application of biocontrol biofilm in agriculture could be an alternative to chemical fertilizers and pest and that could decrease the soil and water pollution.

The biofilms of environmental microbes could be applied for bioremediation. It is suggested that in environment, at site of high concentration of toxic chemical bacteria survive in form of biofilm (Gross et al. 2007). However, the lack of nutrients, application of appropriate microbes/strains and low microbial availability hamper the application in bioremediation. Microbial biofilms are naturally present in environment and execute bioremediation in soil and sediments which is part of nutrient cycling. Although the planktonic bacteria can also metabolize the toxins, however, the planktonic are sensitive for toxicity of pollutants. In contrast, biofilms protects the bacteria and facilitates the degradation of toxins (von Canstein et al. 2002). In addition the poly-microbial biofilm with different species/genus follow different metabolic pathways and produces matrix of different kinds that helps to degrade the environment pollutions (Vu et al. 2009; von Canstein et al. 2002; Boles et al. 2004). The multi-species biofilm found on tidal flats, corroded pipes and streambeds and even on site of infections cooperativities each other. Some specific constituents of EPS help in solubilization of hydrophobic or recalcitrant subtracts that were inaccessible to microbes (Latch et al. 2003; Seo et al. 2009; Wang et al. 2011). The multi-species biofilms of *Burkholderia* sp. NK8 together with *Pseudomonad aeruginosa* PA01 have been used for degradation of chlorinated benzoates (Yoshida et al. 2009).

In this chapter we present a review on the role of biofilms in environment pollution and how biofilm could be utilized for pollution control.

16.2 Microbial Biofilms and Their Characteristics

According to Center for Biofilm Engineering, a bacteria biofilm is defined as "a structured community of bacterial cells enclosed in a self-produced polymeric matrix and adherent to an inert or living surface" (Costerton et al. 1999). The biofilm formation began when free floating bacteria attach to a biotic or abiotic surface and began to excrete a slimy, glue-like substance that anchor various kinds of materials including soil particle, metals, plastics, human tissue and medical implants etc. The initial attachment of bacteria is loose and reversible, and if the colonies remain attached then irreversible attachment occurs and permanent adhesion could take place by adhesion molecules and proteins present on cell surface. Microbial biofilms could be found everywhere in nature. From human/animal tissue to medical indwelling devices, body implants, on plants, in soil, on rocks and on bottom of water. Further-more, the biofilms are important components of food chains in rivers and streams and are grazed upon by the aquatic invertebrates upon which many fish feed. Biofilms could grow in extreme environments such as in the hot, acidic pools at Yellowstone National Park and on glaciers in Antarctica. It is evident that the microbial growth under biofilm state is an ancient and integral component of the prokaryotic life cycle, and is a key factor for survival in diverse environments (Hall-Stoodley et al. 2004). Biofilms formation may be a defensive reaction of microbes in presence of antibiotics. *Escherichia coli* and *Pseudomonas aeruginosa* biofilms has been detected at sub-MIC concentration of aminoglycosides antibiotics (Hoffman et al. 2005). Similarly, *Escherichia coli* and *Pseudomonas aeruginosa* treatment with sub-inhibitory concentrations of tetracycline and cephradine induced biofilm formation and enhance pB10 plasmid transfer rate among the biofilm biomass (Salcedo et al. 2014).

The biofilm bacteria surrounded by EPS are resistance to antibiotics and exhibits 10–1000 time resistance to antibiotic with compare to planktonic cells (Gilbert et al. 2002). The increased resistance in biofilm bacteria is due to combination of different factors: (a) obstruction in the diffusion of antibiotics by biofilm polysaccharides matrix. (b) Presence of resistance or persister cells in the depth of biofilms. (c) The presence of slowing growing or metabolically inactive cells (Stewart and Costerton 2001; Stewart 2002). These properties of biofilm act synergistically along with conventional factors presence in bacteria that give rise to antibiotic resistance genes (ARGs) harbors in antibiotic resistant strains. For example, the biofilms of β-lactamase producing bacteria are more resistance to β-antibiotics such as ampicillin, due to the inherited property of bacteria by virtue of which bacteria inactive the β-lactam antibiotic and the biofilm provide addition resistance bacteria (Anderl et al. 2000). Similarly, in *Pseudomonas aeruginosa* the biofilms gene *ampC* induces in response to exposure to antibiotics such as imipenem (Bagge et al. 2004).

16.3 Biofilms and Human Infection

Microbial biofilms causes many infectious diseases in human. The biofilms related infections are difficult to treat due to antibiotic resistance of biofilms that result in an increase in treatment cost and recovery time. According to National Institutes of Health (NIH), more than 75% of bacteria form biofilms under different environmental conditions. Many bacteria including *Staphylococcus*, *Streptococcus*, *Pseudomonad*, *Haemophilus influenza*, *Moraxella catarrhalis*, *E.coli*, etc. form biofilms on human tissue and on medical devices. *Streptococcus pneumoniae* is a Gram positive bacteria *frequently isolated from biofilms*. *S. pneumoniae* in children, the elderly and in immuno-compromised patients causes severe invasive infections such as pneumonia, septicemia, and meningitis, especially (Jedrzejas 2001; Klugman et al. 2008; van der Poll and Opel 2009; Weimer et al. 2010). It is suggested that initial colonization of pneumococcal in the nasopharynx is a prerequisite to causes infection, the pneumococci persists for months in the nasopharynx, forming specialized structures referred to as biofilms (Bogaert et al. 2004; Simell et al. 2012). The pneumococci can migrate to other anatomic sites from biofilms to cause severe biofilm-associated diseases such as pneumonia and otitis media (Hall-Stoodley et al. 2006; Sanchez et al. 2010). From the lungs of patients with pneumococcal pneumonia or the ear cavity of patients with otitis media, planktonic pneumococci can disperse from the biofilm structure and invade sterile sites, such as the blood stream or brain, to cause lethal bacteremia or meningitis, respectively (Ash and Sheffield 2013; Pichichero 2013; Shak et al. 2013). *Staphylococcus aureus* causes various infections in human including toxic shock syndrome, sepsis, bacteremia, endocarditis, skin infections, respiratory tissue infections and bone joints (Lowy 1998). *Staphylococcus aureus* and *Pseudomonas aeruginosa* are known to cause biofilm-related infections, and methicillin-resistant *S. aureus* (MRSA) has emerged as a clinically relevant pathogen because of its resistance to antibiotics and its ability to form biofilms (Chopra et al. 2015). MRSA and *Pseudomonas aeruginosa* has been isolated from biofilms of chronic suppurative otitis media (CSOM) and chronic middle ear infections (Jung et al. 2009; Kim et al. 2015), chronic otitis media, cholesteatoma, chronic adenoiditis, chronic sinusitis, post-operative trampansomay, and nasal polyposis (Post et al. 2004; Bendouah et al. 2006; Boase et al. 2013). The multi-species of *Staphylococcal aureus* and *Pseudomonas* has been detected in chronic rhinosinusitis (CRS) patients, with enhanced mucosal inflammation, more severe osteitis, higher incidence of recurrent infection (Dong et al. 2014) and post-operative outcomes (Singhal et al. 2011), and post-surgery progression (Bendouah et al. 2006). Many reports suggested that *Pseudomonas* produces metabolites that inhibits or clear the *staphylococcus*, however, in poly-microbial colonization of *Pseudomonas aeruginosa* and *S. aureus*, *Pseudomonas aeruginosa* does not completely inhibit the colonization of *S. aureus*; rather, *S. aureus* employs numerous defense strategies for its survival in the same ecological niche and grows as a small-colony variant (SCV) (Biswas et al. 2009). The poly-microbial colonization result in competitive interactions between *S. aureus* and *Pseudomonas aeruginosa*, results a more persistent and antibiotic-resistant strain with altered colony morphology (namely, SCV) emerges with increased virulence (Nair et al. 2014). Bacteria are affected by the environment factors in which they grow and the interactions present in multispecies biofilms, basic knowledge on several aspects of sociomicrobiology can be gained (Parsek and Greenberg 2005). In human bacteria causes various infection by growing in biofilm mode. In addition, in the multi-species biofilm each species benefit each other for survival and nutrients requirements (Fig. 16.1). For example biofilm formation in oral cavity, it has been reported that different bacterial species in biofilms affect one another positively or negatively. The advantage of cooperative include co-aggregation of cells (Palmer et al. 2003; Yamada et al. 2005; Rickard et al. 2003; Sharma et al. 2005), conjugation (Ghigo 2001), and protection of one or several species from eradication when the biofilm is exposed to antimicrobial compounds (Cowan et al. 2000; Erb et al. 1997; Leriche et al. 2003). These protection or benefits are contributed by a number of factors including enzyme complementation (Shu et al. 2003) and organized spatial distribution of the cells in the biofilm (Cowan et al. 2000; Leriche et al. 2003). The association of bacteria in multi-species biofilm is synergistic results in cooperative biofilm formation by strains that were unable to form a biofilm alone (Filoche et al. 2004; Ghigo 2001; Palmer et al. 2001). However, the negative interaction in multi-species biofilms has also been observed. The negative interaction occurs in species producing

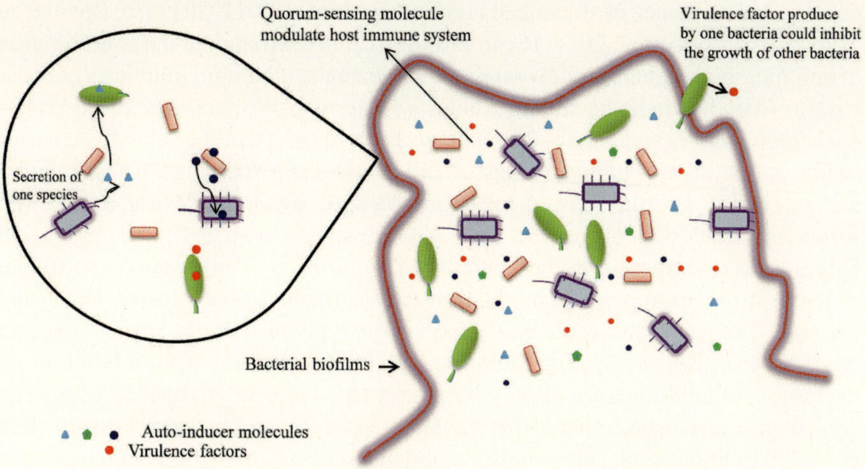

Fig. 16.1 Systematic representation of synergism in multi-species biofilm

antibacterial agents such as bacteriotixins (Rao et al. 2005; Tait and Sutherland 2002) and lowering of pH by one member of consortium (Burne and Marquis 2000; Sissons 1997). Mette et al. (2006), reported synergistic effect in the mixture of the epiphytic bacteria *Microbacterium phyllosphaerae*, *Shewanella japonica*, *Dokdonia donghaensis*, and *Acinetobacter lwoffii* multi-species biofilms. And the bacterial species gain fitness advantages from residing in multispecies biofilm consortia compared to their biology as single-species biofilms (Mette et al. 2006).

Microbial biofilms has been detected on indwelling medical devices including contact lenses, prosthetic heart valves, urinary catheters, intravenous catheters, cerebrospinal fluid shunts, joint prostheses and orthopedic fixation devices, vascular prosthesis, cardiac pacemakers, peritoneal dialysis catheters, breast implants, intrauterine devices biliary tract stents, dentures and in the dental area caries and periodontitis (Abidi et al. 2013; Donlan 2001; Tran et al. 2012; Fux et al. 2006; Donlan 2001; Song et al. 2013; Tollefson et al. 1987; Santos et al. 2011; Dasgupta 2002; Rieger et al. 2013; Donelli et al. 2012; Auler et al. 2010; Murakami et al. 2013).

16.4 Biofilms: A Reservoir of Antibiotic Resistance Genes

Human activities continuously discharge effluents into water reservoirs such as rivers and lakes. These effluents containing toxic chemicals and other pollutants, including industry and hospital waste. As a result the microorganisms inhabiting these water-bodies are exposed to a low but constant concentration of a wide range of chemical pollutants including antibiotics, analgesics, anti-inflammatory and psychiatric drugs, β-blockers, pesticides etc. (Bernier and Surette 2013; Boxall 2014). Many research has demonstrated the effects of pollutants on the composition, activity, and resilience of streambed biofilms (Proia et al. 2011, 2013a, b; Bonnineau et al. 2010; Osorio et al. 2014; Ricart et al. 2010). In environment the concentration of antimicrobials agents are several times less than the minimum inhibitory concentrations (MIC) of most bacterial pathogens (Waksman 1961; Davies 2006; Davies et al. 2006; Davies and Davies 2010). It has been suggest that the sub-MIC concentrations of antibiotics effects variety of cell processes including gene transcription and expression, quorum sensing, inter- or intra-species communication and biofilm formation (Davies 2006; Romero et al. 2011; Sengupta et al. 2013; Andersson and Huges 2014). Furthermore, the low concentration of antibiotics may also trigger different stress responses that might accelerate horizontal gene transfer (HGT) and the spread of antibiotic resistance genes in susceptible bacteria species (Beaber et al. 2004; Miller et al. 2004; Maiques et al. 2006). Bacteria with in biofilms are enclosed and held together closely by matrix that increases frequency of genetic competence and gene transfer that could spread antibiotic resistance genes (Fux et al. 2005). Previously, many studies reported increased conjugation efficiencies in biofilms in compare to that of planktonic bacteria (Ehlers and Bouwer 1999). Indeed, Hausner and Wuertz reported that biofilms increases conjugation rates 1000 times more (Hausner and Wuertz 1999). Molin and Tolker-Nielsen showed that efficiency of gene transfer is correlated with biofilm surface and suggest that high surface/volume ratios favor transformation within or between biofilm population (Molin and Toler-Nielsen 2003). *VanA* gene confers resistance to vancomycin and present in vancomycin-resistant enterococci. In drinking water biofilms Schwartz et al. (2003) detected *vanA* gene in absence of enterococci indicating potential gene transfer from them to autochthonous bacteria in drinking water systems (Schwartz et al. 2003). Similarly, Gillings et al. (2008) reported that 30% biofilms from ground water preatment plants were positive for class 1 integrase gene and only 1–2% bacteria isolated from lake sediments were positive for class 1 integrase gene (Gillings et al. 2008). Farkas et al. (2013) reported that 9.4% isolates of drinking water biofilms were positive for class 1 integrons mostly associated with *Enterobacteriaceae* causing micobial contamination (Farkas et al. 2013).

16.5 Role of Biofilms in Acquisition and Spread of Antibiotic Resistance Genes

The sub-MIC concentration of antibiotics exerts a selective pressure on biofilms bacteria and simulates the emergency of antibiotic resistance (Allen et al. 2010; Andersson and Huges 2014; Marti et al. 2014; Chow et al. 2015). More-over the presence of other pollutants including fertilizers, anti-fouling products, pesticides, organic and inorganic chemicals and heavy metals also contributes in the co-selection of antibiotic resistance because the close location of genes encoding for these resistance phenotypes in the same mobile genetic elements Seiler and Berendonk 2012). The antibiotic sensitive bacteria become resistant through acquiring mutation or by horizontal gene transfer spread by antibiotic resistance determinants. Mobile genetic elements plays and important role in evolution and adaption of bacteria to changes environmental condition and also contributes in horizontal gene transfer (Frost et al. 2005). Mobile genetic elements are short segments of DNA encoding a variety of enzymes and proteins needed for the intracellular mobility within the host genome or intercellular mobility between bacterial cells. In biofilm the bacteria are in close proximity and in exposures to antibiotic and pollutants results in selection and abundance of mobile genetic elements that facilities the spread of antibiotic resistance gene in inter or intra species (Sentchilo et al. 2013). Furthermore, it is well established that antibiotic resistance genes accumulate more in biofilms than in planktonic bacteria. Borjesson et al. (2009) detected high proportion of aminoglycosides and tetracyclines resistance genes in biofilms collected from waste water treatment plant (Borjesson et al. 2009). The relationship between human activity and increase in the presence of antibiotic resistance genes in water bodies has been reported by previously (Winkworth 2013). Winkworth (2013) reported high level of antibiotic resistance gene in biofilms collected from site of near to greater human and dairy farming in compare to biofilms collect from low human activity.

16.6 Biofilm and Bioremediation

Bioremediation is degradation of toxic pollutant from soil, water and air using microorganism (Alexander and Loehr 1992; Prasad and Prasad 2012). In bioremediation eco-friendly diverse microorganisms are employed that could degrade environmental pollutants. In nature various microbes such as bacteria and fungi naturally possess ability for the degradation of the pollutant in which it exists; those properties of microbes can be utilize for bioremediation. More-over, those microbes can be easily genetically manipulated for the degradation of specific pollutants. In bioremediation multi-species bacteria could grow following different metabolic pathway, employing different enzymes and produce different metabolites (Das and Dash 2014). The final product of one species could act as subtract for other results in degradation of complex pollutant. Thus the combination of multi-species could completed degraded the complex pollutants (Fig. 16.2). Furthermore, it has been reported that at site of high toxic concentration of pollutants the bacteria exist in adherent state called biofilms. Biofilms is defined as aggregation of microbes single or multi-species to a biological or inert surface encased in a self-produced matrix comprising of carbohydrates, proteins, extracellular DNA and water (Costerton et al. 1987). The poly-microbial biofilms harbors different microbial species that follow different metabolic degradation pathway, which could be capable of degrading several pollutants either individually or collectively (Gieg et al. 2014; Horemans et al. 2013). The biofilms bacteria are better persist, survive and resist the harsh environmental conditions; however, the planktonic bacteria would be vulnerable to cytotoxicity of environmental condition. Biofilms demonstrated differential gene

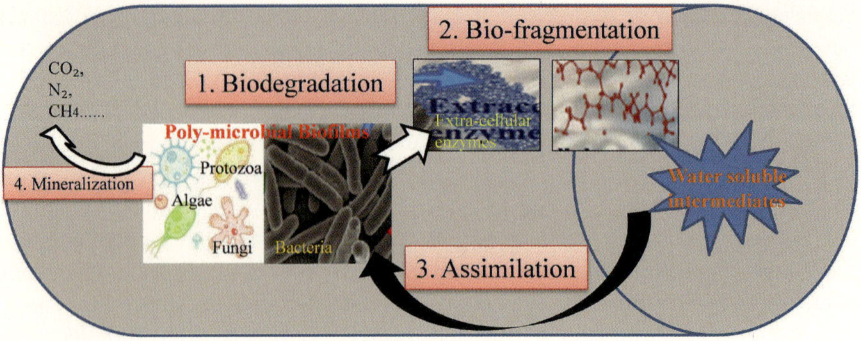

Fig. 16.2 Systematic representation of biodegradation by multi-species biofilms

expressions than counter-part planktonic cells (Yadav et al. 2012). Theses change in gene expression facilities the bacteria to adapt the conditions according to local available concentration of nutrients and oxygen. The alteration in gene expression provides the opportunity to bacteria for the activation of pathway for the degradation of various pollutants. The biofilm formation also alter the phenotype of bacteria such as swimming, swarming, twitching motility, chemotaxis, quorum sensing in soil and in water, that coordination movements increases the biodegradation ability of bacteria (Pratt and Kolter 1999; Lacal et al. 2013). The bacteria with in biofilms are embedded in matrix made of extracellular polymeric substances (EPS) acts as physical barrier and protect microbes (Flemming and Wingender 2010; More et al. 2014). The EPS is made of water, lipids, proteins, carbohydrates, nucleic acid which are secreted by bacteria and/or bounded. Furthermore the composition and structure of EPS varies between species and environmental conditions persist (Branda et al., 2005; Jung et al. 2013; Kreft and Wimpenny 2001; Miqueleto et al. 2010). The bacteria within biofilm are protected against harsh environmental conditions such as acid stress, shear stress, antimicrobial agents, UV damages, desiccation, biocides, solvents, high concentration of toxic chemicals and pollutants and predatory protozoa (Davey and O'Toole 2000; Mah and O'Toole 2001). The biofilm bacteria are also adapted to limited concentration so nutrients and oxygen concentration. These properties of biofilm bacteria make biofilm favorable in compare to planktonic bacteria. Furthermore, the biofilm bacteria are immobilized in matrix and could also immobilized pollutants and facilities degradation (Sutherland 2001).

16.6.1 Advantages of Biofilms in Bioremediation

The microbial biofilm have various advantages in compare to its counterpart planktonic bacteria, such as genetic material exchange, protection from harsh environmental conditions, persistence in different metabolic state, communication with each other and to environment and nutrient availability (Davies et al. 1998; Costerton et al. 1999; Flemming and Wingender 2001). More-over, the biofilms formed with multi-species consists of microorganism originates from one or more kingdoms such as bacteria, archaea bacteria, fungi and algae with varying metabolism and requirements such as electron acceptors/donors (Ferrera et al. 2004; Baker et al. 2009). As a result of multi-species there is co-operativity among microbes for survival during harsh environmental conditions. The bacteria with-in biofilms embedded in EPS secreted by microbes, as the biofilm matures the water channel are formed that helps to transport nutrients and oxygen or other reduced compounds (Picioreanu et al. 2000; Chen et al. 2013a). In multi-species biofilm found on tidal flats, corroded pipes and streambeds and even on site of infections cooperativity among different microbes has been reported. Although, the common constituents of microbial biofilms in multi-species or poly-microbial biofilms are protein, polysaccharides, e-DNA and lipids, however, some specific constituents of EPS help in solubilization of hydrophobic or recalcitrant subtracts that were inaccessible to microbes (Latch et al. 2003; Seo et al. 2009; Wang et al. 2011).

In bioremediation and biotransformation, the microbial biofilms have advantages over planktonic microbes. The planktonic microbes could be inhibited by chemical in question due to higher concentration at contaminated sites. However, the biofilms are tolerated to such toxic and hazardous chemicals. The sessile biofilms or floating biofilms has ability to tolerate changes in environmental conditions such as exposure to high pollutants, antibiotics, nutrients, pH, temperature, salt concertation and water contents (Heipieper et al. 1991; Beveridge et al. 1997; Hall-Stoodley et al. 2004). It is suggested that in heavily contaminated sites and in presence of sub-MIC concentration of antibiotics bacteria predominantly grow in biofilm mode (Gross et al. 2007). And to kill the biofilm bacteria almost 100–1000 times higher concentration antibiotics are required (Mah et al. 2003). In addition the bacteria with in biofilms under stress conditions release membrane vesicles (MVs). It is reported that the *E. coli* MVs could neutralize environmental agents and protect cell lysis, while the *Pseudomonas putida* secretes MVs in response to hydrocarbons alters cell surface and hydrophobicity (Manning and Kuehn 2011; Baumgarten et al. 2012).

16.6.2 Biofilm Mediated Bioremediation of Persistent Organic Pollutants

The organic compounds produced by industries including polycyclic aromatic hydrocarbons (PAHs), polychlorinated byphenyls (PCBs), polychlorinated dibenzo-p -dioxins and difurans (PCDD/Fs), and polychlorinated ethenes (PCEs) are persistent organic pollutants (POP) found in water, sediments and air (Ritter et al. 2002; Roots et al. 2010; Buth et al. 2009). Along with this the fertilizers, herbicides and pesticides used for farming and xenobiotic compounds are important pollutants found in sediments and water (Karlaganis et al. 2001; Lammel and Lohmann 2012). It is suggested that these POP could be metabolized and mineralized by using environmental bacteria (Chua et al. 2001; Zhang et al. 2010; Kataoka and Takagi 2013).

Polycyclic Aromatic Hydrocarbons (PAHs) are absorbed in soil and are insoluble that hinders the bioremediations process. And could enters in food chain and accurate in fatty tissues and causes mutation and cancer. For the successful bioremediation of PAHs, biofilms has been used that increased solubility and transfer of recalcitrant into biotransformation (Johnsen and Karlson 2004). Another obstruct in bioremediation of PAHs is the mixture of PAHs. In environment more than 100 different types of PAHs has been detected. However, the multi-species or mixed biofilms could be used for the bioremediation. In that each species with different metabolism contributes in solubilization and transformation of PAHs (Rodriguez and Bishop 2008). Dehalococcoides bacteria have been used for bioremediation of chlorinated ethenes (Loffler et al. 2013). For the reductively dechlorination of chlorinated PBC dechlorinating chloroflexi has been used (Fagervold et al. 2005). Biofilm has been applied for the degradation of penta-chlorobiphenyls (Macedo et al. 2005). The multi-species biofilms of *Burkholderia* sp. NK8 together with *P. aeruginosa* PA01 have been used for degradation of chlorinated benzoates (Yoshida et al. 2009). The dioxins present in soil have been degraded by apply biofilms of *Comamonas* sp. Strain KD7 (Wang and Oyaizu 2011).

16.6.3 Biofilm and Petroleum Hydrocarbon Bioremediation

The release of petro-chemical in soil is causes soil pollution. The soil remediation using civil-engineering methods are expensive and could not retail soil fertility. Therefore the bioremediation using hydrocarbons degrading microbes along with phyto and rhizo remediation could be employed (Khan et al. 2013; Kuiper et al. 2004; McGuinness and Dowling 2009; Vangronsveld et al. 2009; Gkorezis et al. 2016). The members of hydrocarbon degradation bacteria includes *Bacillus*, *Pseudomonas*, *Brevibacterium*, *Acintobactor*, *Mycobacterium*, *Gordonia*, *Dietzia*, *Burkholderia* and *Aeromicrobium* (Das and Mukherjee 2007; von der Weid et al. 2007; Zhang et al. 2014). These bacteria possess genes encoding enzymes that are required for degrade of petroleum hydrocarbons. For example *Pseudomonas*, *Burkholderia, Rhodococcus* possess alkane hydroxylases (*e.g. alkB* related), *Methylococcus, Methylocella*, or *Methylobacter* possess monooxygenases, *Acinetobacter, Caulobacter* and *Mycobacterium possess* bacterial P450 oxygenase that could metabolize the hydrocarbon and convert them into less or nontoxic forms (Ibrahim et al. 2013). The bioremediation using bacteria biofilm could be applied to non-aqueous phase hydrocarbon degradation (Singh et al. 2006). The biofilms can improve the access of microbes to the surface of hydrocarbon, protect microbes from stress, higher rate gene transfer and improve overall microbial community health (Arutchelvi et al. 2011). Recently, Balseiro-Romero et al. (2017) isolated various bacteria from petroleum hydrocarbon contaminated site and reported hydrocarbon degrading potential (Balseiro-Romero et al. 2017).

16.7 Biofilm and Agriculture

With the increasing use of pesticides for phyto-pathogens control and chemical fertilizers in agriculture increased the soil contamination and accumulation of toxic chemical in soil Horrigan et al. 2002. Furthermore, the toxic chemicals could disseminated and pollute environment, results in human diseases. An alternative route for eco-friendly practice of agriculture could be by using biological control Jordan et al. 2014. A biological control or biocontrol is microorganism that inhibits pathogen growth or produces agents that protect plants and promotes plants growth (Pal and McSpadden Gardener 2006).

In the recent years many researcher highlighted biofilm formation capability and advantages biocontrol agent biofilms (Pandin et al. 2017). Bacteria can colonize on plants and form biofilms on stems, leaves and the rhizosphere of plants, as well as soil particles, mushrooms or organic compost (Ramey et al. 2004; Weyens et al., 2009; Prigent-Combaret et al. 2012). The biofilm formation on plants takes place in following steps (Fig. 16.3). (1) First step involve initial attachment of free floating planktonic bacteria. At this stage bacteria possess locomotor organs. (2) The second stages is the adhesion of bacteria to the substratum and loosing locomotor organs. (3) Then the proliferation of the bacteria takes place. (4) Then the spatial organization of cells and maturation of biofilm takes place and produces exo-polysaccharides. (5) Biofilm ageing or environmental conditions un-favorable for the maintenance of the biofilm results in regulated dispersion of the biofilm.

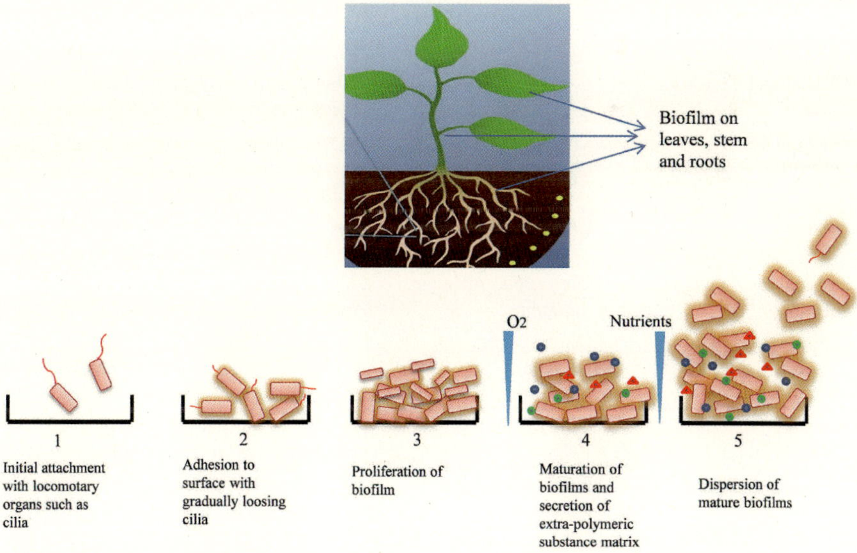

Fig. 16.3 Systematic representation of different stages of biofilm formation and maturation of bacterial biofilm on plants

Table 16.1 Biofilm forming bacteria on plants and their mechanism

Biofilm forming bacteria	Biofilms on host	Effect of biofilms on host
Bacillus atrophaeus	Biofilms on tomato, sugar beet plants	Biofilms produces surfactin and fengycin that are antimicrobial agents and induced systemic resistance
Bacillus amyloliquefaciens SQR9	Cucumber roots	The bacteria biofilms produces antimicrobial agent bacillomycin
Bacillus subtilis	Wheat seeds	The bacteria biofilms on wheat seeds prevent fungal mycelial growth
*Bacillus subtilis*3610	Tomato roots	Bacillus biofilms produces surfactin that is antimicrobial agent
Bacillus subtilis Bs916	Rice stem	Bacillus biofilm produces antimicrobial agent fengycin
Bacillus subtilis UMAF6614	Melon phylloplane	Bacillus biofilms produces Bacillomycin and fengycin are antimicrobial agents
Bacillus subtilis 6051	*Arabidopsis thaliana*	Biofilms of Bacillus subtilis protects Arabidopsis thaliana by producing antimicrobial-agents surfactin
Bacillus amyloliquefaciens SQR9	Maize roots	The biofilm bacteria showed promote plant growth (PGP) activity
Bacillus amyloliquefaciens SQY 162	Tobacco roots	In tobacco plants roots bacillus biofilms produces surfactin and induces systematic resistance
Paenibacillus polymyxa	*Arabidopsis thaliana*	The biofilms give mechanical protection to plants and clear pathogen from site
Paenibacillus polymyxa A26	Wheat seeds	The biofilms clear pathogen from site
Paenibacillus polymyxa B5	*Arabidopsis thaliana*	The biofilms clears pathogen from site of infection
Pseudomonas corrugata CCR04 and CCR80	Pepper roots	The biofilm of pseudomonas inhibits other pathogen by competitive colonization.
*Pseudomonas chlororaphis*PA23	Canola roots, wheat roots	In roots pseudomonas biofilms produces pyrrolnitrin which prevent fungal pathogen infection.
Pseudomonas putida 06909	Citrus roots	*Pseudomonas putida* biofilms protect root by preventing mycelial attachment.
Pseudomonas putida KT2440	Corn roots *Arabidopsis thaliana*	The biofilms of *Pseudomonas putida* induces systematic resistance and enhance plant growth
Pichia kudriavzevii	Pear fruit	*Pichia kudriavzevii* biofilms activates anti-oxidation system
Kloeckera apiculate	Citrus fruit	Biofilms clear pathogens and give mechanical protection to plant
Pseudoalteromonas tunicate	Green macroalga, *Ulvalactuca*	Produces anti-fouling compounds that inhibits colonization

The biofilm formation by plant-pathogen has also been detected in plants. For example, Dickeya dadantii is a gram negative bacteria cause soft rot diseases in a wide range of plant species. The bacteria colonize and form biofilms on chicory leaves and causes disease due to the production of degradative enzymes (Prigent-Combaret et al. 2012; Pandin et al. 2016). However, the Green macroalga, *Ulvalactuca* encourages biofilm formation of marine Gram negative bacteria *Pseudoalteromonas tunicate* (*P. tunicate*). *P. tunicate* is a endophytic bacteria produces anti-fouling compounds that inhibits colonization and biofilm formation (Egan et al. 2002). Similarly, the *Pseudomonas chlororaphis* forma biofilms on wheat rhizosphere and protect from fungal disease (Maddula et al. 2008).

The soil microbes including the biocontrol bacteria forms biofilms and protect various plants (Table 16.1). *Bacillus* species is known for biocontrol activity and produced bacillomycin, fengycin and surfactin lipopeptides. Bacillomycins and fengycins that have antagonistic activity towards fungal and bacterial pathogens of cucurbits. Zeriouh et al. (2014) propose that the biocontrol activity of *Bacillus* is due to coordinated action of the three families of lipopeptides. More-over, *Bacillus subtilis* produces surfactin to trigger biofilm formation on melon phylloplane, which ensures the long-term persistence and the adequate secretion of suppressive lipopeptides, bacillomycins and fengycins, which efficiently target pathogens Zeriouh et al. 2014). Aleti et al. (2016) reported that surfactin A and C with subtle structural differences have varying signal strengths on biofilm formation and root colonization and act specifically on the respective producing strain (Aleti et al. 2016). *Pseudomonas putida* protects the citrus roots against phytopathogen infection by forming biofilms. The bacteria initially colonize on the mycelium of the *Phytophthora parasitica* and feed on its exudates and gradually produces biofilm around the citrus roots that inhibits pathogen growth (Steddom et al. 2002; Ahn et al. 2007). The biofilms of biocontrol agents could protect the host by following mechanism: (a) Biofilms of biocontrol bacteria could prevent stress tolerance (Timmusk et al. 2005; Harriott and Noverr 2009; Pu et al. 2014). (b) The biofilm of biocontrol microbes could inhibit the plant pathogen by producing anti-microbial agents (Bais et al. 2004; Selin et al. 2010; Chen et al. 2013a, b; Sang and Kim 2014; Xu et al. 2014; Zeriouh et al. 2014; Wu et al. 2015; Zhou et al. 2016). (c) The biofilm could produce antagonism and thus clear pathogen and could exert competition for nutrients (Timmusk et al. 2005; Haggag and Timmusk 2008; Pu et al. 2014; Abd El Daim et al. 2015). (d) In multi-species biofilms the metabolite and enzymes could hinder the pathogen growth or the host could be benefited through cooperativity (Hogan et al. 2004; Audrain et al. 2015; Chen et al. 2015). (e) The biofilm bacteria could directly effect the plant physiology, e.g. activation of plant defenses (Wu et al. 2015) and/or stimulation of plant growth (Espinosa-Urgel et al. 2002; Zhang et al. 2015).

References

Abd El Daim I, Häggblom P, Karlsson M, Stenström E, Timmusk S (2015) Paenibacillus polymyxaA26 Sfp-type PPTase inactivation limits bacterial antagonism against Fusarium graminearum but not of F. culmorum in kernel assay. Front Plant Sci 6:368

Abidi SH, Sherwani SK, Siddiqui TR et al (2013) Drug resistance profile and biofilm forming potential of Pseudomonas aeruginosa isolated from contact lenses in Karachi-Pakistan. BMC Ophthalmol 13:57

Agarwal A, Singh KP, Jain A (2010) Medical significance and management of staphylococcal biofilm. FEMS Immunol Med Microbiol 58:147–160

Ahn S-J, Yang C-H, Cooksey DA (2007) Pseudomonas putida 06909 genes expressed during colonization on mycelial surfaces and phenotypic characterization of mutants. J Appl Microbiol 103:120–132

Aleti G, Lehner S, Bacher M, Compant S, Nikolic B, Plesko M, Schuhmacher R, Sessitsch A, Brader G (2016) Surfactin variants mediate species-specific biofilm formation and root colonization in Bacillus. Environ Microbiol 18:2634–2645

Alexander M, Loehr RC (1992) Bioremediation review. Science 258:874

Allen HK, Donato J, Wang HH, Cloud-Hansen KA, Davies J, Handelsman J (2010) Call of the wild: antibiotic resistance genes in natural environments. Nat Rev Microbiol 8:251–259

Andersson DI, Hughes D (2014) Microbiological effects of sublethal levels of antibiotics. Nat Rev Microbiol 12:465–478

Anderl JN, Franklin M, Stewart PS (2000) Role of antibiotic penetration limitation in Klebsiella pneumoniae biofilm resistance to ampicillin and ciprofloxacin. Antimicrob Agents Chemother 44:1818–1824

Arutchelvi J, Joseph C, Doble M (2011) Process optimization for the production of rhamnolipid and formation of biofilm by Pseudomonas aeruginosa CPCL on polypropylene. Biochem Eng J 56(1–2):37–45

Ash SY, Sheffield JVL (2013) Pneumococcus. Med Clin N Am 97:647–666

Audrain B, Farag MA, Ryu C-M, Ghigo J-M (2015) Role of bacterial volatile compounds in bacterial biology. FEMS Microbiol Rev 39:222–233

Auler ME, Morreira D, Rodrigues FF et al (2010) Biofilm formation on intrauterine devices in patients with recurrent vulvovaginal candidiasis. Med Mycol 48(1):211–216

Bagge N, Schuster M, Hentzer M, Ciofu O, Givskov M, Greenberg EP et al (2004) Pseudomonas aeruginosa biofilms exposed to imipenem exhibit changes in global gene expression and β -lactamase and alginate production. Antimicrob Agents Chemother 48:1175–1187

Bais HP, Fall R, Vivanco JM (2004) Biocontrol of Bacillus subtilis against infection of Arabidopsis roots by Pseudomonas syringae is facilitated by biofilm formation and surfactin production. Plant Physiol 134:307–319

Baker BJ, Tyson GW, Goosherst L, Banfield JF (2009) Insights into the diversity of eukaryotes in acid mine drainage biofilm communities. Appl Environ Microbiol 75:2192–2199

Balseiro-Romero M, Gkorezis P, Kidd PS, Van Hamme J, Weyens N, Monterroso C, Vangronsveld J (2017) Characterization and degradation potential of diesel-degrading bacterial strains for application in bioremediation. Int J Phytoremediation:0–0. doi.org/10.1080/15226514.2017.1337065

Baumgarten T, Vazquez J, Bastisch C, Veron W, Feuilloley MG, Nietzsche S, Wick LY, Heipieper HJ (2012) Alkanols and chlorophenols cause different physiological adaptive responses on the level of cell surface properties and membrane vesicle formation in Pseudomonas putida DOT-T1E. Appl Microbiol Biotechnol 93:837–845

Beaber JW, Hochhut B, Waldor MK (2004) SOS response promotes horizontal dissemination of antibiotic resistance genes. Nature 427:72–74

Bendouah Z, Barbeau J, Hamad WA, Desrosiers M (2006) Biofilm formation by Staphylococcus Aureus and Pseudomonas Aeruginosa is associated with an unfavorable evolution after surgery for chronic sinusitis and nasal polyposis. Otolaryngol Head Neck Surg 134(6):991–996

Bernier SP, Surette MG (2013) Concentration-dependent activity of antibiotics in natural environments. Front Microbiol 4:20

Beveridge TJ, Makin SA, Kadurugamuwa JL, Li Z (1997) Interactions between biofilms and the environment. FEMS Microbiol Rev 20:291–303

Biswas L, Biswas R, Schlag M, Bertram R, Götz F (2009) Small-Colony variant selection as a survival strategy for Staphylococcus aureus in the presence of Pseudomonas aeruginosa. Appl Environ Microbiol 75(21):6910–6912

Boase S, Foreman A, Cleland E, Tan L, Melton-Kreft R, Pant H et al (2013) The microbiome of chronic rhinosinusitis: culture, molecular diagnostics and biofilm detection. BMC Infect Dis 13(1):1–9

Bogaert D, van Belkum A, Sluijter M, Luijendijk A, de Groot R, Rümke HC, Verbrugh HA, Hermans PWM (2004) Colonisation by Streptococcus pneumoniae and Staphylococcus aureus in healthy children. Lancet 363:1871–1872

Boles BR, Thoendel M, Singh PK (2004) Self-generated diversity produces"insurance effects" in biofilm communities. PNAS 101:16630–16635

Bonnineau C, Guasch H, Proia L, Ricart M, Geiszinger A, Romaní AM et al (2010) Fluvial biofilms: a pertinent tool to assess β-blockers toxicity. Aquat Toxicol 96:225–233

Börjesson S, Dienues O, Jarnheimer PA, Olsen B, Matussek A, Lindgren PE (2009) Quantification of genes encoding resistance to aminoglycosides, β-lactams and tetracyclines in wastewater environments by real-time PCR. Int J Environ Health Res 19:219–230

Boxall AB (2014) The environmental side effects of medication. EMBO Rep 5:1110–1116

Branda SS, Vik S, Friedman L et al (2005) Biofilms: the matrix revisited. Trends Microbiol 13:20–26

Burmølle M, Webb JS, Rao D, Hansen LH, Sørensen SJ, Kjelleberg S (2006) Enhanced biofilm formation and increased resistance to antimicrobial agents and bacterial invasion are caused by synergistic interactions in multispecies biofilms. Appl Environ Microbiol 72(6):3916–3923

Burne RA, Marquis RE (2000) Alkali production by oral bacteria and protection against dental caries. FEMS Microbiol Lett 193:1–6

Buth JM, Grandbois M, Vikesland PJ, McNeill K, Arnold WA (2009) Aquatic photochemistry of chlorinated triclosan derivatives: potential source of polychlorodibenzo-P-dioxins. Environ Toxicol Chem 28:2555–2563

von Canstein H, Kelly S, Li Y, Wagner-Döbler I (2002) Species diversity improves the efficiency of mercury-reducing biofilms under changing environmental conditions. Appl Environ Microbiol 68:2829–2837

Chen X, Suwarno SR, Chong TH, McDougald D, Kjelleberg S, Cohen Y, Fane AG, Rice SA (2013a) Dynamics of biofilm formation under different nutrient levels and the effect on biofouling of a reverse osmosis membrane system. Biofouling 29:319–330

Chen Y, Yan F, Chai Y, Liu H, Kolter R, Losick R, Guo J (2013b) Biocontrol of tomato wilt disease by Bacillus subtilis isolates from natural environments depends on conserved genes mediating biofilm formation. Environ Microbiol 15:848–864

Chen Y, Gozzi K, Yan F, Chai Y (2015) Acetic acid acts as a volatile signal to stimulate bacterial biofilm formation. MBio 6:e00392

Chopra S, Harjai K, Chhibber S (2015) Antibiotic susceptibility of ica-positive and ica-negative MRSA in different phases of biofilm growth. J Antibiot 68(1):15–22

Chow L, Waldron L, Gillings MR (2015) Potential impacts of aquatic pollutants: sub-clinical antibiotic concentrations induce genome changes and promote antibiotic resistance. Front Microbiol 6:803

Chua H, PH Y, Lo W, Sin SN (2001) The degradation of xenobiotic branched carboxylic acids in anaerobic sediment of the Pearl River in Southern China. Sci Total Environ 266:221–228

Costerton JW, Cheng KJ, Geesey GG et al (1987) Bacterial biofilms in nature and disease. Annu Rev Microbiol 41:435–464

Costerton JW, Stewart PS, Greenberg EP (1999) Bacterial biofilms: a common cause of persistent infections. Science 284:1318–1322

Cowan SE, Gilbert E, Liepmann D, Keasling JD (2000) Commensal interactions in a dual-species biofilm exposed to mixed organic compounds. Appl Environ Microbiol 66:4481–4485

Das S, Dash HR (2014) 1 – Microbial bioremediation: a potential tool for restoration of contaminated areas. In: Das S (ed) Microbial biodegradation and bioremediation. Elsevier, Oxford, pp 1–21

Das K, Mukherjee AK (2007) Crude petroleum-oil biodegradation efficiency of Bacillus subtilis and Pseudomonas aeruginosa strains isolated from a petroleum-oil contaminated soil from North-East India. Bioresour Technol 98(7):1339–1345

Dasgupta MK (2002) Biofilms and infection in dialysis patients. Semin Dial 15(5):338–346

Davey ME, O'Toole GA (2000) Microbial biofilms: from ecology to molecular genetics. Microbiol Mol Biol Rev 64:847–867

Davies J (2006) Are antibiotics naturally antibiotics? J Ind Microbiol Biotechnol 33:496–499

Davies J, Davies D (2010) Origins and evolution of antibiotic resistance. Microbiol Mol Biol Rev 74:417–433

Davies DG, Parsek MR, Pearson JP, Iglewski BH, Costerton JW, Greenberg EP (1998) The involvement of cell-to-cell signals in the development of a bacterial biofilm. Science 280:295–298

Davies J, Spiegelman GB, Yim G (2006) The world of subinhibitory antibiotic concentrations. Curr Opin Microbiol 9:445–453

Donelli G, Vuotto C, Cardines R et al (2012) Biofilm-growing intestinal anaerobic bacteria. FEMS Immunol Med Microbiol 65(2):318–325

Dong D, Yulin Z, Xiao W, Hongyan Z, Jia L, Yan X et al (2014) Correlation between bacterial biofilms and osteitis in patients with chronic rhinosinusitis. Laryngoscope 124(5):1071–1077

Donlan RM (2001) Biofilms and device-associated infections. Emerg Infect Dis 7(2):277–281

Donlan RM (2002) Biofilms: microbial life on surfaces. Emerg Infect Dis 8(9):881–890

Egan S, James S, Holmstrom C et al (2002) Correlation between pigmentation and antifouling compounds produced by Pseudoalteromonas tunicata. Environ Microbiol 4:433–442

Ehlers LJ, Bouwer EJ (1999) Rp4 plasmid transfer among species of Pseudomonas in a biofilm reactor. Water Sci Technol 39:163–171

Erb RW, Eichner CA, Wagner-Dobler I, Timmis KN (1997) Bioprotection of microbial communities from toxic phenol mixtures by a genetically designed pseudomonad. Nat Biotechnol 15:378–382

Espinosa-Urgel M, Kolter R, Ramos J-L (2002) Root colonization by Pseudomonas putida: love at first sight. Microbiology 148:341–343

Fagervold SK, Watts JE, May HD, Sowers KR (2005) Sequential reductive dechlorination of meta-chlorinated polychlorinated biphenyl congeners in sediment microcosms by two different Chloroflexi phylotypes. Appl Environ Microbiol 71:8085–8090

Farkas A, Butiuc-Keul A, Ciatarâks D, Neamktu C, Crvaciunaks C, Podar D et al (2013) Microbiological contamination and resistance genes in biofilms occurring during the drinking water treatment process. Sci Total Environ 443:932–938

Ferrera I, Massana R, Casamayor EO, Balague V, Sanchez O, Pedros-Alio C, Mas J (2004) High-diversity biofilm for the oxidation of sulfide containing effluents. Appl Microbiol Biotechnol 64:726–734

Filoche SK, Anderson SA, Sissons CH (2004) Biofilm growth of Lactobacillus species is promoted by Actinomyces species and Streptococcus mutans. Oral Microbiol Immunol 19:322–326

Flemming HC, Wingender J (2001) Relevance of microbial extracellular polymeric substances (EPSs) – part II: technical aspects. Water Sci Technol 43:9–16

Flemming HC, Wingender J (2010) The biofilm matrix. Nat Rev Microbiol 8:623–633

Frost LS, Leplae R, Summers AO, Toussaint A (2005) Mobile genetic elements: the agents of open source evolution. Nat Rev Microbiol 3:722–732

Fux CA, Costerton JW, Stewart PS, Stoodley P (2005) Survival strategies of infectious biofilms. Trends Microbiol 13:34–40

Fux CA, Quigley M, Worel AM et al (2006) Biofilm-related infections of cerebrospinal fluid shunts. Clin Microbiol Infect 12(4):331–337

Ghigo JM (2001) Natural conjugative plasmids induce bacterial biofilm development. Nature 412:442–445

Gieg LM, Fowler SJ, Berdugo-Clavijo C (2014) Syntrophic biodegradation of hydrocarbon contaminants. Curr Opin Biotechnol 27:21–29

Gilbert P, Maira-Litran T, McBain AJ, Rickard AH, Whyte F (2002) The physiology and collective recalcitrance of microbial biofilm communities. Adv Microb Physiol 46:203–256

Gillings M, Boucher Y, Labbate M, Holmes A, Krishnan S, Holley M et al (2008) The evolution of class 1 integrons and the rise of antibiotic resistance. J Bacteriol 190:5095–5100

Gkorezis P, Daghio M, Franzetti A, Van Hamme JD, Sillen W, Vangronsveld J (2016) The interaction between plants and bacteria in the remediation of petroleum hydrocarbons: an environmental perspective. Front Microbiol 7:1836

Gross R, Hauer B, Otto K, Schmid A (2007) Microbial biofilms: new catalysts for maximizing productivity of long-term biotransformations. Biotechnol Bioeng 98:1123–1134

Haggag WM, Timmusk S (2008) Colonization of peanut roots by biofilm-forming Paenibacillus polymyxa initiates biocontrol against crown rot disease. J Appl Microbiol 104:961–969

Hall-Stoodley L, Costerton JW, Stoodley P (2004) Bacterial biofilms: from the natural environment to infectious diseases. Nat Rev Microbiol 2:95–108

Hall-Stoodley L, Hu F, Gieseke A et al (2006) DIrect detection of bacterial biofilms on the middle-ear mucosa of children with chronic otitis media. JAMA 296:202–211

Harriott MM, Noverr MC (2009) Candida albicans and Staphylococcus aureus form polymicrobial biofilms: effects on antimicrobial resistance. Antimicrob Agents Chemother 53:3914–3922

Hausner M, Wuertz S (1999) High rates of conjugation in bacterial biofilms as determined by quantitative in situ analysis. Appl Environ Microbiol 65:3710–3713

Heipieper HJ, Keweloh H, Rehm HJ (1991) Influence of phenols on growth and membrane permeability of free and immobilized Escherichia coli. Appl Environ Microbiol 57:1213–1217

Hengzhuang W, Wu H, Ciofu O et al (2011) Pharmacokinetics/pharmacodynamics of colistin and imipenem on mucoid and nonmucoid Pseudomonas aeruginosa biofilms. Antimicrob Agents Chemother 55(9):4469–4474

Hengzhuang W, Wu H, Ciofu O et al (2012) In vivo pharmacokinetics/pharmacodynamics of colistin and imipenem in Pseudomonas aeruginosa biofilm infection. Antimicrob Agents Chemother 56(5):2683–2690

Hoffman LR, D'Argenio DA, MacCoss MJ, Zhang Z, Jones RA, Miller SI (2005) Aminoglycoside antibiotics induce bacterial biofilm formation. Nature 436:1171–1175

Hogan DA, Vik A, Kolter R (2004) A Pseudomonas aeruginosa quorum-sensing molecule influences Candida albicans morphology. Mol Microbiol 54:1212–1223

Hoiby N, Ciofu O, Johansen HK et al (2011) The clinical impact of bacterial biofilms. Int J Oral Sci 3(2):55–65

Horemans B, Breugelmans P, Hofkens J et al (2013) Environmental dissolved organic matter governs biofilm formation and subsequent linuron degradation activity of a linuron-degrading bacterial consortium. Appl Environ Microbiol 79:4534–4542

Horrigan L, Lawrence RS, Walker P (2002) How sustainable agriculture can address the environmental and human health harms of industrial agriculture. Environ Health Perspect 110:445–456

Ibrahim ML, Ijah UJJ, Manga SB, Bilbis LS, Umar S (2013) Production and partial characterization of biosurfactant produced by crude oil degrading bacteria. Int Biodeter Biodegr 81:28–34

Jedrzejas MJ (2001) Pneumococcal virulence factors: structure and function. Microbiol Mol Biol Rev 65:187–207

Johnsen AR, Karlson U (2004) Evaluation of bacterial strategies to promote the bioavailability of polycyclic aromatic hydrocarbons. Appl Microbiol Biotechnol 63:452–459

Jordan K, Dalmasso M, Zentek J, Mader A, Bruggeman G, Wallace J et al (2014) Microbes versus microbes: control of pathogens in the food chain. J Sci Food Agric 94:3079–3089

Jung H, Lee SK, Cha S-H, Byun JY, Park MS, Yeo SG (2009) Current bacteriology of chronic otitis media with effusion: high rate of nosocomial infection and decreased antibiotic sensitivity. J Infect 59(5):308–316

Jung JH, Choi NY, Lee SY (2013) Biofilm formation and exopolysaccharide (EPS) production by Cronobacter sakazakii depending on environmental conditions. Food Microbiol 34:70–80

Karlaganis G, Marioni R, Sieber I, Weber A (2001) The elaboration of the "Stockholm convention" on persistent organic pollutants (POPs): a negotiation process fraught with obstacles and opportunities. Environ Sci Pollut Res Int 8:216–221

Kataoka R, Takagi K (2013) Biodegradability and biodegradation pathways of endosulfan and endosulfan sulfate. Appl Microbiol Biotechnol 97:3285–3292

Khan S, Afzal M, Iqbal S, Khan QM (2013) Plant–bacteria partnerships for the remediation of hydrocarbon contaminated soils. Chemosphere 90(4):1317–1332

Kim SH, Kim MG, Kim SS, Cha SH, Yeo SG (2015) Change in detection rate of methicillin-resistant Staphylococcus aureus and Pseudomonas aeruginosa and their antibiotic sensitivities in patients with chronic suppurative otitis media. J Int Adv Otol 11:151–156

Klugman KP, Madhi SA, Albrich WC (2008) Novel approaches to the identification of Streptococcus pneumoniae as the cause of community-acquired pneumonia. Clin Infect Dis 47:S202–S206

Kreft JU, Wimpenny JW (2001) Effect of EPS on biofilm structure and function as revealed by an individual-based model of biofilm growth. Water Sci Technol 43:135–141

Kuiper I, Lagendijk EL, Bloemberg GV, Lugtenberg BJJ (2004) Rhizoremediation: a beneficial plant-microbe interaction. Mol Plant Microbe Interact 17(1):6–15

Lacal J, Reyes-Darias JA, García-Fontana C et al (2013) Tactic responses to pollutants and their potential to increase biodegradation efficiency. J Appl Microbiol 114:923–933

Lammel G, Lohmann R (2012) Identifying the research needs in the global assessment of toxic compounds 10 years after the signature of the Stockholm convention. Environ Sci Pollut Res Int 19:1873–1874

Latch DE, Packer JL, Arnold WA, McNeill K (2003) Photochemical conversion of triclosan to 2,8-dichlorodibenzo-p-dioxin in aqueous solution. J Photochem Photobiol A Chem 158:63–66

Leriche V, Briandet R, Carpentier B (2003) Ecology of mixed biofilms subjected daily to a chlorinated alkaline solution: spatial distribution of bacterial species suggests a protective effect of one species to another. Environ Microbiol 5(1):64–71

Löffler FE, Ritalahti KM, Zinder SH (2013) Dehalococcoides and reductive dechlorination of chlorinated solvents. In: Stroo HF et al (eds) Bioaugmentation for groundwater remediation, vol 5. Springer, New York, pp 39–88

Lowy FD (1998) Staphylococcus aureus infections. N Engl J Med 339:520–532

Macedo AJ, Kuhlicke U, Neu TR, Timmis KN, Abraham WR (2005) Three stages of a biofilm community developing at the liquid–liquid interface between polychlorinated biphenyls and water. Appl Environ Microbiol 71:7301–7309

Mack D, Rohde H, Harris LG, Davies AP, Horstkotte MA, Knobloch JK-M (2006) Biofilm formation in medical device-related infection. Int J Artif Organs 29(4):343–359

Maddula VS, Pierson EA, Pierson LS (2008) Altering the ratio of phenazines in Pseudomonas chlororaphis (aureofaciens) strain 30-84: effects on biofilm formation and pathogen inhibition. J Bacteriol 190:2759–2766

Mah TF, O'Toole GA (2001) Mechanisms of biofilm resistance to antimicrobial agents. Trends Microbiol 9:34–39

Mah TF, Pitts B, Pellock B, Walker GC, Stewart PS, O'Toole GA (2003) A genetic basis for Pseudomonas aeruginosa biofilm antibiotic resistance. Nature 426(6964):306–310

Maiques E, Úbeda C, Campoy S, Lasa I, Novick RP, Barbé J et al (2006) β-lactam antibiotics induce the SOS response and horizontal transfer of virulence factors in Staphylococcus aureus. Appl Environ Microbiol 188:2726–2729

Maki DG, Kluger DM, Crnich CJ (2006) The risk of bloodstream infection in adults with different intravascular devices: a systematic review of 200 published prospective studies. Mayo Clin Proc 81:1159–1171

Manning AJ, Kuehn MJ (2011) Contribution of bacterial outer membrane vesicles to innate bacterial defense. BMC Microbiol 11:258

Marti E, Variatza E, Balcazar JL (2014) The role of aquatic ecosystems as reservoirs of antibiotic resistance. Trends Microbiol 22:36–41

McGuinness M, Dowling D (2009) Plant-associated bacterial degradation of toxic organic compounds in soil. Int J Environ Res Public Health 6(8):2226–2247

Miller C, Thomsen LE, Gaggero C, Mosseri R, Ingmer H, Cohen SN (2004) SOS response induction by beta-lactams and bacterial defense against antibiotic lethality. Science 305:1629–1631

Miqueleto AP, Dolosic CC, Pozzi E et al (2010) Influence of carbon sources and C/N ratio on EPS production in anaerobic sequencing batch biofilm reactors for wastewater treatment. Bioresour Technol 101:1324–1330

Molin S, Tolker-Nielsen T (2003) Gene transfer occurs with enhanced efficiency in biofilms and induces enhanced stabilization of the biofilm structure. Curr Opin Biotechnol 14:255–261

More TT, Yadav JSS, Yan S et al (2014) Extracellular polymeric substances of bacteria and their potential environmental applications. J Environ Manage 144:1–25

Murakami M, Nishi Y, Seto K et al (2013) Dry mouth and denture plaque microflora in complete denture and palatal obturator prosthesis wearers. Gerodontology 32:188. https://doi.org/10.1111/ger.12073

Nair N, Biswas R, Götz F, Biswas L (2014) Impact of Staphylococcus aureus on pathogenesis in polymicrobial infections. Infect Immun 82(6):2162–2169

Osorio V, Proia L, Ricart M, Pérez S, Ginebreda A, Cortina JL et al (2014) Hydrological variation modulates pharmaceutical levels and biofilm responses in a Mediterranean river. Sci Total Environ 472:1052–1061

Otto M (2010) Looking toward basic science for potential drug discovery targets against community-associated MRSA. Med Res Rev 30:1–22

Pal KK, McSpadden Gardener B (2006) Biological control of plant pathogens. Plant Health Instr. https://doi.org/10.1094/PHI-A-2006-1117-02

Palmer RJ Jr, Kazmerzak K, Hansen MC, Kolenbrander PE (2001) Mutualism versus independence: strategies of mixed-species oral biofilms in vitro using saliva as the sole nutrient source. Infect Immun 69:5794–5804

Palmer RJ Jr, Gordon SM, Cisar JO, Kolenbrander PE (2003) Coaggregation-mediated interactions of streptococci and actinomyces detected in initial human dental plaque. J Bacteriol 185:3400–3409

Pandin C, Caroff M, Condemine G (2016) Antimicrobial peptide resistance genes in the plant pathogen Dickeya dadantii. Appl Environ Microbiol 82:6423–6430

Pandin C, Le Coq D, Canette A, Aymerich S, Briandet R (2017) Should the biofilm mode of life be taken into consideration for microbial biocontrol agents? Microbial Biotechnol, n/a-n/a 10:719. https://doi.org/10.1111/1751-7915.12693

Parsek MR, Greenberg EP (2005) Sociomicrobiology: the connections between quorum sensing and biofilms. Trends Microbiol 13:27–33

Pichichero ME (2013) Otitis media. Pediatr Clin North Am 60:391–407

Picioreanu C, Van Loosdrecht MC, Heijnen JJ (2000) Effect of diffusive and convective substrate transport on biofilm structure formation: a two-dimensional modeling study. Biotechnol Bioeng 69:504–515

Post JC, Stoodley P, Hall–Stoodley L, Ehrlich GD (2004) The role of biofilms in otolaryngologic infections. Curr Opin Otolaryngol Head Neck Surg 12(3):185–190

Prasad MN, Prasad R (2012) Nature's cure for cleanup of contaminated environment—a review of bioremediation strategies. Rev Environ Health 27:181–189

Pratt LA, Kolter R (1999) Genetic analyses of bacterial biofilm formation. Curr Opin Microbiol 2:598–603

Prigent-Combaret C, Zghidi-Abouzid O, Effantin G, Lejeune P, Reverchon S, Nasser W (2012) The nucleoid-associated protein Fis directly modulates the synthesis of cellulose, an essential component of pellicle–biofilms in the phytopathogenic bacterium Dickeya dadantii. Mol Microbiol 86:172–186

Proia L, Morin S, Peipoch M, Romaní AM, Sabater S (2011) Resistance and recovery of river biofilms receiving short pulses of Triclosan and Diuron. Sci Total Environ 409:3129–3137

Proia L, Osorio V, Soley S, Köck-Schulmeyer M, Pérez S, Barceló D et al (2013a) Effects of pesticides and pharmaceuticals on biofilms in a highly impacted river. Environ Pollut 178:220–228
Proia L, Lupini G, Osorio V, Pérez S, Barceló D, Schwartz T et al (2013b) Response of biofilm bacterial communities to antibiotic pollutants in a Mediterranean river. Chemosphere 92:1126–1135
Pu L, Jingfan F, Kai C, Chao-an L, Yunjiang C (2014) Phenylethanol promotes adhesion and biofilm formation of the antagonistic yeast Kloeckera apiculata for the control of blue mold on citrus. FEMS Yeast Res 14:536–546
Ramey BE, Koutsoudis M, von Bodman SB, Fuqua C (2004) Biofilm formation in plant-microbe associations. Curr Opin Microbiol 7:602–609
Rao D, Webb JS, Kjelleberg S (2005) Competitive interactions in mixed-species biofilms containing the marine bacterium Pseudoalteromonas tunicata. Appl Environ Microbiol 71:1729–1736
Ricart M, Guasch H, Alberch M, Barceló D, Bonnineau C, Geiszingerb A et al (2010) Triclosan persistence through wastewater treatment plants and its potential toxic effects on river biofilms. Aquat Toxicol 100:346–353
Rickard AH, Gilbert P, High NJ, Kolenbrander PE, Handley PS (2003) Bacterial coaggregation: an integral process in the development of multi-species biofilms. Trends Microbiol 11:94–100
Rieger UM, Mesina J, Kalbermatten DF et al (2013) Bacterial biofilms and capsular contracture in patients with breast implants. Br J Surg 100(6):768–774
Ritter L, Solomon K, Sibley P, Hall K, Keen P, Mattu G, Linton B (2002) Sources, pathways, and relative risks of contaminants in surface water and groundwater: a perspective prepared for the Walkerton inquiry. J Toxic Environ Health A 65:1–142
Rodriguez S, Bishop P (2008) Enhancing the biodegradation of polycyclic aromatic hydrocarbons: effects of nonionic surfactant addition on biofilm function and structure. J Environ Eng 134:505–512
Romero D, Traxler MF, López D, Kolter R (2011) Antibiotics as signal molecules. Chem Rev 111:5492–5505
Roots O, Roose A, Kull A, Holoubek I, Cupr P, Klanova J (2010) Distribution pattern of PCBs, HCB and PeCB using passive air and soil sampling in Estonia. Environ Sci Pollut Res Int 17:740–749
Salcedo DE, Lee JH, Ha UH, Kim SP (2014) Theeffectsofantibiotics on the biofilm formation and antibiotic resistance gene transfer. Desalin Water Treat 54:3582–3588
Sanchez CJ, Shivshankar P, Stol K, Trakhtenbroit S, Sullam PM, Sauer K, Hermans PWM, Orihuela CJ (2010) The pneumococcal serine-rich repeat protein is an intra-species bacterial adhesin that promotes bacterial aggregation *in vivo* and in biofilms. PLoS Pathog 6:e1001044
Sang MK, Kim KD (2014) Biocontrol activity and root colonization by Pseudomonas corrugatastrains CCR04 and CCR80 against Phytophthora blight of pepper. BioControl 59:437–448
Santos AP, Watanabe E, Andrade D (2011) Biofilm on artificial pacemaker: fiction or reality? Arq Bras Cardiol 97(5):e113–e120
Schwartz T, Kohnen W, Jansen B, Obst U (2003) Detection of antibiotic-resistant bacteria and their resistance genes in wastewater, surface water, and drinking water biofilms. FEMS Microbiol Ecol 43:325–335
Seiler C, Berendonk TU (2012) Heavy metal driven co-selection of antibiotic resistance in soil and water bodies impacted by agriculture and aquaculture. Front Microbiol 3:399
Selin C, Habibian R, Poritsanos N, Athukorala SNP, Fernando D, de Kievit TR (2010) Phenazines are not essential for Pseudomonas chlororaphis PA23 biocontrol of Sclerotinia sclerotiorum, but do play a role in biofilm formation. FEMS Microbiol Ecol 71:73–83
Sengupta S, Chattopadhyay MK, Grossart HP (2013) The multifaceted roles of antibiotics and antibiotic resistance in nature. Front Microbiol 4:47
Sentchilo V, Mayer AP, Guy L, Miyazaki R, Green Tringe S, Barry K et al (2013) Community-wide plasmid gene mobilization and selection. ISME J 7:1173–1186. https://doi.org/10.1038/ismej.2013.13
Seo Y, Lee WH, Sorial G, Bishop PL (2009) The application of a mulch biofilm barrier for surfactant enhanced polycyclic aromatic hydrocarbon bioremediation. Environ Pollut 157:95–101

Shak JR, Vidal JE, Klugman KP (2013) Influence of bacterial interactions on pneumococcal colonization of the nasopharynx. Trends Microbiol 21:129–135

Sharma A, Inagaki S, Sigurdson W, Kuramitsu HK (2005) Synergy between Tannerella forsythia and Fusobacterium nucleatum in biofilm formation. Oral Microbiol Immunol 20:39–42

Shu M, Browngardt CM, Chen YY, Burne RA (2003) Role of urease enzymes in stability of a 10-species oral biofilm consortium cultivated in a constant-depth film fermentor. Infect Immun 71:7188–7192

Simell B, Auranen K, Käyhty H, Goldblatt D, Dagan R, O'Brien KL (2012) The fundamental link between pneumococcal carriage and disease. Expert Rev Vaccines 11:841–855

Singh R, Paul D, Jain RK (2006) Biofilms: implications in bioremediation. Trends Microbiol 14(9):389–397

Singhal D, Foreman A, Bardy J-J, Wormald P-J (2011) Staphylococcus aureus biofilms. Laryngoscope 121(7):1578–1583

Sissons CH (1997) Artificial dental plaque biofilm model systems. Adv Dent Res 11:110–126

Song Z, Borgwardt L, Høiby N et al (2013) Prosthesis infections after orthopedic joint replacement: the possible role of bacterial biofilms. Orthop Rev (Pavia) 5(2):65–71

Steddom K, Menge JA, Crowley D, Borneman J (2002) Effect of repetitive applications of the biocontrol bacterium Pseudomonas putida 06909-rif/nal on citrus soil microbial communities. Phytopathology 92:857–862

Stewart PS (2002) Mechanisms of antibiotic resistance in bacterial biofilms. Int J Med Microbiol 292:107–113

Stewart PS, Costerton JW (2001) Antibiotic resistance of bacteria in biofilms. Lancet 358:135–138

Sutherland IW (2001) The biofilm matrix--an immobilized but dynamic microbial environment. Trends Microbiol 9:222–227

Tait K, Sutherland IW (2002) Antagonistic interactions amongst bacteriocin-producing enteric bacteria in dual species biofilms. J Appl Microbiol 93:345–352

Thornton RB, Wiertsema SP, Kirkham L-AS, Rigby PJ, Vijayasekaran S et al (2013) Neutrophil extracellular traps and bacterial biofilms in middle ear effusion of children with recurrent acute otitis media – a potential treatment target. PLoS One 8:e53837

Timmusk S, Grantcharova N, Wagner EGH (2005) Paenibacillus polymyxa invades plant roots and forms biofilms. Appl Environ Microbiol 71:7292–7300

Tollefson DF, Bandyk DF, Kaebnick HW et al (1987) Surface biofilm disruption. Enhanced recovery of microorganisms from vascular prostheses. Arch Surg 122(1):38–43

Tran PL, Lowry N, Campbell T et al (2012) An organoselenium compound inhibits Staphylococcus aureus biofilms on hemodialysis catheters in vivo. Antimicrob Agents Chemother 56(2):972–978

van der Poll T, Opal SM (2009) Pathogenesis, treatment, and prevention of pneumococcal pneumonia. Lancet 374:1543–1556

Vangronsveld J, Herzig R, Weyens N, Boulet J, Adriaensen K, Ruttens A, Thewys T, Vassilev A, Meers E, Nehnevajova E, van der Lelie D, Mench M (2009) Phytoremediation of contaminated soils and groundwater: lessons from the field. Environ Sci Pollut Res Int 16(7):765–794

von der Weid I, Marques JM, Cunha CD, Lippi RK, dos Santos SCC, Rosado AS, Lins U, Seldin L (2007) Identification and biodegradation potential of a novel strain of Dietzia cinnamea isolated from a petroleum-contaminated tropical soil. Syst Appl Microbiol 30(4):331–339

Vu B, Chen M, Crawford RJ, Ivanova EP (2009) Bacterial extracellular polysaccharides involved in biofilm formation. Molecules 14:2535–2554

Waksman SA (1961) The role of antibiotics in nature. Perspect Biol Med 4:271–287

Wang Y, Oyaizu H (2011) Enhanced remediation of dioxins-spiked soil by a plant–microbe system using a dibenzofuran-degrading Comamonas sp. and Trifolium repens. Chemosphere 85:1109–1114

Wang LH, Sun D, Hu X, Yu L, Wang HJ (2011) Relationship between bacterial biofilm and clinical features of patients with chronic rhinosinusitis. Eur Arch Otorhinolaryngol 269(1):155–163

Weimer KED, Armbruster CE, Juneau RA, Hong W, Pang B, Swords WE (2010) Coinfection with Haemophilus influenzae promotes pneumococcal biofilm formation during experimental otitis media and impedes the progression of pneumococcal disease. J Infect Dis 202:1068–1075

Weyens N, van der Lelie D, Taghavi S, Newman L, Vangronsveld J (2009) Exploiting plant–microbe partnerships to improve biomass production and remediation. Trends Biotechnol 27(10):591–598

Winkworth CL (2013) Antibiotic resistance genes in freshwater biofilms along a whole river. J Water Health 11:186–197

Wu K, Fang Z, Guo R, Pan B, Shi W, Yuan S et al (2015) Pectin enhances bio-control efficacy by inducing colonization and secretion of secondary metabolites by Bacillus amyloliquefaciens SQY 162 in the rhizosphere of tobacco. PLoS One 10:e0127418

Xu Z, Zhang R, Wang D, Qiu M, Feng H, Zhang N, Shen Q (2014) Enhanced control of cucumber wilt disease by Bacillus amyloliquefaciens SQR9 by altering the regulation of its DegU phosphorylation. Appl Environ Microbiol 80:2941–2950

Yadav MK, Kwon SK, Cho CG, Park S-W, Chae S-W, Song J-J (2012) Gene expression profile of early in vitro biofilms of Streptococcus pneumoniae. Microbiol Immunol 56:621–629

Yamada M, Ikegami A, Kuramitsu HK (2005) Synergistic biofilm formation by Treponema denticola and Porphyromonas gingivalis. FEMS Microbiol Lett 250:271–277

Yoshida S, Ogawa N, Fujii T, Tsushima S (2009) Enhanced biofilm formation and 3-chlorobenzoate degrading activity by the bacterial consortium of Burkholderia sp. NK8 and Pseudomonas aeruginosa PAO1. J Appl Microbiol 106:790–800

Zeriouh H, de Vicente A, Pérez-García A, Romero D (2014) Surfactin triggers biofilm formation of Bacillus subtilis in melon phylloplane and contributes to the biocontrol activity. Environ Microbiol 16:2196–2211

Zhang W, Wang H, Zhang R, XZ Y, Qian PY, Wong MH (2010) Bacterial communities in PAH contaminated soils at an electronic waste processing center in China. Ecotoxicology 19:96–104

Zhang X, Liu X, Wang Q, Chen X, Li H, Wei J, Xu G (2014) Diesel degradation potential of endophytic bacteria isolated from Scirpus triqueter. Int Biodeter Biodegr 87:99–105

Zhang N, Yang D, Wang D, Miao Y, Shao J, Zhou X et al (2015) Whole transcriptomic analysis of the plant-beneficial rhizobacterium Bacillus amyloliquefaciens SQR9 during enhanced biofilm formation regulated by maize root exudates. BMC Genomics 16:685

Zhou H, Luo C, Fang X, Xiang Y, Wang X, Zhang R, Chen Z (2016) Loss of gltb inhibits biofilm formation and biocontrol efficiency of Bacillus subtilis Bs916 by altering the production of γ-polyglutamate and three lipopeptides. PLoS One 11:1–20

Chapter 17
Sustainable Bioelectricity Generation from Living Plants

Mohnish Borker, T.V. Suchithra, M. Srinivas, and S. Jayaraj

17.1 Introduction

Depletion of natural resources with a rapidly changing climate has forced mankind to develop and exploit new possibilities in obtaining sustainable and reliable energy sources. Renewable technologies like solar, hydro, wind and bioenergy have already been implemented in daily life. Market share of biofuels, such as bioethanol, biodiesel and bioelectricity is increasing (Eisentraut 2010). However bio energy is not always sustainable. Deforestation and competition with food production on arable land are the two frequent disadvantages (Haveliek and Schneider 2011). Thus, many countries focus on increased usage of renewable energy, as it helps curb difficulties faced due to greenhouse gas emissions and reducing global warming by replacing conventional energy sources (Panswar et al. 2011; Arent et al. 2011; Swift-Hook 2013).When renewable energy sources are considered for electricity generation, they provide benefits for managerial techniques for the economic growth, as well as for environment (Varun et al. 2009). The various bioenergy sources discovered till date are: biodiesel, biohydrogen, bioethanol, biogas, microbial fuel cell and plant microbial fuel cells.

Depending on the source of bioenergy, the biofuels are classified as 1st generation biofuels that are obtained from edible crop seeds, 2nd generation biofuels which are obtained from non-edible seeds or energy crops, the 3rd generation biofuels which are algae based biofuels and the 4th generation biofuels that are advanced biofuels (Daroch et al. 2013). In 1911, a new source for bioenergy generation was discovered which exploits microorganisms for electricity production. They are the Microbial fuel cells (MFCs).

M. Borker • M. Srinivas • S. Jayaraj
Department of Mechanical Engineering, National Institute of Technology Calicut, Kozhikode, Kerala, India

T.V. Suchithra (✉)
School of Biotechnology, National Institute of Technology Calicut, Kozhikode, Kerala, India
e-mail: drsuchithratv@nitc.ac.in

J.K. Patra et al. (eds.), *Microbial Biotechnology*,
https://doi.org/10.1007/978-981-10-6847-8_17

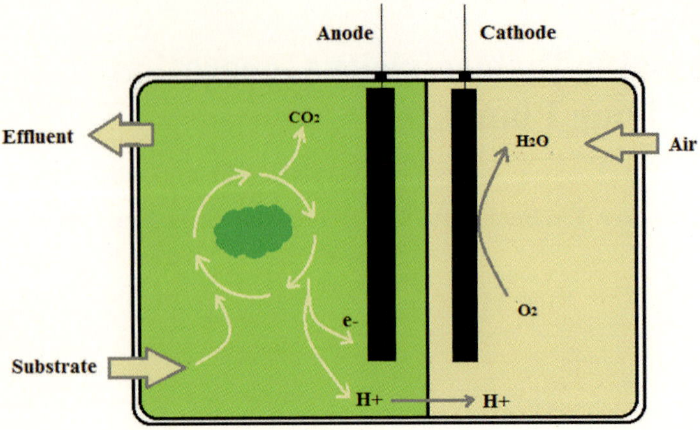

Fig. 17.1 Basic configuration of Microbial fuel cell

17.1.1 Microbial Fuel Cells

Microbial fuel cell is a biological fuel cell that utilises the microorganisms to generate electricity. These microbes breakdown the organic compounds contained in the electrolyte system into electrons and protons (Liu et al. 2004). A general setup of an MFC consists of two chambers separated by a membrane. The anode is placed in the anode chamber where the necessary fuel is supplied. Oxidation reaction takes place where the organic compounds are broken into electrons and H^+ protons (Park and Gregory 2000). An oxidizer is supplied to the cathode chamber where a reduction reaction takes place near the cathode. Figure 17.1 gives a brief setup of the microbial fuel cell.

Unlike other conventional fuel cells, MFCs utilise biological matter as its nutrient and energy source (Chaudhuri and Lovley 2003). In general MFCs can be classified as mediated or mediator-less cells based on the addition of electron carriers in the anode chamber. Mediator-less fuel cell has better control on the fuel cell and provides a higher efficiency potential (Liu et al. 2004). MFC has many possible advantages over conventional fuel cells, which is the reason why increasing amount of research is being conducted on it. Recent advances in MFCs show that plants are introduced in the electrolyte system to boost the current generation.

17.1.2 Plant Microbial Fuel Cell (PMFC)

Plants release a considerable amount of organic compounds ($C_6H_{12}O_6$) during the process of photosynthesis. About 50–60% of the fixed carbon is transferred from the leaves to the roots, depending on the type of plant, the growth structure and the surrounding environmental conditions (Lynch and Whipps 1990). Bacteria which are

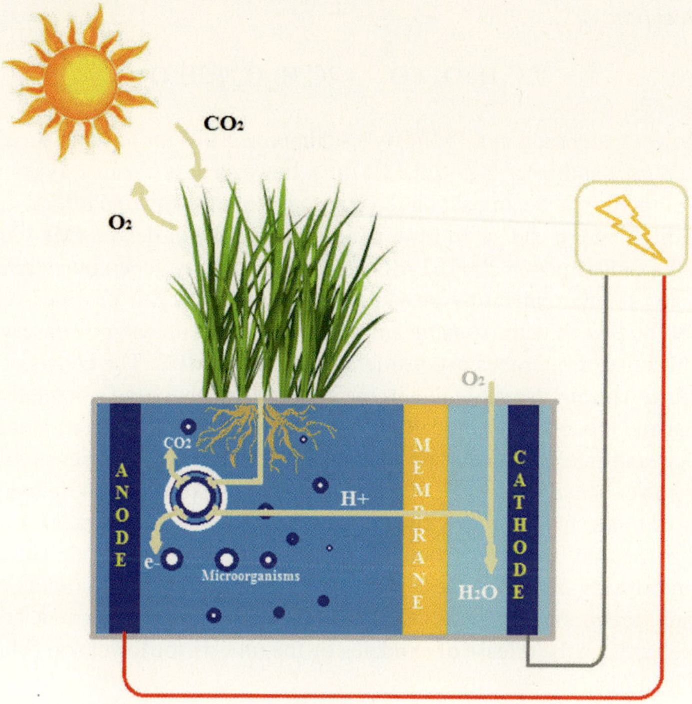

Fig. 17.2 Basic design of a Plant microbial fuel cell

electrochemically active, present around the roots breakdown this organic matter, releasing electrons. Electricity is generated when an electron acceptor or an electrode is placed in the vicinity of these bacteria. The electrode material with a higher potential is used as an anode. The plant is allowed to grow in the anode region where the anodic environment is made favourable for the plant growth. Figure 17.2 provides the basic design of a plant microbial fuel cell.

Electricity generation takes place in two steps. (i) In the anode chamber, electrons are released wherein they are taken up by the anode and transferred to the cathode by an external circuit through a load (Timberlake 2009). (ii) The electrons combine with the protons that permeate through the membrane forming water in the cathode region.

Equations 17.1, 17.2 and 17.3 gives the anode and cathode reactions as follows:

Anode:

$$2C_6H_{12}O_6 \rightarrow 2C_6H_{10}O_6 + 4H^+ + 4\bar{e} \quad (17.1)$$

Cathode:

$$O_2 + 4H^+ + 4\bar{e} \rightarrow 2H_2O \quad (17.2)$$

Net Reaction:

$$2C_6H_{12}O_6 + O_2 \rightarrow 2C_6H_{10}O_6 + 2H_2O \tag{17.3}$$

The power generation in a PMFC is less (approx. 10 W/m^2) as compared to wind (5–7.7 MW/m^2) and solar (4.5–7.5 MW/m^2); however it is cheaper as compared to the both of them.But, the impact on the environment of both wind turbines and solar panels is high and at the same time is a source of social debate (Mc Gowan and Connors 2000; Kazmerski 2006). PMFCs aim at electricity generation while sustaining the natural environment by providing in-situ bioenergy. It provides continuous electricity for a week without using any substrates and with energy conservation.

The diversity of microbes is considerably high in PMFC. The conversion of rhizodeposits to current is mainly due to the interactions between fermentative microorganisms and the electrochemically active bacteria. Optimization of plant species, cell design and nutrient media are essential to enhance the current production in terms of power density (Lu et al. 2015). Demarcating exudates and rhizodeposits in plants is difficult, hence, root exudates are defined as all the organic substances released by healthy and intact roots of the plant into the lower soil. The root exudates comprise of carbohydrates, amino acids, amides, aliphatic acids, aromatic acids, fatty acids, and vitamins (Grayston et al. 1997). The nutrient acquisition mainly depends on the release of exudates by the roots in the lower soil (Marschener 1998) (Fig. 17.3).

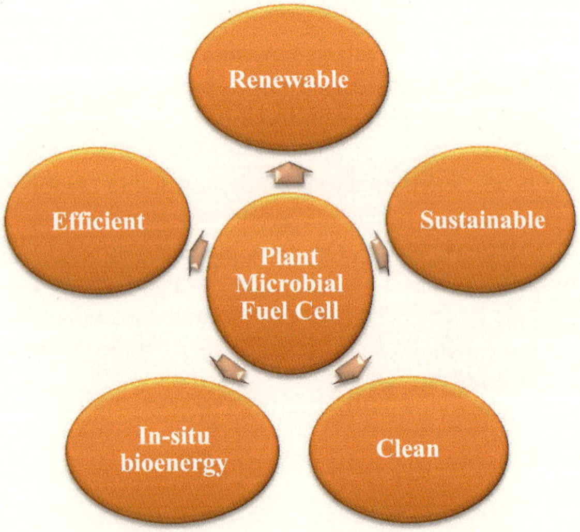

Fig. 17.3 Diverse advantages of a plant MFC

17.2 Material Selection

17.2.1 Selection of Plants for PMFC

The efficiencies of energy conversion are highly depending on the plant species. Hence Selection of plants has to be done precisely. Flowering plants consume more organic matter for pollen formation, hence are substituted by marshland grasses. Table 17.1 show various plants suitable for PMFC.

Plants like *Spathyphilum* (peace lily) have a flower like formation, but are known to withstand soaked soil for a good duration. Examples of such plants under study are those of *Spartina anglica*, *Glyceria maxima*, *Oryza sativa*, *Arundinella anomala* and *Musa acuminate.*

17.2.2 Selection of Electrodes

The electrodes can provide support for the growth of biocatalyst and connection. They act as electron acceptors completing the circuit. Hence, selection of the electrodes surface is very crucial. Just like Plant free MFC, electrode material with higher potential behaves as an anode. Table 17.2 gives a list of the possible materials of electrodes and other components.

Proper placing of the anode and cathode is crucial as the distance between the anode and cathode affects the proton diffusion between the two (Cheng et al. 2006). Also, cathode modification with catalyst accelerates oxygen reduction (Tender et al. 2002). In PMFCs the potential of the cell is dependent on the maximum theoretical voltage of the cell, current density and the internal resistance of the cell, all with respect to the geometric area of the anode or the membrane. Acetate is generally added to supply substrate to the microbe culture in the cell to enhance electricity generation (Kaku et al. 2008).

Table 17.1 Plants for Plant Microbial Fuel Cell

	Plant	Substrate	Application
1.	*Oryza sativa* (Rice paddy)	Glucose	Plant/microbe cooperation for electricity generation in a rice paddy field, Microbial fuel cell generating electricity from Rhizodeposits of Rice plants
2.	*Spartina anglica*		New plant growth medium for increased power output of the pant microbial fuel cell,
3.	*Glyceria maxima*		Electricity generation by a novel design tubular plant microbial fuel cell Green Electricity production with living plants and bacteria in a fuel cell
4.	*Arundinella anomala*		Concurrent bioelectricity and biomass production in three plant microbial fuel cells

Table 17.2 Components of a PMFC

	Item	Material
1.	Anode	Graphite, graphite felt, carbon paper, carbon-cloth, graphite granules
2.	Cathode	Graphite, graphite felt, carbon paper, carbon-cloth
3.	Anodic chamber	Glass, polycarbonate, Plexiglas
4.	Cathodic chamber	Glass, polycarbonate, Plexiglas
5.	Membranes	Proton exchange membrane: Nafion, Ultrex, polyethylene

17.3 Classification of PMFC

PMFCs are generally of two types, single chambered and dual chambered. The most basic form of a plant MFC is the 'sediment type PMFC' which consists of a single chamber incorporating both the electrodes in the same chamber. The dual chamber PMFC consists of an anode and a cathode chamber separated by a proton exchange membrane. Figure 17.4 explains the various classifications of PMFC models.

17.3.1 Sediment PMFC

A sediment PMFC consists only of one chamber that maintains both the anode and the cathode. Figure 17.5 depicts the basic model of the sediment PMFCs. The cathode is however exposed to the atmosphere near the top soil (Letebvre et al. 2008). The absence of the proton exchange membrane makes the system very cost effective. But the rate of current generation is very low due to high diffusion rates of the electrons and proton (Chiao et al. 2006).

17.3.2 Rooftop PMFC

An advanced form of a PMFC is a rooftop system, which combines the advantages of green roofs with that of electricity generation (Helder et al. 2013). Rooftop PMFCs are a type of single chamber PMFCs wherein they incorporate the electrode-plant assembly onto the household roofs. As the area of application is large and to make it cost affective, the use of a membrane is eliminated. As seen in Fig. 17.6, it generally employs grass species which grow in extensive covers.

Other advantages of a rooftop PMFC is that it prevents rainwater runoff, provides higher aesthetic value, preserves biodiversity, improves the oxygen content in the atmosphere and decreases the temperature within cities (Strik et al. 2011).

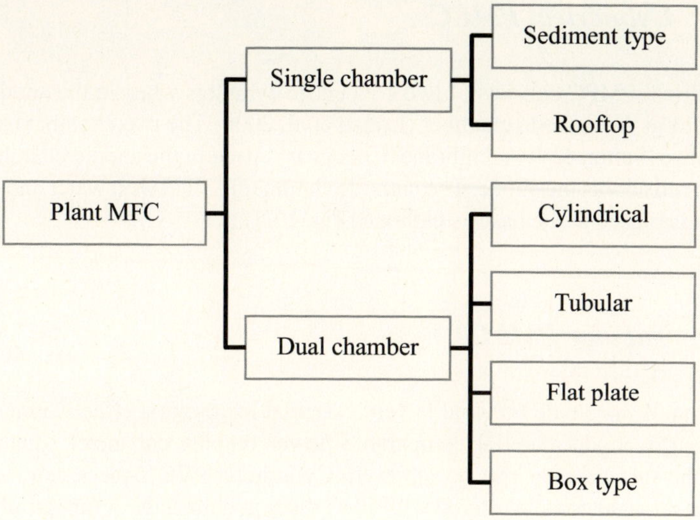

Fig. 17.4 Classifications of PMFC models

Fig. 17.5 Sediment PMFC

Fig. 17.6 Rooftop plant MFC

17.3.3 Cylindrical PMFC

A cylindrical PMFC consists of two concentric cylinders wherein the anode chamber is held in the cathode chamber (Logan et al. 2006).The two chambers are separated with a Nafion 117A membrane. Cocopeat is used in the anodic chamber and a graphite rod acts as the anode. The cathode chamber is filled with water and a graphite sheet is placed at the base as cathode (Fig. 17.7).

17.3.4 Tubular PMFC

Reduction of the anode material is very essential for the cost effective models. On reducing the anode material comparable power outputs per meter square of the membrane surface areas can be achieved. Cylindrical PVC T-piece can be used to create an air tight container fixed with PVC discs at both ends. A tubular ultrafiltration membrane is held through the centre of both discs. Anode is placed outside the membrane in the tube while the cathode is placed inside the ultrafiltration membrane (Timmers et al. 2013). Figure 17.8 depicts the tubular PMFC design. The system can be tried for anode materials of graphite felt and graphite granules.

Fig. 17.7 Cylindrical PMFC

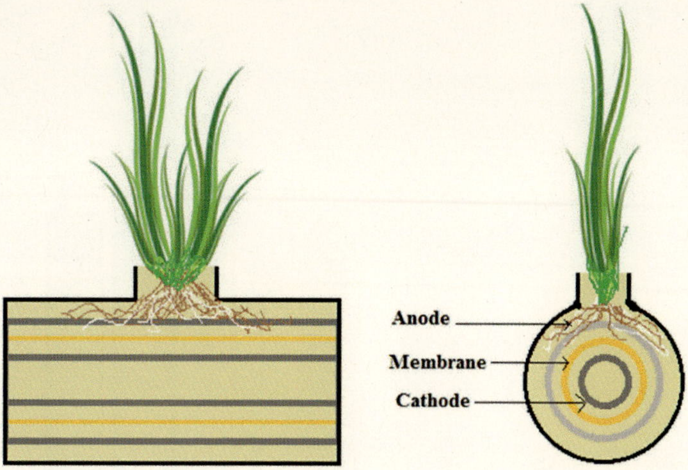

Fig. 17.8 Dual chamber tubular PMFC

17.3.5 Flat Plate PMFC

A flat porous plate PMFC can be constructed using acrylic sheets. The anode is made from three graphite felt layers, stacked with plastic rings in between them. The three layers are connected using golden wire as current collector. Cathode consists of one layer of graphite felt and catholyte is constantly circulated through the cathode. System is controlled by external resistances ranging from 500 to 1000 Ω (Park and Gregory 2000; Helder et al. 2012) (Fig. 17.9).

17.3.6 Dual Chambered Box Type PMFC

In a box type dual chambered PMFC, the anode and cathode chambers are separated with a Nafion 117A membrane. Both the anode and cathode chambers are of box shape having equal dimensions. The desired plant is potted along with the anode in the anode chamber, while still water is maintained in the cathode chamber. Graphite sheets and cocopeat-water mixture can be used as electrode material and anolyte respectively (Fig. 17.10).

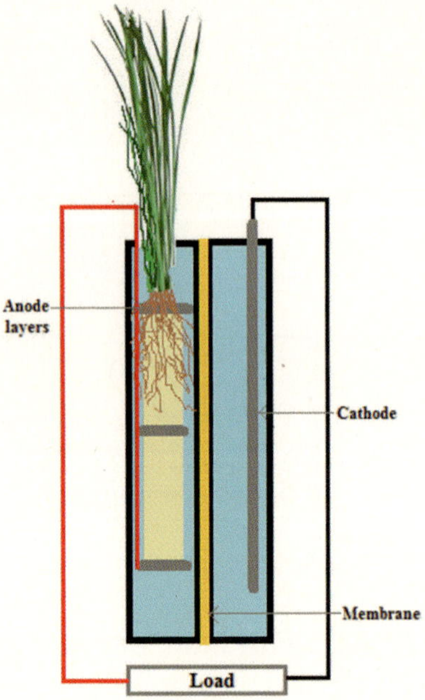

Fig. 17.9 Flat plate PMFC

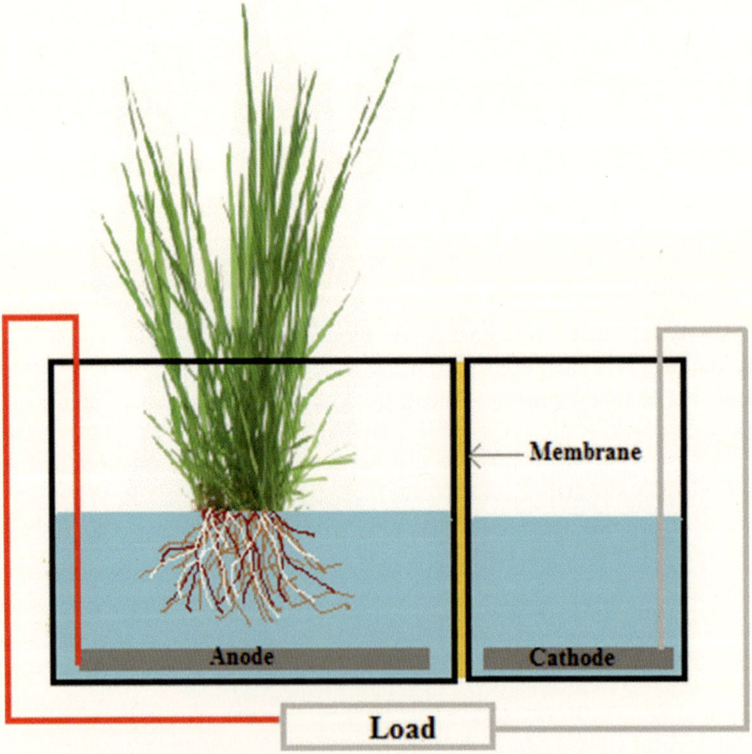

Fig. 17.10 Dual chamber box type PMFC

17.4 Measurement Techniques

For considering a PMFC in an application, it is necessary to determine the performance parameters. The following are the measured parameters of the plant in beaker sediment MFC system

V_{oc}: Open Circuit Voltage (Volts)
I_{sc}: Short circuit Current (mA/m^2GA)
V: Generated Voltage (Volts)
R_{ext}: External Resistance (Ω)

The three main performance parameters are the current density (I), the internal resistance (R_{int}) and the power density (P) of the PMFC. All these parameters are evaluated with respect to the anode geometric area (GA). Anode which comprises of mostly carbon cloth or graphite sheets is expensive. Transforming the PMFC into a cost efficient and easy accessible system is a prime motive. Hence, the performance parameters are evaluated with respect to the GA of the anode (Strik et al. 2008). Equations 17.4, 17.5 and 17.6 provide the performance parameters necessary for the evaluation of the PMFC.

Current density (mA/m^2GA)

$$I = \frac{V}{A_{anode} R_{Ext}} \tag{17.4}$$

Internal Resistance (Ω/m^2 GA)

$$R_{\text{int}} = \frac{(Voc - V)}{I} \tag{17.5}$$

Power density (mW/m^2GA)

$$P = V \times I \tag{17.6}$$

17.5 Polarization Curves

Based on the current densities and the power densities, polarization curves were obtained for different values of resistances like 50*Ω*, 100*Ω*, 500*Ω*, 1000*Ω*, 1500*Ω* and 2000*Ω*. The polarization curves are necessary, for they determine the maximum power density of the system. From the maximum power density, one can obtain the internal resistance of the system. The internal resistance of the PMFC is the external resistance corresponding to the maximum power density in the polarization curve. In order to enhance the performance of the system, the internal resistance needs to be reduced by maintaining a constant resistance (value maintained to that of internal resistance) across the circuit.

17.6 Model Reviews

MFC performance of PMFCs varies with its difference in its design (Table 17.3). A preliminary MFC performance test for the voltage it can generate and the current produced for different resistances is essential for the selection of suitable plant for a PMFC setup. This data provides us with the plants individual power density with respect to the anode GA. Current and power densities are measured in terms of the anode GA, so as to determine the systems dependence on the anode material (Strik et al. 2008).

Determining the internal resistance is quite crucial. Polarization curves are drawn with the current and voltages at different resistance values for the systems. Internal resistance can be obtained from polarisation curves, which is equivalent to the external resistance at max power density. N. Kaku, et al. obtained the polarisation curve for rice paddy electricity generation system, with max power density of 5.75 mW/m^2 (Kaku et al. 2008).

An initial incubation period of 50–100 days or an open cell voltage of 0.4 V, whichever is attained earlier, is necessary for producing an electrical current. This is mainly due to various reasons such as omission of nutrients, release of oxygen, conducted through the 'aerenchyma', scavenging the electrons collected at the anode or lack of an adapted anodic microbial consortium (Schamphelaire et al. 2008). Experimental evidences show that bright sunny days give higher output compared to dull rainy days (Kaku et al. 2008). The amount of biomass produced above ground determines the functioning ability of the plant in the PMFC system. Experimental data by M. Helder et al. determines how the above ground biomass

Table 17.3 Power efficiencies of plant MFCs

Model	Plant	Current density (mA/m^2)	Power density (mW/m^2)	References
Single chamber sediment type	*O. sativa*	24	5.75	Kaku (2008)
Anode: Graphite felt				
Cathode: Graphite felt				
Dual chamber cylindrical plant MFC	*S. anglica*	39	222	Helder (2012)
Anode: Graphite granules + electrode	*A. anomala*	31	22	
Cathode: Graphite felt				
Dual chamber tubular plant MFC	*Glyceria maxima*	25.8	60	Timmers (2013)
Anode: Graphite felt				
Cathode: Graphite felt				
Flat plate plant MFC	*S. anglica*	2.08×10^3	679	Wetser (2015)
Anode: Graphite felt				
Cathode: Graphite felt				

values in form of total leave and stem length increases during the time of study (Helder et al. 2010).

A novel design tubular PMFC was tested by R. Timmers et al. (2013). The design mostly concentrated on reducing the anode material by incorporating it in the form of a tubular roll. The design achieved a maximum power density of 60 mW/m^2 during initial polarization. K. Wetser et al. (2015) used a dual cathode chamber flat plate PMFC system for salt water species of *S.anglica* (Wetser et al. 2015).

17.7 Conclusion

Recent advances and extensive research has been carried out in developing the technique of PMFC. Innovative techniques were involved to improve the power density and match the current crop based electricity systems. Four designs were discussed depicting their functional parameters. Power density as high as 222 mW/m^2 anode GA was obtained for a dual chamber cylindrical PMFC. Also, reducing the anode material in a tubular design, power density of 60 mW/m^2 is achieved. Yet, further fundamental research and technological integration is needed to display the full plant power electricity potential. After which complete environmental and economic analysis can be done.

References

Arent DJ, Wise A, Gelman R (2011) The status and prospects of renewable energy for combating global warming. Energy Econ 33:584–593

Chaudhuri SK, Lovley DR (2003) Electricity generation by direct oxidation of glucose in mediatorless microbial fuel cells. Nat Biotechnol 21:1229–1234

Cheng S, Liu H, Logan BE (2006) Increased power generation in a continuous flow MFC with advective flow through the porous anode and reduced electrode spacing. Environ Sci Technol 40:2426–2432

Chiao M, Blam K, Lewei L (2006) Micro machined microbial and photosynthetic fuel cells. J Micromechanical Micro Eng 16:2547–2553

Daroch M, Geng S, Wang G (2013) Recent advances in liquid biofuel production from algal feed stocks. Appl Energy 102:1371–1381

Eisentraut A (2010) Sustainable production of second generation biofuels. IEA Energy papers, France, vol 1, pp 1–221

Grayston SJ, Vaughan D, Jones D (1997) Rhizosphere carbon flow in trees, in comparison with annual plants: the importance of root exudation and its impact on microbial activity and nutrient availability. Appl Soil Ecol 5:29–56

Havliek P, Schneider UA (2011) Global land-use implications of first and second generation biofuel targets. Energy Policy 39:5690–5702

Helder M, Strik D, Hamelers H (2010) Concurrent bioelectricity and biomass production in three plant microbial fuel cells. J Bioresource Technol 101:3541–3547

Helder M, Strik DP, Hamelers HV, Kuijken RCP, Buisman CJ (2012) Year round performance of the flat-plate plant-microbial fuel cell. Bioresour Technol 104:417–423

Helder M, Strik D, Timmers R (2013) Resilience of roof-top plant-microbial fuel cells during Dutch winter. J Biomass and Bioenergy 51:1–7

Kaku N, Yoneawa N, Kodama Y (2008) Plant/microbe cooperation for electricity generation in a rice paddy field. J Appl Microbiol Biotechnol 79:43–49

Kazmerski LL (2006) Solar photovoltaic R&D at the tipping point: a 2005 technology overview. J Electron Spectrosc Relat Phenom 150:105–135

Lefebvre O, Al-Mamun A, Ooi WK, Tang Z, Chua DH, Ng HY (2008) An insight into cathode options for microbial fuel cells. Water Sci Technol 57:2031–2037

Liu H, Ramnarayanan R, Logan R (2004) Production of electricity during wastewater treatment using a single chamber microbial fuel cell. Environ Sci Technol 38:2281–2285

Logan BE, Hamelers HWM, Rozendal R, Schröder U (2006) Microbial fuel cells: methodology and technology. Environ Sci Technol 40:5181–5192

Lu L, Xing D, Ren Z (2015) Microbial community structure accompanied with electricity production in a constructed wetland plant MFC. J Bioresource Biotechnol 195:115–121

Lynch JM, Whipps JM (1990) Substrate flow in the rhizosphere. Plant and Soil 129:1–10

Marschener H (1998) Role of root growth, arbuscular mycorrhiza, and root exudates for the efficiency in nutrient acquisition. Field Crop Res 56:203–207

McGowan JG, Connors SR (2000) Windpower: a turn of the century review. Annu Rev Energy Environ 25:147–197

Panswar NL, Kaushik SC, Kothari S (2011) Role of renewable energy sources in environmental protection: a review. Renew Sustain Energy Rev 15:1513–1524

Park DH, Gregory Z (2000) Electricity generation in microbial fuel cells using neutral red as an Electronophore. Appl Environ Microbiol 66:1292–1297

Schamphelaire LVD, Bossche H, Dand S (2008) Microbial fuel cell generating electricity from Rhizodeposits of Rice plants. J Environ Sci Technol 42:3053–3058

Strik DPBTB, Hamelers HVM, Snel JFH, Buisman CJN (2008) Green electricity production with living plants and bacteria in a fuel cell. Int J Energy Resources 32:870–876

Strik DPBTB, Timmers RA, Helder M, Steinbusch KJJ, Hamelers HVM, Buisman CJN (2011) Microbial solar cells: applying photosynthetic and electrochemically active organisms. Trends Biotechnol 29:41–49

Swift-Hook DT (2013) The case for renewables apart from global warming. Renew Energy 49:147–150

Tender LM, Reimers CE, Stecher HA, Holmes DE, Bond DR (2002) Harnessing microbially generated power on the seafloor. Nat Biotechnol 20:821–825

Timberlake KC (2009) Basic chemistry, 3rd edn. Prentice Hall, Upper Saddle River

Timmers R, Strik D, Hamelers H (2013) Electricity generation by a novel design tubular plant microbial fuel cell. J Biomass Bioenergy 51:60–67

Varun, Parakash R, Bhat IK (2009) Energy, economics and environmental impacts of renewable energy systems. Renew Sustain Energy Rev 13:2716–2721

Wetser K, Sudirjo E, Buisman C (2015) Electricity generation by a plant microbial fuel cell with an integrated oxygen reducing biocathode. J Appl Energy 137:151–157

Chapter 18
Biosurfactants: An Agent to Keep Environment Clean

Padma Mahanti, Sanjeet Kumar, and Jayanta Kumar Patra

18.1 Introduction

Earlier the advance methods of detection of inorganic and organic contaminants of soil and water had allowed for only high ranges (per billion levels). Even those detection, many heavy metals and organic compounds was properly not detected. After the detection, the remediation was carried out via soil washing or *in situ* flushing. This process is called scientifically solublization and it performed with only water or water with additives but some organic contaminants like trychloro ethylene, polycyclic aromatic hydrocarbons and polychlorinated biphenyls was difficult to remove. For the above mentioned problems encouraged to the researchers to produce an agent for permanent removal or conversion in degradable form (Vijayakumar and Sarvana 2015). As a result, an amphiphilic compounds were produced that reduce the free energy of the systems by replacing the bulk molecules of higher energy at an interface called surfactants. They are produced having sound potential to mobilize the contaminants, lower surface tension, increase solubility, detergency power, wetting ability and foaming capacity (Desai and Banat 1997). In general, they are used to save energy and energy costs. They are used to remove metals with addition of organic solvents, chelating agents, acid and bases. They also play a functional key in the agricultural sectors which is directly related to the

P. Mahanti
Directorate of Environment and Climate Change, Trivandrum, Kerala, India

S. Kumar (✉)
Ambika Prasad Research Foundation, Odisha, India
e-mail: sanjeet.biotech@gmail.com

J.K. Patra (✉)
Research Institute of Biotechnology and Medical Converged Science, Dongguk University, Goyang-si, Gyeonggi-do, South Korea
e-mail: jkpatra@dongguk.edu

J.K. Patra et al. (eds.), *Microbial Biotechnology*,
https://doi.org/10.1007/978-981-10-6847-8_18

environment. In the year of 2004, Deleu & Paquot reported that about 0.2 million tons of surfactants are used in crop protection.

The modern era is the day of "Green Concept". Therefore, to enhance the potential and minimize the limitation of surfactants with green technology concept, Biosurfactants came in existence. They are combinations of low molecular weight surface-active substances produced by the microorganisms (bacteria, yeast and fungi). They are grouped as glycolipids, lipopeptides, phospholipids, fatty acids, neutral lipids, polymeric and particulate compounds (Thavasi 2011). A key character of BF is a hydrophilic-lipophilic balance which is the indicative the portion of hydrophilic and hydrophobic constituents in surface-active substances. Due the unique structure of BF, it increase the surface area of hydrophobic water-insoluble substances, increase the water bioavailability of such substances and change the character of the bacterial cell membrane. They are active at extreme temperatures, pH and salinity as well, and can be produced from industrial wastes and from by-products and this make them to allow utilizing waste substrates without or less polluting effect. They possess hydrophobic and hydrophilic characters which help to aggregate at interface between fluids with different polarities. In comparison to chemically synthesized surfactants, they are environment friendly, biodegradable, biocompatibility, digestability, less toxic and non-hazardous. The three prime uses of BF are, 1: used to increase the surface area of hydrophobic substrate, 2: used to increase the bioavailability of hydrophobic substrate through solubilization, 3: used to regulate the attachment and removal of microorganisms from the surface. Due to above all mentioned properties, BF are used in diverse industries like organic chemical industries, petroleum, petrochemicals, mining, metallurgy, agrochemicals, fertilizers, foods, beverages, cosmetic industries, pharmaceuticals and Crop protection. They also used as emulsifiers, demulsifiers, foaming agents, wetting agents, detergents, spreading agents and ingredients of nutraceutical. Keeping this in view an attempt has been taken to gather the complete information on BF and their application on environmental issues created by the contamination.

18.2 Key Properties of Biosurfectants

Biosurfactants (BFs) have unique and distnict key characters. Hence, it is better than chemically synthesized surfactants. The major key characters of BFs are discussed below in detail (Fig. 18.1):

(a) *Less toxicity:*

It is the prime key character which is fit in the "Green Concept". The BFs are generally less / non-toxic bio products for the removal of contaminant to keep environment clean. Many researchers has proved it by various experiments such as in the year of 1991, Poremba et al., demonstrated that chemical derived surfactants showed highest toxic then BFs produced from. Flasz et al. (1998) and Cavalero and Cooper (2003) also demonstrated that BFs produced from *Pseudomonas aeruginosa*

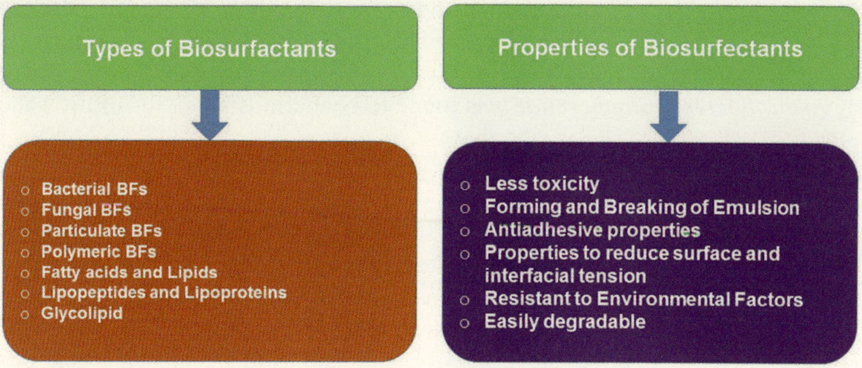

Fig. 18.1 Types and properties of different biosurfactants

and *Candida bombicola* respectively were non-toxic and non-mutagenic then chemically synthesized surfactants like sophorolipids.

(b) Forming and Breaking of Emulsion:

BFs show more stability with emulsion for months to years. They also act as good emulsifier or de-emulsifiers. In the year of 1993, Velikonja and Kosaric showed stability characters with oil-in water/water-in-oil emulsions. In the year 1985, Cirigliano and Carman described a water-soluble emulsifier Liposan synthesized by *Candida lipolytica*.

(c) Antiadhesive properties:

Sometimes group of micro-organisms or some organic matter colonized on any surface and such accumulated stuffs harm the host surface. These bacterial / organic accumulated surface is called biofilm (Hood and Zottola 1995). The above colonolization was affected by various factors such as bacterial strains, hydrophobicity and the electrical charges, physiological conditions and types of extracellular polymers produced by which help cells to anchor to surface (Zottola 1994). In above situations, BFs act as antiadehsive agents. In the year of (2012), Chakrabarti demonstrated that BFs produced from *Streptococcus thermophilus* slow down the colonolization of other Thermophilic strains of *Streptococcus* over steel and BFs from *Pseudomonas fluorescens* inhibited the attachment of *Listeria monocytogens* on steel surface.

(d) Properties to reduce surface and interfacial tension:

BFs are very much effective and efficient in reducing surface and Interfacial tension. In the year of 1981, Cooper et al., reported BFs produced from *B. subtilis* and in the year 1985, Syldatk et al., reported from *Pseudomonas aeruginosa*.

(e) Resistant to Environmental Factors:

BFs are mostly resistant to the environmental factors such as temperature and pH. In the year of 1990, McInerney et al. reported that BFs produced from

Bacillus licheniformis was resistant to temperature up to 50 °C, pH between 4.5 to 9.0 and NaCl & Ca concentration up to 50 and 25 g L^{-1} respectively while in the year of 2004, Singh & Cameotra documented that BFs produced by *Arthrobacter protophermiae* was found to be thermostable (30–100 °C) and pH (2 to 12).

(f) Easily degradable:

Now-days everyone want "Green" & "Biodegradable" concept to keep the environment healthy. Always chemical surfactants create environment problems hence BFs are in high demand due to its biodegradability potential. BFs easily degraded as compared to synthetic surfactants (Mohan et al. 2006). In the year 2008, Lee et al., controlled the bloom of marine algae, Cochlodinium using the biodegradable BFs Sophorolipid with the removal efficiency of 90% in 30 min treatment (Vijayakumar and Sarvana 2015).

18.3 Types of Biosurfactants

BFs are produced from the micro-organisms and therefore they are classified as per the microbial origin and biochemical compositions as compared to chemically synthesized surfactants are classified as per the polarity index. The most major types of BFs are following (Fig. 18.1):

(i) Bacterial BFs
(ii) Fungal BFs
(iii) Particulate BFs
(iv) Polymeric BFs
(v) Fatty acids and Lipids
(vi) Lipopeptides and Lipoproteins
(vii) Glycolipid

18.3.1 Bacterial BFs

As diverse organic compounds are used as a carbon source by the microbes for their growth, they become the vector of the diffusion into the microbial cell by producing number of substances called BFs. Some specific micro-organisms are able to change the cell wall structure using non-ionic or Lipopolysacharides BFs such as non-ionic trechlose Corbynomycolates produced by *Rhodococcus erythropolic* and from some *Mycobacterium* species and *Arthrobacter* species (Ristau and Wanger 1983; Kilburn and Takayama 1981; Kretschmer et al. 1982; Vijayakumar and Sarvana 2015). In the year of 1982, Kretschmer et al. reported lipopolysaccharides having some properties produced from *Acinetobacter* species.

18.3.2 Fungal BFs

BFs produced from bacterial strains is not cheaper so researcher are screening the low cost BFs. For this, production of BFs from fungal species are sound. The most common fungi as a source of BFs are *Candida batistae, Candida bombicola, Candida lipolytica, Candida ishiwade, Aspergillus ustus, Trichosporon ashii* etc. (Rufino et al. 2007; Chandran and Das 2010; Aljendro et al. 2011; Casas et al. 1997; Sarubbo et al. 2007; Thanomsub et al. 2004).

18.3.3 Particulate BFs

Particulate BFs are the BFs form the extracellular membrane vesicles and play an important role in hydrocarbon (alkane) uptake by the cells. In the year of 1979 and 2012 respectively Kaeppeli and Finnerty & Chakarbarati reported a vesicles having diameter 20–30 nm with 1.158 cubic gcm of buoyant density produced by *Acinetobacter* species strain HO_1-N.

18.3.4 Polymeric BFs

Some BFs are the combinations of many biomolecules called polymeric BFs. Emulsion is an example of polymeric BFs and it is an effective emulsifying agent for hydrocarbon in water. These include emulsan, liposan, alasan, lipomanan etc. According to Hatha (2007) a very low concentration (0.001–0.01%) of emulsan emulsify hydrocarbons in water. *Candida lipolytica* also produces polymeric BFs like liposan (83% carbohydrate and 17% protein) (Cooper and Paddock 1984). A potent extracellular polyanionic amphipathic heteropolysaccharide is produced by *Acinetobacter calcoaceticus* RAG-1 and acts as efficient emulsifier (Rosenberg et al. 1979). In the year of 2012, Chakrabarti reported polymeric BFs called Liposan, is an extracellular water-soluble emulsifier synthesized by *Candida lipolytica* which is composed of 83% carbohydrate and 17% of protein.

18.3.5 BFs with Fatty Acid

During the production of n-alkane, micro organisms produce large quantities of fatty acids and lipids BFs such as *Acinetobacter* species produced 1-N, Phosphatidyl ethanolamine-rich vesicles which form optically clear micro-emulsions of alkanes in water (Vijayakumar and Sarvana 2015). Gautam and Tyagi reported that N-phosphatidylethanolamine produced by *Acinetobacter* spp. was able to emulsify alkanes dissolved in water. The length of aliphatic chain of fatty acid determined the hydrophilicity and lipophilicity of biosurfactant. (Gautam and Tyagi 2006).

18.3.6 Glycolipid

In the year 1949, Jarvis and Johnson described modern BFs as they are Carbohydrates linked to long-chain aliphatic acids or hydrocarboxyaliphatic acids by an ester group. Hence BFs are mainly Glycolipids. A glycolipid is defined as sugar moiety attached to a chain of n- numbers of aliphatic acids through an ester bond. The glycan parts are mostly mono- or di-hexoses attached lipid moieties. Commonly rhamnose are observed to be present in the biosurfactants produced by the *Pseudomonas* genera. Some best known Glycolipids BFs are:

(i) *Rhamnolipids*: it is most popular and studied BFs are produced by *Pseudomonas aeroginosa.* Mainly they are a combination of rhamnose molecules with hydroxydacanolic acids.
(ii) *Trehalolipids*: In the year of 1978, Asselineau and Asselineau documented Trehalose lipids as BFs produced from *Rhodococcus erthropolis* and *Arthobacter* species.
(iii) *Sophorlipids:* Sophorlipids BFs consist of a dimeric carbohydrate sophorose linked to a long-chain hydroxyl fatty acid by glycosidic linkage (Vijayakumar and Sarvana 2015). They are generally group of six to nine different hydrophobics (Gautam and Tyagi 2006).

18.3.7 BFs with Polypeptide

The BFs, where they biochemically lipid molecules attached with polypeptide chain. Lipopeptides are biosurfactants consisting of a lipid molecule attached to a polypeptide chain. Some of the best examples include surfactin, arthrofacin, iturin, serrawetin W2 etc. (Priya and Usharani 2009; Tang et al. 2010).

These BFs are in highly demand for pharmacological research as well as environmental work. Two most popular BFs are:

(i) *Lichenysins*: These BFs is produced from the bacteria *Bacillus licheniformis* which show excellent stability with high temperature, pH and salinity. In the year of 1990, McInerney et al. a BF called Lichenysin which reduce the surface and interfacial tension.
(ii) *Surfactin*: It is unique BFs in structure and widely used. It is a combination of a seven amino-acid ring structure attached with a fatty acid chain by Lactone linkage (Arima et al. 1968; Vijayakumar and Sarvana 2015). The β–sheet structure of Surfactin, resembles with a horse saddle in both aqueous and air interface. Its various applications like recovery of oil, removal of heavy metals from soil, antifungal activity, antibacterial property, and cytotoxic effects.

18.4 Environment Applications of BFs

BFs have sound applications to keep environment clean. The prime uses in environmental stuffs are (Table 18.1; Fig. 18.2):

18.4.1 Petroleum and Oil Contamination

Now-days oil and petroleum is a big problem for environment in general and for wetlands ecosystem, water bodies, riverine ecology, marine ecosystem and beach ecology in particular. Oil and petroleum contamination creates big hazards for the flora, fauna, micro-climate and bio-chemo-physio status of the environment. To address the problems and getting the stuffs for restoration of healthy environment, BFs play an important role. The work of BFs in biodegradation of hydrocarbon is started from the experiments of Itoh and Suzuki in the year 1972. They showed that hydrogen culture media stimulated the growth of Rhamnolipid producing strain of *Pseudomonas aeruginosa*. Research showed that BFs are used to enhance the water solubility and increase the displacement of oily substances (Chang et al. 2008). In the year of 2009, Olivera et al. documented that *Alcanivorax* and *Cycloclasticus* genera are highly specialized hydrocarbons degraders in marine ecology and *Alcanivorax borkumensis* utilizes aliphatic hydrocarbons as its main carbon source for growth and produce an anionic glucose lipid BFs. Vijayakumar and Sarvana (2015) reported that Gordonia species BS29 grows on aliphatic hydrocarbons as sole carbon source has found to produce bioemulsion which effectively degrade

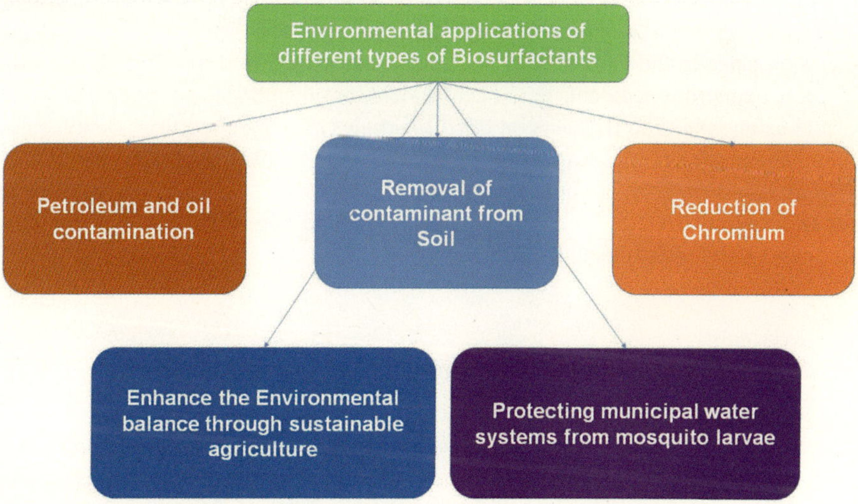

Fig. 18.2 Environmental applications of different types of buisurfactants

Table 18.1 Major Bisurfactants and their source organisms

Sl. No.	Biosurfactans	Source organisms
1.	Carbohydrate –lipid	*Debarymyces polmorphus*
2.	Diglycosyl diglycerides	*Lactobacillus fermentum*
3.	Glycolipid	*Candida ishiwade*
4.	Glycolipoprotein	*Aspergillus ustus*
5.	Ornithine Lipids	*Thiobacillus thiooxidans*
6.	Polyol lipids	*Rhodotorula glutinis*
7.	Protein-Lipopolysaccharide-complex	*Candida lipolytica*
8.	Protein PA	*Pseudomonas aeruginosa*
9.	Rhamnolipids	*Pseudomonas chleroraphis*
10.	Serrawettin	*Serratia marcescens*

crude oil. In the year of 2008, Whang and his group showed that *Pseudomonas aeruginosa* and *Bacilius subtilis* produce rhamnolipid and surfactin increase solubility and bioavailability of petrochemical mixture.

Petroleum industry is one of the best non-renewable energy resources for major transportation systems in a country. However they are also equally major environmental pollutants. Spillage of petroleum products into the water bodies not only contaminates the water system but also brings toxicity to a major population of organisms living in water bodies. Previewing this, a lot of attention has been paid to remediate such oil spillage accidents and following it innovative techniques and technologies are being implemented. Application of BFs is one such innovative technology to remediate the oil spillage.

A major accident that took place in 2011, November (Sedco 706 oil rig, Brazil) causing a leakage of total of 5943 L spreading over 163 km^2. Yet another oil spillage that took place in the Gulf of Mexico in 2010 caused by the explosion of an oil rig in USA. Numerous accidents have been evident for oil spillage and have been remediated using BFs. The BFs acts on the oil particles causing distortion of molecules (emulsification) thus leading to breakdown of oil. A novel approach which uses *Pseudomonas putida* commonly recognized as superbug was found to be suitable for restoration of water bodies from oil spillage and continues to one of the best findings till date (Zhang et al. 2005).

18.4.2 Degradation of Toxic Pollutants

The biologically degradation or bioremediation usually consists of the application of Nitrogenous and phosphorus fertilizers, adjusting the pH and water content, supplying air and adding bacteria. In this process, the addition of BFs as emulsifiers is advantageous when bacterial growth is slow or when the pollutants consist PAHs. In the above case, BFs play a sound role to stimulate the bioremediation process. Increase of bioemulsifier as a BFs concentration in the time of bioremediation would increase the emulsifying characters. In the year of 1995, Navon-Venezia et al. justified this approach. Most of hydrocarbons exist in strongly absorbed forms when they are introduced into environment. So, their removal efficiency might be limited in low mass transfer phases. In this case, addition of BFs as solubilization agents to the system enhance the bioavailability of low solubility and highly sorptive compounds (Shin et al. 2004; Vijaykumar and Sarvana 2015).

18.4.3 Bioavailability Potential

The low water solubility of a potent pollutant, Polycyclic Aromatic Hydrocarbons is believed to limit their availability to microorganism which is potential problems for bioremediation of contaminated sites. It has been assumed that BFs would enhance the bioavailability of hydrophobic compounds. In the year of 1996, reported this application of BFs to clean environment using *Pseudomonas stutzeri* while in the year 1998 using *Mycobacterium* species and similar experiments using *Mycobacterium* species by Willumsen et al. in the year 2001.

18.4.4 Potential to Increase Surface Area

BFs have character to increase the surface area of hydrophobic water insoluble substrate. In the year 1999, Sekelsky and Shreve showed the role of BFs in keeping environment healthy using the BFs.

18.4.5 Removal of Contaminant from Soil

Soil is the integral part of environment. The cultivated and wild land is affected by presence of organic and inorganic pollutants which create negative impacts on both types of land. For the removal of pollutants to make soil healthy, need bioremediation. In the year of 2006, Sun et al. reported that the BFs produced from the microorganisms are excellent for removal of hydrocarbons and heavy metals. It was also observed that desorption of hydrophobic pollutants tightly bound to soil particles is accelerated by BFs. BFs would also enhance the degradation of certain chemical insecticides which are accumulated in the soil (Zhang et al. 2011; Sharma et al. 2009; Wattanaphon et al. 2008; White et al. 2006; Nielsen and Sorensen 2003; Sachdev and Cameotra 2013). There are many research have cited in the literature which justify that BFs play a sound role in improving the health of soil by the process of soil remediation and also on pesticidebiodegradation by surfactin and degradation of chlorinated hydrocarbon by Glycolipids (Mata-Sandoval et al. 2001). BFs produced from *Lactobacillus pentosus* has demonstrated reduction by 58.6% to 62.8% of octane hydrocarbon from soil (Moldes et al. 2011; Sachdev and Cameotra 2013). BFs are very good against heavy metals degradation. Heavy metals pollution originates from excessive use of metal salt based fungicides, sewage and sludge amendments applied on the cultivated land. In the limited amount, these pollutants serve as essential micronutrients and are required for various actions in plant metabolisms but in higher concentration, they damage plant in the form of root tissue necrosis and purpling of foliage. For this problem BFs act as an agent to degrade the pollutants from the soil through bioremediation. In the year 2006 and 2011, Kassab & Roane and Pacwa-Plociniczak et al. demonstrated that BFs produced by *Pseudomonas* species, *Bacillus* species and *Acinetobacter* species are used for removal of heavy metals from contaminated soil and even acceleration of pesticides degradation. Some BFs like Rhamnolipid and Surfactin are known as for the removal of heavy metals such as Ni, Cd, Mg, Mn, Ca, Ba, Li, Cu and Zn (Herman et al. 1995; Mulligan et al. 2001; Nielson and Sorensen 2003; Mulligan and Wang 2004; Sachdev and Cameotra 2013).

18.4.6 Enhance the Environmental Balance Through Sustainable Agriculture

Sustainable agriculture is the prime key for the balancing of ecosystem and healthy environment of the locality. BFs produced from microbes have antimicrobial potential against the pathogens of agricultural crop as well as a common plant species. Hence, they are considered to a promising biocontrol molecule for achieving sustainable agriculture. Several reports are cited in the literature on this prime applications of BFs such as Nihorimbere et al. (2011) showed that BFs produced from Rhizo-bacteria has antagonist properties. Krzyzanowska et al. (2012) showed that BFs produced from *Pseudomonas* and *Bacillus* species exhibited biocontrol of soft root causing *Pectobacterium* and *Dickeya* species while Kim et al. (2011) reported that BFs from a strain of *Pseudomonas* have sound insecticidal activity against Green peach aphid (*Myzus persicae*). Kruijit et al. (2009) has also reported that plant growth-promoting *Pseudomonas putida* produces BFs that can cause lysis of zoospores of the oomycete pathogen (*Phytophthora capsici*), causative agent of dumping-off in cucumber. The lipopeptide BFs produced by *Bacillus* species exhibit growth inhibition of phytopathogenic fungi Fusarium, Aspergillus species and Biopolaris sorokiniana. Such BFs could used as bioacontrol agents (Velho et al. 2011; Sachdev and Cameotra 2013). *Brevibacillus brevis* strain HOB1 produce surfactin isoform and this lipopeptide BFs has demonstrated sound antimicrobial properties (Haddad 2008; Sachdev and Cameotra 2013). In the year of Hultberg et al. reported that fluorescent pseudomonad's with the BFs producing ability may inhibit the growth of fungal pathogens like *Pythium ultimum, Fusarium oxysporum, Phytophthora cryptogea*. In the year of 2003, Andersen et al. showed that BFs produced from *Pseudomonas* species terminate the growth of *Rhizoctonia solani* and *Phythium ultimum* which causes several plant diseases. Whereas Kim et al. (2010) showed that *Colletotrichum gloeosporides*, causative agent for anthracnose on papaya leaves is reported to be controlled by BFs producing *Bacillus subtilis*. The above all reports reveal that BFs play an important role in sustainable agriculture which is directly proportional to the balanced ecosystem and healthy environment.

18.4.6.1 Control of Invasive Species and Weeds

Invasive species and weeds create hazard for the biodiversity and environment. It destroys the native flora which reflects on the balancing of ecosystems. In this case, BFs play an important role control the weeds and invasive flora. In the year of 2002, Boyette et al. reported that when BFs are used with the combination of a fungus, *Myrothecium verrucaria*, the combination destroy the weed species.

18.4.7 Reduction of Chromium

Chromium is the 7th most rich element on earth and can be found in 9 different oxidation states. Amongst the most common are the Cr (III) and Cr (VI) in nature. Cr (VI)is considered to be carcinogenic and is considered as an environmental hazard and on other side Cr (III) is required to animals as nutrition. The difference on part of both the states of Cr is due to their solubility. A thorough study on the detoxification of Cr (VI) in presence of negatively charged biosurfactant was carried out by Massara et al. They reported that the biosurfactants could enhance the removal of Cr from kaolinite was significant. They also studied it by using 2 factorial experiments. Over a period of 24 days, near about 100% Cr was reduced which significantly suggested that rhamnolipids are of beneficial use (Massara et al. 2007).

18.4.8 Protecting Municipal Water Systems from Mosquito Larvae

Mosquitoes are one of the deadliest vectors causing a large mortality (WHO reports 500 million) throughout the world (Nabar and Lokegaonkar 2015). The increase in the population density has become more prone to the mosquitoes due to increased rate of municipal wastes caused by people. Chemical industries are in urge to produce more of chemical mosquitocides and the increased resistivity of mosquitoes has brought serious concerns in the scientific community. Microbial mosquitocides can be considered as alternative to the chemical ones. Thus microbial insecticides can be considered as alternatives to chemical insecticides. A study by Nabar & Lokegaonkar reported that bacteria producing larvicidal metabolites from various environments are highly effective. The tested against larvae of *Culex* and *Aedes moquitoes* and found that 100% mortality of the mosquito larvae caused at very low concentration of secondary metabolites. Microbially produced metabolites are highly capable due their selective toxicity and ready decomposability and are economic friendly. The high utility of BFs against mosquito larvae may be due to surface tension reducing capacity which forces larvae sink into the water and thus death occurs.

18.5 New BFs as an Agent to Keep Environment Clean

As BFs are much important in many industries and sectors in general and to keep environment clean in particular, many advance research produced some new BFs which has sound applications on environmental problems (Mulligan 2009). Such studies are discussed in detail: (1) Nayak et al. (2009) identified a *Pseudoxanhomonas* species PNK-04 strain which produced Rhamnolipids. The Rhamnolipids were a mixture of mono-and di-rhamnolipids units. Emulsification ability was high compared to synthetic surfactants. They were also able to enhance biodegradation of 2-chlorobenzoic acid, 3-chlorobenzoic acid and 1-methyl apphthalene through solubility enhancement; (2) Somayeh et al. (2008) isolated a BF-producing bacteria from oil contaminated soil and water. The bacteria were identified as *Pseudomonas aeruginosa*. They showed that surface tension was less than 40 mN/m; (3) In the year of 2009, Martin et al. showed that 99% of the aromatic polyaromatic hydrocarbons up to 3 than 4 rings were not effectively remediated.

18.6 Peculiar Findings and Conclusion

The study revealed that Biosurfactants are extracellular compounds produced by microorganisms. BFs are surface active molecules having hydrophobic and hydrophilic moieties as their constituents which allow BFs to interact at interface and reduce the surface tension. They are classified based on the origin and biochemical compositions. They have unique and distinct features which make them a sound agent to keep environment clean. They are used in many environmental sectors, majors are followings:

1. Bioremediation of oil, heavy metals and pesticides
2. Enhanced oil recovery process
3. Antimicrobial agents

4. Accessibility or bioavailability of the pollutants to the microbes involved in biodegradation process
5. Heavy metal degradations etc.

Despite the benevolent properties of BFs, there are many challenges that remain unsolved in the field of BFs. Need more work on BFs as a tools for the Environmental problems. Also need a commercial production of BFs in low cost. There are following necessary aspects for advanced research on BFs to make them a versatile eco-friendly biomolecules for keeping environment healthy and sustainable development:

a. Advance research on BFs properties mainly degradation efficiency for the environmental problems.
b. Evaluation of toxicity levels of each BF on the biotic components of Environment.
c. Isolation of new and safe BFs based on green concept of the society and environment.
d. Strategies for the production of cost effective techniques

References

Aljendro CS, Humberto HS, Maria JF (2011) Producation of glycolipids with antimicrobial activity by *Ustilago maydis* FBD 12 in submerged culture. African J Microbiol Res 5:2512–2523

Arima K, Kakinuma A, Tamura G (1968) Surfactin, a crystalline peptidelipid surfactant produced by Bacillus subtilis: isolation, characterization, and its inhibition of fibrin clot formation. Biochem Biophys Res Commun 3:488–494

Asselineau C, Asselineau J (1978) Trehalose-containing glycolipids. Prog Chem Fats Other Lipids 16:59–99

Boyette CD, Walker HL, Abbas HK (2002) Biological control of kudzu (*Pueraria lobata*): with an isolate of *Myrothecium verrucaria*. Biocontrol Sci Tech 12:75–82

Casas JA, De Lara SG, Garcia F (1997) Optimization of a synthetic medium for *Candida bombicola* growth using factorial design of experiments. Enzyme Microb Technol 21:221–229

Cavalero DA, Cooper DG (2003) The effect of medium composition on the structure and physical state of sophorolipids produced by *Candida bombicola* ATCC 22214. J Biotechnol 103:31–41

Chakrabarti S (2012) Bacterial biosurfactants: characterization, antimicrobial and metal remediation properties. PhD. Thesis, National Institute of Technology

Chandran P, Das N (2010) Biosurfactant production and diesel oil degradation by yeast species *Trichosporon asahii* isolated from petroleum hydrocarbon contaminated soil. Int J Eng Sci Technol 2:6942–6953

Chang MW, Holoman TP, Yi H (2008) Molecular characterization of surfactant-driven microbial community changes in aerobic phenanthrene-degrading cultures under methanogenic conditions. Biotechnol Lett 30:1595–1601

Cirigliano MC, Carman GM (1985) Purification and characterization of liposan, a biomulsifier from *Candida lipolytica*. Appl Environ Microbiol 50:846–850

Cooper DG, Paddock DA (1984) Production of a biosurfactant from *Torulopsis bombicola*. Appl Environ Microbiol 47:173–176

Cooper DG, Macdonald CR, Duff SJB, Kosaric N (1981) Enhanced production of surfactin from *Bacillus subtilis* by continuous product removal and metal cation additions. Appl Environ Microbiol 42:408–412

Desai JD, Banat IM (1997) Microbial production of surfactants and their commercial potential. Microbiol Mol Biol Rev 61:47–64

Flasz A, Rocha CA, Mosquera B, Sajo C (1998) A comparative study of the toxicity of a synthetic surfactants and one produced by *Pseudomonas aeruginosa* ATCC 55925. Med Sci Res 26:181–185

Gautam KK, Tyagi VK (2006) Microbial surfactants: a review. J Oleo Sci 55:155–166

Haddad NI (2008) Isolation and characterization of a biosurfactant producing strain, brevibacilis brevis HOB1. J Ind Microbiol Biotechnol 35:1597–1604

Hatha AAM (2007) Microbial biosurfactants–review. J Mar Atmos Res 3:1–17

Herman DC, Artiola JF, Miller RA (1995) Removal of cadmium, lead and zinc from soil by a rhamnoid biosurfactant. Environ Sci Technol 29:2280–2285

Hood SK, Zottola EA (1995) Biofilms in food processing. Food Control 6:9–18

Itoh S, Suzuki T (1972) Effect of rhamnolipids on growth of Pseudomonas aeruginosa mutant deficient in n-parafin-utilizing ability. Agric Biol Chem 36:2233–2235

Jarvis FG, Johnson MJ (1949) A glycolipid produced by Pseudomonas aeruginosa. J Am Chem Soc 71:4124–4126

Kaeppeli O, Finnerty WR (1979) Partition of alkane by an extracellular vesicle derived from hexadecane-grown *Acinetobactor*. J Bacteriol 140:707–712

Kassab DM, Roane TM (2006) Differential responses of a mine tailings *Pseudomonas* isolate to cadmimum and lead exposures. Biodegradation 17:379–387

Kilburn JO, Takayama K (1981) Effects of ethambutol on accumulation and secretion of trehalose mycolates and free mycolic acid in *Mycobacterium smegmatis*. Antimicrob Agents Chemother 20:401–404

Kim PI, Ryu J, Kim YH, Chi YT (2010) Production of biosurfactant lipopeptide Iturin A. fengycin and surfaction A from Bacillus subtilis CMB32 for control of Colletotrichum glocosporides. J Microbiol Biotechnol 20:138–145

Kim SK, Kim YC, Lee S, Kim JC, Yun MY, Kim IS (2011) Insecticidal activity of rhamnolipid isolated from Pseudomonas sp. EP-3 against green peach aphid (Myzus persicae). J Agric Food Chem 59:934–938

Kretschmer A, Bock H, Wagner F (1982) Chemical and physical characterization of interfacial-active lipids from *Rhodococcuc erythropolis* grown on n-alkane. Appl Environ Microbiol 44:864–870

Kruijit M, Tran H, Raaijmakers JM (2009) Functional, genetic and chemical characterization of biosurfactants produced by plant growth-promoting *Pseudomonas putida* 267. J Appl Microbiol 107:546–556

Krzyzanowska DM, Patrykus M, Golanowska M, Polonis K, Gwizdek A, Lojkowska E, Jafra S (2012) Rhizospore bacteria as potential biocontrol agents against soft rot caused by various *Pectobacterium* and *Dickeya* spp. Strains. J Plant Pathol 94(2):367–368

Lee YJ, Choi JK, Kim EK, Youn SH, Yang EJ (2008) Field experiments on mitigation of harmful algal bloom using a Sophorolipid: Yellow clay mixture and effects on marine plankton. Harmful Algae 7:154–162

Martin VG, Kalil SJ, Alberto, Costa V (2009) In situ bioremediation using biosurfactant produced by solid state fermentation. World J Microbiol Biotechnol 25:843–851

Mata-Sandoval JC, Karns J, Torrents A (2001) Influence of rhamnolipids and Triton X-100 on the biodegradation of three pesticides in aqueous and soil slurries. J Agric Food Chem 49:3296–3303

McInerney MJM, Javaheri M, Nagle DP (1990) Properties of the biosurfactant produced by *Bacillus licheniformis* strain JF-2. J Ind Microbiol Biotechnol 5:95–101

Mohan PK, Nakhla G, Yanful EK (2006) Biokinetics of biodegradation of surfactants under aerobic, anoxic and anaerobic conditions. Water Res 40:533–540

Moldes AB, Paradelo R, Rubinos D, Devesa-Rey R, Cruz JM, Barral MT (2011) Ex-situ treatment of hydrocarbon-contaminated soil using biosurfactants from *Lactobacillus pentosus*. J Agric Food Chem 59:9443–9447

Mulligan CN (2009) Recent advances in the environmental applications of biosurfactants. Curr Opin Colloid Interface Sci 14:372–378

Mulligan CN, Wang S (2004) Remediation of a heavy metal contaminated soil by a rhanolipid foam. In: Thomas HR, Yangt RN (eds) Geoenvironmental engineering. Integrated management of groundwater and contaminated land. Thomas Telford, London, pp 544–551

Mulligan CN, Yong RN, Gibbs BF (2001) Heavy metal removal from sediments for biosurfactants. J Hazard Mater 85:111–125

Nabar BM, Lokegaonkar S (2015) Larvicidal activity of microbial metabolites extracted from extremophiles against vector mosquitoes. Int J Mosquito Res 2:161–165

Navon-Venezia SZ, Zosim A, Gottlieb A, Legmann R, Carmeli S, Ron EZ, Roserberg E (1995) Alasan, a new bioemulsifer from *Acinetobacter radioresistens*. Appl Environ Microbiol 61:3240–3244

Nayak AS, Vijaykumar MH, Karegoudar TB (2009) Characterization of Biosurfactants produced by *Pseudoxanthomonas* sp. PNK-04 and its application in bioremediation. Int Biodeterior Biodegrad 63:73–79

Nielsen TH, Sorensen J (2003) Producation of cyclic lipopeptides by Pseudomonas fluorescens strains in bulk soil and in the sugar beet rhizosphore. Appl Environ Microbiol 69:861–868

Nihorimbere V, Marc O, Smargiassi M, Thonart P (2011) Beneficial effect of the rhizosphore microbial community for plant growth and health. Biotechnol Agron Soc Environ 15:327–337

Olivera NL, Nievas ML, Lozanda M, Prado G, Dionisi HM, Sineriz F (2009) Isolation and characterization of biosurfactants potential of natural bilge waste microflora. J Ind Microbiol Biotechnol 30:542–548

Pacwa-Plociniczak P, Piotrowska SZ, Cameotra SS (2011) Environmental applications of biosurfactants: recent advances. Int J Mol Sci 12:633–654

Poremba KW, Gunkel S, Lang S, Wagner F (1991) Toxicity testing of synthetic and biogenic surfactants on marine microorganisms. Environ Toxicol Water Qual 6:157–163

Priya T, Usharani G (2009) Comparative study for biosurfactant production by using *Bacillus subtilis* and *Pseudomonas aeruginosa*. Bot Res Int 2:284–287

Ristu E, Wanger F (1983) Formation of novel trehalose lipids from *Rhodocuccus erythropolis* under growth RR limiting conditions. Biotechnol Lett 5:95–100

Rosenberg E, Zuckerberg A, Rubinovitz C, Gutnick DL (1979) Emulsifier of *Arthrobacter* RAG-1: isolation and emulsifying properties. Appl Environ Microbiol 37:402–408

Rufino RD, Sarubbo LA, Takaki GM (2007) Enhancement of stability of biosurfactants produced by *Candida lipolytica* using industrial residue as substrate. World J Microbiol Biotechnol 23:729–734

Sachdev DP, Cameotra SS (2013) Biosurfactants in agriculture. Appl Microbiol Biotechnol 97:1005–1016

Sarubbo LM, Farias CBB, Takaki GMC (2007) Co-utilization of canola oil and glucose on the production of a surfactant by *Candida lipolytica*. Curr Microbiol 54:68–73

Sekelsky AM, Shreve GS (1999) Kinetic model of biosurfactants enchanced hexadecane biodegra dation by *Pseudomonus aeruginosa*. Biotechnol Bioeng 63:401–409

Sharma S, Singh P, Raj M, Chadha BS, Saini HS (2009) Aqueous phase partitioning of hexachlorocyclohexane (HCH) isolates by biosurfactants produced by *Pseudomonas aerginosa* WH-2. J Hazard Mater 171:1178–1182

Shin KH, Kim KW, Seagren EA (2004) Combined effects of pH and biosurfactants addition on solublization and biodegradation of phenanthrene. Appl Microbiol Biotechnol 65:336–343

Singh P, Cameotra SS (2004) Potential applications of microbial surfactants in biomedical sciences. Trends Biotechnol 22:142–146

Somayeh C, Abbas AS, Nouhi AS (2008) Study the role of isolated bacteria from oil contaminated soil in bioremediation. J Bacteriol 1365:678–707

Sun X, Wu L, Luo Y (2006) Application of organic agents in remediation of heavy metals-contaminated soil. Ying Yong Sheng Tai Xue Bao 17:1123–1128

Syldatk C, Lang S, Wagner F, Wray V, Witte L (1985) Chemical and physical characterization of four interfacial-active rhamnolipids from *Pseudomonas* species. DSM 2874 grown on n-alkane. Zeitschrift Naturforschung C 40:51–60

Tang JS, Zhao F, Gao H, Dai Y, Yao ZH, Hong K et al (2010) Characterization and online detection of surfactin isomers based on HPLC-MS n analyses and their inhibitory effects on the overproduction of nitric oxide and the release of TNF and IL-6 in LPS-induced macrophages. Mar Drugs 8:2605–2618

Thanomsub B, Watcharachaipong T, Chotelersak K, Arunrattiyakorn P, Nitoda T, Kanzaki H (2004) Monoacylglycerols: Glycolipids biosurfactants produced by a thermotolerant yeast, *Candida ishiwadae*. J Appl Microbiol 96:588–592

Thavasi R (2011) Microbial biosurfactants: from an environmental applications point of view. J Bioremediation Biodegradation 2(5):1000104e

Velho RV, Medina LF, Segalin J, Brandelli A (2011) Production of lipopeptides among Bacillus strains showing growth inhibition of phytopathogenic fungi. Folia Microbial (Praha) 56:297–303

Velikonja J, Kosaric N (1993) Biosurfactants in food applications. In: Kosaric N, Sukun FV (eds) Biosurfactants production: properties: applications. CRC Press, New York, pp 419–448

Vijayakumar S, Sarvana V (2015) Biosurfactants-types, sources and applications. Res J Microbiol 10(5):181–192

Wattanaphon IIT, Kerdsin A, Thammacharoen C, Sangvanich P, Vangnai AS (2008) A biosurfactant from *Burkholderia cenocepacia* BSP3 and its enhancement of pesticide solubilization. J Appl Microbiol 105:416–423

Whang LM, Liu PWG, Ma CC, Cheng SS (2008) Application of biosurfactants, rhamnolipid and surfaction, for enhanced biodegradation of diesel-contaminated water and soil. J Hazard Mater 151:155–163

White JC, Parrish ZD, Gent MP, Lannucci BW, Eitzer BD, Isleyen M, Mattina MI (2006) Soil amendments, plant age, and intercropping impact p, p- DDE bioavailability to *Cucurbita pepo*. J Environ Qual 35:992–1000

Willumsen P, Nielsen J, Karlson U (2001) Degradation of phenanthrene-analogue azaarenes by *Mycobacterium gilvum* strain LB307T under aerobic conditions. Appl Microbiol Biotechnol 56:539–544

Zhang G, Wu Y, Qian X, Meng Q (2005) Biodegradation of crude oil by *Pseudomonas aeruginosa* in the presence of rhamnolipids. J Zhejiang Univ Sci B6:725–730

Zhang C, Wang S, Yan Y (2011) Isomerization and biodegradation of beta-cypermethrin by *Pseudomonas aeruginosa* CH7 with biosurfactant production. Bioresources Technology 102:7139–7146

Zottola EA (1994) Microbial attachment and biofilm formation: a new problem for the food industry? Food Technol 48:107–114

Chapter 19
Bioprospecting of Endophytes for Agricultural and Environmental Sustainability

Sabuj Sahoo, Sarmistha Sarangi, and Rout George Kerry

19.1 Biodiversity of Endophytes

Endophytes are the most diverse group of microorganisms found in virtually every plant on earth. They are widely spread in various environment conditions. Although endophytes commonly refers to mostly fungi or bacteria it has already been recognized in marine algae (Smith et al. 1989; Stanley 1992) and mosses and ferns (Petrini et al. 1992; Raviraja et al. 1996). Endophytes were reported to be chiefly isolated from plants ranging from palm to large trees, marine grasses and lichens. The biodiversity of endophytes are mainly classified into three categories such as fungal, bacterial and algal endophytes.

19.1.1 Fungal Endophytes

Fungal endophytes are most commonly studied among other endophytes. Fungi are plant like organism that lack chlorophyll, true roots, stems and leaves. An endophytic fungus lives in inter or intra cellular spaces in stem, petiole and leaves of a plant for a part or whole of their life cycle. Most endophytes isolated belong to ascomycetes and their anamorph basidiomycetes. The total biodiversity of fungal endophytes may be classified into two major categories. These include the

S. Sahoo (✉)
Post Graduate Department of Biotechnology, Utkal University, Bhubaneswar, Odisha, India
e-mail: sabujbiotech@utkaluniversity.ac.in; sabujbiotech@rediffmail.com

S. Sarangi
ICAR-Central Rice Research Institute, Cuttack, Odisha, India

R.G. Kerry
P.G. Department of Biotechnology, Academy of Management & Information Technology, Khurda, Odisha, India

J.K. Patra et al. (eds.), *Microbial Biotechnology*,
https://doi.org/10.1007/978-981-10-6847-8_19

Table 19.1 Symbiotic criteria used to characterize fungal endophytic classes

	Clavicipitacious	Nonclavicipitaceous		
Criteria	Class-1	Class-2	Class-3	Classs-4
Host range	Narrow	Broad	Broad	Broad
Tissue colonized	Shoot and rhizome	Shoot, root and rhizome	Shoot	Root
Transmission	Vertical and horizontal	Vertical and horizontal	Horizontal	Horizontal
In plantacolonization	Extensive	Extensive	Limited	Extensive
In planta biodiversity	Low	Low	High	Unknown

Table 19.2 Fungal endophytes with their host range

Fungal endophytes	Phyllum	Class	Order	Host range	Colonized part
Colletotrichum	Ascomycota	Sordariomycetes	Glomerellales	Wide	Shoots & roots
Curvularia	Ascomycota	Dothideomycetes	Pleosporales	Wide	Shoots & roots
Epichloë	Ascomycota	Sordariomycetes	Hypocreales	Grasses	Shoots
Fusarium	Ascomycota	Sordariomycetes	Hypocreales	Wide	Shoots & roots
Neotyphodium	Ascomycota	Sordariomycetes	Hypocreales	Grasses	Shoots
Piriformospora	Basidiomycota	Agaricomycetes	Sebacinales	Wide	Roots
Serendipita	Basidiomycota	Agaricomycetes	Sebacinales	Wide	Roots

clavicipitaceous (CE), which infect some grasses confined to cool regions and the non-clavicipitaceous endophytes (NCE), which are widely distributed and found in asymptomatic tissues of non vascular plants, conifers, ferns and angiosperms. However, NCE are reported to be restricted to Ascomycota and Basidicomycota groups (Jalgaonwala et al. 2011; Bhardwaj and Agrawal 2014). Symbiotic criteria used to characterize fungal endophytic classes are shown in Table 19.1.

Typically CE occurs on plant shoots where they form systemic intercellular infections. CE are fastidious in culture and are restricted to some grasses that grow in warm and cool season (Bischoff and White 2005). NCE are primarily ascomycetes fungi recovered from land plants, terrestrial eco systems ranging from agro biosystems to biomes, that are widespread from tropic to tundra (Arnold 2007). The examples of fungal and bacterial endophytes with their host range is shown in Tables 19.2 and 19.3 respectively.

19.1.2 Bacterial Endophytes

Bacterial endophytes are the second most studied endophytes after fungi. More than 16 phyla of bacterial endophytes belonging to about 200 genera were reported, most of which belong to the phyla Actinobacteria, Proteobacteria and Firmicutes

Table 19.3 Bacterial endophytes with their host range

Bacterial endophytes	Examples	Host plant species	Colonizing area
Proteobacteria	*Erwinia, Pseudomonas*	Alfalfa (*Medicago sativa* L.)	Roots
γ -proteobacteria	*Pseudomona, Serratia*	Black pepper (*Pipernigrum* L.)	Roots
β-proteobacteria	β-proteobacteria	Grape (*Vitis* sp.)	Stems
Proteobacteria	*Proteobacteria*	Radish (*Raphanus sativus* L.)	Leaves and roots
Actinobacteria	*Corynebacterium*	Sugar beet (*Beta vulgaris* L.)	Roots
β-proteobacteria	*Burkholderia cepacia*	Wheat (*Triticum aestivum* L.)	Roots
α-proteobacteria	*Erwinia., Agrobacterium*	Soybean (*Glycine max* (L.) Merr.	Stems, leaves, roots

(Golinska et al. 2015). Gram positive and negative bacterial species majorly contribute to the diversity of bacterial endophytes viz., *Achromobacter*, *Acinetobacter*, *Agrobacterium*, *Bacillus*, *Brevibacterium*, *Microbacterium*, *Pseudomonas*, *Xanthomonas* etc (Sun et al. 2013). These endophytes resides intracellularly in root and shoot cells of plants. In thisintracellular form bacteria loss cell wall but continue to divide and metabolize. These wall-less intracellular forms of bacteria are called L-forms (Allan et al. 2009). Actinomyces are prokaryotic organisms belonging to the phylum actinobacteria resembling mycelium like fungus and are spore producers (Chaudhary et al. 2013).

19.1.3 Algal Endophytes

Algae are simple chlorophyll bearing vascular plants. Algae are eukaryotic organisms that have no roots, stems or leaves. A number of endophytes are now known that grows within seaweeds and algae (Andrew et al. 2013). One such example is *Ulvella leptochaete*, which has recently been discovered from host algae including Cladophora and Laurentia from India. Recently a team led by Dr. Felix Bast, Assistant Professor, Central University of Punjab (2016), Bathinda able to discover the microscopic endophytic algae within its host seaweeds. The research involved analysis of green seaweeds, *Cladophora glomerata* collected from Calicut, Kerala and red seaweeds, *Laurentia obtusa* collected from Mandapam, Tamil Nadu.

19.2 Interaction Between the Endophytes and Their Host Plants

The endophytes, which are a major part of plant micro-ecosystems, found in healthier tissues are in an endosymbiotic relationship with the plants they invade. In a greater view it is found that these endophytes not only enhance the tolerance of the plant, in biotic or abiotic stress helps but also improves the quality as well as the quantity of the secondary phytocompounds that are used as bioactive compounds or drugs (Jia et al. 2016). Due the development of sophisticated new versatile pseudo branches of life-sciences such as proteomics, genomics and their subdivided branches such as metabolomics, trancriptomics, it became much easier to determine molecular interaction at the most nano level (Rout and Sahoo 2013; Pervez et al. 2016; Abdurakhmonov 2016). On the basis of these advanced modes of evaluations it is found that the endophytic population may greatly be affected by the age of the plant, genetic variations in the plant, or environmental background of the host plant that lies in it (Jia et al. 2016). Taking the benefits into count, further bioengineer the endophyte with desired efficiency can be designed and incorporated into the host plant to produce the required quantity of the primary, secondary phytocompounds or drugs. Apart from the growth of the host plant, proper uptake of the nutrients by the host plant the disease resistance of the host plant against pathogen can also be improved (Mei and Flinn 2010).

19.3 Transmission of Endophytes

Endophytes which lives inside plants for a part or whole of their life cycle can be transmitted from parents to offspring or between two individual plants in a community. There are basically two types of transmission.

19.3.1 Vertical Transmission

Vertical transmission is a method of direct transfer from parents to offspring. Vertically transmitted fungal endophytes are typically considered clonal and transmit via fungal hyphae penetrating the embryo within the host's seeds (e.g., seed transmitting forms of *Epichloe*) (White et al. 1993; David and Manish 2011).

19.3.2 Horizontal Transmission

Horizontal transmission is among individual plants in a community. Reproduction through asexual or sexual spores leads to horizontal transmission, where endophytes may spread between plants in a population or community (Mariusz et al. 2014). Most fungi are terrestial, growing as hyphae and producing thick walled non motile spores. Spores are single-celled propagules which separate from the organism and can get dispersed. In both sexual and asexual transmission fungi produce spores that disperse among plants via air or through animals.

19.3.3 Transmission of Fungal Endophytes

A fungal hyphae is a long branching filamentous structure required for its vegetative growth. In vertical transmission endophyte is found in the embryo of infected seed. As the seed germinates, the endophyte grows into emerging leaf and finally grows up the stem and into the seed head of the reproductive plant.

19.3.4 Transmission of Bacterial Endophytes

Mostly bacterial endophytes are originated from rhizosphere or phyllosphere, however some are transmitted through seeds Rhizosphere is the narrow region of the soil that is directly influenced by root secretions and associated soil microorganisms. Endophytic bacteria have been isolated from monocotyledaneous [*Miscanthus giganteus*, *Iris pseudacorus*, *Phragmites australis*, *Lythrum salicaria* and *Cladium mariscus*], dicotyledonous [*Gosspium hirsutum*, *Cucumis sativis* and *Beta vulgaris*] to herbaceous plants. The root system secrets chemical exudates (flavinoids, amino acids, carbohydrates, organic acids). Bacterial endophytes attracted towards these chemicals and by chemotactic movement it gets attached to the root. Various molecules like cellulose, pectinase, superoxide dismutase etc. secreted by bacteria to colonize plants. Finally bacteria enter through roots and spread to other parts. Plant roots exude many organic compounds that stimulates microbial growth and can have major impact on the composition of rhizosphere microbiome (Grayston et al. 1998; Miethling et al. 2000).

Endophyte distribution within plants depends on a combination of ability to colonize and allocation of plant resources. Root endophytes often colonize and penetrate the epidermis at sites of lateral root emergence, below the root hair zone, and in root cracks (Dong et al. 2003; Compant et al. 2005; Zakria et al. 2007). These colonizers are capable of establishing populations both inter- and intracellularly (Hurek et al. 1994; Zakria et al. 2007). After initial colonization, some endophytes can move to other areas of the plant by entering the vascular tissues and spreading systemically (Compant et al. 2005; Zakria et al. 2007; David and Manish 2011). Using endophytes labeled with green-fluorescent-protein (GFP), David and Manish (2011) demonstrated the transport of the endophytes from seeds into plant roots and tissues, and endophytes injected into stems moved into the roots and rhizosphere, suggesting that there may be a continuing movement of organisms throughout the root microbiome.

19.4 Endophytes for Environment and Agriculture Sustainability

Endopytes which live a part or full of their life cycle inside the host plant and secrete wide variety of compounds essential for the growth of plants and protection from environmental conditions. Bioactive compounds are of enormous importance to plants as well as human are produced by endophytes. The phytocompounds secreted by the endophytes acts as biocontrol agent by exerting protective action of the plants from repeated grazing of herbivores on same plant. The compounds produced by endophytes are of immense importance as antibiotics, drugs or medicinal compounds useful for food industry and the compounds of high relevance in research. Endohytes are involved in phytoremediation, biodegradation, and nutrient cycling and thus reduces use of pesticides in agriculture and protect our environment from hazardioaus chemicals. Summarising the profound applications of endophytes and its impact on plants, human and environment, they have been proved to be a boon and not a ban and have potential to sustain the agriculture in a better way. Various application of endophytes by means of which it promote plant growth as well as sustain the environment and agriculture are given below.

19.5 Applications of Endophytes

The efficiency of endophytes to produce novel bioactive compounds with unique structures and bioactivities have proven to be helpful in agricultural sustainability, environmental conservation and ecotoxicological importance is currently being explored to their maximum extent. In lieu of a huge reservoir, these vast potentialities of secondary products, as well as their exploitation for agricultural and environmental benefaction, are scanty. The current development in endophytic research is mainly focused on evaluating endophytic microbial populations inhabiting plants, which enhances plant growth, disease resistance and the ability to tolerate or withstand the external environment. It can be assumed that these humble researches will emerge as a boon and would certainly leave a wide impact on agricultural science, environmental sustainability as well as for the welfare of mankind through their personification in technological advancement in phytological technologies. The impact of endophytes that enhances plant growth, disease resistance, agricultural and environmental sustainability as well as its ecotoxicological importance is shown in Table 19.4.

19.5.1 Nutrient Cycling

Nutrient cycling is an important process of balancing of the nutrients and make it available for each ecosystem components. The nutrients are made available by the degradation of biomass by saprophytes and recycled into the environment thus

Table 19.4 Impact of endophytes in agricultural and environmental sustainability

Function	Plants	Endophytic organism	Bioactive compounds/	Function	References
		Bacteria	Phytocompounds		
Phytostimulation	*Phragmites communis*, *Potamogeton crispus*, *Nymphaea tetragona* and *Najas marina*	*Aeromonas caviae*, *A. salmonicida*, *Bacillus niacin*, *B. aryabhattai*, *B. sphaericus*, *B. simplex*, *Deltia tsuruhatensis*, *Paenibacillus lautus*, *Enterobacter hormaechei*, *Flavobacterium oceanosedimentum*, *Klebsiella terrigena*, *Lactococcus lactis*, *Microbacterium flavescens*, *Paenibacillus barcinonensis*, *Pantoea agglomerans*, *Pseudomonas jessenii*, *P. taiwanensis*, *P. xanthomarina*, *Staphylococcus epidermidis*	Aromatic phytocompounds	Biodegradation (Chlorpyrifos Fenpropathrin Naphthalene Bifenthrin) and P-solubilization	Chen et al. (2012)
	Arabidopsis thaliana, *Platycladus orientalis*	*Phyllobacterium brassicacearum*, *B. subtilis*	Abscisic acid (ABA)	Decrement of leaf transpiration (*A. thaliana*), stomatal conductance (*P. orientalis*)	Bresson et al. (2013); Liu et al. (2013)
	Glycine max	*P. putida*,	Gibberellin	Growth promotion	Armada et al. (2014)
	Lavandula dentate	*B. thuringiensis*	Indole-3-acetic acid (IAA)		
	Vitis vinifera L cv.	*Bacillus*, *Mirococcus* and *Pantoa* genera	Ammonia, IAA, IAA-like molecules, siderophores and lytic enzyme	Phosphate solubilization and growth promotion	Baldan et al. (2015)

(continued)

Table 19.4 (continued)

Function	Plants	Endophytic organism	Bioactive compounds/ Phytocompounds	Function	References
		Bacteria			
	Z. mays L., *Solanum lycopersicum* L., *Citrullus lanatus* Thunb.	*B. cereus, B. licheniformis, B. pumilus, B. simplex, Bacillus* sp., *B. thuringiensis, Burkholderia gladioli, B. gladioli* pv. *Alliicola, Paracoccus halophilus, Stenotrophomonas maltophilia,* and *Stenotrophomonas* sp.	Growth hormones	Growth promotion	Xia et al. (2015)
	Phaseolus vulgaris L.	*Trichoderma polysporum, Trichoderma atroviridae* and *Trichoderma harzianum*	Proteolytic enzymes and phosphate solubilization factors	Phytoactivation and/or phytoinhibition	Pierre et al. (2016
	A. thaliana	*Kosakonia radicincitans*	20S proteasome alpha-3 subunit	Growth promotion	Witzel et al. (2017)
	Simmondsia chinensis	*Bacillus* sp., *Streptomyces* sp., *Methylobacterium aminovorans, Rhodococcus pyridinivorans* and *Oceanobacillus kimchi*	Growth hormones	Growth promotion	Perez-Rosales et al. (2017)
		Fungus			
	Suaeda japonica	*Penicillium* sp.	Bioactive gibberellins (GAs) and other inactive GAs	Enhanced seed germination	You et al. (2012)
	Panax ginseng	*Aspergillus* sp., *Cladosporium* sp., *Engyodontium* sp., *Fusarium* sp., *Penicillium* sp., *Plectosphaerella* sp., *Verticillium* sp. and *Ascomycete* sp.	Triterpenoid saponins and ginsenosides	Root growth and protection	Wu et al. (2013)
	Tinospora cordifolia and *Calotropis procera*	*Phoma* sp.	Growth hormones	Growth promotion	Kedar et al. (2014)

	Alstonia boonei, *Enantia chlorantha*, *Kigelia africana*	*Aspergilus niger, Macrophomina* sp., *Trichoderma* sp., *Penicillium citrinum*, *P. nigricans*	Secondary metabolites	Plant protection	Tolulope et al. (2015)
	Phaseolus vulgaris	*Trichoderma atroviridae*, *T. polysporum* and *T. harzianum*	Proteolytic enzymes, phosphate solubilization factors, active volatile and non-volatile metabolites	Enhanced seed germination	Pierre et al. (2016)
	Brassica campestris	*Mucor* sp.	Enhanced IAA, ACC deaminase and solubilize phosphate secretion	Plant growth promotion	Zahoor et al. (2017)
		Actinomycetes			
	Z. mays L.,	*Microbispora*, *Streptomyces* and *Streptosporangium*	–	Plant growth promotion and protection	de Araujo et al. (2000)
	Aloe vera, *Mentha arvensis* and *Ocimum sanctum*	*Streptomyces albosporus*, *S. aureus*, *S. cinereus, S. globisporus*, *S. griseofuscus*, *S. roseosporus*, *S. viridis*, *S. griseorubruviolaceus*, *Actinopolyspora* sp., *Micromonospora* sp., *Saccharopolyspora*sp.	Enhanced IAA, siderophore production	Plant growth promotion, phosphate solubilization	Gangwar et al. (2014)

(continued)

Table 19.4 (continued)

Function	Plants	Endophytic organism	Bioactive compounds/	Function	References
		Bacteria	Phytocompounds		
	Saccharum officinarum	*Streptomyces* sp. and *Micromonospora* sp.	Enhanced IAA, siderophore production	Plant growth promotion, and biosolubilizing activities of insoluble phosphate and leonardite,	Sinma et al. (2015)
	Triticumae stivum	*Streptomyces nobilis, S. kunmingensis, S. mutabilis,*	Enhanced IAA production, siderophore, and HCN and ACC deaminase production	Plant growth promotion, phosphate solubilization	Anwar et al. (2016)
	Solanum lycopersicum	*S. enissocaesilis, S. djakartensis*			
	Dracaena cochinchinensis Lour.	*Streptomyces* sp., *Nocardiopsis* sp., *Brevibacterium* sp., *Microbacterium sp.*, *Tsukamurella* sp., *Arthrobacter* sp., *Brachybacterium* sp., *Nocardia* sp., *Rhodococcus* sp., *Kocuria* sp., *Nocardioides* sp. and *Pseudonocardia* sp.	–	Plant growth promotion and protection	Salam et al. (2017)
Phyto-immobilization		**Bacteria**			
	Albizia lebbeck	*Bacillus* sp., *Rhizobium* sp., *Pseudomonas* sp., *Xanthomonas* sp., *Salinococcus* sp. and *Marinomonas* sp.	–	Phytoaccumulation of cromium	Manikandan et al. (2015)

	Sedum plumbizincicola	*Achromobacter piechaudii*	–	Resistance to cadmium, zinc and lead and phosphate solubilization	Ma et al. (2016)
	Leptochloa fusca	*Pantoea stewartii*, *Microbacterium arborescens* and *Enterobacter* sp.	–	Bioaugmentation, plant growth promotion, removal of both organic and inorganic pollutants	Ashraf et al. (2017)
		Fungus			
	Festuca arundinacea and *F. pratensis*	*Neotyphodium*	–	Enhanced cadmium tolerance and partial immobilization	Soleimani et al. (2010a)
	G. max L.	*Exophiala*, *Metarhizium*, *Promicromonospora* and *P. funiculosum*	–	Enhanced cupper tolerance	Khan and Lee (2013)
	Triticum	*Arbuscular Mycorrhizal Fungus* (AMF)	–	Enhanced cupper zinc, iron and manganese tolerance	Khan et al. (2014)
	Z. mays	*Exophiala pisciphila*	–	Enhanced cadmium tolerance	Wang et al. (2016)
	Oryza sativa	*Piriformospora indica*	–	Modulation of the anti-oxidative enzyme system, enhanced arsenic tolerance	Mohd et al. (2017)

(continued)

Table 19.4 (continued)

Function	Plants	Endophytic organism	Bioactive compounds/ Phytocompounds	Function	References
		Bacteria			
Phytotransformation		**Bacteria**			
	Achillea millefolium, Dactylis glomerata, Solidag ocanadensis and *Trifolium aureum*	*Microbacterium foliorum* and *Plantibacter flavus*	–	Petroleum tolerance and degredation	Lumactud et al. (2016)
Bioremediation	*Lolium multiflorum*	*Psudomonas*	Alkanes	Diesel degradation	Andria et al. (2009)
	Zea maize	*Burkholderia cepacia*	Phenol, toluene	Petroleum tolerance and degredation	Wang et al. (2010)
	Poplar	*Psudomonas putida*	Trichloroethylene	Co-contamination of nickel and trichloroethylene tolerance and degredation	Weyens et al. (2015)
	Alium macrostemon	*Enterobacter*	Pyrenes	Pyrenes tolerance and degredation	Sheng et al. (2008)

become accessible to living system. Endophytes have been reported to have important role in biodegradation of the litter of the inhabitating host plant (Muller et al. 2001; Kumaresan and Suryanarayanan 2002; Osono 2003, 2006; Korkama-Rajala et al. 2008; Fukasawa et al. 2009; Osono and Hirose 2009; Promputtha et al. 2010). The endophytes colonize with in the plants initially, during biodegradation of litter, and is facilitated through antagonistic interaction of the saprophytes (Thormann et al. 2003; Fryar et al. 2001; Terekhova and Semenova 2005).

19.5.2 Plant Growth Promotion by Endophytes

Endophytes orchestrate many ubiquitous roles in the plant growth by colonizing the internal tissue in almost every part of the plant (Santoyo et al. 2016). Moreover, these endophytes could efficiently execute this spectacular attribution of plant growth enhancement via certain important interrelated mechanisms like phytostimulation, phytoimmobilization, phytostabilization, phytotransformation, phytovolatilization, phytofilterastion, biofertilization and biocontrol (Conesa et al. 2012). Phytostimulation is a direct manifestation of plant growth promotion through the up-regulation of phytohormones directly or indirectly (Bloemberg and Lugtenberg 2001). Likewise, phytoimmobilization is another remediation technology where contaminants are effectually removed by the means of plants, the intake and sequential release into the soil is further followed by immobilization in either a mineral-emended soil or a geomate commonly known as mineral-containingmat (Arthur et al. 2005). Removal of the contaminants from the location is quite sophisticated therefore stabilization provides another important and effective mode of technical amelioration strategy adapted by nature for self-sustenance (Perez-de-Mora et al. 2011). Together phytoimmobilization and phytostabilization leads to phytotransformation or otherwise known as phytodegradation (sequestration and/or compartmentalization) phytocompounds involve the transformation or deterioration of intricate organic molecules into unadorned or simple molecular form that could be further absorbed into plant tissue in necessary (Etim 2012). The efficient degradation of contaminants or the xenobiotics, the subsequent release is sometimes regarded as phytoextraction/phytovolatilization where they are efficiently removed out of the plant system into the atmosphere (Limmer and Burken 2016). Phytofiltration is a fascinating scheme tailored by the plants where, with the help of certain surface anchorage phytocompounds, contaminants or nutrients are either absorbed/bound to the nonliving part of the plant (biosorptation), or to the root (Rhizofiltration), or to the seedlings (Blastofiltration) (Conesa et al. 2012). Though there is tremendous advantage in the usage of this phenomenon for the agricultural remediation but till date the advancement of research in the field is scanty. Biofertilazation on the other hand is one of the most growing, demanding, advanced and widely researched area of agriculture and environmental science. It is a technique where living microorganisms effectively aggrandize the nutrients and mineral uptake of the plants by means of establishing a symbiotic relationship with the plant and simultaneously sustains the dynamic nature of the soil (Roychowdhury et al. 2017). And finally, the protection of plant from pathogens before or after the harvest by the help of endophytic microbes is another growing sector of agricultural research where the endophytes produce a bioactive compound or induces the host plant to produce a specific and/or a group of phytoactive compound that inhibits the growth and survivability of the pathogens (Eljounaidi et al. 2016; Shahzad et al. 2017).

19.5.3 Bioremediation/Biodegradation

The method of removal of wastes and hazardous pollutants from the environment employing micro-organisms is known as bioremediation. Endophytes have the potential to breakdown complex compounds. The role of endophytes in bioremediation (Mastretta et al. 2009) resulted in improved biomass production under conditions of stress due to cadmium, thus can withstand higher cadmium concentration when compared to uninoculated plants. The impact of endophytes that enhances plant growth by bioremediation is shown in Table 19.4.

19.5.4 Phytostimulation

Up-regulation or down-regulation of phytocompounds specifically growth hormones and other bioactive compounds plays an indispensable role in plant growth and development. These phytohormones and bioactive compounds are generally self-induced within the plants in coordination with the time and certain abiotic factors which influences plant sustenance within that particular biosphere. But in some case the modulation of these phytohormones or other bioactive phytocompounds is induced by biotic factors living within the plant in a symbiotic relation as endophytes. These endophytes or endo-rhizosphers are currently been exploited for their capacity to produce exopolysaccharides, growth hormones (Indole-3-acetic acid), phyto-enzymes (1-Aminocyclopropane-1-carboxylic acid [ACC] deaminase), volatile compounds, osmoregulatory hormone and antioxidants (Table 19.4) (Vurukonda et al. 2016).

Phytohormones such as auxins, cytokinins and gibberellic acids are produced by bacterial endophytes (Bloemberg and Lugtenberg 2001). The most highly studied example of phytostimulation involves lowering plant hormone ethylene levels by 1-aminocyclopropane-1- carboxylate deaminase (ACC). Several endophytesthat release ACC deaminase have been shown to increase plant growth including *Arthrobacter* spp. and *Bacillus* spp. in pepper plants (*Capsicum annuum*) as well as *Pseudomonas putida* and *Rhodococcus* spp. in peas (*Pisum sativum*) (Sziderics et al. 2007; Belimov et al. 2001). Higher amounts of bioactive GA3. GA4 and GA7 that induced maximum plant growth in rice and soyabean varieties by *Cladosporium sphaerosperrmum,* a fungal endophyte isolated from the roots of *Glycine max* (L)

Merr (Humayun et al. 2009). Endophytes also facilitate uptake of important nutrients from soil, water and organic matter for growth and development of plants. *Phyllobacterium brassicacearum* and *Bacillus subtilis* have been evaluated for enhancing the abscisic acid (ABA) in *Arabidopsis thaliana* and *Platycladus orientalis* respectively. The up-regulation ABA hormone resulted in decrement of leaf transpiration in *A. thaliana* while increment stomatal conductance in *P. orientalis* (Bresson et al. 2013; Liu et al. 2013). *P. putida* embellished the growth of *Glycine max* by inducing the secretion of the hormone gibberellins. In *Lavandula dentate* the hormone indole-3-acetic acid (IAA) was enhanced by *B. thuringiensis*, which further improved K-contents and proline with simultaneously decrease glutathione reductase and ascorbate peroxidase (Armada et al. 2014). The production of IAA was also enhanced in *Triticum* by a group of microbes namely *Rhizobium leguminosarum*, *R. phaseoli* and *Mesorhizobium ciceri* (Hussain et al. 2014). In a different study a total of 377 bacterial endophytes were isolated form *Vitis vinifera* L cv. and were tested for plant growth promoting (PGP) abilities along with their effect of *A. thaliana* was also evaluated. It was found that the endophytes could able to promote ammonia production, phosphate solubilization, IAA and IAA-like molecules biosynthesis, siderophores and lytic enzyme secretion. Further twelve effective endophytes mainly belonging to *Bacillus*, *Mirococcus* and *Pantoa* genera were specifically selected for further studies (Baldan et al. 2015). Again 12 root endophytes from different strains of *Solanum lycopersicum* mostly belonging to species of *Pseudomonas, Rhizobium, Rhodococcus and Agrobacterium were isolated and analyzed for their PGP properties. Improvement in the production of organic acids, IAA, ACC deaminase and* siderophores was observed. The impact was further verified on *A. thaliana* root growth using vertical agar plate assay (Abbamondi et al. 2016). In a most recent study, 101 endophytic bacteria were isolated from *Simmondsia chinensis* root of which 8 endophytic bacteria belonging to *Bacillus* sp., *Streptomyces* sp., *Methylobacterium aminovorans*, *Rhodococcus pyridinivorans* and *Oceanobacillus kimchi* were the partial sequencing of 16S rDNA gene. Later it was found that of these 8 endophytic bacterial species only two endophytes namely *R. pyridinivorans* and *O. kimchi* showed efficient *PGP properties* (Perez-Rosales et al. 2017).

Endophytic fungus also imparts phytostimulation by modulating the production of bioactive phytocompounds within their host plant species. This phenomenon was shown in a study where 35 fungal endophytes were isolated from a halophyte Suaeda japonica and were identified by internal transcribed spacer (ITS). Their PGP ability was verified by their treatment with Waito-c rice seedling and moreover, their secondary metabolites such as bioactive gibberellins (GAs) and other inactive GAs were detected by HPLC and GC-MS SIM analysis (You et al. 2012). In *Panax ginseng* 38 strains of endophitic fungus belong to *Aspergillus*, *Cladosporium*, *Engyodontium*, *Fusarium*, *Penicillium*, *Plectosphaerella*, *Verticillium* and *Ascomycete* species were isolated and investigated for their capacity to produce saponins and ginsenosides which were detected by HPLC (Wu et al. 2013). *Phoma* species isolated from two medicinal plants namely *Tinospora cordifolia* and *Calotropis procera* was evaluated for its PGP activity on *Zea mays*, where it was observed that the fungus indeed showed the ability to promote the plant growth (Kedar et al. 2014). *Trichoderma* endophytes specifically *T. atroviridae*, *T. polysporum* and *T. harzianum* isolated from the roots of *Phaseolus vulgaris* could able to synthesize and release proteolytic enzymes and phosphate solubilization factors. Furthermore most of the active volatile and non-volatile metabolites that had certain stimulatory or inhibitory impact on *P. vulgaris* seed germination were perticularly released by *T. polysporum* and *T. harzianum* (Pierre et al. 2016). A recent study proves that IAA, ACC deaminase and solubilize phosphate secretion enhanced in the plant *Brassica campestris* by *Mucor* species which was identified by 18S and 28S rRNA ITS 1 and 4 sequence homology (Zahoor et al. 2017).

Apart from endophytic bacteria and fungus other microorganisms such as algae, acticomycetes, protozoa and cyanobacteria also mediate phytostimulation and contribute in host plant growth and development (Table 19.4) (Vejan et al. 2016). A study on *Nostoc* in crop plants like *Oryza sativa* and *Triticum* under axenic conditions improves the growth and development by modulating the hormones such as IAA (Hussain et al. 2015). Overall it can be stated that endophytic microbes are an effective means of phytostimulants.

19.5.5 Phytoimmobilization

In the remediation process where in the use of phytocomponents are at the verge of being highly exploited and could be further accelerated in conjugation with the symbiotic endophytes like bacteria, fungus and many other related microorganisms (Ma et al. 2011). Previously substantiation convey, bacterial endophytes as an effective tool that could be implemented in regulation of physiological changes including immobilization of osmolytes, certain micronutrients along with osmotic acclimatization, stabilization of membrane ion conductivity which is directly or indirectly linked with changes in the membrane phospholipid composition (Compant et al. 2005). Phytoimmobilization and transformation ultimately leads to increased ability to sustain or tolerate the stress induced by the abiotic component or factor. Plants like *Elsholtzia splendens* and *Commelina communis* which could tolerate or withstand copper concentration, have been reported to encounter an increase in dry weight of the root as well as the shoot when inoculated with endophytic bacteria in comparison to the control (Sun et al. 2010). On the other hand endophytes that are genetically modified convey an additional channel for phyto-associated neutralization of the contaminants and there by ameliorating the stress induced by these contaminants in the site (Divya and Kumar 2011). On the basis 16S rRNA gene sequencing phylogenic analysis reveled of about 118 isolates comprising of 17 proteobacterial genera in the chromium treated plant sample of *Albizia lebbeck*. The proteobacterial genera commonly comprised of *Bacillus*, *Rhizobium*, *Pseudomonas*, *Xanthomonas*, *Salinococcus* and *Marinomonas* (Manikandan et al. 2015). There are certain multi-metal resistance endophytes, one of such endophyte is *Achromobacter piechaudii* a bacterium isolated from *Sedum plumbizincicola* and characterized by morphological features, biochemical and 16S rDNA sequencing and phylogenic analysis. The bacterial strain portrayed increased level of resistance to cadmium, zinc and lead along with other salutary properties such as solubilization of phosphors and production of IAA and lastly presence of the strain significantly increased the availability of cadmium, zinc and lead in soil (Ma et al. 2016). Effective bioremediation can be generally achieved by constructed wetland vegetated with *Leptochloa fusca*. It was found that, when a combination of three endophytic bacteria namely *Pantoea stewartii*, *Microbacterium arborescens* and *Enterobacter* species used for bioaugmentation, the consortium of the bacteria could enhance the growth of *L. fusca* and simultaneously contributed to the removal of both organic and inorganic pollutants and also ameliorated the toxicity in the constructed wetland (Ashraf et al. 2017).

Fungus on the other side also plays an important role in phytoimmobilization, which is a direct representation of the capacity to tolerate and immobilize pollutants by concurrently increasing the amount of biomass (Sudha et al. 2016). Currently studies are focused on evaluating the capacity of endophytic fungus to induce metal tolerance and partial immobilization in plants. One such example can be cited for the endophytic fungus *Neotyphodium* that enhanced cadmium tolerance and its partial immobilization in infected plants named *Festuca arundinacea* and *F. pratensis*. Additionally it was observed that photochemical efficacy of photosystem II increased, indicating the reduction of cadmium stress (Soleimani et al. 2010a, b). Species of *Exophiala*, *Metarhizium*, *Promicromonospora* and *Pencillium* also showed increased tolerance to metal stress of copper and cadmium. But it was *Pencillium funiculosum* that highly ameliorated biomass yield, chlorophyll and total protein contents in its host plant *Glycine max* L. under Cu stress (Khan and Lee 2013). *Arbuscular Mycorrhizal Fungus* (AMF) is a group of mycorrhizal fungus that could penetrate roots through cortical cells of the vascular plants. In *Triticum* the grain and shoot yield was increased under different concentration of Zn, Cu, Fe and Mn when inoculated with AMF compared to un-inoculated *Triticum* (Khan et al. 2014). Dark Septate Endophytes (DSEs) are asexual chlamydospores ascomycetous fungi that inhabits within the living plant root in a symbiotic relationship. One of such endophyte is *Exophiala pisciphila*, which regulates physiological response in *Z. mays* under soil cadmium stress. The mechanism is not clear how such tolerance is achieved by *Z. mays* by the help DSE (Wang et al. 2016). The immobilization of soluble arsenic is another interesting feature displayed by these endophytic fungus. *Piriformospora indica* is one of such fungus the colony of which was isolated from *Oryza sativa* root and was evaluated for its impact on the plant. Primarily it was found that hyper-colonization of the fungus that ultimately alleviated biomass density, root amelioration, number of chlorophyll and stabilization of oxidoredox status by the modulation of the antioxidative enzyme system which finally protects the plants photo-system under stress induced due to hyper-concentration of arsenic (Mohd et al. 2017). The impact of endophytes that enhances plant growth by phytoimmobilization is shown in Table 19.4.

19.5.6 Phyto-Transformation

In nature, plants commonly perform this phenomenon to neutralize soil pollutants and furthermore this property could be enhanced by the help of endophytes. Mostly endophytic bacteria, fungus, actinomycites and up to some extent algae are widely being explored for their contribution to plant life, in phytotransformation (Shakoor et al. 2017). Presently the persistent organic pollutants like pesticides, explosives, industrial byproducts and other xenobiotics are the major source of abiotic stress.

The role of bacterial endophytes in the metabolism of toxic xenobiotics has already been described successfully in the phytotransformation of toluene and other organic pollutants into intermediate metabolites that could efficiently be used both by the plant and associated interacting micro-organisms (Aken et al. 2010). *Burkholderia fungorum*, a bacterial strain generally present in poplar root tissue, was isolated from oil refinery discharge that had the capability to transform dibenzothiophene, phenanthrene, naphthalene, fluorine and their removal (Andreolli et al. 2013). *Prosopis juliflora* was found to harborcertain endophytic bacterial species such as *Aerococcus*, *Bacillus* and *Staphylococcus* which was confirmed by 16S rRNA gene sequencing phylogenic analysis. The ability to ionize toxic heavy metals such as chromium, cadmium, copper, lead and zinc were also evaluated along with the capacity to promote plant growth was confirmed when inoculated in *Lolium multiflorum* L. (Khan et al. 2015). Four different plants, *Achillea millefolium*, *Dactylis glomerata*, *Solidag ocanadensis* and *Trifolium aureum* could able to grow bounteously in the soil contaminated with petroleum. It was later found that there were about 190 endophytic bacterial species that support these plants in phytotransformation of hydrocarbons. 16S rDNA sequencing showed the presence of *Microbacterium foliorum* and *Plantibacter flavus* in all the plants (Lumactud et al. 2016).

Endophytic fungus also contributes in phytotransformation of contaminants like bacterial endophytes and improves plant's fitness to withstand environmental stress. Alleviation of stress induced due to salt accumulation was achieved by a GAs producing basidiomycetous endophytic fungus *Porostereum spadiceum* (Hamayun et al. 2017). There are many other endophytic fungus that performs similar functions like *Penicillium minioluteum*, *P. funiculosum*, *Metarhizium anisopliae*, *Beauveria bassiana*, *Mucor* sp. etc. (Khan et al. 2011a, b; Greenfield et al. 2016; Zahoor et al. 2017). The impact of endophytes that enhances plant growth by phytotransformation is shown in Table 19.4.

19.5.7 Phytovolatilization

The phenomenon through which a plant can completely remove contaminants from the site and release them into atmosphere in a volatile form can be termed as phytovolatilization. This phenomenon is highly exploited in phytotechnology programs in standardization plant growth and survivability (Schiavon and Pilon-Smits 2017). Plants simultaneously interacts with diversified classes of chemical compounds including both organic and inorganic through either direct or indirect phytovolatilization (Limmer and Burken 2016; Schiavon and Pilon-Smits 2017). Volatilization of metals such as As, Se, Hg and organic compounds such as petroleum hydrocarbons are now archived by certain endophytic bacterium. Some of such endophytic bacterium includes *Pseudomonas aeruginosa*, *P. putida*, *P. stutzeri*, *Rhodococcus wratislaviensis*, *Acinetobacter* sp., *Burkholderia* sp., *Gordonia* sp., *Dietzia* sp., *Gordonia* sp., *Mycobacterium* sp., *Nocardioides* sp., *Novosphingobium* sp., *Ochrobactrum* sp., *Polaromonas* sp., *Rhodococcus* sp., *Sphingomonas* sp. etc. which are currently being explored for their ability to degrade organic compounds (Gkorezis et al. 2017). Certain endophytic fungus are also involved in phytovolatilization of organic and inorganic compounds, the most common of them are *Alternaria alternate*, *Cladosporium cladosporioides*, *Cochliobolus sativus*, *Fusarium oxysporum*, *Muscodor yucatanensis*, *Talaromyces wortmannii*, *Trichoderma viride* etc (Ningxiao et al. 2016). The impact of endophytes that enhances plant growth environmental sustainability by phytovolatilization is shown in Table 19.4.

19.5.8 Biofertilization

As previously mentioned biofertilization is a technique where living microorganisms in co-ordination with nutrients and minerals present in the surrounding, enhances the absorption properties of the plant without altering dynamic nature of the soil (Roychowdhury et al. 2017). The promotion of plant growth by increasing the accessibility or supply of major nutrients is termed biofertilization (Bashan 1998). The biofertilizers basically supply nitrogen, phosphorous, potassium along with certain ion scavenging molecules like siderophores and exopolysaccharides (Saha et al. 2016; Sanlibaba and Çakmak 2016). A well-studied form of biofertilization is nitrogen fixation, which is the conversion of atmospheric nitrogen to ammonia. Several plant growth promoting bacterial endophytes have been extensively evaluated for their efficiency to fix nitrogen including *Azospirillum*sp (Hill and Crossman 1983), *Pantoea agglomerans* (Verma et al. 2001) and *Azoarcus* sp. (Hurek et al. 2002).

In the present scenario much significance is given to biofertilizers in comparison to the conventional fertilizers due to its ecofriendly nature. Generally these biofertilizers can be differentiated into azotobacter, phosphate solubilizers or rhyzobium on the basis of the major type of microorganisms they harbor and/or their solubilization property. Almost all the microorganisms in these ecofriendly fertilizers are symbiotic in nature moreover these organisms also act as endophytes in some cases.

Biofertilizers that harbors azotobacter that is a genus of motile oval bacteria belongs to a family of *Azotobacteriaceae* which are aerobic and heterotrophic in nature includes bacterial species like *A. beijerinckii*, *A. chroococcum*, *A. insignis*, *A.*

macrocytogenes and *A. vinelandii*. All of these bacterial endophytes are commonly found in crop plants like rice, wheat, maize etc. and are already have been proven to be efficient in crop improvement as well as sustenance of fertility of soil (Dursun et al. 2010; Roychowdhury et al. 2017). Inorganic and organic compounds solubilization is a distinct property bestowed by these biofertilizers is remarkable. Most common of these solubilization properties is phosphate solubilization where the bacterial species such as *Azospirillum* sp., *Pseudomonas* sp., *Bacillus* sp., *Proteous* sp. along with certain fungal species like *Aspergillus flavus*, *A. niger*. *A. ochraceus*, *A. sydawi*, *A. terreus*, *A. versicolor*, *Chaetomium globosum*, *Fusarium* sp., *Mucor* sp., *Penicillium* sp. etc. are involved in solubilization of insoluble inorganic phosphate such as hydroxyapatite, rock phospahate tricalcium phosphate and dicalcium phosphate, into ions (Selvi et al. 2017; Roychowdhury et al. 2017). Lastly, biofertilizers, composed rhizobacterial species that could fix atmospheric nitrogen to the soil or to the root nodules are under vigorous study. Common rhizobacterial species includes *Bacillus megaterium*, *Bradyrhizobium japonicum*, *Rhizobium*, *Bradyrhizobium*, *Stenotrophomonas rhizophila* etc. (Rajeswari et al. 2017; Tarekegn and Kibret 2017).

19.5.9 Biocontrol

The promotion of plant growth through protection from phytopathogens is known as biocontrol. The use of synthetic chemicals for controlling plant diseases is the major risk factor raised against ecological and environmental niches. The search for an ecofriendly way to fight against these diseases is the major public and research concern. This concern for a sustainable means has paved the way for an alternative approach that is, the potential use of endophytes, as biocontrol (Eljounaidi et al. 2016). Some of the recently evaluated endophytic bacterial species such as *Achromobacter piechaudii*, *Enterobacter cloacae*, *Erwinia persicina*, *Pantoea agglomerans*, *P. fluorescens*, *Serratia plymuthica*, *S. marcescens*, *B. subtilis*, *S. iquefaciens*, *B. amyloliquefaciens*, *Paenibacillus* sp., *Stenotrophomonas* sp., *Enterobacter* sp. etc. have shown promising results in against various plant diseases such as Verticillium, Fusarium, Eggplant and Verticillium wilt etc. (Eljounaidi et al. 2016; Shahzad et al. 2017; Egamberdieva et al. 2017). The main and common mechanism exploited by thes endophytes is the elevation of certain growth hormones, induced systemic resistance, signal interference, production of antimicrobial proteins, siderophores, antibiotics and inhibitory compounds (Eljounaidi et al. 2016).

Siderophore produced by a microorganism can bind iron with high specificity and affinity making iron unavilable for other microorganism; thereby limiting their growth. It may stimulate plant growth directly by increasing availability of iron in the soil sorrounding the roots or indirectly by competitively inhibiting the growth of plant pathogens with less efficient iron uptake system..

Siderophores, such as pyochelin and salicylic acid, chelate iron and can indirectly contribute to disease control by competing with phytopathogens for trace metals (Duffy and Defago 1999). Antimicrobial metabolites produced by plant growth promoting bacterial endophytes such as 2, 4 diacetylphloroglucinol (DAPG) can enhance disease suppression in plants. Endophytic microorganisms are regarded as an effective biocontrol agent, alternative to chemical control. *Beauveria bassiana* an endophytic fungi, was reported to control the borer insects in coffee seedlings, thus acts as an entomopathogen (Posada and Vega 2006) and sorghum (Tefera and Vidal 2009). The fungal pathogen *Botrytis cinerea* causes severe rotting in tomatoes during storage and shelf life can be well antogonised by endophytic bacteria *Bacillus subtilis* isolated from *Spaeranskiatuberculata* (Bge.) Baill (Wang et al. 2009). A new strain of *Burkholderia pyrrocinia* and *B. cepacia*, were identified as potential biocontrol agent against poplar canker (Ren et al. 2011). Bioactive compounds from endophytes and their use against pathogenic micro-organisms is shown in Table 19.5.

Other endophytic microorganisms like fungus and actinomycites have also been identified to play an indispensible role as a biocontrol in various cases. Some such endophytic fungus with biocontrol activity against woolly aphid, fusarium wilt, against various soil born bacterial pathogen and pests are *Gibberella fujikuroi*, *Aspergillus tubingensis*, *A. flavus*, *Trichoderma koningiopsis*, *Galactomyces geotrichum*, *P. simplicissimum*, *P. ochrochloron*, *Eupenicillium javanicum* (Potshangbam et al. 2017). Some endophytes and their use against pathogenic microorganisms are shown in Table 19.5.

19.6 Impact of Endophytes on Bioactive Compounds of Host Plant

Advancement proteomics, genomics studies have revolutionized the present prospective of evaluation biomolecules and their regulation, modulation as well as their impact on or within the host, both in active and in-active state. These advanced studies includes certain sophisticated instrumentations and certain gel- based approach such as 2D Gel Electrophoresis coupled with Edman Sequencing, Matrix- Assisted Laser Desorption/ionisation Time Of Flight (MALDI- TOF), Surface- Enhanced Laser Desorption/Ionization Time of Flight (MS), Electrospray Ionization (ESI)-MS/MS analysis, Multidimensional Protein Identification Technology (MudPIT), Isobaric Tags for Relative and Absolute Quantification (iTRAQ), Isotope- Coded Affinity Tag (ICAT) and Tendem Mass Tags (TMT) (Ahsan et al. 2009; Hu et al. 2015). The bioactive compounds of the host plants such as enzymes are one of the most significant phytocompounds that regulates almost every aspect of plant life. Much work has been done on the evaluation of enzymatic activity and their up or down-regulation based on the impact of the endophytic microbes the host plant harbor (Castro et al. 2014). Form these evaluation it can be speculated that all phytoenzymes that have a direct or indirect heterogeneous impact on plants growth and

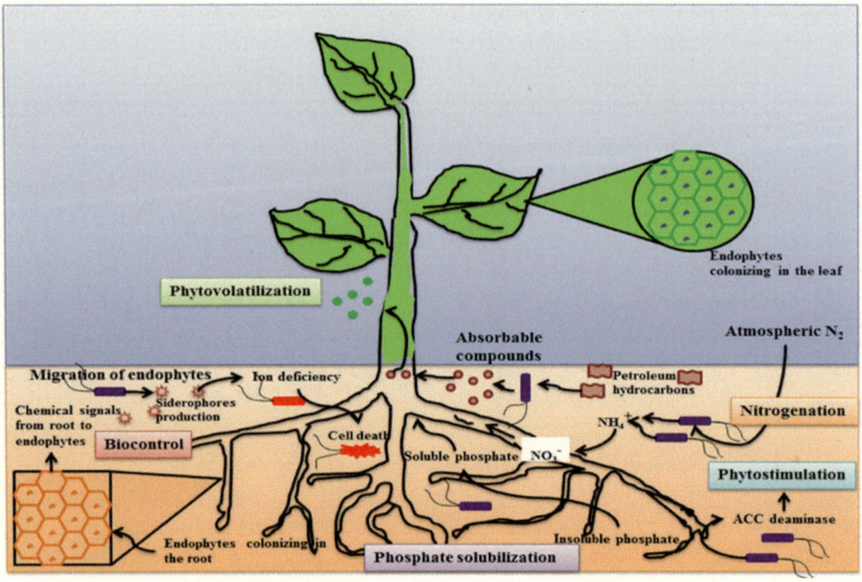

Fig. 19.1 Mechanism, interaction and beneficial effect, of endophytes

survivable under biotic and/or abiotic stress possesses four basic enzymatic activity that is proteolytic, amilolytic/endoglucanase, lipolytic and esterasic (Table 19.5) (Carrim et al. 2006; Castro et al. 2014). The mechanism, interaction and beneficial effect, of endophytes is summarized in Fig. 19.1.

19.7 Extracellular Enzymes from Endophytes

Endophytes exhibit a complex network interact in association with the host plants were widely studied as inexhaustible sources of new bioactive natural products. Enzymes of the endophytes degrade the polysaccharides available in the host plants. Fungal strains, isolated from various plants of medicinal importance viz., *Alpinia calcarata*, *Bixa orellana*, *Calophyiium inophyllum* and *Catharanthus roseus* have been reported to produce enzymes such as amylase, cellulose, laccase, lipase, pectinase, xylanase, −1, 4- glucan, phosphotases and proteinase e, −1, 4- glucan, without lyase, phosphotases and proteinase and protease extracellularly. The hydrolytic enzymes produced through endophytes differs from species to species and depends on the interactions with host and their ecological factors (Sunitha et al. 2013).

Table 19.5 Source of bioactive compounds from endophytes and their use against pathogenic microorganisms

Source of endophytes	Bioactive compounds	Cure against pathogen	Mode of endophyte transmission of the pathogen	Reference
Boesenbergia rotunda Streptomyces coelicolor	Munumbicins	*Escherichia coli*	Ground meats, raw or under pasteurized milk	Golinska et al. (2015)
Chloridium sp.	Javanicin	*Pseudomonas* sp.	Contaminated water or surgical instruments	Jalgaonwala et al. (2011)
Cytonaema sp.	Cytonic acids A and B	*Human cytomegalovirus*	Shellfish, berries or contaminated water	Bhardwaj and Agrawal (2014)
Diaporthe helianthi	Fabatin, tyrosol	*Enterococcus hirae*	Nosocomial infection	Godstime et al. (2014)
Fusarium proliferatum	Kakadumyci, beauvericin	*Listeria monocytogenes*	Raw or under pasteurized milk	Golinska et al. (2015)
Streptomyces hygroscopicus	Clethramycin	*Cryptococcus neoformans*	Lettuce harvested from tropical regions	
Streptomyces sp.	Kakadumycin A, hypericin	*Shigella* sp.	Contaminated food, water a	Golinska et al. (2015); Joseph and Priya (2011)
Streptomyces tsusimaensis	Valinomycin	*Corona virus*	Food or water contaminated with infected fecal matter	Alvin et al. (2014)
Thottea grandiflora	Streptomycin	*Bacillus cereus*	Uncooked meat and raw milk	Joseph and Priya (2011)
Xylaria sp. *Ginkgo biloba*	Sordaricin 7	*Yersinia enterocolitica*	Swine meat and meat products, milk and dairy products	Joseph and Priya (2011)
Fusarium proliferatum	amino-4-methylcoumarin, Beauvericin	*Yersinia enterocolitica*	Swine meat and meat products, milk and dairy products	Joseph and Priya (2011)

19.8 Conclusion

Endophytes are capable of synthesizing number of bioactive metabolites which are mostly used as effective drugs against various diseases and are having profound impact on agricultural and environmental sustainability. These secondary metabolites were categorised into various functional groups, alkaloids, benzopyranones, flavonoids, phenolicsacids, quinones, steroids, saponins, tannins, tertaralones, xanthones and many others (Schulz et al. 2002; Strobel and Daisy 2003; Jalgaonwala et al. 2011; Pimentel et al. 2011; Godstime et al. 2014). Source of bioactive compounds from endophytes is presented in Table 19.5.

References

Abbamondi GR, Tommonaro G, Weyens N (2016) Plant growth-promoting effects of rhizospheric and endophytic bacteria associated with different tomato cultivars and new tomato hybrids. Chem Biol Technol Agric 3(1):1–10

Abdurakhmonov IY (2016) Genomics era for plants and crop species – advances made and needed tasks ahead. In: Abdurakhmonov IY plant genomics. InTech, Croatia, pp 3–26

Ahsan N, Renaut J, Komatsu S (2009) Recent developments in the application of proteomics to the analysis of plant responses to heavy metals. Proteomics 9:2602–2621

Aken BV, Correa PA, Schnoor JL (2010) Phytoremediation of polychlorinated biphenyls: new trends and promises. Environ Sci Technol 44(8):2767–2776

Alvin A, Kristin I, Miller B, Neilan A (2014) Exploring the potential of endophytes from medicinal plants as sources of antimycobacterial compounds. Microbiol Res 169:483–495

Allan EJ, Hoischen C, Gumpert J (2009) Bacterial L-Forms. Adv Appl microbial 68:1–39

Andreolli M, Lampis S, Poli M et al (2013) Endophytic *Burkholderia fungorum* DBT1 can improve phytoremediation efficiency of polycyclic aromatic hydrocarbons. Chemosphere 92:688–694

Andrew JF, Katelyn TE, Joseph S, Forward E et al (2013) *Macroalgal endophytes from the* atlantic coast of Canada: a potential source of antibiotic natural products? Microorganisms 1(1):175–187

Andria V, Reichenauer TG, Sessitsch A (2009) Expression of alkane monooxygenase (alkB) genes by plant-associated bacteria in the rhizosphere and endosphere of Italian ryegrass (*Lolium multiflorum* L.) grown in diesel contaminated soil. Environ Pollut 157:3347–3350

Anwar S, Ali B, Sajid I (2016) Screening of rhizospheric actinomycetes for various *in-vitro* and *in-vivo* Plant Growth Promoting (PGP) traits and for agroactive compounds. Front Microbiol 7(1334):1–11

Armada E, Roldan A, Azcon R (2014) Differential activity of autochthonousbacteria in controlling drought stress in native Lavandula and Salvia plants species under drought conditions in natural arid soil. Microb Ecol 67:410–420

Arnold AE (2007) Understanding the diversity of foliar fungal endophytes: progress, challenges, and frontiers. Fungal Biol Rev 21:51–66

Arthur EL, Rice PJ, Rice PJ et al (2005) Phytoremediation—an overview. Crit Rev Plant Sci 24(2):109–122

Ashraf S, Afzal M, Naveed M (2017) Endophytic bacteria enhance remediation of tannery effluent in constructed wetlands vegetated with *Leptochloa fusca*. Int J Phytoremediation. https://doi.org/10.1080/15226514.2017.1337072

Baldan E, Nigris S, Romualdi C et al (2015) Beneficial bacteria isolated from grapevine inner tissues shape *Arabidopsis thaliana* roots. PLoS One 10(10):e0140252

Bashan Y (1998) Inoculants of plant growth-promoting bacteria for use in agriculture. Science 16:729–770

Belimov AA, Safronova VI, Sergeyeva TA et al (2001) Characterization of plant growth promoting rhizobacteria isolated from polluted soils and containing 1-aminocyclopropane-1-carboxylate deaminase. Can J Microbiol 47:642–652

Bhardwaj A, Agrawal P (2014) A review fungal endophytes: as a store house of bioactive compound. World. J Pharm Pharm Sci 3:228–237

Bischoff JF, White JF Jr (2005) Evolutionary development of the *Clavicipitaceae*. In: Dighton J, White JF, Oudemans P (eds) The fungal community: its organization and role in the ecosystem. Taylor & Francis, Boca Rato, pp 505–518

Bloemberg GV, Lugtenberg BJJ (2001) Molecular basis of Plant growth promotion and biocontrol by rhizobacteria. Curr Opin Plant Biol 4:343–350

Bresson J, Varoquaux F, Bontpart T et al (2013) The P.G.P.R., strain *Phyllobacterium brassicacearum* STM196 induces a reproductive delay and physiological changes that result in improved drought tolerance in Arabidopsis. New Phytol 200:558–569

Carrim AJI, Barbosa EC, Vieira JDG (2006) Enzymatic Activity of Endophytic Bacterial Isolates of *Jacaranda decurrens* Cham. (Carobinha-do-campo). Braz Arch Biol Technol 49(3):353–359

Castro RA, Quecine MC, Lacava PT et al (2014) Isolation and enzyme bioprospection of endophytic bacteria associated with plants of Brazilian mangrove ecosystem. SpringerPlus 3:382

Chaudhary HS, Soni B, Shrivastava AR, Shrivastava S (2013) Diversity and versatility of actinomycites and its role in antibiotic production. Int J Pharm Sci 3:S83–S94

Chen W-M, Tang Y-Q, Mori K, X-L W (2012) Distribution of culturable endophytic bacteria in aquatic plants and their potential for bioremediation in polluted waters. Aquat Biol 15:99–110

Compant S, Reiter B, Sessitsch A et al (2005) Endophytic colonization of Vitis vinifera L. by plant growth-promoting bacterium *Burkholderia* sp. strain PsJN. Appl Environ Microbiol 71:1685–1693

Conesa HM, Evangelou MWH, Robinson BH, Schulin R (2012) A critical view of current state of phytotechnologies to remediate soils: still a promising tool? Sci World J 2012:1–10

David J-M, Manish NR (2011) Conservation and diversity of seed associated endophytes in Zea across boundaries of evolution, ethnography and ecology. PLoS One 6(6):e20396

De Araujo JM, Da Silva AC, Azevedo JL (2000) Isolation of endophytic actinomycetes from roots and leaves of maize (*Zea mays* L.) Braz Arch Biol Technol 43. https://doi.org/10.1590/S1516-89132000000400016

Divya B, Kumar MD (2011) Plant-Microbe interaction with enhanced bioremediation. Res J Biotechnol 6:72–79

Dong Y, Iniguez AL, Triplett EW (2003) Quantitative assessments of the host range and strain specifi city of endophytic colonization by *Klebsiella pneumonia* 342. Plant and Soil 257:49–59

Duffy BK, Defago G (1999) Environmental factors modulating antibiotic and siderophore biosynthesis by Pseudomonas fluorescens biocontrol strains. Appl Environ Microbiol 65:2429–2438

Dursun A, Ekinci M, Donmez MF (2010) Effects of foliar application of plant growth promoting bacterium on chemical contents, yield and growth of tomato (*Lycopersicon esculentum* L.) and cucumber (*Cucumis sativus* L.) Pak J Bot 42(5):3349–3356

Egamberdieva D, Wirth S, Behrendt U et al (2017) Antimicrobial activity of medicinal plants correlates with the proportion of antagonistic endophytes. Front Microbiol 8(199):1–11

Eljounaidi K, Lee SK, Bae H (2016) Bacterial endophytes as potential biocontrol agents of vascular wilt diseases – review and future prospects. Bio Cont 103:62–68

Etim EE (2012) Phytoremediation and its mechanisms: a review. Int J Environ Bioenergy 2(3):120–136

Fryar SC, Yuen TK, Hyde KD, Hodgkiss IJ (2001) The influence of competition between tropical fungi on wood colonization in streams. Microb Ecol 41(3):245–251

Fukasawa Y, Osono T, Takeda H (2009) Effects of attack of saprobic fungi on twig litter decomposition by endophytic fungi. Ecol Res 24(5):1067–1073

Gangwar M, Dogra S, Gupta UP, Kharwar RN (2014) Diversity and biopotential of endophytic actinomycetes from three medicinal plants in India. Afr J Microbiol Res 8(2):184–191

Gkorezis P, Daghio M, Franzetti A et al (2017) The interaction between plants and bacteria in the remediation of petroleum hydrocarbons: an environmental perspective. Front Microbiol 7(1836):1–27

Godstime OC, Enwa FO, Augustina JO, Christopher EO (2014) Mechanisms of antimicrobial actions of phytochemicals against enteric pathogens – a review. J Pharm Chem Biol Sci 2:77–85

Golinska P, Wypij M, Agarkar GM et al (2015) Endophytic actinobacteria of medicinal plants: diversity and bioactivity. Antonie Van Leeuwenhoek 108:267–289

Grayston SJ, Ang SW, Ampbell CDC, Dwards ACE (1998) Selective influence of plant species on microbial diversity in the rhizosphere. Soil Biol Biochem 30:369–378

Greenfield M, Gómez-Jimenez MI, Ortiz V (2016) *Beauveria bassiana* and *Metarhizium anisopliae* endophytically colonize cassava roots following soil drench inoculation. Biol Control 95:40–48

Hamayun M, Hussain A, Khan SA (2017) Gibberellins producing endophytic fungus *Porostereum spadiceum* AGH786 rescues growth of salt affected soybean. Front Microbiol 8(686):1–13

Hill A, Crossman SM (1983) Characterization of N_2 -fixing bacteria associated with sweet potato roots. Can J Microbiol 29:860–862

Hu J, Rampitsch C, Bykova NV (2015) Advances in plant proteomics toward improvement of crop productivity and stress resistance. Front Plant Sci 6(209):1–15

Humayun M, Afzal Khan S, Ahmad N et al (2009) *Cladosporium sphaerospermum* as a new plant growth-promoting endophyte from the roots of *Glycine max* L. Merr. World J Microbiol Biotechnol 25(4):627–632

Hurek T, Reinhold-Hurek B, Montagu MV, Kellenberger E (1994) Root colonization and systemic spreading of *Azoarcus*strain BH72 in grasses. J Bacteriol 176:1913–1923

Hurek T, Handley LL, Reinhold-Hurek B, Piche Y (2002) Azoarcusgrass endophytes contribute fixed nitrogen to the plant in an un-culturable state. Mol Plant Microbe Interact 15:233–242

Hussain MB, Zahir ZA, Asghar HN, Asghar M (2014) Exopolysaccharides producing rhizobia ameliorate drought stress in wheat. Int J Agric Biol 16:3–13

Hussain A, Shah ST, Rahman H (2015) Effect of IAA on in vitro growth and colonization of Nostoc in plant roots. Front Plant Sci 6(46):1–9

Jalgaonwala RE, Mohite BV, Mahajan RT (2011) Natural products from plant associated endophytic fungi. J Microbiol Biotechnol Res 1:21–32

Jia M, Chen L, Xin HL et al (2016) A friendly relationship between endophytic fungi and medicinal plants: a systematic review. Front Microbiol 7:906

Joseph B, Priya RM (2011) Bioactive compounds from endophytes and their potential in pharmaceutical effect: a review. Am J Biochem Mol Biol 1:291–309

Kedar A, Rathod D, Yadav A (2014) Endophytic *Phoma* sp. isolated from medicinal plants promote the growth of *Zea mays*. Nusantara bioscie 6:132–139

Khan AL, Lee I-J (2013) Endophytic *Penicillium funiculosum* LHL06 secretes gibberellin that reprograms *Glycine max* L. growth during copper stress. BMC Plant Biol 13:86

Khan AL, Hamayun M, Ahmad N et al (2011a) Salinity stress resistance offered by endophytic fungal interaction between *Penicillium minioluteum* LHL09 and *Glycine max* L. J Microbiol Biotechnol 21:893–902

Khan AL, Hamayun M, Kim Y-H et al (2011b) Ameliorative symbiosis of endophyte (*Penicillium funiculosum* LHL06) under salt stress elevated plant growth of *Glycine max* L. Plant Physiol Biochem 49:852e861

Khan A, Sharif M, Ali A et al (2014) Potential of AM fungi in phytoremediation of heavy metals and effect on yield of wheat crop. Am J Plant Sci 5:1578–1586

Khan MU, Sessitsch A, Harris M et al (2015) Cr-resistant rhizo and endophytic bacteria associated with *Prosopis juliflora* and their potential as phytoremediation enhancing agents in metal degraded soils. Front Plant Sci 5(755):1–10

Korkama-Rajala T, Muller MM, Pennanen T (2008) Decomposition and fungi of needle litter from slow- and fast-growing Norway spruce (*Picea abies*) clones. Microb Ecol 56(1):76–89

Kumaresan V, Suryanarayanan TS (2002) Endophyte assemblages in young, mature and senescent leaves of Rhizophora apiculata: evidence for the role of endophytes in mangrove litter degradation. Fungal Divers 9:81–91

Limmer M, Burken J (2016) Phytovolatilization of organic contaminants. Environ Sci Technol 50(13):6632–6643

Liu F, Xing S, Ma H et al (2013) Cytokinin producing, plant growth promoting rhizobacteria that confer resistance to drought stress in Platycladusorientalis container seedlings. Appl Microbiol Biotechnol 97:9155–9164

Lumactud R, Shen SY, Lau M, Fulthorpe R (2016) Bacterial endophytes isolated from plants in natural oil seep soils with chronic hydrocarbon contamination. Front Microbiol 7(755):1–10

Ma Y, Prasad MNV, Rajkumar M, Freitas H (2011) Plant growth promoting rhizobacteria and endophytes accelerate phytoremediation of metalliferous soils. Biotechnol Adv 29:248–258

Ma Y, Zhang C, Rui S, Oliveira RS et al (2016) Bioaugmentation with endophytic bacterium E6S homologous to *Achromobacter piechaudii* enhances metal rhizoaccumulation in host *Sedum plumbizincicola*. Front Plant Sci 7(75):1–9

Manikandan M, Kannan V, Mendoza OH et al (2015) The contribution of endophytic bacteria to *Albizia lebbeck*-mediated phytoremediation of tannery effluent contaminated soil. Int J Phytoremediation 18(1):77–86

Mariusz T, Marshall SB, James FW (2014) Epichloo spp. associated with grasses: new insights on life cycles, dissemination and evolution. Mycologia 106(2):181–201

Mastretta C, Taghavi S, van der Lelie D et al (2009) Endophyticbacteria from seeds of Nicotiana tabacum can reduce cadmiumphytotoxicity. Int J Phytoremediation 11:251–267

Mei C, Flinn BS (2010) The use of beneficial microbial endophytes for plant biomass and stress tolerance improvement. Recent Pat Biotechnol 4(1):81–95

Miethling R, Wieland G, Backhaus H, Tebbe C (2000) Variation of microbial rhizosphere communities in response to crop species, soil origin, and inoculation with *Sinorhizobium meliloti* L33. Microb Ecol 41:43–56

Mohd S, Shukla J, Kushwaha AS et al (2017) Endophytic fungi *Piriformospora indica* mediated protection of host from arsenic toxicity. Front Microbiol 8(754):1–14

Muller MM, Valjakka R, Suokko A, Hantula J (2001) Diversity of endophytic fungi of single Norway spruce needles and their role as pioneer decomposers. Microb Ecol 10(7):1801–1810

Ningxiao LI, Alfiky A, Vaughan MM, Kang S (2016) Stop and smell the fungi: Fungal volatile metabolites are overlooked signals involved in fungal interaction with plants. Fungal Biol Rev 30(3):134–144

Osono T (2003) Effects of prior decomposition of beech leaf litter by phyllosphere fungi on substrate utilization by fungal decomposers. Mycoscience 44(1):41–45

Osono T (2006) Role of phyllosphere fungi of forest trees in the development of decomposer fungal communities and decomposition processes of leaf litter. Can J Microbiol 52(8):701–716

Osono T, Hirose D (2009) Effects of prior decomposition of Camellia japonica leaf litter by an endophytic fungus on the subsequent decomposition by fungal colonizers. Mycoscience 50(1):52–55

Perez-de-Mora A, Madejon P, Burgos P (2011) Phytostabilization of semiarid soils residually contaminated with trace elements using by-products: sustainability and risks. Environ Pollut 159:3018–3027

Perez-Rosales E, Alcaraz-Melendez L, Puente ME (2017) Isolation and characterization of endophytic bacteria associated with roots of jojoba (*Simmondsia chinensis* (Link) Schneid). Curr Sci 112:396–401

Pervez Z, Alam MS, Islam MS (2016) First report of bacterial soft rot of Aloe Vera (*Aloe barbadensis*) caused by *Pectobacterium chrysanthemi* in Bangladesh. J Plant Pathol Microbiol 7(12):e110

Petrini O, Fisher PJ, Petrini LE (1992) Fungal endophyte of bracken (Pteridium aquilinum), withsome reflections on their use in biological control. Sydowia 44:282–293

Pierre E, Louise NW, Marie TKR, Valere TFP (2016) Integrated assessment of phytostimulation and biocontrol potential of endophytic *Trichoderma* spp against common bean (*Phaseolus vulgaris* L.) root rot fungi complex in centre region, cameroon. Int J Pure App Biosci 4(4):50–68

Pimentel MR, Molina G, Dionisio AP et al (2011) Use of endophytes to obtain bioactive compounds and their application in biotransformation process. Biotechnol Res Int 2011:576286

Posada F, Vega FE (2006) Inoculation and colonization of coffee seedlings (*Coffea arabica* L.) with the fungal entomopathogen *Beauveria bassiana* (Ascomycota: Hypocreales). Mycoscience 47(5):284–289

Potshangbam M, Devi SI, Sahoo D, Strobel GA (2017) Functional characterization of endophytic fungal community associated with *Oryza sativa* L. and *Zea mays* L. Front Microbiol 8(325):1–15

Promputtha I, Hyde KD, McKenzie EHC et al (2010) Can leaf degrading enzymes provide evidence that endophytic fungi becoming saprobes? Fungal Divers 41:89–99

Rajeswari P, Aishwaryaalakshmi B, Jeyagowri C (2017) Isolation, identification and screening of *Rhizobium* for plant growth promotion. Int. J Appl Res 3(1):732–733

Raviraja NS, Sridhar KR, Barlocher F (1996) Endophytic aquatic hyphomycetes of roots ofplantation crops and ferns from India. Sydowia 48:152–160

Ren JH, Ye JR, Liu H et al (2011) Isolation and characterization of a new *Burkholderia pyrrocinia* strain JK-SH007 as a potential biocontrol agent. World J Microbiol Biotechnol 27(9):2203–2215

Rout JR, Sahoo SL (2013) Antioxidant enzyme gene expression in response to copper stress in *Withania somnifera* L. Plant Growth Regul 71(1):95–99

Roychowdhury D, Mondal S, Banerjee SK (2017) The effect of biofertilizers and the effect of vermicompost on the cultivation and productivity of maize – a review. Adv Crop Sci Tech 5(1):1–4

Saha M, Sarkar S, Sarkar B et al (2016) Microbial siderophores and their potential applications: a review. Environ Sci Pollut Res 23(5):3984–3999

Salam N, Khieu T-N, Liu M-J (2017) Endophytic actinobacteria associated with *Dracaena cochinchinensis* Lour.: Isolation, diversity, and their cytotoxic activities. Biomed Res Int 2017:1308563

Sanlibaba P, Çakmak GA (2016) Exopolysaccharides production by lactic acid bacteria. Appl Microbiol 2(2):1–5

Santoyo G, Moreno-Hagelsieb G, Orozco-Mosqueda MC, Glick BR (2016) Plant growth-promoting bacterial endophytes. Microbiol Res 183:92–99

Schiavon M, Pilon-Smits EAH (2017) Selenium biofortification and phytoremediation phytotechnologies: a review. J Environ Qual 46:10–19

Schulz B, Boyle C, Draeger S et al (2002) Endophytic fungi: a source of novel biologically active secondary metabolites. Mycol Res 106:996–1004

Selvi KB, Paul JJA, Vijaya V, Saraswathi K (2017) Analyzing the efficacy of phosphate solubilizing microorganisms by enrichment culture techniques. Biochem Mol Biol J 3:1

Shahzad R, Khan AL, Bilal S (2017) Plant growth-promoting endophytic bacteria versus pathogenic infections: an example of *Bacillus amyloliquefaciens* RWL-1 and *Fusarium oxysporum* f. sp. lycopersici in tomato. PeerJ 5:e3107

Shakoor A, Abdullah M, Sarfraz R et al (2017) A comprehensive review on phytoremediation of cadmium (Cd) by mustard (*Brassica juncea* L.) and sunflower (*Helianthus annuus* L.) J Biol Environ Sci 10(3):88–98

Sheng X, Chen X, He L (2008) Characteristics of an endophytic pyrene-degrading bacterium of *Enterobacter* sp. 12J1 from *Allium macrostemon* Bunge. Int Biodeter Biodegr 62:88–95

Sinma K, Nurak T, Khucharoenphaisan K (2015) Potentiality of endophytic actinomycetes isolated from sugar cane. KMITL Sci Tech J 15:88–97

Smith CS, Chand T, Harris RF, Andrews JH (1989) Colonization of a submersed aquatic plant, eurasian water milfoil (*Myriophyllum spicatum*), by fungi under controlled conditions. Appl Environ Microbiol 55:2326–2332

Soleimani M, Hajabbasi MA, Afyuni M (2010a) Effect of endophytic fungi on cadmium tolerance and bioaccumulation by *Festuca arundinacea* and *Festuca pratensis*. Int J Phytoremediation 12(6):535–549

Soleimani M, Afyuni M, Hajabbasi MA (2010b) Phytoremediation of an aged petroleum contaminated soil using endophyte infected and non infected grasses. Chemosphere 81:1084–1090

Stanley SJ (1992) Observations on the seasonal occurrence of marine endophytic and parasitic fungi. Can J Bot 70:2089–2096

Strobel GA, Daisy B (2003) Bioprospecting of microbial endophytes and their natural products. Microbiol Mol Biol Rev 67:491–502

Sudha V, Govindaraj R, Baskar K (2016) Biological properties of endophytic fungi. Braz Arch Biol Technol 59:e16150436

Sun LN, Zhang YF, He LY et al (2010) Genetic diversity and characterization of heavy metal-resistant-endophytic bacteria from two copper-tolerant plant species on copper mine wasteland. Bioresour Technol 101:501–509

Sun H, He Y, Xiao Q et al (2013) Isolation, characterization and antimicrobial activity of endophytic bacteria from *Polygonum cuspidatum*. Afr J Microbiol Res 7:1496–1504

Sunitha VH, Devi DN, Srinivas C (2013) Extracellular enzymatic activity of endophytic fungal-strains isolated from medicinal plants. World J Agri Sci 9(1):01–09

Sziderics AH, Asche FR, Trognitz F et al (2007) Bacterial endophytes contribute to abiotic stress adaptation in pepper plants (*Capsicum annuum*L.) Can J Microbiol 53:1195–1202

Tarekegn MA, Kibret K (2017) Effects of rhizobium, nitrogen and phosphorus fertilizers on growth, nodulation, yield and yield attributes of soybean at Pawe Northwestern Ethiopia. WSN 67(2):201–218

Tefera T, Vidal S (2009) Effect of inoculation method and plant growth medium on endophytic colonization of sorghum by the entomopathogenic fungus *Beauveria bassiana*. Bio Control 54(5):663–669

Terekhova VA, Semenova TA (2005) The structure of micromycete communities and their synecologic interactions with basidiomycetes during plant debris decomposition. Microbiology 74(1):91–96

Thormann MN, Currah RS, Bayley SE (2003) Succession of microfungal assemblages in decomposing peat land plants. Plant and Soil 250(2):323–333

Tolulope RA, Adeyemi AI, Erute MA, Abiodun TS (2015) Isolation and screening of endophytic fungi from three plants used in traditional medicine in Nigeria for antimicrobial activity. Int J Green Pharm 9:58–62

Vejan P, Abdullah R, Khadiran T (2016) Role of plant growth promoting rhizobacteria inagricultural sustainability – a review. Molecules 21(573):1–17

Verma SCC, Ladha JKK, Tripathi AKK (2001) Evaluationof plant growth promoting and colonization ability of endophytic diazotrophs from deep water rice. J Biotechnol 91:127–141

Vurukonda SSKP, Vardharajula S, Shrivastava M, Ali Sk Z (2016) Enhancement of drought stress tolerance in crops by plant growth promoting rhizobacteria. Microbiol Res 184:13–24

Wang S, Hu T, Jiao Y et al (2009) Isolation and characterization of *Bacillus subtilis* EB-28, an endophytic bacterium strain displaying biocontrol activity against *Botrytis cinerea* Pers. Front Agric China 3(3):247–252

Wang Y, Li H, Zhao W et al (2010) Induction of toluene degradation and growth promotion in corn and wheat by horizontal gene transfer within endophytic bacteria. Soil Biol Biochem 42:1051–1057

Wang J-L, Li T, Liu G-Y et al (2016) Unraveling the role of dark septate endophyte (DSE) colonizing maize (*Zea mays*) under cadmium stress: physiological, cytological and genic aspects. Sci Rep 6:22028

Weyens N, Beckers B, Schellingen K et al (2015) The Potential of the Ni-Resistant TCE-Degrading *Pseudomonas putida* W619-TCE to reduce phytotoxicity and improve phytoremediation efficiency of poplar cuttings on A Ni-TCE co- contamination. Int J Phytoremediation 17(1):40–48

White JF, Morgan-Jones G, Morrow AC (1993) Taxonomy, life cycle, reproduction and detection of Acremonium endophytes. Agric Ecosyst Environ 44(1–4):13–37

Witzel K, Ustun S, Schreiner M et al (2017) A proteomic approach suggests unbalanced proteasome functioning induced by the growth-promoting bacterium *Kosakonia radicincitans* in *Arabidopsis*. Front Plant Sci 8(661):1–10

Wu H, Yang H-Y, You X-L, Li Y-H (2013) Diversity of endophytic fungi from roots of *Panax ginseng* and their saponin yield capacities. SpringerPlus 2:107

Xia Y, Bolt SD, Dreyer J, Scott D, Williams MA (2015) Characterization of culturable bacterial endophytes and their capacity to promote plant growth from plants grown using organic or conventional practices. Front Plant Sci 6(490):1–10

Xiao X, Luo S, Zeng G et al (2010) Biosorpiton of cadmium by endophytic fungus (EF) *Microsphaeropsis* sp. LSE10 isolated from cadmium hyperaccumulator *Solanum nigrum* L. BioresourTechnol 101:1668–1674

You Y-H, Yoon H, Kang S-M et al (2012) Fungal diversity and plant growth promotion of endophytic fungi from six halophytes in Suncheon bay. J Microbiol Biotechnol 22(11):1549–1556

Zahoor M, Irshad M, Rahman H et al (2017) Alleviation of heavy metal toxicity and phytostimulation of *Brassica campestris* L. by endophytic *Mucor* sp. MHR-7. Ecotoxicol Environ Saf 142:139–149

Zakria M, Njoloma J, Saeki Y, Akao S (2007) Colonization and nitrogen-fixing ability of *Herbaspirillum*sp. strain B501 gfp1 and assessment of its growth-promoting ability in cultivated rice. Microbiol Environ 22:197–206

Chapter 20
Tools and Techniques for Genetic Engineering of Bio-Prospective Microorganisms

Balasubramani S.P. and Vanitha Ramesh

20.1 Introduction

Even before understanding the complete biology, humans have been exploiting microorganisms for agriculture and food. Inter-cropping with leguminous plants for nitrogen fixing by soil and preparation of wine, bread, and cheese are examples. Humans have been using microbes without even having much knowledge on their biology. Genetically modified organisms (GMO's) have been used by humans for centuries. Application of GMO's has been mainly in food production, which includes commonly consumed foods like bread, cheese and wine. Wild populations of several of the bacteria and yeasts have got genetic diversity and individual traits through natural selection or induced mutagenesis. These changes lead to development of useful characteristics for large scale industrial or environmental applications.

Microorganisms play an important role in food production, either in food fermentation or in food spoilage. Several of the enzymes and secondary metabolites produced by micro-organisms are also used food production and processing. Endogenous micro-organisms present in the food preparation are sufficient to induce the fermentative process. But to impart uniformity and predictability, starter culture is introduced, which completes the fermentation process. Most industrial fermentations are performed this way. Generally, bacteria, moulds and yeasts are used in food fermentation. The most widely used organisms are lactic acid bacteria (LAB) and yeasts (*Sacchromyces cerevisiae*). To improve the process and products obtained

B. S.P. (✉)
School of Integrative Health Sciences, Trans-Disciplinary University, Bangalore, India
e-mail: balasubramani.sp@tdu.edu.in

V. Ramesh
Department of Genetics, Indian Academy Degree College (Autonomous), Bangalore, India

J.K. Patra et al. (eds.), *Microbial Biotechnology*,
https://doi.org/10.1007/978-981-10-6847-8_20

through fermentations these organisms were subjected to genetic modification methods like enrichment selection, mutagenesis, conjugation and protoplast fusion.

Conventionally, genetic modification of LAB was performed employing chemical- or ultra-violet-induced mutagenesis. Mutants with superior characteristics were obtained by followed by enrichment or selection process. Alternately, natural methods of genetic exchange like conjugation can also occur between strains of LAB (Steenson and Klaenhammer 1987).

Protoplast fusion is another common method performed to facilitate recombination between two desired strains with unique characteristics. This method was originally used in mapping the bacterial genome but the same method has been shown to successfully produce LAB strains with desired characteristics (Patnaik et al. 2002). Yeast strains that can produce a variety of biochemical substrates for use in the fermentation process has also been developed using protoplast fusion (Pina et al. 1986).

In order to obtain desirable traits in the starter cultures people have been looking at different strategies. One way is to identify organisms with desirable traits and enrich them by conducting small-scale fermentations. This approach is far from practical and also does not guarantee higher productivity. Low throughput is also a limiting factor in the success rate of this approach. The alternate approach is to use the molecular genetics tools and genetic engineering. In the post-genomics era modifying genetic make-up for imparting desired characters has become very attractive and efficient way to bio-prospect. Whole genome sequence for several important bacteria and fungi are already available in public databases. The advantage of the genome based methods is their precision with which strains can be engineered.

20.2 Transformation of Competent Cells

Transformation in bacterial cells was first discovered by Frederick Griffith with *Streptococcus pneumoniae* in 1928. However, the transforming principle was identified by DNA in 1944 by Oswald T. Avery, Colin M. MacLeod, and Maclyn McCarty. These findings were important milestones in the elucidation of the molecular nature of genes.

The process by which genetic alteration of cells take place through direct uptake, incorporation and expression of exogenous DNA is called transformation. During this process, DNA is obtained from the surrounding environment through the cell membrane (Fig. 20.1).

In some of the species of bacteria, transformation is a natural process to adopt to the environment. Bacterial transformation can also be effected by artificial means. Environmental stress, starvation, cell density etc., can also induce the microbial cells to be competent and induce transformation. The artificial methods for transformation of bacteria by introducing recombinant DNA into the recipient cells can be facilitated by using chemical agents, enzymes, or electroporation.

Transformation of microbial strains for exploitation in industry and environmental process has become much simpler and several commercial strains are available

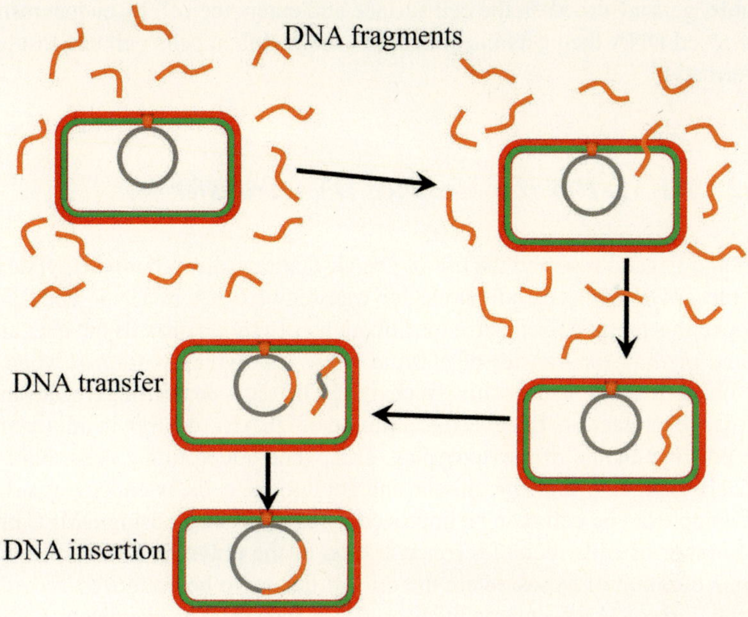

Fig. 20.1 Schematic representation of bacterial transformation.Competent bacterial cells obtain DNA from the surrounding environment

for dedicated purposes. Academic research has also been facilitating the development of transformed micro-organisms by developing tools and techniques to study the mechanisms of transformation and also in application of transformants.

Cell membrane is a protective layer which separates internal and external environments of the cell and allows selective permeability of substances to move across. It is composed of amphipathic molecules. DNA being an anionic hydrophilic polymer, macromolecule cannot easily cross the physical barrier of membrane and enter the cells unless assisted. Moreover, cells have nuclease that can degrade DNA from external matrices. In order to facilitate the transfer of DNA into the cell, certain specific synthetic compounds are used. In general, this process aims at altering the cell membrane, widen the pore size, inactivate DNA degrading enzymes and allow direct transfer of DNA into the cells. Some of the chemicals used in bacterial transformation are calcium phosphate, DEAE dextran, cationic lipid, polymers like poly-L-lysine (PLL), polyphosphoester, chitosan and dendrimers.

20.2.1 Calcium Phosphate Mediated DNA Transfer

This method was first described in 1973 by Graham and van der Eb, while analyzing the infectivity of adenoviral DNA. While the exact principle of this method is not clear, it is hypothesized that DNA/calcium phosphate co-precipitate, becomes insoluble, gets adsorbed on the cell surface and enters the cell by endocytosis. The endocytosed DNA then gets integrated into the recipient cells genome resulting in transformation.

20.2.2 DEAE-Dextran Mediated DNA Transfer

Vaheri and Pagano reported the use of DEAE-dextran (diethylaminoethyl dextran), a soluble polycationic carbohydrate for transfer of DNA in 1965. They initially developed this method to enhance the infectivity of viruses on cells but later adapted the same method for transfer of plasmid DNA. Apolyplex is formed when negatively charged DNA and positively charged DEAE – dextran aggregate through electrostatic interaction. Excess concentration of DEAE – dextran in mixture results in net positive charge of the apolyplex. This when added to the cells, bind to the negatively charged plasma membrane and get into the cells by endocytosis. Uptake of the complexed by cells can be improved by osmotic shock using DMSO or glycerol. Number of cells required, concentration of the polymer, DNA concentration and time duration of exposure are the critical factors to be optimized according to each cell during DAEA-dextran mediated DNA transfer. The chemicals used in this process are toxic to cells at higher concentrations, so a dose response study with the cells to be transformed should be performed before the actual experiment. However, this is a simple, inexpensive and sensitive method for transient transformation.

20.2.3 Lipofection

Lipofection uses artificial phospholipid vesicles called liposomes for the delivery of DNA or other molecules into the cells. These vesicles resemble the multi-lamellar or unilamellar membrane of the cells with a size range of 0.1–10 μm or 20–25 nm respectively. This method for DNA transformation was first described in 1965. The DNA- liposomes complex can be formed through membrane-membrane fusion or by endocytosis. This liposome complex has ability fuse with the protoplasts of the cells, gets internalized and then release the contents into the cell. In general, most of the bacteria and yeast protoplasts are susceptible to lipofection method. Liposomes are classified as cationic and pH-sensitive.

20.2.3.1 Cationic Liposomes

Positively charged or cationic liposomes associate with the negatively charged DNA molecules by electrostatic interactions forming a stable complex. To avoid toxic nature of the cationic liposomes, generally, neutral liposomes are used as DNA carriers or helpers. This complex is denoted as lipoplex. Some of the commonly used neutral co-lipids are dioleoylphosphatidyl ethanolamine (DOPE) or dioleoylphosphatidyl choline (DOPC). Positive charge of the liposome complex facilitates binding to the negatively charged cell membrane which is then endocytosed into the cell and then into nucleus. The concentration of lipid and ratio of lipid to DNA used in the formation of lipolex determines the efficiency of cationic liposome gene transfer method.

20.2.3.2 Negatively Charged Liposomes

This gene transfer process is not very effective. The electrostatic forces between phosphate backbone of DNA and negatively charged groups of the lipids in liposome repel each other and does not allow them to form a complex. Sometimes, DNA molecules even get entrapped within the aqueous interior of these liposomes. Divalent cations like Ca^{2+}, Mg^{2+}, Mn^{2+} and Ba^{2+} are used to neutralize the mutual electrostatic repulsion and to facilitate formation of lipoplex complex. In general, anionic lipoplexes are physiologically safe but they are pH sensitive and destabilize at low pH.

Delivery of genes by liposome into the nucleus is still poorly understood. Cationic liposomes are more efficient in gene delivery than that of pH sensitive liposomes. Liposome mediated gene transfer are easy, economic, efficient, minimal toxicity to the cell and DNA, tractable process and adaptable to high-throughput systems. However, this process cannot be applied to all cell types.

20.2.4 Microinjection

DNA microinjection involves delivery of foreign DNA into a living cell through a fine glass micropipette. This method was first proposed in the early nineteenth century by Dr. Marshall A. Barber. DNA of interest is delivered into the recipient cell using a glass micropipette tip of 0.5 mm diameter by observing under a powerful microscope. The cell membrane is pierced with the help of micropipette to deliver the DNA into the cytoplasm. The introduced DNA may express extra-chromosomally or gets integrated with the genome through homologous recombination. Some of the advantages of microinjection are precise delivery of DNA directly into a single cell, easy identification of transformed cells upon injection of dye along with DNA.

20.2.5 Particle Bombardment

Sanford and colleagues at Cornell University, USA developed this method in 1987. Particle gun, micro projectile bombardment and particle acceleration are some of the other terms used to denote this gene transfer method. This process uses high-velocity micro projectiles to deliver DNA or other substances into cells. This is one of the commonly followed method for gene transfer. The biolistic gun works on the principle of conservation of momentum. It uses the helium gas passage in the cylinder with a range of velocities required for transformation of DNA to various cell types. The equipment consists of a bombardment chamber consisting of a plastic rupture disk. Below the rupture disk is the macro carrier which is loaded with micro carriers, which are micro pellets made of gold or tungsten micro pellets coated with DNA for transformation. Bombardment chamber also consists of an outlet for vacuum creation.

To ensure sterility, the apparatus is placed in laminar flow while carrying out the procedure. The recipient cells are placed in the apparatus and a stopping screen is placed between the cells and micro carrier assembly. High pressure helium allows the plastic rupture disk to rupture and propelling the macro carrier and micro carriers with DNA. Presence of the stopping screen prevents the passage of macro projectiles and allows the DNA coated micro pellets to pass through it to the cell and deliver DNA.

The method is simple and convenient by involving gold microcarrier coated with DNA or RNA, loading sample cartridges, pointing the nozzle and firing the device. No pre-processing of the cell is required and intact cells are used. It is possible to manipulate of genome of sub-cellular organelles by employing this method. This method also does not involve any of the potentially harmful viruses or toxic chemical treatment as gene delivery vehicle. It is possible to place DNA or RNA in precise locations within the cell. However, establishment of specialized infrastructure may be involving cost. Technical disadvantages include the possibility of cell damage, random integration of DNA into the cell and chances of multiple copy insertions leading to gene silencing.

20.2.6 Sonoporation

The process of using ultrasound for inducing temporary permeabilization of the cell membrane is called sonoporation. The pores can be used to transport DNA, drugs or other therapeutic compounds from the extracellular environment. The compound is trapped inside the cells after ultrasound exposure. Acoustic cavitation of micro bubbles is employed the for enhancing the delivery of large molecules like DNA. Injection and ultrasound treatment allows the micro bubbles complex with DNA to be delivered into the target cells. This method does not cause any damage to the cells at the same time it is very simple and highly efficient for gene transfer. Cell viability is one of the important factors to be considered to opt this method for gene transfer, as in rare cases, high exposure to low-frequency (<MHz) ultrasounds result in rupture and death of cells.

20.3 Protoplast Fusion

Protoplast refers to cells in which the cell wall is removed. Cytoplasmic membrane forms the other most covering in protoplasts. They can be generated by treating with specific lytic enzymes. Protoplast fusion (PF) has been used as a strategy to genetically engineer microorganisms for more than past 75 years. It's a physical phenomenon by which two or more protoplasts will be fused spontaneously or by induction. Fusion of protoplasts using polyethylene glycol (PEG) in *Bacillus megaterium* (Fodor and Alföldi 1976), *B. subtilis* (Schaeffer et al. 1976) and in *Streptomyces* (Hopwood et al. 1977) are the initial reports describing this method. The success of PF largely depends upon the type of organism used. In general fusion of protoplasts work well with gram-positive organisms and fungi when compared to the gram-negative due to the presence of outer membrane. PF is one of the simple and important tools for gene transfer in microorganisms. This method generates genetically improved inter-specific or intra-specific hybrid strains of microorganisms to improve fitness or for industrial exploitation. The hybrid protoplast will contain fused nuclei from both the parents and heteroplasoic cytoplasm.

20.3.1 Methods of Protoplast Fusion

(a) **Spontaneous fusion**

Removal of the cell wall and the enlarged plasmodesmata can allow the movement of organelles into the neighbouring cells. Thus, simple physical contact of protoplasts induces spontaneous fusion. This process is usually intra-specific and result in formation of homokaryons.

(b) **Induced fusion**

The negative charge on the surface of protoplast around the outside of plasma membrane does not allow to fuse with each other. So, they are induced to form the hybrids. Generally, the fusion is induced by mechanical, chemical or electric stimulation.

Mechanical fusion is facilitated by bringing the selected protoplasts into intimate contact under a microscope. Micromanipulator and perfusion micropipette fitted with 1 mm tip is used to hold and induce fusion of the protoplasts.

Chemicals such as sodium or potassium nitrate ($NaNo_3$/KNo_3) (Power et al. 1970), polyethylene glycol (PEG) (Hansen and Stadler 1977), calcium ions (Ca^{++}) at high alkaline pH (about 10.5) (Blow et al. 1979) are fusogens and are used to induce protoplast fusion. These chemical fusogens facilitate adhesion of isolated protoplasts, lead to agglutination and finally make them fuse. Intergenic fusion is possible by this method. Less expensive and non-specificity are advantages of using chemical fusogens, but some of these chemicals can be cytotoxic at higher concentrations.

Srinivas and Panda (1997) fused protoplasts of *Trichoderma reesei* QM9414 and *Saccharomyces cerrevesei* NCIM 3288 to convert cellulose to ethanol by single step process. The key enzyme involved in this process was found to be endoglucanase. Similarly, protoplast of tetracycline resistant (Tetr) and erythromycin resistant (Eryr) *Lactobacillus fermentatum* 604 with was fused with *L. fermentatum* 605 using PEG for transfer of antibiotic resistance (Iwata et al. 1986). Coupling sodium nitrate with sucrose solution and PEG with DMSO (10–15%) (Klebe and Mancuso 1982) or concanvalin A or sea water (Betina et al. 1973) are said to increase the efficiency of fusogen.

Fusing of protoplasts in the presence of electric current is called electrofusion. This is performed by placing two glass capillary microelectrodes in contact with the protoplasts. The first step in this method uses mild electrical stimulation (10 kv m^{-3}) to generate dielectrophoretic pole within the protoplast suspension leading to formation of pearl chain arrangement to fuse protoplasts. Subsequent application of high intensity electric impulse (100 kv m^{-3}) for some microseconds results in disruption of cell membrane and induce fusion of protoplasts. This process is said be 100% efficient in generating protoplast fusion, reproducible and non-cytotoxic, but requires sophisticated instrumentation (Finaz et al. 1984).

20.3.2 Disruption of Cell Wall for Generating Protoplasts

Generally bacterial and fungal cell walls can be degraded using lysozyme (Fodor and Alfoldi 1976). Some of the fugal species requires mixture of enzymes due to their complex cell wall structure. In such cases, mixture of glucanase and chitinase is used (Jogdand 2001). A combination of lysozyme and achromopeptidase are used to remove the cell wall of *Streptomyces* (Narayanswamy 1994). In case of filamentous mycelial organisms like *Streptomyces*, the organism can also be cultured in medium containing high concentrations of glycine (0.8–2%). Glycine replaces the D-alanine in the peptidoglycan layer and interfere in the cross linking thereby reducing the integrity of cell wall (Hammes et al. 1973). Some researchers have also simultaneously used glycine incorporation to the medium and lysozyme treatment. Lysostaphin is used in case of *Staphylococcus sp.* (Novick et al. 1980). For some of the gram-negative bacteria like *E. coli*, lysozyme was used with EDTA (Weiss 1976). Protoplasts of *Providencia alcalifaciens* were prepared by treating the cells with glycine-lysozyme-EDTA method (Coetzee et al. 1979). Use of antibiotic fosfomycin, a cell was synthesis inhibitor have been also used to prepare protoplasts of *E. coli* and *Serratiamarcescens*(Rodicio et al. 1978).

20.3.3 Protoplast Regeneration

Specialised medium and conditions are required for regeneration of hybrid protoplasts. For e.g., *B. megaterium* protoplasts require soft agar overlay on hypertonic medium (Fodor and Alföldi 1976), but *B. subtilis* can be plated directly on the surface of the medium (Schaeffer et al. 1976). Dehydrated culture plate (about 20%) enhanced the regeneration of *Streptomyces fradiae* protoplasts (Hopwood 1981). Temperature is one of the important factors determining the regeneration of protoplasts. Regeneration frequencies of protoplasts are estimated by relating colony counts on regeneration plates to the number of organisms in the original culture from which the protoplasts were prepared or by using haemocytometer counts of protoplasts.

20.4 Electroporation

Electroporation was developed a method of gene transfer during the late 1980s. This method employs electrical currents to disrupt cell membrane, create pores in the cell envelope so that DNA from an external source can enter (Luchansky et al. 1988). Owing to its simplicity in the procedure, this is the most widely used gene transfer method. However, it lacks efficiency in many different species.

Cell organelles have different levels polarisation. Electroporation or electropermeabilisation is a method in which the cells are exposed to a strong electric current for short duration and alter their polarisation. Strong electric current supplied to the cell membrane shifts the position of lipids and creates reversible local disorganisation and form pore channels. These pore channels are large enough to allow the influx and efflux of macromolecules. Permeability of DNA, protein or DNA-protein complex are facilitated through these pores. Electroporation is performed by passing 200–400 mV/cm across the suspended cells in an electroporation cuvette. Following electroporation, cells must be handled carefully until they divide, produce new cells and express the transformed genes.

Creating 1 V gradient across the ~1 μm diameter microbial cell ie., 10,000 $cm^{-1}$for about 10 ms for transformation of bacterial cells is much complex and costlier than the higher cells. Square wave generators are said to produce transformation efficiency of 10^6 transformants per μg of DNA.

In general, transformation efficiency of electroporation is better than the protoplast fusion and other chemical treatments. Additional treatment with PEG, dithiothreitol (DTT), lithium acetate or dimethyl sulfoxide (DMSO) is said to further improve the transformation efficiency of electroporation. These treatment parameters differ from species to species. Growth phase and density of the cells, cell wall structure and thickness, electroporation settings also determine the efficiency of this process.

Each organism has its own voltage range requirement. *Lactobacillus bulgaricus* cells have maximum transformation efficiency at 2.5 kV cm^{-1} but cannot withstand at electric field of 2.5 kV cm^{-1}. But, organisms like *Streptococcus thermophilus* show maximum efficiency at 10 kV cm^{-1} and *E. coli* at 12 kV cm^{-1}. Probably, membrane composition, cell diameter and cell-wall composition determine the amount of electric current required for electroporation. Transformation efficiency is also determined by the medium in which the cells are suspended. Concentrated sucrose is a preferred medium. Cells freezed in 15% glycerol or sucrose show better transformation efficiency. Physical barriers like presence of capsular polysaccharide with *Klebsiella pneumoniae* affect the transformation frequency.

20.5 Conjugation

Bacterial conjugation is a horizontal gene transfer process from a donor cell bearing one or more conjugative plasmids to a plasmid-free recipient cell (Fig. 20.2). Conjugative plasmids in most of the bacterial species can be transferred to distantly related or even unrelated microorganisms or to even eukaryotic host cells. This is one of the chief mechanisms by which antibiotic resistance spreads among pathogenic bacteria. In both Gram-negative and Gram-positive bacteria, to transfer the plasmid, both the partner cells need to be close physical contact and the transfer is mediated by plasmid-encoded proteins, which provide transfer (tra) functions.

During conjugation a special type of replication process in which one copy of plasmid DNA molecule remains in the donor while the other copy is transferred to the recipient. Replication process is initiated by creating a single-stranded nick at the plasmid's origin of transfer (oriT). Several genes present in plasmid are involved in the conjugation process, including the genes required to form a mating pair between donor and recipient. Some of the bacterial species might show unique mechanisms in conjugation, for eg., in the Gram-positive bacteria of the genus Streptomyces (i) replication of the plasmid occurs in the donor and a double stranded plasmid is transferred to the recipient, and (ii) protein encoded by a single plasmid is sufficient for conjugative transfer of plasmid between mating cells.

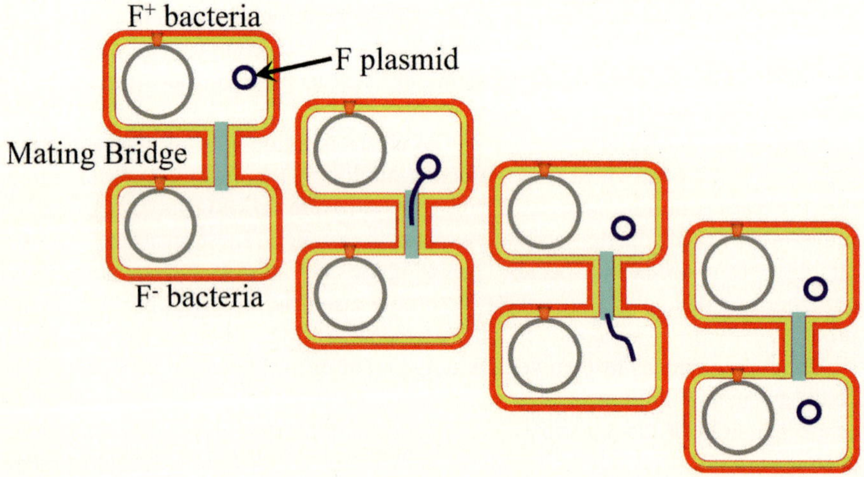

Fig. 20.2 Schematic representation of conjugation in bacteria. Bacteria can transfer genetic material from one cell to another through this process

20.6 Transduction

Transduction is a process of by which genes are introduced into a host cell's genome using viruses as carriers. Viruses have ability to deliver nucleic acids directly into cells and, they are capable of activating transcription and replication machinery. The gene of interest is packed into virus particle and allowed to infect the target cells. Following infection, the gene interest with the viral genome enters the host cells through a receptor- mediated process. The introduced gene may persist as an independent episome or integrate with the host genome and can express the desired character.

LAB was genetically engineered by introducing genes with the help of a bacteriophage (Bierkland and Holo 1993). Transduction may end up abortive as in some cases the introduced genes may be deleted for unknown reasons.

Viral vectors that carry and introduce cellular nucleic acids to recipient cells on infection are called transducing particles. Generally, bacteriophages act as transducing viruses as they carry parts of bacterial or plasmid genome from one bacterial host to another. In some phages the DNA gets integrated with the viral chromosome in the particle and is described as specialized transduction. While in few other transducing phages, the DNA from the host origin is present as independent expressing entity and is denoted as generalized transduction. Retroviruses transfer RNA into the recipient cell.

The viruses may transfer any region of the donor chromosome to the recipient in generalized transduction. But, in specialized transduction, always the same segment is carried into the recipient.

20.7 Clustered Regularly Interspaced Short Palindromic Repeats (CRISPR)

CRISPR or Clustered Regularly Interspaced Short Palindromic Repeats is the most recent genetic engineering technique.

CRISPR is a family of DNA sequences in bacteria which is obtained from viruses that have earlier attacked the bacterium and other mobile genetic elements. These prokaryotic DNA segments contain short, repetitive base sequences separated by 'spacers'. The processed RNA transcribed from the repeat-spacer array generate small crRNA that specify the target sequences (also known as protospacers) cleaved by CRISPR system. Protospacer-adjacent motif (PAM) which is present immediately downstream to the target region facilitates the cleaving. Cas9 enzyme responsible for crRNA biogenesis and targeting is coded by CRISPR-associated (*cas*) genes which flank the repeat-spacer array. Cas9 is an endonuclease which is guided by crRNA to specify the site of cleavage in double-stranded (ds)DNA. During the processing of the crRNA precursor loading of the crRNA guide onto Cas9 occurs. This process also requires a small tracrRNA or antisense RNA to the precursor and

RNAse III. The CRISPR sequences play an important role in bacterial defense mechanism. They help the host bacterium to detect and destroy DNA from further attacks by similar viruses. These sequences were initially discovered in the *E. coli* genome in 1987.

Breakthrough was achieved when this bacterial immune mechanism was adopted to edit the genome of prokaryote and eukaryotic organisms (Gilbert et al. 2013; Jinek et al. 2013). Presently, the CRISPR/Cas9 system forms the basis of a genome editing technology known as that allows permanent modification of genes within organisms. Unlike the other genome editing tools which are not site specific or require sophisticated protein engineering, CRISPR/Cas9 only requires a short crRNA guide. Due to its simplicity and adaptability, CRISPR has rapidly become one of the most popular approaches for genome engineering.

The CRISPR genome editing tool consists of two components. They are:

(i) Cas9, endonuclease enzyme that can cut the two strands of DNA at a specific location of interest in the genome. A gene can then be inserted or removed in the cut position.
(ii) Guide RNA (gRNA), is a short synthetic pre-designed RNA consisting of a 'scaffold' sequence for binding of Cas9 and a user-defined ~20 nucleotide 'spacer' or 'targeting' sequence which defines the target genomic region to be modified.

Thus, one can specifically determine the genomic target of Cas9 by changing the targeting sequence present in the gRNA.

This technology enables geneticists and medical researchers to edit parts of the genome by removing, adding or altering sections of the DNA sequence (Fig. 20.3). It is currently the simplest, most versatile and precise method of genetic manipulation. It is faster, cheaper and more accurate than previous techniques of editing DNA and has a wide range of potential applications. CRISPR was originally employed to "knock-out" target genes in various cell types and organisms, but modifications to the Cas9 enzyme have extended the application of CRISPR to selectively activate or repress target genes, purify specific regions of DNA, and even image DNA in live cells using fluorescence microscopy. Furthermore, the ease of generating gRNAs makes CRISPR one of the most scalable genome editing technologies and has been recently utilized for genome-wide screens.

Initial experiments were performed with *S. thermophilus*, exposed to predatory phages to test if exogenous phage DNA can be incorporated into the CRISPR repeats of the bacterial genome. Genes involved in vital process like Cas (CRISPR associated) genes, which code for polymerases, nucleases, and helicases, were knocked down to determine their roles in this CRISPR process.

At present, CRISPR/Cas9 specificity is determined by the gRNA which consists of a specific 20 base sequence. Essentially, the gRNA sequence need to have complementarity to the target sequence in the gene to be edited. However, it may not be possible for all 20 bases to match for the guide RNA to be able to bind. There may be case that the gRNA may bind to less complementary sites of the genome other than intended site. This would enable the Cas9 enzyme to cut at the wrong site and

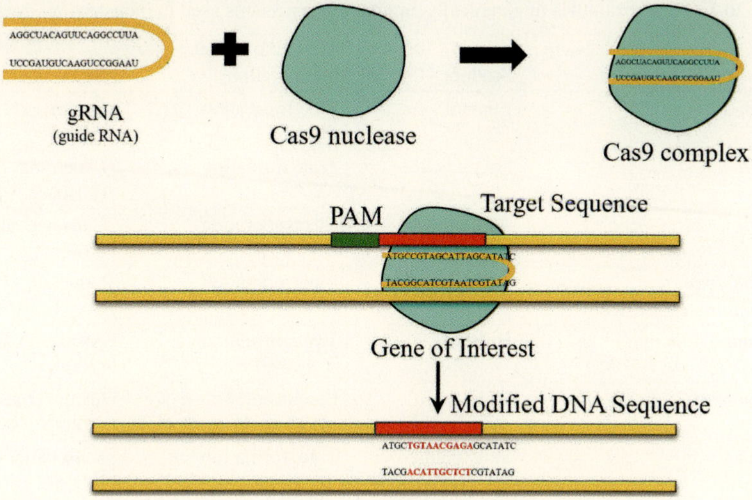

Fig. 20.3 Schematic representation of CRISPR/Cas9 genome editing.Combination of Cas9 endonuclease with guiding RNA, can edit genome of organisms at pre-determined sites

end up introducing mutation in the wrong location which might have deleterious consequences. Scientists are still keen to find a way to ensure that the CRISPR-Cas9 binds and cuts accurately. This could be achieved by the design of better, more specific guide RNAs using our knowledge of the DNA sequence of the genome and the 'off-target' behaviour of different versions of the Cas9-gRNA complex.

Possibilities of using Cas9 enzyme that will only cut a single strand of the target DNA rather than the double strand is also being considered. In such cases two Cas9 enzymes and two guide RNAs have to be in the same location of the genome for editing. This reduces the probability of the cut being made in the wrong place and improve specificity.

20.8 Conclusions

This chapter has reviewed some of the commonly used genetic modification techniques employed in improvement of microorganisms used in environment and health. A list of examples where these methods have been used to improve the efficiency of micro-organisms employed in environment, industry and health is presented in Tables 20.1, 20.2 and 20.3. This is just a glimpse of the methods available. There are several other methods followed for genetic improvement of the strains which are organism and species specific. The methods described in this chapter are commonly followed with a variety of organisms.

Table 20.1 Some examples of genetically engineered organisms used in industrial production

Organism	Product	Method of genetic modification	Reference
E. coli	Ethanol	Electroporation	Durnin et al. (2009)
E. coli	Lactic acid	Transformation	Mazumdar et al. (2010)
E. coli	Acetate	Electroporation	Causey et al. (2003)
Mannheimia succiniciproducens	Succinic acid	Electroporation	Lee et al. (2006)
C. acetobutylicum	Butanol	Transformation	Alsaker et al. (2004)
Lactobacillus plantarum	Lactic acid	Electroporation	Okano et al. (2009)
E. coli	Fatty acid ethyl esters (FAEEs)	Transformation	Kalscheuer et al. (2006)
Kluyveromyces lactis	Xylonic acid	Transformation	Nygard et al. (2011)
Pichia kudriavzevii	Xylonic acid	Transformation	Toivari et al. (2013)
S. cerevisiae	Xylonic acid	Transformation	Toivari et al. (2012)

Table 20.2 Some examples of genetically engineered organisms in bio-remediation of environment

Organism	Compound to be degraded	Reference
E. coli	Atrazine	Strong et al. (2000)
Pseudomonas fluorescence	Napthalene degradation	Sayler and Ripp (2000)
Pseudomonas sp.	PCB	Erickson and Mondello (1993)
P. putida	4-ethyl benzoate	Ramos et al. (1987)
Thauera aromatica	Toluene degradation	Coschigano (2002)
C. testosteroni	o-, p-monochlorobiphenyls	Hrywna et al. (1999)
P. pseudoalcaligenes	PCB, benzene, toluene	Suyama et al. (1996)
E. coli	TCE, toluene	Winter et al. (1989)
A. eutrophus	Nonpolar narcotics	Layton et al. (1999)
Burkholderia cepacia	Toluene	Taghavi et al. (2005)

Table 20.3 Some examples of genetically engineered organisms used in human health

Organism	Product	Method	Reference
E. coli	Erythromycin analogs	Electroporation	Zhang et al. (2015)
	Macrolide 6-deoxyerythromycin D	Mutagenesis	Lee et al. (2011)
	Calf chymosin	Transformation	Emtage et al. (1983)
	Trypsin	Transformation	Vasquez et al. (1989)
	Insulin	Transformation	Chen et al. (1995)
	Gonadotropin releasing hormone	Transformation	Xu et al. (2006)
	Human growth hormone	Transformation	Seeburg et al. (1978)
	Human parathyroid hormone	Transformation	Rabbani et al. (1988)
	IGF-1	Transformation	Kim and Lee (1996)
	MAB	Transformation and genetic engineering	Spadiut et al. (2014)
Streptomyces coelicolor	Amidated polyketide	Protoplast transformation	Zhang et al. (2006)
Streptomyces clavuligerus	Clavulanic acid	Mutagenesis and genetic engineering	Paradkar (2013)
Streptomyces lividans	Daptomycin	Protoplast transformation	Penn et al. (2006)
Streptomyces roseosporus	Daptomycin derivates	Transformation	Miao et al. (2005)
Saccharopolyspora erythrae	Erythromycin A	Transformation	Donadio et al. (1996)
Saccharomyces cerevisiae	IGF-1	Transformation	Bayne et al. (1988)
Aspergillus niger	Secondary metabolites	Transformation	Richter et al. (2014)
Pichia pastoris	MAB	Electroporation	Li et al. (2006)

References

Alsaker KV, Spitzer TR, Papoutsakis ET (2004) Transcriptional analysis of *spo0A* overexpression in *Clostridium acetobutylicum* and its effect on the Cell's response to Butanol stress. J Bacteriol 186(7):1959–1971. https://doi.org/10.1128/JB.186.7.1959-1971.2004

Bayne ML, Applebaum J, Chicchi GG, Hayes NS, Green BG, Cascieri MA (1988) Expression, purification and characterization of recombinant human insulin-like growth factor I in yeast. Gene 66(2):235–244

Betina V, Bálint S, Hajnická V, Nadová A (1973) Diphasic production of secondary metabolites by PenicilliumnotatumWestling S-52. I. Biosynthesis of eitrinin. Folia Microbiol (Praha) 18(1):40–48

Bierkland NK, Holo H (1993) Transduction of a plasmid carrying the cohesive end region from Lactococcus lactis bacteriophage LC3. Appl Environ Microbiol 59(6):1966–1968

Blow AM, Botham GM, Lucy JA (1979) Calcium ions and cell fusion. Effects of chemicalfusogens on the permeability of erythrocytes to calcium and other ions. Biochem J 182(2):555–563

Causey TB, Zhou S, Shanmugam KT, Ingram LO (2003) Engineering themetabolism of Escherichia Coli W3110 for the conversion of sugar to redox-neutral and oxidized products: homoacetate production. Proc Natl Acad Sci U S A 100:825–832

Chen JQ, Zhang HT, MH H, Tang JG (1995) Production of human insulin in an E. Coli system with met-Lys-human proinsulin as the expressed precursor. Appl Biochem Biotechnol 55(1):5–15. PubMed PMID: 7486987

Coetzee JN, Sirgel FA, Lecatsas G (1979) Genetic recombination in fused spheroplastsof providence alcalifaciens. J Gen Microbiol 114(2):313–322

Coschigano PW (2002) Construction and characterization of insertion/deletion mutations of the tutF, tutD, and tutG genes of Thaueraaromatica strain T1. FEMS MicrobiolLett 217(1):37–42

Donadio S, Staver MJ, Katz L (1996) Erythromycin production in Saccharopolysporaerythraea does not require a functional propionyl-CoA carboxylase. Mol Microbiol 19(5):977–984

Durnin G, Clomburg J, Yeates Z, Alvarez PJ, Zygourakis K, Campbell P, Gonzalez R (2009) Understanding and harnessing the microaerobic metabolism of glycerol in Escherichia Coli. Biotechnol Bioeng 103(1):148–161. https://doi.org/10.1002/bit.22246

Emtage JS, Angal S, Doel MT et al (1983) Synthesis of calf prochymosin (prorennin) in Escherichia Coli. Proc Natl Acad Sci U S A 80(12):3671–3675

Erickson BD, Mondello FJ (1993) Enhanced biodegradation of polychlorinated biphenyls after site-directed mutagenesis of a biphenyl dioxygenase gene. Appl Environ Microbiol 59(11):3858–3862

Finaz C, Lefevre A, Teissié J (1984) Electrofusion. A new, highly efficienttechnique for generating somatic cell hybrids. Exp Cell Res 150(2):477–482

Fodor K, Alföldi L (1976) Fusion of protoplasts of bacillus megaterium. Proc Natl Acad Sci U S A 73(6):2147–2150

Gilbert LA, Larson MH, Morsut L, Liu Z, Brar GA, Torres SE, Stern-Ginossar N, Brandman O, Whitehead EH, Doudna JA, Lim WA, Weissman JS, Qi LS (2013) CRISPR-mediated modular RNA-guided regulation of transcription in eukaryotes. Cell 154(2):442–451. https://doi.org/10.1016/j.cell.2013.06.044

Hammes W, Schleifer KH, Kandler O (1973) Mode of action of glycine on thebiosynthesis of peptidoglycan. J Bacteriol 116(2):1029–1053

Hansen D, Stadler J (1977) Increased polyethylene glycol-mediated fusion competence in mitotic cells of a mouse lymphoid cell line. Somatic Cell Genet 3(5):471–482

Hopwood DA (1981) Genetic studies with bacterial protoplasts. Annu Rev Microbiol 35:237–272

Hopwood DA, Wright HM, Bibb MJ, Cohen SN (1977) Genetic recombination throughprotoplast fusion in Streptomyces. Nature 268(5616):171–174

Hrywna Y, Tsoi TV, Maltseva OV, Quensen JF, Tiedje JM (1999) Construction and characterization of two recombinant bacteria that grow on *ortho-* and *para-*substituted chlorobiphenyls. Appl Environ Microbiol 65:2163–2169

Iwata M, Mada M, Ishiwa H (1986) Protoplast fusion of *Lactobacillus fermentum*. Appl Environ Microbiol 52:392–393

Jinek M, East A, Cheng A, Lin S, Ma E, Doudna J (2013) RNA-programmed genome editing in human cells. Elife 2:e00471. https://doi.org/10.7554/eLife.00471

Jogdand SN (2001) Protoplast technology. Gene biotechnology, 3rd edn. Himalaya Publishing house, Mumbai, pp 171–186

Kalscheuer R, Stölting T, Steinbüchel A (2006) Microdiesel: Escherichia Coli engineered for fuel production. Microbiology 152(Pt 9):2529–2536

Kim SO, Lee YI (1996) High-level expression and simple purification of recombinant human insulin-like growth factor I. J Biotechnol 48(1):97–105

Klebe RJ, Mancuso MG (1982) Enhancement of polyethylene glycol-mediated cellhybridization by inducers of erythroleukemia cell differentiation. Somatic Cell Genet 8(6):723–730

Layton AC, Gregory B, Schultz TW, Sayler GS (1999) Validation of genetically engineered bioluminescent surfactant resistant bacteria as toxicity assessment tools. Ecotoxicol Environ Saf 43:222

Lee SJ, Song H, Lee SY (2006) Genome-based metabolic engineering of *Mannheimiasucciniciproducens* for succinic acid production. Appl Environ Microbiol 72(3):1939–1948. https://doi.org/10.1128/AEM.72.3.1939-1948.2006

Lee HY, Harvey CJB, Cane DE, Khosla C (2011) Improved precursor directed biosynthesis in *E. coli* via directed evolution. J Antibiot 64(1):59–64. https://doi.org/10.1038/ja.2010.129

Li H, Sethuraman N, Stadheim TA, Zha D, Prinz B, Ballew N, Bobrowicz P, Choi BK, Cook WJ, Cukan M, Houston-Cummings NR, Davidson R, Gong B, Hamilton SR, Hoopes JP, Jiang Y, Kim N, Mansfield R, Nett JH, Rios S, Strawbridge R, Wildt S, Gerngross TU (2006) Optimization of humanized IgGs in glycoengineered Pichiapastoris. Nat Biotechnol 24(2):210–215

Luchansky JB, Muriana PM, Klaenhammer TR (1988) Application of electroporation for transfer of plasmid DNA to Lactobacillus, Lactococcus, Leuconostoc, Listeria, Pediococcus, Bacillus, Staphylococcus, Enterococcus, and Propionibacterium. Mol Microbiol 2(5):637–646

Mazumdar S, Clomburg JM, Gonzalez R (2010) Escherichia Coli strains engineered for homofermentative production of D-lactic acid from glycerol. Appl Environ Microbiol 76:4327–4336

Miao V, Coëffet-Legal MF, Brian P, Brost R, Penn J, Whiting A, Martin S, Ford R, Parr I, Bouchard M, Silva CJ, Wrigley SK, Baltz RH (2005) Daptomycin biosynthesis in Streptomyces roseosporus: cloning and analysis of the gene cluster and revision of peptide stereochemistry. Microbiology 151(Pt 5):1507–1523

Narayanswamy S (1994) Plant cells and tissue cultures. Plant protoplast: isolation, culture and fusion. TATA MCGraw Hill Publishing Company, New Delhi, pp 391–469

Novick R, Sanchez-Rivas C, Gruss A, Edelman I (1980) Involvement of the cellenvelope in plasmidmaintenance: plasmid curing during the regeneration ofprotoplasts. Plasmid 3(3):348–358

Nygård Y, Toivari MH, Penttilä M, Ruohonen L, Wiebe MG (2011) Bioconversion of d-xylose to d-xylonate with Kluyveromyceslactis. Metab Eng 13(4):383–391. https://doi.org/10.1016/j.ymben.2011.04.001

Okano K, Zhang Q, Shinkawa S, Yoshida S, Tanaka T, Fukuda H, Kondo A (2009) Efficient production of optically pure D-lactic acid from raw corn starch by using a genetically modified L-lactate dehydrogenase gene-deficient and alpha-amylase-secreting lactobacillus plantarum strain. Appl Environ Microbiol 75(2):462–467. https://doi.org/10.1128/AEM.01514-08

Paradkar A (2013) Clavulanic acid production by Streptomyces clavuligerus: biogenesis, regulation and strain improvement. J Antibiot (Tokyo) 66(7):411–420. https://doi.org/10.1038/ja.2013.26

Patnaik RS, Louie S, Gavrilovic V, Perry K, Stemmer WPC, Ryan CM, del Cardayre S (2002) Genome shuffling of Lactobacillus for improved acid tolerance. Nat Biotechnol 20(7):707–712

Penn J, Li X, Whiting A, Latif M, Gibson T, Silva CJ, Brian P, Davies J, Miao V, Wrigley SK, Baltz RH (2006) Heterologous production of daptomycin in Streptomyces lividans. J Ind Microbiol Biotechnol 33(2):121–128

Pina A, Calderon IL, Benitex T (1986) Intergeneric hybrids of Saccharomyces cerevisiae and Zygosaccharomyces fermentati obtained by protoplast fusion. Appl Environ Microbiol 51(5):995–1003

Power JB, Cummins SE, Cocking EC (1970) Fusion of isolated plant protoplasts. Nature (Lond) 225:1016–1018

Rabbani SA, Yasuda T, Bennett HP, Sung WL, Zahab DM, Tam CS, Goltzman D, Hendy GN (1988) Recombinant human parathyroid hormone synthesized in Escherichia Coli. Purification and characterization. J Biol Chem 263(3):1307–1313

Ramos JL, Wasserfallen A, Rose K, Timmis KN (1987) Redesigning metabolic routes: manipulation of TOL plasmid pathway for catabolism of alkylbenzoates. Science 235:593–596

Richter L, Wanka F, Boecker S, Storm D, Kurt T, Vural Ö, Submuth R, Meyer V (2014) Engineering of Aspergillus niger for the production of secondary metabolites. Fungal Biol Biotech 1:4. https://doi.org/10.1186/s40694-014-0004-9

Rodicio MR, Manzanal MB, Hardisson C (1978) Protoplast-like structures formationfrom two species of Enterobacteriaceae by fosfomycin treatment. Arch Microbiol 118(2):219–221

Sayler GS, Ripp S (2000) Field applications of genetically engineered microorganisms for bioremediation processes. Curr Opin Biotechnol 11(3):286–289

Schaeffer P, Cami B, Hotchkiss RD (1976) Fusion of bacterial protoplasts. Proc Natl Acad Sci U S A 73(6):2151–2155

Seeburg PH, Shine J, Martial JA, Ivarie RD, Morris JA, Ullrich A, Baxter JD, Goodman HM (1978) Synthesis of growth hormone by bacteria. Nature 276(5690):795–798. PubMed PMID: 364320

Spadiut O, Capone S, Krainer F, Glieder A, Herwig C (2014) Microbials for the production of monoclonal antibodies and antibody fragments. Trends Biotechnol 32(1):54–60. https://doi.org/10.1016/j.tibtech.2013.10.002

Srinivas R, Panda T (1997) Localization of carboxymethylcellulase in the intergeneric fusants of TrichodermmareeseiQM 9414 and Saccharomyces cereviseeNCIM 3288. Bioprocess Biosyst Eng 18:71–73

Steenson LR, Klaenhammer TR (1987) Conjugal transfer of plasmid DNA between Streptococci immobilized in calcium alginate gel beads. Appl Environ Microbiol 53(4):898–900

Strong LC, McTavish H, Sadowsky MJ, Wackett LP (2000) Field-scale remediation of atrazine-contaminated soil using recombinant Escherichia Coli expressing atrazine chlorohydrolase. Environ Microbiol 2(1):91–98

Suyama A, Iwakiri R, Kimura N, Nishi A, Nakamura K, Furukawa K (1996) Engineering hybrid pseudomonads capable of utilizing a wide range of aromatic hydrocarbons and of efficient degradation of trichloroethylene. J Bacteriol 178(14):4039–4046

Taghavi S, Barac T, Greenberg B, Borremans B, Vangronsveld J, van der Lelie D (2005) Horizontal gene transfer to endogenous endophytic bacteria from poplar improves phytoremediation of toluene. Appl Environ Microbiol 71(12):8500–8505

Toivari M, Nygård Y, Kumpula EP, Vehkomäki ML, Benčina M, Valkonen M, Maaheimo H, Andberg M, Koivula A, Ruohonen L, Penttilä M, Wiebe MG (2012) Metabolic engineering of Saccharomyces cerevisiae for bioconversion of D-xylose to D-xylonate. Metab Eng 14(4):427–436. https://doi.org/10.1016/j.ymben.2012.03.002

Toivari M, Vehkomäki ML, Nygård Y, Penttilä M, Ruohonen L, Wiebe MG (2013) Low pH D-xylonate production with Pichiakudriavzevii. Bioresour Technol 133:555–562. https://doi.org/10.1016/j.biortech.2013.01.157

Vasquez JR, Evnin LB, Higaki JN, Craik CS (1989) An expression system for trypsin. J Cell Biochem 39(3):265–276. PubMed PMID: 2651464

Weiss RL (1976) Protoplast formation in Escherichia Coli. J Bacteriol 128(2):668–670

Winter RB, Yen K-M, Ensley BD (1989) Efficient degradation of trichloroethylene by a recombinant Escherichia Coli. Biotechnology 7:282–285

Xu J, Li W, Wu J, Zhang Y, Zhu Z, Liu J, Hu Z (2006) Stability of plasmid and expression of a recombinant gonadotropin-releasing hormone (GnRH) vaccine in Escherichia Coli. Appl Microbiol Biotechnol 73(4):780–788. Epub 2006 Aug 9. PubMed PMID: 16896597

Zhang W, Ames BD, Tsai S-C, Tang Y (2006) Engineered biosynthesis of a novel AmidatedPolyketide, using the Malonamyl-specific initiation module from the Oxytetracycline Polyketide Synthase. Appl Environ Microbiol 72(4):2573–2580. https://doi.org/10.1128/AEM.72.4.2573-2580.2006

Zhang G, Li Y, Fang L, Pfeifer BA (2015) Tailoring pathway modularity in the biosynthesis of erythromycin analogsheterologously engineered in E. Coli. Sci Adv 1(4):e1500077. https://doi.org/10.1126/sciadv.1500077

Index

J.K. Patra et al. (eds.), *Microbial Biotechnology*,
https://doi.org/10.1007/978-981-10-6847-8

P

R

S

T

Printed by Printforce, the Netherlands